Environmental Hydrology and Water Management

Environmental Hydrology and Water Management

Edited by **Sarah Luck**

SYRAWOOD
PUBLISHING HOUSE
New York

Published by Syrawood Publishing House,
750 Third Avenue, 9th Floor,
New York, NY 10017, USA
www.syrawoodpublishinghouse.com

Environmental Hydrology and Water Management
Edited by Sarah Luck

International Standard Book Number: 978-1-68286-267-4 (Hardback)

Printed in China.

Contents

Preface

The purpose of the book is to provide a glimpse into the dynamics and to present opinions and studies of some of the scientists engaged in the development of new ideas in the field from very different standpoints. This book will prove useful to students and researchers owing to its high content quality.

Environmental hydrology is a significant field of study that is concerned with the distribution and movement of water across the planet. It plays a prominent role in sustainable management of various water resources. The book aims to explore significant topics like physico-chemical processes behind hydro-climatology, evaluation of various water resources and their interaction around the world, impact of human activities on water quality and distribution, etc. Students, researchers, and others associated with the fields of environmental hydrology and water management will benefit alike from this book.

At the end, I would like to appreciate all the efforts made by the authors in completing their chapters professionally. I express my deepest gratitude to all of them for contributing to this book by sharing their valuable works. A special thanks to my family and friends for their constant support in this journey.

Editor

1

On inclusion of water resource management in Earth system models – Part 1: Problem definition and representation of water demand

A. Nazemi and H. S. Wheater

Global Institute for Water Security, University of Saskatchewan, 11 Innovation Boulevard, Saskatoon, SK, S7N 3H5, Canada

Correspondence to: A. Nazemi (ali.nazemi@usask.ca)

Abstract. Human activities have caused various changes to the Earth system, and hence the interconnections between human activities and the Earth system should be recognized and reflected in models that simulate Earth system processes. One key anthropogenic activity is water resource management, which determines the dynamics of human–water interactions in time and space and controls human livelihoods and economy, including energy and food production. There are immediate needs to include water resource management in Earth system models. First, the extent of human water requirements is increasing rapidly at the global scale and it is crucial to analyze the possible imbalance between water demands and supply under various scenarios of climate change and across various temporal and spatial scales. Second, recent observations show that human–water interactions, manifested through water resource management, can substantially alter the terrestrial water cycle, affect land–atmospheric feedbacks and may further interact with climate and contribute to sea-level change. Due to the importance of water resource management in determining the future of the global water and climate cycles, the World Climate Research Program's Global Energy and Water Exchanges project (WRCP-GEWEX) has recently identified gaps in describing human–water interactions as one of the grand challenges in Earth system modeling (GEWEX, 2012). Here, we divide water resource management into two interdependent elements, related firstly to water demand and secondly to water supply and allocation. In this paper, we survey the current literature on how various components of water demand have been included in large-scale models, in particular land surface and global hydrological models. Issues of water supply and allocation are addressed in a companion paper. The available algorithms to represent the dominant demands are classified based on the demand type, mode of simulation and underlying modeling assumptions. We discuss the pros and cons of available algorithms, address various sources of uncertainty and highlight limitations in current applications. We conclude that current capability of large-scale models to represent human water demands is rather limited, particularly with respect to future projections and coupled land–atmospheric simulations. To fill these gaps, the available models, algorithms and data for representing various water demands should be systematically tested, intercompared and improved. In particular, human water demands should be considered in conjunction with water supply and allocation, particularly in the face of water scarcity and unknown future climate.

1 Background and scope

1.1 Large-scale modeling – an introduction to land-surface and global hydrological models

The Earth system is an integrated system that unifies the physical processes at the Earth's surface. These processes include a wide range of feedbacks and interactions between and within the atmosphere, land and oceans and cover the global cycles of climate, water and carbon that support planetary life (e.g., Schellnhuber, 1999; Kump et al., 2010). From the advent of digital computers, Earth system models have been a key tool to identify past changes and to predict the future of planet Earth. These models normally include sub-models that represent various functions of the land, atmosphere and oceans (Claussen, 2001; Schlosser et al., 2007). A crucial sub-model in Earth system models is the land-surface model (LSM) that represents the land portion of the

Earth system. LSMs contain interconnected computational modules that characterize physical processes related to soil, vegetation and water over a gridded mesh, and account for their influences on water, energy and, increasingly, carbon exchanges. A wide range of LSMs is currently available, and these can be differentiated based on how, and to what extent, different land-surface processes are represented; nonetheless, a LSM should explicitly or implicitly include the dynamics of these processes, and account for their drivers at various temporal and spatial scales (see Trenberth, 1992; Sellers, 1992).

The importance of representing the terrestrial water cycle in LSMs is well-established (see Pitman, 2003, and references therein), and there has been progressive development of LSMs in representing various components of the hydrologic cycle, such as soil moisture, vegetation, snowmelt and evaporation. In early LSMs, hydrology was conceptualized as a simple lumped bucket model (Manabe, 1969), but this representation has progressively been improved by including more complexity and explicit physics in canopy, soil moisture and runoff calculations (see Deardorff, 1978; Dickinson, 1983, 1984; Sellers et al., 1986, 1994, 1996a; Nicholson, 1988; Pitman et al., 1990). Despite these improvements, major limitations and uncertainties remain in the hydrological simulations, causing systematic bias in water and energy balance calculations. These deficiencies have been attributed (in part) to unrealistic assumptions and incomplete parameterizations of catchment response in LSMs (Soulis et al., 2000; Music and Caya, 2007; Sulis et al., 2011). Further attempts, therefore, have focused on including catchment-scale runoff generation and routing processes (e.g., Miller et al., 1994; Hagemann and Dümenil, 1997; Oki and Sud, 1998; Oleson et al., 2008; Lawrence et al., 2011). These components determine the hydrological response at the larger scales and have been frequently used in large-scale hydrological models, so-called global hydrologic models (GHMs). Similar to LSMs, GHMs are gridded large-scale models; however, they are typically simpler in structure and focus on representing the water cycle rather than other land-surface processes (such as the energy and carbon cycles). LSMs have been applied frequently in regional and global modeling (e.g., Liang et al., 1994; Pietroniro et al., 2007; Adam et al., 2007; Livneh et al., 2011) and compared to GHMs (see Haddeland et al., 2011). At this stage of research, however, both LSMs and GHMs are still imperfect and incomplete, as current simulations cannot match recent hydrological observations (see Lawrence et al., 2012).

1.2 Modeling human–water interactions

While external forcing, mainly the energy flux from the Sun, is the main driver of the Earth system, internal disturbances such as volcanic eruptions, wildfires and human activities can substantially affect the natural Earth system cycles (Vitousek et al., 1997; Trenberth and Dai, 2007; Bowman et al., 2009). In particular, post-industrial human activities, from the mid-20th century onwards, have severely perturbed the Earth system (Crutzen and Steffen, 2003; Crutzen, 2006). This has initiated a new geological epoch, informally termed the "Anthropocene", in which it is recognized that the natural processes within the land surface system are highly controlled and regulated by humans (see McNeil, 2000; Steffen et al., 2007, 2011). Accordingly, Earth system models should address feedbacks and interactions between the natural Earth system and the anthroposphere, which includes human cultural and socio-economic activities (Schellnhuber, 1998, 1999; Claussen, 2001). The terrestrial water cycle is one set of Earth system processes that is greatly perturbed by human activities; it also is of critical importance in determining human health, safety and livelihoods, as well as local, regional and global economies (e.g., Nilsson et al., 2005). However, although some anthropogenic effects, such as the emission of greenhouse gases and land-use change, have been incorporated in LSMs (e.g., Lenton, 2000; Zhao et al., 2001; Karl and Trenberth, 2003; Brovkin et al., 2006; Solomon et al., 2009), less effort has been made to represent human–water interactions (e.g., Trenberth and Asrar, 2012; Lawrence et al., 2012; Oki et al., 2013). This can be a major reason for current deficiencies in hydrological performance of large-scale modes (i.e., LSMs and/or GHMs). In fact, large-scale models still widely assume that human effects on the terrestrial water cycle can be ignored. This assumption is highly questionable and can result in the neglect of important hydrologic processes (see Gleick et al., 2013).

Human–water interactions include a wide spectrum of anthropogenic interventions, including land-use change and water resource management. During the past century, human water consumption has increased more than 6-fold, with around 5, 18 and 10 times increase in agricultural, industrial and municipal consumption, respectively (see Shiklomanov, 1993, 1997, 2000). Supplying such intensive demands has required large changes in the natural water cycle – which can be even more than the effects of warming climate (see Haddeland et al., 2014), and is associated with major environmental water stress at the global scale. Smakhtin et al. (2004) concluded that over 1.4 billion people currently live in river basins with high environmental water stress and this number will increase as water withdrawals grow. For instance, surface-water withdrawals for supplying human needs decrease downstream flows, often substantially, and result in seasonal decline in flows of major rivers such as the Colorado River (e.g., Cayan et al., 2010). Similarly, dam operations considerably change the timing, volume, peak and the age of natural streamflow and reduce inputs to wetlands, lakes and seas (e.g., Vörösmarty et al., 1997, 2005; Vörösmarty and Sahagian, 2000; Meybeck, 2003; Tang et al., 2010). This is associated with some extreme effects, such as the death of the Aral Sea (e.g., Precoda, 1991; Small et al., 2001) and aggressive decline in the area of Lake Urmia in Iran (Aghakouchak et al., 2014). In parallel, groundwater abstractions are associated with declining groundwater levels, reduced baseflow contributions and loss of wetlands. For instance, current assessments reveal significant groundwater depletion in some areas of the globe, such as Indian peninsula, the US Midwest, and Iran (Giordano, 2009; Rodell et al., 2009; Gleeson et al., 2012; Döll et al., 2014). Without considering human withdrawals, these changes in surface-water and groundwater availability cannot be captured by large-scale models. It should be noted that human activities have large effects on water quality as well. For instance, extensive groundwater pumping is also associated with potential long-term contamination, for example by salt-water intrusion (Sophocleous, 2002; Antonellini et al., 2008), and nutrient pollution of surface and groundwater is an outstanding global challenge. These water quality impacts, however, remain beyond the scope of this survey.

As human life and water availability are tightly interconnected (see Sivapalan et al., 2012), current and future changes in the water availability are not only important for Earth system modeling, but are also of major importance to human society, and these issues can be explored to a large extent with large-scale models. Although human water use still accounts for a small proportion of total water on and below the surface (see Oki and Kanae, 2006), total human withdrawals currently include around 26 % of terrestrial evaporation and 54 % of the accessible surface runoff that is geographically and temporally available (Postel et al., 1996). There are already major water scarcity issues across highly populated regions of the globe (e.g., Falkenmark, 2013; Schiermeier, 2014), which raise fundamental concerns about how future demand should be supplied, particularly considering climate change (e.g., Arnell, 1999, 2004; Tao et al., 2003; Döll, 2009; Taylor et al., 2013; Hanasaki et al., 2013a, b; Wada et al., 2013; Milano et al., 2013; Mehta et al., 2013; Schewe et al., 2014). Such important threats to water security necessitate a detailed understanding of water availability and demand in time and space; and therefore large-scale models are required for impact assessments.

Apart from the hydrologic and water security relevance discussed above, human–water interactions can have broader implications for the water cycle and affect climate, although these issues are yet to be fully explored, and remain in some cases controversial. For instance, irrigation can disturb the "natural" atmospheric boundary conditions (e.g., Sacks et al., 2009; Destouni et al., 2010; Gerten et al., 2011; Pokhrel et al., 2012; Hossain et al., 2012; Guimberteau et al., 2012; Dadson et al., 2013). At this stage of model development, the available quantitative understanding of these land–atmospheric implications is limited. To explore these issues it is necessary to include these processes in coupled land–atmospheric models, and this requires explicit representation of relevant human–water interactions within LSM computational schemes. Moreover, the return flows from human usage, entering the seas and oceans, can affect salinity and temperature and consequently impact their circulation patterns (e.g., Rohling and Bryden, 1992; Skliris and Lascaratos, 2004; Vargas-Yàñez et al., 2010). This is of particular concern for closed oceans and the polar environment, where a change in freshwater input can modify the oceanic circulations and thus feedback on continental rainfall (Polcher, 2014). However as noted above, issues related to water quality remain beyond the scope of our survey.

1.3 Aim and scope of this survey

The aim of our survey is to consider the associated scientific and data challenges, the state of current practice, and directions for future research around including human effects on the terrestrial water cycle. In this paper and a companion paper (hereafter Nazemi and Wheater, 2015), we focus on human–water activities manifested through water resource management and note that this is subject to operational and policy constraints. We only consider water quantity aspects of water resource management, which we define as a suite of anthropogenic activities related to storage, abstraction and redistribution of available water sources for various human demands. Although a fully coupled representation of water resource management in Earth system models is not currently available, important progress is being made, and more generally a body of literature is gradually shaping around describing different aspects of water resource management in large-scale models, in particular within the context of GHMs. Nonetheless, there are still fundamental obstacles in including water resource management within large-scale models.

First, a fundamental principle in Earth system models as well as LSMs and GHMs is the conservation of water. To represent water resource management, therefore, it is necessary to fully capture water in a coupled human–natural system. To achieve this (i) modeling complexity should be increased, (ii) process representations related to both natural and anthropogenic systems should be improved and (iii) modeling capability should be extended to new domains (see Polcher, 2014, for an in-depth discussion). For instance, a large proportion of human demand is supplied by groundwater, which is often absent or crudely represented in both LSMs and GHMs and is widely considered disjoint from other elements of the Earth system such as climate.

Second, multiple factors affect water resource management at the larger scales, such as climate, hydrology, landcover and socio-economy as well as land and environment management. Moreover, real-world management decisions often include cultural values and political concerns (Gober and Wheater, 2014). These various influences are so far considered in isolation and the interactions among them are widely unseen (e.g., Beddington, 2013).

Third, there is considerable lack of regional and global data concerning the actual use and operation of water resources systems, and therefore large-scale models cannot be properly tuned or validated. This major limitation, for instance, has led the research community to use estimated demand as a surrogate for actual use. Lack of data about human operations can also introduce large uncertainty into simulations of terrestrial storage and runoff. For instance Gao et al. (2012) noted that the "[. . .] results from global reservoir simulations are questionable" as "there are no direct observations of reservoir storage".

Fourth, there is a major gap between the scope of local operational water resource models and large-scale applications and research needs. Essentially, the scale at which local water resource management takes place is often within the sub-grid resolution of current large-scale models, which requires narrowing the resolution in large-scale models for explicit representation (see Wood et al., 2011) or adding more sub-grid heterogeneity into grid calculations for implicit parameterization. In addition, there is (and will increasingly be) competition between various water demands which requires allocation decisions. At this stage of model development, however, it is still unclear how operational policies should best be reflected at larger scales. At the local scale, detailed information on physical and operational systems as well as climate and water supply conditions are available (or can be generated as scenarios; see, e.g., Nazemi et al., 2002, 2013; Nazemi and Wheater, 2014a, b) and the competition between demands is often reflected as an optimization problem. As the simulation scale moves from local and small basin scales to regional and global scales, the data availability degrades considerably and the high level of calculations within optimization algorithms cannot be maintained, due to computational barrier.

Conceptually, water resource management at larger scales can be seen as an integration of two fully interactive elements, related to water demand as well as water supply and allocation: water demand is constrained by water availability and drives water allocation, which results in extraction from water sources and determines the extent of change in hydrological elements of the land surface. Moreover, as noted briefly above, perturbations in the terrestrial water cycle due to water resource management can further interact with other elements of the Earth system, particularly with climate (see Fig. 1). To assess the impacts of water resource management on land-surface processes and associated feedbacks with climate, the elements of water demand and water allocation should be described using computational algorithms and included in large-scale models. For the purpose of our survey, and reflecting the state of algorithm development and data availability, we focus in this paper only on the representation of water demand, and in the Nazemi and Wheater (2015) on water supply and allocation. Here, we classify human water demands under two general categories, namely irrigative and non-irrigative, and further divide non-irrigative demands into municipal, industrial, environmental, energy-related, and livestock water needs. This is useful to put current algorithms and modeling applications into context. Accordingly, we discuss how these demands are characterized using various computational algorithms. As will be

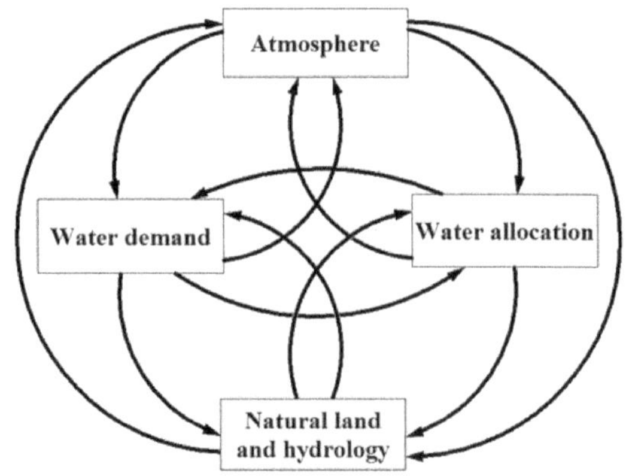

Figure 1. Water resource management as an integration of water demand and water allocation and its interactions with natural land-surface and climate.

shown later in this paper, human demands are mainly quantified either using downscaling (i.e., top-down approaches) or through direct modeling at the grid scale (i.e., bottom-up approaches). Depending on the type of application, the algorithms can be included in a wide range of large-scale models. Throughout our review, we consider both offline and online implications of water demand. Offline simulations assess the effects of water demand on land-surface processes without considering the associated feedbacks to the climate system, but can be linked to atmospheric driving variables to simulate land-surface and/or hydrological responses to climate and water resource management. Online models also account for the effects of water demand on land–atmospheric feedbacks and are further coupled with climate models. This is done by considering the effects of water demand on the dynamics of land-surface variables and updating the surface boundary conditions in climate models (Verseghy, 1991, 2000; Verseghy et al., 1993). Online applications are also termed in the LSM community as coupled land–atmospheric simulations (e.g., Entekhabi and Eagleson, 1989; Noilhan and Planton, 1989) and are more computationally demanding compared to offline simulations. While offline models include both LSMs and GHMs, it should be noted that GHMs cannot be used for online applications as they do not account for the energy balance and therefore cannot fully represent land–atmosphere feedbacks.

The structure of this paper is as follows: in Sect. 2 we highlight the impacts of irrigative and non-irrigative water demands on the terrestrial water cycle and land–atmospheric feedbacks. Sections 3 and 4 provide an overview of available representations of irrigative and non-irrigative demands at larger scales, respectively. In Sect. 5, we briefly explore state-of-the-art applications and highlight current limitations and uncertainties in estimating current and future water demand and associated online and offline impacts. We further discuss current gaps in Sect. 6 and provide some suggestions for future developments. Finally, Sect. 7 summarizes this first part of our survey and outlines our main findings with respect to representing human water demand.

2 Types of human demand and their impacts on the water cycle

Human water demands can be divided into irrigative and non-irrigative categories. Irrigation is the dominant human water use and has significantly intensified since the 1950s, due to population growth and technological development (Steffen et al., 2011). This has major importance for global food security, as it produces approximately 40 % of the world's food (Abdullah, 2006). Currently, around 25 % of harvested crop area is irrigated (Portmann et al., 2010). This accounts for some 90 % of water consumption at the global scale (Döll et al., 2009; Siebert et al., 2010), which is around 70 % of the total water withdrawals from surface and groundwater resources (Wisser et al., 2008; Gerten and Rost, 2010). Clearly supplying such a large water demand can severely disturb the "natural condition" by decreasing streamflow volume (e.g., Meybeck, 2003; Gaybullaev et al., 2012; Lai et al., 2014) and groundwater levels (e.g., Rodell et al., 2009; Gleeson et al., 2012; Wada et al., 2010, 2012, 2014; Döll et al., 2014). Currently, surface water is the main supplier of global irrigative needs, accounting for 57 % of the total consumptive irrigation use at the global scale (Siebert et al., 2010).

Apart from driving hydrological changes, irrigation-induced changes in soil moisture can affect land surface–atmosphere feedbacks (see Eltahir, 1998). Pokhrel et al. (2012) showed that increased soil water content through irrigation substantially enhances evapotranspiration, and therefore transforms the surface energy balance. Evapotranspiration due to irrigation leads to cooling of the land surface (e.g., Haddeland et al., 2006; Saeed et al., 2009; Destouni et al., 2010), as well as enhanced cloud cover and chance of convective precipitation (e.g., Moore and Rojstaczer, 2001; Douglas et al., 2009; Harding and Snyder, 2012a, b; Qian et al., 2013). Irrigation may also alter regional circulation patterns due to temperature difference between irrigated areas and neighboring regions (e.g., DeAngelis et al., 2010; Wei et al., 2013). Over highly irrigated regions, this can mask important climate change signals. Gerten et al. (2011), for instance, showed that the irrigation in South Asia has offset the increasing temperature in the region.

Non-irrigative water demands include municipal and industrial uses, energy-related withdrawals, other agricultural uses, such as livestock, as well as designated environmental water uses, which can be an important constraint on water management. Non-irrigative demands contribute a lesser proportion to total human water use at the global scale. This proportion, however, has significant spatial variability (Vassolo and Döll, 2005; Flörke et al., 2013) as regional differences in population, income, life style and technological developments can alter the extent of non-irrigative demand significantly (e.g., Alcamo et al., 2003; Flörke and Alcamo, 2004; Hejazi et al., 2013a). However, while irrigation is predominantly a consumptive water use, only a small portion of the non-irrigative withdrawal is consumptive (e.g., Hanasaki et al., 2013a). Non-irrigative withdrawals, therefore, partially or totally return to surface-water or groundwater systems with varying degrees of time lag. Still, this can considerably perturb the streamflow regime (e.g., Maybeck, 2003; Förster and Lilliestam, 2010). Non-irrigative water demands are currently on a rapid incline due to growing population and industrial development. This can increase water stress in both time and space (Hejazi et al. 2013a, b, c, d). As non-irrigative demands are mainly non-consumptive, they are less likely to change the energy balance and/or perturb the atmospheric moisture condition significantly and therefore they are less relevant to land–atmospheric interactions. However, changing timing of flows can have significant local effects, for example on wetland inundation. Similarly, for some large-scale mining activities in which the extent of water withdrawals is considerable, the associated changes in soil moisture and land cover can be potentially relevant to land–atmospheric feedbacks. To the best of our knowledge, such online considerations for non-irrigative withdrawals have not yet been explored in the literature.

3 Available representations of irrigative demand in large-scale models

Irrigation is an important element of water resource management and has been explored more in depth than non-irrigative demands. To simplify our presentation, we classify current representations with respect to the scale (regional vs. global) and/or mode of simulation (offline vs. online). Tables 1 and 2 summarize representative examples of offline simulations at both regional (Table 1) and global (Table 2) scales. Table 3 presents some online examples. In brief, current online applications have mainly been performed at rather fine temporal and spatial resolutions with shorter simulation periods than offline representations. In contrast, a wide spectrum of host models (i.e., large-scale models in which the irrigation algorithm is embedded), as well as forcing and land-use data, has been used in current offline examples (see Tables 1 and 2). Model resolutions in offline applications can vary in time from 1 h (e.g., Leng et al., 2013) to 1 day (e.g., Haddeland et al., 2007) with a grid size ranging from a few kilometers (e.g., Siebert and Döll, 2010; Nakayama and Shankman, 2013) to a few hundred kilometers (e.g., Gueneau et al., 2012) in space. Moreover, offline irrigation demand calculations have already been performed globally under future climate conditions.

Table 1. Representative examples including regional irrigation in large-scale models (offline mode).

Reference	Irrigation data	Irrigation demand	Region	Host model	Forcing	Temporal resolution	Spatial resolution
Haddeland et al. (2006)	Döll and Siebert (2002)	Difference between current soil moisture content and minimum of FAO Penman–Monteith crop-specific evapotranspiration and soil moisture content at field capacity.	Colorado (USA) and Mekong (east Asia)	VIC (Liang et al., 1994)	Adam and Lettenmaier (2003); Maurer et al. (2002)	3 h	$0.5^\circ \times 0.5^\circ$
Haddeland et al. (2007)	Siebert et al. (2005)	Haddeland et al. (2006)	North America and Asia	VIC (Liang et al., 1994)	Maurer et al. (2002)	24 h	$0.5^\circ \times 0.5^\circ$
Gueneau et al. (2012)	GAEZ (IIASA/FAO, 2012); FRIS (USDA, 2008)	Difference between actual and potential evapotranspiration based on Farmer et al. (2011). Crop growth and irrigation losses included.	USA	CLM3.5 (Oleson et al., 2004, 2008)	NCC (Ngo-Duc et al., 2005b)	6 h	$2.5^\circ \times 2.5^\circ$
Leng et al. (2013)	MODIS (Ozdogan and Gutman, 2008); NASS (USDA, 2002)	Difference between current and ideal soil moisture content based on CLM4CNcrop crop growth model of CLM4 (Levis and Sacks, 2011; Levis et al., 2012).	Contermi-nous USA	CLM4 (Lawrence et al., 2011)	NLDAS (Cosgrove et al., 2003)	1 h	$0.125^\circ \times 0.125^\circ$
Nakayama and Shankman (2013)	Liu et al. (2010)	Difference between current soil moisture content and soil moisture at the field capacity.	Changjing, Yellow River basins (China)	NICE (Nakayama, 2011)	ECMWF (http://www.ecmwf.int/en/forecasts/datasets)	6 h	10 km × 10 km
Voisin et al. (2013)	Crop area projections in Chaturvedi et al. (2013a, b).	Downscaling GCAM model estimations (Wise and Calvin, 2011; Wise et al., 2009a) using methods of Hejazi et al. (2013a), Siebert and Döll (2008) and Hanasaki et al. (2013a, b).	US Mid-west	SCLM-MOSART (Lawrence et al., 2011; Li et al., 2013); Tesfa et al. (2014)	CASCaDE (http://cascade.wr.usgs.gov)	1 h	$0.125^\circ \times 0.125^\circ$

Table 2. Representative examples including global irrigation in large-scale models (offline mode).

Reference	Irrigation data	Irrigation demand	Host model	Forcing	Temporal resolution	Spatial resolution
Döll and Siebert (2002)	Döll and Siebert (2000)	Difference between Smith (1992) effective rainfall and Priestley and Taylor (1972) crop-specific potential evapotranspiration and Allen et al. (1998) multipliers.	WaterGAP (Alcamo et al., 2003)	CRU TS 1.0 (New et al., 1999, 2000)	24 h	$0.5° \times 0.5°$
de Rosnay et al. (2003)*	Döll and Siebert (2002)	Difference between effective rainfall and FAO potential evapotranspiration (Allen et al., 1998) without considering irrigation efficiency.	ORCHIDEE (Ducoudré et al., 1993)	ISLSCP-I (Sellers et al., 1996b)	24 h	$1° \times 1°$
Hanasaki et al. (2006)	Döll and Siebert (2000)	Similar to Döll and Siebert (2002). Reference evaporation is based on FAO Penman–Monteith.	TRIP (Oki and Sud, 1998)	ISLSCP-I (Sellers et al., 1996b)	24 h	$0.5° \times 0.5°$
Wisser et al. (2008)	Siebert et al. (2005, 2007); GIAM (Thenkabail et al., 2009)	Similar to Haddeland et al. (2006) using Allen et al. (1998) procedure.	WBM (Vörösmarty et al., 1998)	CRU TS 2.1 (Mitchell and Jones, 2005); NCEP (Kalnay et al., 1996)	24 h	$0.5° \times 0.5°$
Rost et al. (2008, 2009)	Siebert et al. (2007)	Difference between available plant moisture and an updated Priestley and Taylor (1972) potential evaporation based on potential canopy conductance of carbon and water (Sitch et al., 2003).	LPJmL (Bondeau et al., 2007)	CRU TS 2.1 (Mitchell and Jones, 2005)	24 h	$0.5° \times 0.5°$
Hanasaki et al. (2008a, b)	Döll and Siebert (2000)	Difference between current and 75 % of field capacity. Irrigation applied 30 days prior to planting. Detailed crop growth representation based on SWIM (Krysanova et al., 1998).	H08 (Hanasaki et al., 2008a, b)	NCEP-DOE (Kanamitsu et al., 2002); GSWP-2 (Zhao and Dirmeyer, 2003)	24 h	$1° \times 1°$
Siebert and Döll (2010)	MIRCA2000 (Portmann et al., 2010)	Difference between actual and crop-dependent reference evapotranspiration computed according to Priestley and Taylor (1972). Crop coefficients obtained from Allen et al. (1998).	GCWM (Siebert and Döll, 2008)	CRU TS 2.1 (Mitchell and Jones, 2005)	24 h	$0.08° \times 0.08°$
Wada et al. (2011, 2012)	MIRCA2000 (Portmann et al., 2010)	Difference between actual and potential transpiration according to van Beek et al. (2011), using Priestley and Taylor (1972) for calculating crop-specific and transpiration (Allen et al., 1998).	PCR-GLOBWB (van Beek et al., 2011)	CRU TS 1.0 (New et al., 1999, 2000)	24 h	$0.5° \times 0.5°$
Pokhrel et al. (2012)	Siebert et al. (2007)	Procedure of Hanasaki et al. (2008a, b). Crop calendar is based on potential evapotranspiration (Allen et al., 1998).	MASTIRO (Takata et al., 2003)	Kim et al. (2009); GPCC (Rudolf et al., 2005)	6 h	$1° \times 1°$
Wada et al. (2014)	MIRCA2000 (Portmann et al., 2010)	Constant 50 mm surface-water depth for paddy irrigation until 20 days before harvesting. For non-paddy areas, the difference between current and ideal plant available moisture at field capacity with dynamic root zone.	PCR-GLOBWB (van Beek et al., 2011)	ERA-Interim (Dee et al., 2011); MERRA (http://gmao.gsfc.nasa.gov/merra/)	24 h	$0.5° \times 0.5°$

* The simulation is performed globally but the results are analyzed only over the Indian peninsula.

Table 3. Representative examples including irrigation in coupled land-surface models (online mode).

Reference	Irrigation data	Irrigation demand	Region	Host LSM	Climate model	Temporal resolution	Spatial resolution
Adegoke et al. (2003)	LandSat (http://landsat.gsfc.nasa.gov/)	Target soil moisture deficit (difference between actual and saturated soil moisture).	High Plains (USA)	LEAF-2 (Walko et al., 2000)	RAMS (Pielke et al., 1992)	30 s nested in 1 min	10 km × 10 km nested in 40 km × 40 km
Sacks et al. (2009)	FAO-AQUASTAT (http://www.fao.org/nr/water/aquastat/main/index.stm)	AQUASTAT irrigated water uses applied at constant rate when LAI exceeds 80 % of the maximum annual value.	Global	CLM3.5 (Oleson et al., 2008)	CAM (Collins et al., 2004, 2006)	20 min	2.8° × 2.8°
Sorooshian et al. (2011)	CIMIS-MODIS (http://wwwcimis.water.ca.gov/)	Target soil moisture deficit (irrigation starts when the soil moisture drops below a maximum depletion threshold beyond which the plant in stressed (a percentage of field capacity, depending on the crop) and continues to field capacity)	California Central Valley (USA)	Noah (Ek et al., 2003)	NCAR-MM5 (Chen and Dudhia, 2001a, b)	30 min 1 h	4 km × 4 km 12 km × 12 km 36 km × 36 km
Harding and Snyder (2012a, b)	MODIS (Friedl et al., 2002; Ozdogan and Gutman, 2008); NASS (USDA, 2002)	Target soil moisture deficit (difference between actual and saturated soil moisture to depth of 2 m).	Great Plains (USA)	Noah (Ek et al., 2003)	WRF (Skamarock et al., 2005)	30 and 25 s	10 km × 10 km
Guimberteau et al. (2012)	Döll and Siebert (2002)	Difference between potential transpiration and the net water amount kept by the soil (i.e., the difference between precipitation reaching the soil and total runoff).	Global	ORCHIDEE (Ducoudré et al., 1993)	LMDZ4 (Hourdin et al., 2006)	30 min	2.5° × 1.25°
Qian et al. (2013)	MODIS (Ozdogan and Gutman, 2008; Ozdogan et al., 2010)	Similar to Sorooshian et al. (2011). Based on Ozdogan et al. (2010), moisture threshold is fixed at 50 % of filed capacity. Roots grow based on the greenness index.	Southern Great Plains (USA)	Noah (Ek et al., 2003)	WRF (Skamarock et al., 2005)	3 h	12 km × 12 km

3.1 Framework and general procedure

Irrigated lands normally introduce heterogeneity into the computational grids of LSMs and GHMs. Such sub-grid heterogeneity can be represented as an additional "tile", similar to forested land, bare soil and snow cover (Polcher et al., 2011). Essentially, irrigation algorithms are required to estimate the irrigation demand, and accordingly irrigative water use, at the grid scale. Here we refer to the irrigation demand as the water required for ideal crop growth in addition to the available water from precipitation. To simulate the grid-based irrigation demand, crop type and the extent of irrigated regions and growing seasons should be first identified. The location and area of irrigation districts and the associated crop types can be extracted from regional and global data sets (e.g., USDA, 2002, 2008; Siebert et al., 2005, 2007; Portmann et al., 2010) and/or remotely sensed data (e.g., Adegoke et al., 2003; Qian et al., 2013). There are two general approaches for identifying growing seasons. The choice of these options depends on the level of detail in the host model. In simpler models, where no energy-balance calculation is available (i.e., GHMs), crops can grow when and where simple temperature- and precipitation-based criteria are met (e.g., Döll and Siebert, 2002). In more detailed models (i.e., LSMs) the optimal growing season can be identified based on biophysical conditions of crop growth and/or soil water, canopy and energy balance conditions to estimate the cropping period that is necessary to obtain mature and optimal plant biomass (e.g., Rost et al., 2008; Pokhrel et al., 2012). Both approaches are subject to uncertainty. On one hand, models with fixed crop calendars ignore inter-annual variability in growing seasons. On the other hand, even models with fully dynamic crop growth algorithms may misrepresent the seasonality. After the growing season is identified, the irrigation demands (and under some assumptions, actual irrigation withdrawals) at each simulation time step can be calculated. A variety of top-down and bottom-up procedures are available for calculating the irrigation demand in large-scale models and are reviewed further below. If the irrigation demand is completely fulfilled, then the actual evapo(transpi)ration would be equal to crop-specific evapo(transpi)ration under standard conditions (see Allen et al., 1998). In offline applications, the irrigation rate can perturb soil moisture content, evaporation, deep percolation and runoff in irrigated tiles (e.g., Hanasaki et al., 2008a, b; Wada et al., 2011, 2012, 2014). In online applications, the vertical vapor and heat fluxes need also to be considered. The total fluxes for each grid can be then calculated as the sum of the flux contributions from irrigated and non-irrigated portions of the grid (e.g., Haddeland et al., 2006; Pokhrel et al., 2012), and can be further introduced to climate models as coupled surface boundary conditions (e.g., Sorooshian et al., 2011; Harding and Snyder, 2012a, b).

3.2 Top-down algorithms for calculating irrigation demand

In top-down approaches, the irrigation demand is not directly calculated, but estimated based on downscaling information available at coarser scales, often at national or geopolitical scales. Such information is based on census-based inventories (e.g., Sacks et al., 2009) or socio-economic model outputs (e.g., Voisin et al., 2013). Top-down approaches are highly influenced by the availability of global data on water use, such as FAO's Information System on Water and Agriculture (AQUASTAT; http://www.fao.org/nr/water/aquastat/main/index.stm), which provides annual inventory data on national (and in some cases also sub-national) scales, and has been further extended to include socio-economic model outputs. An example of such a model is the Global Change Assessment Model (GCAM; Wise et al., 2009a, b; Wise and Calvin 2011), which estimates agricultural production based on socio-economic variables, from which the irrigation water use is indirectly calculated using the water required for each crop per unit of land. Downscaling is performed mainly using land-use, technological and/or socio-economic proxies. There are various sources of uncertainty associated with top-down algorithms. First, both inventory and model-based products have major limitations due to their spatial and temporal scales as irrigation practices are highly variable within a country and a typical year. Moreover, the quality of both census and model-based products is poor. For instance, there are inconsistencies between census data and data quality varies from country to country (see Portman et al., 2010, for a detailed discussion). Also, socio-economic models widely ignore water availability constraints (Hejazi et al., 2013d). As a result, calculation of irrigation demand is mainly pursued through bottom-up schemes.

3.3 Bottom-up algorithms for calculating irrigation demand

In contrast to top-down schemes, bottom-up approaches estimate the irrigation demand directly at the grid scale by mimicking the optimal crop growth for irrigated tiles. Despite major limitations due to the heterogeneity in soil and crops, bottom-up algorithms have been widely used in the literature. These algorithms include a range of modeling assumptions; however, they are all centered around estimation of an ideal crop water requirement, i.e., where there is no water deficit. This requirement is based on estimation of "potential evapo(transpi)ration", which characterizes the atmospheric moisture deficit (Hobbins et al., 2008). There are multiple approaches to estimate the potential evapo(transpi)ration, and the estimates obtained may vary considerably. LSMs typically include detailed energy balance calculations and resolve the diurnal cycle; therefore, they can directly calculate potential and actual evaporations (see Milly, 1992; Barella-Ortiz et al., 2013, for a detailed description). Alternative approaches adopt a variety of methods, and are heavily influenced by FAO's guidelines for calculating irrigation water requirements (see Allen et al., 1998). These approaches are mainly used in GHMs, where the evapotranspiration is calculated for a reference crop and corrected as a function of crop type and development stage using a set of empirical coefficients. Various methods are used to characterize the reference evapotranspiration, such as FAO Penman–Monteith (Allen et al., 1998), Priestley and Taylor (1972) and modified Hargreaves (Farmer et al., 2011) to name a few (see McKenney and Rosenberg, 1993, for more examples). The choice of appropriate formulation for reference evapotranspiration is rather arbitrary and depends largely on the data availability as well as the level of detail supported in the host model. It should be noted that, due to the difference in estimation of evaporation, incorporating FAO's guidelines for estimation irrigation demand in LSMs can introduce inconsistencies with the evaporation estimated by the model at various timescales, particularly over dry regions where the irrigation is likely to occur (Polcher, 2014).

Here we briefly explain the currently available bottom-up algorithms, from the more simple to the more comprehensive algorithms, and highlight their strengths and weaknesses.

In the most simple bottom-up representations, the irrigation demand at every time step is the water required to bring the soil moisture at the root zone to saturation (e.g., Lobell et al., 2006; Harding and Snyder, 2012a, b), which describes an extreme demand condition and clearly overestimates the actual irrigation water requirement (Sacks et al., 2009). In a more realistic but still naïve representation, the soil moisture requirement during the growing season is considered to be the field capacity (e.g., Nakayama and Shankman, 2013); therefore, the irrigation water need is the water required to bring the soil moisture to field capacity. The description of the irrigation demand based on the field capacity can also overestimate the actual water requirements, as the evaporation often reaches potential level before the soil reaches field capacity. The threshold at which the evaporation reaches potential evaporation is crop-dependent, but often considered as a constant value in large-scale models. As an offline example, Hanasaki et al. (2008a) assumed that paddy and non-paddy crops require soil moisture content of 100 or 75 % of the field capacity at the root zone with constant depth at the global scale. Yoshikawa et al. (2014) later updated the assumption for non-paddy soil moisture requirement and used 60 % of field capacity, referring to the requirement for wheat. This is again rather unrealistic as (1) by assuming a constant percentage of the field capacity for all crop types, the diversity in crop water requirement is ignored; and (2) a constant root zone depth at the global scale can result in misestimating the irrigation demand. There are attempts to address these limitations. For instance, Sorooshian et al. (2011) assumed that the required soil moisture content can change for each grid based on the dominant crop. Leng et al. (2013) and Qian et al. (2013) implemented root growth in their irrigation demand algorithm to avoid overestimation of demand due to a constant root zone. It should be noted that calculating the root growth is also subject to uncertainty; however, associated limitations remain beyond the scope of this paper.

More realistic definitions of irrigation water demand are based on the difference between the crop-dependent potential evapotranspiration and available crop water. This definition has been widely used in global irrigation demand projections (see Table 2). In earlier examples (e.g., Döll and Siebert, 2002; de Rosnay et al., 2003), crop development is described by constant monthly multipliers for potential evapotranspiration and the effective rainfall is used as a surrogate for available crop water. In more advanced algorithms, the correction factors are considered as functions of daily climate, stage of vegetation and root growth. Moreover, actual evapotranspiration or soil moisture content can be used instead of effective rainfall (Haddeland et al., 2006, 2007; Gueneau et al., 2012). There are two key limitations associated with this approach to simulation of irrigation demands. First, FAO's definition of irrigation water requirement considers both transpiration from crop and evaporation from soil. It has been noted that this quantification may result in overestimating the irrigation demand and may not properly represent the dynamics of vegetation (Polcher et al., 2011). Second, it is assumed that crop growth is a function of water availability only; therefore, the effects of other drivers such as CO_2 on photosynthesis are wholly ignored.

Some efforts try to overcome these limitations by defining irrigation demand based on potential transpiration instead of potential evapotranspiration (e.g., Wada et al., 2011, 2012), in conjunction with models that have more comprehensive vegetation schemes. Potential transpiration is the transpiration that would occur if the crop is not water stressed. Potential transpiration takes into account CO_2 fertilization effects and can represent the adaptation of the plants to climatic conditions and/or crop growth cycles, if the host model is equipped with relevant calculations (Guimberteau et al., 2012); therefore, this approach is mainly used in LSMs with detailed consideration of vegetation growth. As an example, Rost et al. (2008) coupled a transpiration deficit algorithm with the Lund-Potsdam-Jena managed Land scheme (LPJmL; Bondeau et al., 2007), which has a detailed vegetation growth module based on carbon and water availability (see Sitch et al., 2003; Gerten et al., 2004). The crop water limitation was calculated based on the atmospheric water deficit, soil moisture, plant hydraulic states as well as the CO_2 effects. Considering the effects of both carbon and water in vegetation can provide a basis for explicit linkage between CO_2 emission, crop growth and irrigation water requirement. This would be important for future predictions under increasing CO_2 effects. Moreover, some recent simulations showed that the irrigation requirement changes if a dynamic growth model is used; and this can improve the partitioning of latent heat flux, which is relevant to online applications (e.g., Lu, 2013). Nonetheless, it should be noted that the success of potential transpiration algorithm depends strongly on the way various tiles are treated at the grid scale. Normally, LSMs can define multiple crops at the grid scale and can distinguish the various water needs across different tiles within a grid. If potential transpiration is implemented consistently with sub-grid soil moisture divisions, then the water taken from the irrigated tiles optimizes photosynthesis and is only evaporated by the crops and not used by other surface types (e.g., bare soil, non-irrigated crops). In contrast, if all tiles share the same soil moisture reservoir at the grid scale, irrigation will increase the soil moisture and evaporation and therefore reduce water stress over the whole grid.

3.4 Projection of irrigative demand

From water and food security perspectives, particularly under various global change scenarios, it is crucial to investigate future irrigation demand and assess various possibilities for irrigation deficit. Climate model projections under IPCC emission scenarios (IPCC, 2000) have been widely used to force bottom-up irrigation demand algorithms (e.g., Arnell, 1999; Wada et al., 2013; Rosenzweig et al., 2014). Efforts have been also made to include intermediate socio-economic scenarios that can be matched to current climate change scenarios (see, e.g., Arnell, 2004; Fischer et al., 2007; Alcamo et al., 2007). For irrigation, intermediate scenarios describe changes in irrigated areas, irrigation efficiency and crop type, using empirical approaches. For example, Hanasaki et al. (2013a) recently proposed intermediate scenarios based on newly developed Shared Socio-economic Pathways (SSPs; Kriegler et al., 2012; see also Moss et al., 2010), which are consistent with Representative Concentration Pathways (RCPs; Meinshausen et al., 2011; K. E. Taylor et al., 2012). Constructing intermediate scenarios using empirical procedures, however, is uncertain as mechanisms that link irrigation expansion to socio-economic factors are not fully known and current empirical relationships can contain large uncertainties. More dynamic linkage between irrigation expansion and socio-economic drivers can be provided by coupled socio-economy–energy–carbon models. One emerging model of such a kind is GCAM, which has been recently implemented for simulating the future expansions in irrigation areas and demands (Hejazi et al., 2013b, c, d) as well as policy implications for irrigation water requirements (e.g., Chaturvedi et al., 2013a, b). Although these models can represent the dynamic effects of various drivers on irrigation, they remain uncertain as their simulations are rather coarse and do not incorporate water availability constraints. There are emerging efforts to avoid this limitation by linking the irrigation demand to climate, economy and water management constraints. This can result in prediction of regions in which irrigation can be developed and sustained considering changing climate, water availability, water price and water management infrastructure (see Nassopoulos et al., 2008, 2012). Such approaches however have not been applied at larger regional and global scales.

4 Available representations of non-irrigative demand

4.1 Forms and drivers of non-irrigative demand

Non-irrigative water demands relate to a wide range of environmental, municipal, industrial and energy-related uses, as well as other agricultural water needs (e.g., livestock), and include both consumptive and non-consumptive withdrawals. Among these, livestock water demand is assumed fully consumptive, and can be estimated by livestock number and demand per livestock head (e.g., Wada et al., 2011; Strzepek et al., 2012b; Hejazi et al., 2013d). Wada et al. (2014) made a further improvement by estimating daily livestock requirements at $0.5° \times 0.5°$ spatial resolution using livestock data of Steinfeld et al. (2006). Daily demand was considered as a function of daily temperature.

In contrast to livestock water demand, environmental flow needs can be considered as a fully non-consumptive need, required to protect rivers' health and aquatic life. Considering the extent of environmental degradation at the global scale, accounting for environmental flow needs becomes more and more relevant and should be considered as an integral part of water resource management at larger scales. Tharme (2003) made an extensive review of available methodologies for estimating environmental flow needs and identified more than 200 methodologies based on various hydrological, hydraulic rating, habitat simulation and holistic guidelines at the river basin scale. There are also some recent trends to involve scientists, water-resource managers and stakeholders to analyze available hydrological information and convert them into ecologically based and socially acceptable goals for estimating the environmental flow needs (see Poff et al., 2009). Such procedures however are widely dependent on the availability of relevant information and, therefore, cannot be easily implemented in large-scale models. Currently, implementation of environmental flow needs in large-scale models remains rather limited and simplistic and these needs are often calculated based on generic rules. For instance, Smakhtin et al. (2004) assigned thresholds for fair (Q_{90}), natural (Q_{50}) and good (Q_{75}) natural flow conditions. Shirakawa (2004, 2005, referenced from Hanasaki et al., 2008a) distinguished between two factors, i.e., minimum and perturbation flow requirements, which can also accommodate transient streamflow conditions. Currently, the perturbation flow requirements are often ignored in large-scale models and the environmental needs are estimated as a minimum flow threshold (often Q_{90} or 10 % of mean annual), which should be maintained in the river reaches (e.g., Hanasaki et al., 2008a; Döll et al., 2009; Strzepek et al., 2010, 2012b; Blanc et al., 2013). Other rules have been also suggested. For instance, Haddeland et al. (2006) considered a 7-day consecutive low flow with a 10-year recurrence period as the environmental flow requirement. Although these rules are easily implementable for larger regions and global scales, they widely ignore natural system complexity and the local policy context and can contribute to misunderstanding of the extent of environmental water stress (Arthington et al., 2006).

At this stage of model development, municipal, industrial and energy-related water demands are considered as the most dominant forms of non-irrigative uses. These demands are estimated using complex functions of socio-economic and technological factors, with high variability in time and space. Population is the most significant factor driving these withdrawals (e.g., Alcamo et al., 2003; Hanasaki et al., 2008a; Wada et al., 2014). National gross domestic product (GDP) is also a strong factor (e.g., Gleick, 1996; Cole, 2004; Wada et al., 2011). Although higher GDP may trigger more municipal water use per capita (Alcamo et al., 2007), Hughes et al. (2010) showed that, in general, water uses per capita are greater in developing than developed countries due to low-tech water delivery and industrialization. Strzepek et al. (2010) argued that industrial water use increases with the level of resource industry and decreases when a country moves toward the service sector. Industrial technology is another important factor for non-irrigative use as the extent of both consumptive and non-consumptive uses can significantly change based on the type of technology. Macknick et al. (2011), for instance, provided estimates of total water withdrawals and consumption for most electricity generation technologies within the US. Comparing to recirculating cooling technology, they noted that once-through cooling requires 10 to 100 times more water withdrawal per unit of electric generation. However, the latter consumes less than half of the water consumed by recirculating cooling technology. Climate can be another important factor controlling both consumptive and non-consumptive withdrawals (e.g., Wada et al., 2011, 2014; Hejazi et al., 2013a; Voisin et al., 2013), but has often been ignored as an explicit driver of non-irrigative water demand.

4.2 Top-down algorithms for estimation of grid-based non-irrigative withdrawals

Unlike irrigation demand, top-down approaches have been widely used for non-irrigative withdrawals to transfer national or geopolitical data to basin or grid scales. Various downscaling procedures have been suggested, based on different proxies (see Table 4). These top-down schemes are heavily influenced by the availability of national and global data sets and the downscaling algorithms within the Water – Global Assessment and Prognosis scheme, which is a global water budget and use model (WaterGAP; Alcamo et al., 1997, 2003, 2007). Currently, the availability of different global information sources has provided the opportunity to generate gridded products from different sources. As an example, Hanasaki et al. (2008a) merged the FAO-AQUASTAT data with population distributions and national boundary information from Columbia University (CIAT, 2005) and the consumptive ratios of Shiklomanov (2000) to come up with gridded industrial and municipal water withdrawals and uses at the global scale. More detailed information on various industrial uses resulted in breaking down the industrial withdrawals into their components. For instance, Vassolo and Döll (2005) distinguished between industrial water uses related to thermoelectric power generation and manufacturing production. Temporal disaggregation of annual withdrawals, however, has received much less attention. Recently Wada et al. (2011, 2014) and Voisin et al. (2013) developed simple algorithms to disaggregate annual data to monthly and daily estimates (see Table 5).

4.3 Projection of non-irrigative demand

Characterizing the past and future evolution of non-irrigative demands is required to understand the mechanisms controlling water use and water allocation. Current projections have coarse temporal and spatial resolution and describe non-irrigative demands as functions of socio-economic and technological developments (e.g., Davies et al., 2013; Blanc et al., 2013; Hejazi et al., 2013b, d; Voisin et al., 2013). These changes can be characterized by intermediate socio-economic and technological scenarios, as briefly explained above for irrigation expansion (see Sect. 3.4). The projected demands can be further downscaled using various proxy variables, as explained in Sect. 4.2. Table 6 summarizes some representative efforts, which can be classified as explicit and implicit algorithms. In explicit algorithms, changes in water withdrawals are directly described as functions of changes in socio-economy, technology and water price using simple parametric structures (e.g., Strzepek et al., 2012b; Flörke et al., 2013; Hanasaki et al., 2013a; Hejazi et al., 2013a). The parameters can be assigned using the available global and regional data. In implicit procedures, first the production (or population) is estimated based on integrated economy and population models or prescribed scenarios. By considering the amount of water withdrawal per unit of production (or population) and accounting for technological and/or socio-economic shifts, water withdrawals are consequently projected.

5 State of large-scale modeling applications

The algorithms reviewed in Sects. 3 and 4 have had a wide range of online and offline applications. In comparison to offline applications, online simulations are still at a relatively early stage of development; they typically only include irrigation, mainly implemented at regional scale and under current conditions, and present rather contradictory results. Offline applications in contrast include both irrigative and non-irrigative demands, performed under current and future con-

Table 4. Representative examples calculating grid-based non-irrigative demands using downscaling of coarse-scale estimates.

Reference	Estimated demand	Downscaling procedure	Data support	Targeted resolution
Alcamo et al. (2003)	Domestic	Distributing country-level withdrawals based on population, ratio of rural to urban population (constant for each country) and percentage of population with access to drinking water	Population (van Woerden et al., 1995); access to drinking water (WRI, 1998)	0.5° × 0.5° (global)
	Industrial	Downscaling countywide industrial withdrawals based on proportion of urban population	Population (van Woerden et al., 1995)	
Vassolo and Döll (2005)	Thermoelectric cooling	Calculating the gridded data for power production based on downscaling global estimates. Allocating constant flow to each unit of production according to type of cooling system.	World Electric Power Plants Data Set (http://www.platts.com)	0.5° × 0.5° (global)
	Manufacturing	Estimating countrywide sectoral production volumes along with water intensity for each unit of production in each sector. Downscaling total demand to the grid scale based on city nighttime light.	Industrial production volumes (UN, 1997; CIA, 2001); sectoral intensity (Shiklomanov, 2000; WRI, 2000); night city light pollution (US Air Force, www.ngdc.noaa.gov/dmsp)	
Hanaskai et al. (2008a)	Domestic and industrial	Countrywide data downscaled to grid scale by weighting population and national boundary information, further converted to water consumption estimates.	AQUASTAT countrywide withdrawals, population and national boundaries (CIAT, 2005); ratio of consumption to withdrawal (Shiklomanov, 2000).	1° × 1° (global)
Hejazi et al. (2013b)	Municipal and industrial	Demand estimates of GCAM model (http://wiki.umd.edu/gcam) downscaled as a function of population. Population density assumed static in time.	Global population density data based on WWDR-II and methodology of Wada et al. (2011, 2013a)	0.5° × 0.5° (global)

Table 5. Representative examples for disaggregating annual non-irrigative demand into monthly estimates.

Reference	Estimated demand	Disaggregation procedure	Data support
Wada et al. (2011, 2014)	Municipal and livestock	Downscaling annual demand to monthly fluctuations as a function of temperature	CRU (New et al., 1999, 2000)
Voisin et al. (2013)	Electrical	Dividing electrical use into industry, transportation and building sectors. Assuming uniform distribution for industry and transportation uses and capturing the monthly fluctuations in building use based on heating/cooling degree days.	CASCaDE (http://cascade.wr.usgs.gov)

Table 6. Representative examples for projection of non-irrigative water demands using socio-economic variables.

Reference	Simulated demands	Simulation procedure	Temporal resolution	Spatial resolution
Alcamo et al. (2003)	Domestic and industrial	Explicit simulation of change in industrial and domestic withdrawal as functions of usage intensity and technological change. Usage intensities are functions of GDP.	Annual	Countrywide
Strzepek et al. (2012b)	Municipal and industrial	Explicit simulation of change in municipal water use as a function of population and per capita income. Industrial water use considered as a function of water use per capita and GDP considering growth rate and climatic and water availability factors.	Annual	Assessment sub-regions (global)
Flörke et al. (2013)	Domestic and industrial	Explicit simulation of domestic demand using Alcamo et al. (2003) with parameterization based on HYDE (http://themasites.pbl.nl/tridion/en/themasites/hyde/) and UNEP (http://www.unep.org/) data sets. Technological change influenced electrical demand. Manufacturing water use computed as a function of baseline structural intensity and rates of manufacturing gross value and technological change.	Annual	Countrywide (global)
Davies et al. (2013)	Electrical	Implicit simulation – changes in regional cooling system shares estimated based on shift from wet to dry cooling technologies. Reductions in water withdrawal and consumptions estimated based on level of technological change.	Annual	Geopolitical regions (global)
Hanasaki et al. (2013a)	Industrial and municipal	Explicit simulation of industrial withdrawal as a function of electricity production and water intensity which decreases linearly in time. Municipal water use calculated as a function of population and change in municipal intensity, varying based on GDP.	Five-year interval	Countrywide
Blanc et al. (2013)	Electrical, domestic, industrial and mining	Electrical demand projected implicitly using ReEDS (Short et al., 2009) and integration with USREP model (Rausch and Mowers, 2013). Water withdrawal and consumption to meet electrical demand estimated using Strzepek et al. (2012a). Other demands categorized into three groups: public supply, self-supply and mining supply and simulated explicitly. Public supply considered as a function of population and GDP per capita. Self-supply considered as function of sectoral GDP. Mining supply considered as a function of mining's GDP.	Annual	Assessment sub-regions (US)
Hejazi et al. (2013a)	Municipal	Withdrawal per capita explicitly determined as a function of GDP per capita, water price and technological development. Technological development considered as a function of operational efficiency, which further determines extent of water use.	Annual	Geopolitical regions (global)
Hejazi et al. (2013b, d)	Industrial	Manufacturing water demand is explicitly simulated based on population and GDP. Water demand for primary energy scaled by amount of fuel production and water demand for secondary energy.	Annual	Geopolitical regions (global)
Wada et al. (2014)	Industrial and municipal	Industrial and municipal withdrawal taken from WWDR-II data set (Shiklomanov, 1997; Vörösmarty et al., 2005) and backcasted explicitly using economic and technological proxies. Net municipal water demand calculated as a function of fraction of urban to total population and recycling ratio.	Annual	Countrywide (global)

ditions, and provide relatively more consistent results. Here, we briefly summarize recent applications and highlight the limitations in current simulations.

5.1 Online representation

Recent studies have shown that including irrigation in coupled land-surface schemes can generally improve climate simulations. With respect to regional temperature, for instance, Saeed et al. (2009) showed that representing irrigation activities over northwestern India and Pakistan can reduce climate model simulation bias by 5° K. It should be noted, however, that there are still large disagreements in quantifying the effects of irrigation on regional and global temperature (see, e.g., Boucher et al., 2004 vs. Lobell et al., 2006), mainly attributed to the difference in the implemented irrigation demand calculations. Sacks et al. (2009) tried to overcome the limitations in demand algorithms by downscaling the AQUASTAT irrigative water use data to the grid scale. They concluded that irrigation has significant importance for regional temperature, but at global scale the temperature cooling in some regions due to irrigation is canceled by temperature warming in some other areas due to climate, land-cover and circulation changes. There are, however, some limitations in their study, as the irrigation demand did not vary between years and they applied irrigation only when the LAI is around 80 % of the annual LAI. These assumptions can result in large uncertainty.

Irrigation-induced precipitation has been studied for quite some time and irrigation has been shown to have a significant effect on local and regional precipitation patterns (e.g., Barnston and Schickedanz, 1984; Moore and Rojstaczer, 2001). For instance despite regional decline, Tuinenberg et al. (2011) found a positive precipitation trend in climate stations located in the irrigated regions of the southern Asia. Lucas-Picher et al. (2011) tested four climate models and argued that lack of representation of irrigation is the main reason for precipitation bias over the Indian monsoon area. Guimberteau et al. (2012) showed that irrigation can also affect the onset of mean monsoon date over the Indian peninsula, leading to a significant decrease in precipitation during May to July. Nonetheless, there are still large disagreements in (1) identifying the dominant mechanisms that drive the irrigation-induced precipitation; and (2) estimating the amount and spatial extension of change in precipitation. DeAngelis et al. (2010) noted that the growing season precipitation increased in the Great Plains of the US during the 20th century as a result of intensive irrigation. Using vapor tracking analysis, they indicated that evaporation from irrigated lands adds to downwind precipitation, which increases as the evaporation increases. Harding and Snyder (2012a, b), however, noted that the extent of effects on precipitation also depend on the antecedent soil moisture. They argued that, in low soil moisture conditions, further irrigation can result in suppression of regional precipitation. Guimberteau et al. (2012) argued that these contrasting results might be due to differences in local moisture, where the irrigation is applied. Based on a 30-year simulation, they showed an increase in summer precipitation over the arid western region of the Mississippi River basin in association with enhanced evapotranspiration. However, a decrease in precipitation was identified over the wet eastern part of the basin. These results, however, are based on only one set of models and the coarse grid resolution might degrade the quality of simulations – see the discussion below. With respect to the scale of disturbance, Sorooshian et al. (2011) showed that irrigation over California's Central Valley significantly decreases local temperature and increases local precipitation; however, they argued that the effects of irrigation do not expand far from the place where irrigation takes place. In contrast, Lo and Famiglietti (2013) argued that irrigation in California's Central Valley intensifies the water cycle in the southwestern US and can increase the flow in the Colorado River.

There are two main limitations associated with available simulations of irrigation-induced rainfall discussed above. First, in most of the online studies, water availability is not a constraint. As a result, the water balance is not closed and they simply analyze whether evaporation increase can enhance atmospheric moisture convergence or not. This can be considered as a major limitation as the available water can control the extent of irrigation (and consequently evaporation) and stabilize the associated feedback processes. Second, it is known that sharp landscape contrasts (i.e., transitions between wet and cool as well as dry and hot areas) critically affect rainfall formation (e.g., Taylor 2009; C. M. Taylor et al., 2012). Although irrigation can create such transitions due to enhanced evaporation and decreased surface temperature, current LSMs are generally unable to generate the atmospheric perturbations due to these transitions (Polcher, 2014). Due to these limitations, the results of current sensitivity analyses should be considered with caution.

Online simulations under future climate change are limited and have been performed mainly at regional scales. Gerten et al. (2011) used a nested regional climate model to dynamically downscale the future simulations of a global climate model over southern Asia and considered two modes of simulation: with or without irrigation. They concluded that including irrigation can result in roughly half of the temperature increase predicted without representing irrigation. With respect to future precipitation, simulation with and without irrigation both showed a decrease in precipitation over northern India and increase in precipitation over the southern peninsula; the latter was enhanced with irrigation. They noted that the increase in precipitation cannot be seen if the global-scale simulations are not dynamically downscaled. This highlights the importance of including irrigation schemes in regional climate models for dynamic downscaling of future climate change scenarios.

In summary, despite current limitations and differences in the host climate and LSM models, irrigation demand algorithms and simulation settings, significant feedback effects are associated with irrigation. Large uncertainties, however, exist in current coupled irrigation–land-surface–climate modeling, which emphasizes the need for more research in this area.

5.2 Offline representation

Offline representation of water demands is more common, and a wide variety of GHMs and LSMs in conjunction with different demand algorithms have been used to simulate the dynamics of water demand under both current and future conditions. The available global simulations under current conditions are compared and summarized in Wada et al. (2014) and Chaturvedi et al. (2013a, b) for irrigative demands and in Alcamo et al. (2003) and Hejazi et al. (2013b) for total water consumption. Although incorporating water demand calculations can generally result in more realistic river discharge simulations (see Ngo-Duc et al., 2005a, b, 2007), current simulations exhibit large differences in estimates of water demand and use at countrywide, continental and global scales. This can be referred to the differences in data support, demand calculation schemes and host models – see the discussion of Sect. 6.

Normally, future projections of water demands include more uncertainty than simulation of current conditions as they are also conditioned on uncertain climate futures and/or socio-economic and technological scenarios. Considering future climate projections, with or without considering irrigation expansion, irrigation demand algorithms have mainly projected increase in irrigation demand under climate change scenarios. As an early example, Fischer et al. (2007) estimated irrigation water requirement as a function of both projected irrigated land and climate change from 1990 to 2080. They showed that the impact of climate change on increasing irrigation water requirement could be nearly as large as the changes initiated by socio-economic developments. There are, however, two sets of uncertainty associated with future projections of irrigation demand. First, gridded climate products have significant deficiencies in representing current and future climate, particularly with respect to precipitation (e.g., Lorenz and Kunstmann, 2012; Grey et al., 2013). This can further propagate to estimation of irrigation demand at the sub-grid scale. Second, there are large disagreements between irrigation demand projections with respect to different climate model simulation, irrigation algorithms and host large-scale models. One possible approach to account for these uncertainties would be using a multi-model approach, as recommended by Gosling et al. (2011) and Haddeland et al. (2011, 2014) and implemented to some extent by Wada et al. (2013) and Rosenzweig et al. (2014). Based on the latest IPCC climate scenarios (K. E. Taylor et al., 2012), these studies generally concluded that a significant increase in future demand is likely, with possibly 1-month or more shift in the peak irrigation demand in mid-latitude regions (Wada et al., 2013), but large uncertainties are associated with the predictions (see Rosenzweig et al., 2014). Moreover, both studies noted that CO_2 increases might have beneficial effects on crop transpiration efficiency, if other factors are not limiting (see also Gerten et al., 2011; Konzmann et al., 2013). Nonetheless, it still remains unclear whether increased transpiration efficiency is canceled out by increased transpiration due to increasing biomass and plant growth. More studies, therefore, are required in this direction (see Gerten, 2013). This is a context for which LSMs can offer an ideal platform as they have the explicit modules required for considering dynamic interactions of carbon, vegetation and water – see the discussion of Sect. 6.

Similar conclusions were obtained with respect to non-irrigative demands. Alcamo et al. (2007) and Hejazi et al. (2013d) showed that increasing domestic and industrial water uses, if not controlled, can be a major threat for water supply. There are, however, large discrepancies between different projections of non-irrigative demands (Gleick, 2003), in which the divergence between modeling results becomes more highlighted as the projection horizon increases (see Davis et al. (2013) for electrical demand and associated water use). These uncertainties can be referred to limitations in current data availability for supporting robust and reliable projections, differences in socio-economic and technological scenarios, as well as some underlying assumptions in demand calculation algorithms, which can limit their efficiency in future simulations.

As the current global potential for expanding water demand is rather limited (Rost et al., 2009; Gerten and Rost, 2010), adaptation and mitigation strategies are required to moderate human water demands. In such cases prescribed "policy" scenarios can be introduced into large-scale models for impact assessment. Using this approach, it has been shown that mitigation can significantly decrease future global water demand. For example, Hanasaki et al. (2013a) showed approximately 7-fold and 2.5-fold variation in industrial and municipal demands, depending on the SSP considered. The effects of mitigation, however, have large regional variation. For irrigative demands, Fischer et al. (2007) showed that some regions may be negatively affected by mitigation actions, which depend on specific combinations of CO_2 changes that affect crop water requirement and projected precipitation and temperature changes. Kyle et al. (2013) showed that applying CO_2 mitigation policies can result in high deployment of other high-tech solutions for electrical generation (e.g., solar power) that have low water requirements. Hejazi et al. (2013c) further showed that taxation can be an important factor in mitigating the effect of water scarcity by regulating more water efficient options for irrigation. Hejazi et al. (2013a) further showed the possibility of a slight decrease in municipal withdrawals in the year 2100 under a high-tech scenario, despite significant population growth. Davies et al. (2013) showed similar results for electricity water withdrawals if high-tech solutions are employed. Large-scale models also showed that promoting international trade can be a strong adaptation option for controlling regional demand, in which water-limited regions can import water-expensive products from other areas (e.g., Siebert and Döll, 2010; Hanasaki et al., 2010; Konar et al., 2013). Assessment of trade scenarios and water footprinting, however, needs detailed tracking of the water cycle (see Chenoweth et al., 2014) and is highly dependent on how reasonable the human demands and production, as well as water availability and water allocation, are described in time and space. Such a level of accuracy is currently not available and therefore the assessments remain widely uncertain.

In summary, current offline projections agree on large impacts of future change in climate, socio-economy and technology on water demands and the importance of adaptation and mitigation strategies for managing future water security threats. Available projections, however, are rather limited and suffer from major sources of uncertainty, which is revealed by large discrepancies between different simulation products under current and future conditions. We now turn to discuss these gaps in more detail and identify the research needs and priorities.

6 Discussion

Major gaps remain in the current capability in modeling water demands and understanding their online and offline impacts on the terrestrial water cycle and human livelihoods. These gaps are partially due to inherent complexity in modeling Earth system processes, which is more significant in coupled simulation modes. Apart from various computational barriers, one main challenge in online simulations is the uncertainty associated with coupling land and atmospheric models, as given a unique land-surface boundary condition, the simulations obtained by different climate models can be divergent (Koster et al., 2004; Pitman et al., 2009; Dadson et al., 2013). Another major challenge for coupled irrigation–land-surface–climate simulations is the choice of appropriate temporal and spatial resolutions, at which the relevant physical processes and feedbacks between land and atmosphere should be represented and described. Ideally, the optimal modeling resolution should be identified based on physical realism; nonetheless, the choice of resolution in coupled simulations is mainly constrained by computational resources, data availability and the complexity supported by the LSMs. If these are not limiting factors, it has been shown that finer temporal and spatial resolutions can improve online representation of irrigation. For instance, using six different combinations of temporal/spatial resolutions, Sorooshian et al. (2011) concluded that spatial and temporal resolution in coupled irrigation–land–climate models can significantly change both temperature and precipitation simulations over irrigated grids and a fine level of detail is required for representing the physical processes controlling the feedbacks between irrigation and atmosphere. However, these findings remain regionally and seasonally dependent and are closely linked to the level of complexity supported in the considered irrigation parameterization and host model. It should be noted that, by increasing the spatial resolution, more processes need to be included in order to ensure water conservation within the model and that can further complicate the issues related to water availability – see the discussion below. The effects of fine modeling resolution seem to be in general less significant in offline runs, as far as the evaporation calculation is consistent with estimation of crop water requirements and each crop is supplied by a unique moisture reservoir. Compton and Best (2011) conducted offline global simulations and showed that fine spatial resolution has little importance on long-term modeling of evaporation and runoff; however, the temporal resolution does change the mean evaporation/runoff balance. The issues around modeling resolution are explored further in Nazemi and Wheater (2015).

Large uncertainties are also associated with offline human water demand simulations under current and future conditions. Lissner et al. (2012), for instance, noticed significant difference in terms of water demand per capita between the simulated products of WaterGAP and reported AQUASTAT data. These uncertainties are mainly related to (i) available data support, (ii) demand calculation algorithms and (iii) host models. These sources are widely connected and cannot be easily addressed and quantified independently. Here we briefly discuss these sources and propose few directions for future developments.

1. Uncertainty in current data support: primarily, there are considerable uncertainties across the input and forcing data required for executing large-scale models. Generally, large-scale models discussed in this paper are forced and initialized using various data sources that are developed and maintained independently. This results in major inconsistencies, particularly at the grid scale, where it is often the case that information coming from different sources does not match each other (e.g., soil properties do not fit to land use). Typically, modelers fix these issues by applying simple rules or assumptions; however, inconstancies in personal judgments can highly affect the quality of simulations at the local and regional scales (see Bormann et al., 2011, for a local study). Major uncertainties are also associated with the data required for executing demand calculation algorithms. Siebert et al. (2005) noted that even the locations of irrigation districts are uncertain in many regions and sub-grid variability of crops within irrigated areas is not generally available. Wisser et al. (2008) argued that major uncertainties are associated with forcing, irrigation and crop maps, and this can result in large differences between simulations of irrigation water requirement. Another source of data uncertainty is the generally sparse information on irrigation techniques.

This can be important for understanding the amount of water loss and thus estimating the actual irrigation use and evaporation (see, e.g., Evan and Zaitchik, 2008). The issues around data support apply to non-irrigative demands as well. For the case of water use for electricity generation in the US, Macknick et al. (2011) noted that "federal data sets on water use in power plants have numerous gaps and methodological inconsistencies." Data uncertainty can propagate into structural and parametric identification during model development and can further extend to future projections. The availability of different sources of global and regional data has resulted in emergence of various data sets, with varying degrees of quality, which can potentially support demand calculation algorithms. At this stage of research, the various data sets have not been systematically compared with respect to their uncertainty and the associated effects on demand simulations. This is a major need for future exploration.

2. Uncertainty in demand calculation algorithms: this includes both irrigative and non-irrigative demands.

    a. Irrigative demand: limitations in current algorithms mainly include the uncertainty in describing the crop moisture requirements in time and space and constraining the irrigation to water availability. If the irrigation is limited by the water available at the grid scale, then the quality of simulation is hindered by the ability of the host model to describe water allocation from surface and groundwater resources (see Nazemi and Wheater, 2015). In addition, current bottom-up algorithms do not appropriately consider plant-specific water requirements at the sub-grid scale due to missing soil and crop diversity. This can result in misestimating the irrigation demand. In the best situation, where the same assumption is used for the calculation of the crop evaporation and the irrigation demand, the uncertainty of the irrigative demand is the same as evaporation, but this can still vary greatly across various host models. Considering future simulations, widely used irrigation demand estimates based on FAO guidelines often require several input variables (see, e.g., Farmer et al., 2011 and Hejazi et al., 2013b, for simplifications), and given the need for downscaling of climate variables for future simulations, these can be outperformed by simpler models (e.g., Vörösmarty et al., 1998; Wisser et al., 2010). At the current stage of research, different methods for calculating irrigative demand have not yet been fully intercompared to identify appropriate algorithms with respect to region, climate and type of crops. This can be considered as an important need for further research. Another avenue for future development is improving the demand simulations using data assimilation and model calibration. These opportunities will be discussed further in Nazemi and Wheater (2015).

    b. Non-irrigative demand: the current offline modeling capability is generally temporally coarse, and available downscaling and projection algorithms mainly do not account for seasonal variations in water demand. There are also parametric and structural uncertainties in functional mappings that link water demand to socio-economic and technological proxies due to limitations in available data as well as the diversity and spatiotemporal variability in non-irrigative demands. At this stage, it is not fully understood how these uncertainties propagate into future projections considering additional uncertainty in future climate and socio-economic scenarios. Developing robust downscaling and projection algorithms for estimation of non-irrigative demands therefore is an important need for future development.

3. Uncertainty in host models: host models can add substantial uncertainty to demand simulations, particularly for irrigation. As noted in Sect. 3, the calculation of irrigation demand involves solving the soil water balance at every simulation time step and this is determined by how the relevant natural processes, such as actual evapotranspiration and soil moisture, are parameterized in the host model. Haddeland et al. (2011) showed major differences in the global simulations obtained from six LSMs and five GHMs due to differences in underlying assumptions, process representations, and related parameterizations. It is also shown that considering feedback effects between irrigation and atmosphere can considerably change potential evaporation (e.g., Blyth and Jacobs, 2011; Lu, 2013); therefore offline irrigation demand simulations based on GHMs might be biased as they inherently ignore climate feedbacks. Moreover, GHMs often cannot represent important processes such as the effects of increased carbon concentration on irrigation demand. This limitation may result in major deficiencies in simulating climate change scenarios as CO_2 increases can significantly change vegetation dynamics (e.g., Prudhomme et al., 2014), which can further alter the evaporation and runoff regimes (Gerten et al., 2004). From this perspective, it can be concluded that online LSMs are superior to GHMs with respect to simulations under increasing CO_2 concentration and future water stress, as they often include many of the required computational components for investigating interactions between climate, carbon, vegetation and water cycles. Efforts are however needed to transfer recent demand calculation algorithms developed in the context of GHMs into LSMs. In addition, although it has been argued that the uncertainties in host models are more

significant than in climate forcing (e.g., Wada et al., 2013), uncertainties in irrigation algorithms and large-scale host models have not been fully disaggregated and distinguished. This requires a "mix and match" of multiple demand algorithms with multiple host models to conduct a systematic intercomparison and sensitivity analysis. This can be considered as an important research direction – see Nazemi and Wheater (2015).

7 Summary and concluding remarks

The terrestrial water cycle has been greatly affected in time and space by human activities during the recent past, to the extent that the current geological era has been named the "Anthropocene". Anthropogenic activities, therefore, are required to be represented in models that are used for impact assessments, large-scale hydrological modeling and land–atmosphere feedback representations. Current human–water interactions are mainly manifested through water resource management, which can be further broken down into two interacting components, related to water demand as well as water supply and allocation. In this paper we considered the representation of water demand in large-scale models. Water demand was further divided into irrigative and non-irrigative categories. We summarized current demand calculation algorithms based on type of demand, modeling procedure and underlying assumptions. Current applications were overviewed; and limitations in knowledge were identified and discussed. Considering current gaps in representing the anthropogenic demands in large-scale models, three main directions are suggested for future developments. These include (1) systematic intercomparisons between different data sets, demand algorithms and host models and associated uncertainties with respect to different geographic regions as well as various socio-economic and climate conditions; (2) developing improved algorithms for calculating both irrigative and non-irrigative demands in time and space considering data limitations as well as diversity and spatiotemporal variability in human demand; and finally (3) transferring the algorithms developed in the context of GHMs to LSMs for (a) improved irrigation demand calculation under increasing CO_2 effects; and (b) further coupled studies with climate models to address various scientific questions with respect to interactions between carbon, irrigation and climate under climate change conditions. Apart from these immediate research needs, efforts are also required to link with socio-economic and energy models to have a full understanding of the dynamic interactions between natural and anthropogenic drivers of human water availability, demand and consumption (Calvin et al., 2013). This seems to be more of a long-term development due to the limitations in current demand algorithms, LSMs as well as socio-economic and energy models.

As a final remark, it must be noted that the effects of water demand on both the terrestrial water cycle and water security cannot be fully studied unless considered in conjunction with water supply and allocation, which determine the extent of human intervention in water cycle. This is particularly important for future predictions, as increasing water scarcity is a major limiting factor for water demand and can substantially increase competition over available water sources. In Nazemi and Wheater (2015), we review how water supply and allocation have been represented at larger scales and been integrated with various water demands and natural land-surface processes at grid and sub-grid scales.

Acknowledgements. The first author would like to thank Amir Aghakouchak for his valuable inputs during the early stages of this survey. The authors gratefully acknowledge the constructive comments from the editor, Jan Polcher, Ingjerd Haddeland, Jason Evans and two anonymous reviewers, which have enabled us to make significant improvements to this paper. Financial support for this study was provided by the Canada Excellence Research Chair in Water Security at the University of Saskatchewan.

Edited by: W. Buytaert

References

Abdullah, K. B.: Use of water and land for food security and environmental sustainability, Irrig. Drain., 55, 219–222, doi:10.1002/ird.254, 2006.

Adam, J. C. and Lettenmaier, D. P.: Adjustment of global gridded precipitation for systematic bias, J. Geophys. Res., 108, 4257, doi:10.1029/2002JD002499, 2003.

Adam, J. C., Haddeland, I., Su, F., and Lettenmaier, D. P.: Simulation of reservoir influences on annual and seasonal streamflow changes for the Lena, Yenisei and Ob' rivers, J. Geophys. Res.-Atmos., 112, D24114, doi:10.1029/2007JD008525, 2007.

Adegoke, J. O., Pielke Sr., R. A., Eastman, J., Mahmood, R., and Hubbard, K. G.: Impact of irrigation on midsummer surface fluxes and temperature under dry synoptic conditions: A regional atmospheric model study of the US High Plains, Mon. Weather Rev., 131, 556–564, 2003.

AghaKouchak, A., Norouzi, H.-R., Madani, K., Mirchi, A., Azarderakhsh, M., Nazemi, A., Nasrollahi, N., Farahmand, A.-R., Mehran, A., and Hasanzadeh, E.: Aral Sea syndrome desiccates Lake Urmia: Call for action, Journal of Great Lakes Research, doi:10.1016/j.jglr.2014.12.007, in press, 2014.

Alcamo, J., Döll, P., Kaspar, F., and Siebert, S.: Global change and global scenarios of water use and availability: an application of WaterGAP 1.0, Center for Environmental Systems Research (CESR), University of Kassel, Germany, available at: http://www.usf.uni-kassel.de/usf/archiv/dokumente/projekte/watergap.teil1.pdf (last access: 6 May 2014), 1997.

Alcamo, J., Döll, P., Henrichs, T., Kaspar, F., Lehner, B., Rösch, T., and Siebert, S.: Development and testing of the WaterGAP 2 global model of water use and availability, Hydrolog. Sci. J., 48, 317–337, 2003.

Alcamo, J., Flörke, M., and Märker, M.: Future long-term changes in global water resources driven by socio-economic and climatic changes, Hydrolog. Sci. J., 52, 247–275, 2007.

Allen, R. G., Pereira, L. S., Raes, D., and Smith, M.: Crop evapotranspiration-Guidelines for computing crop water requirements-FAO Irrigation and drainage paper 56, FAO, Rome, http://www.engr.scu.edu/~emaurer/classes/ceng140_watres/handouts/FAO_56_Evapotranspiration.pdf (last access: 6 May 2014), 1998.

Antonellini, M., Mollema, P., Giambastiani, B., Bishop, K., Caruso, L., Minchio, A., Pellegrini, L., Sabia, M., Ulazzi, E., and Gabbianelli, G.: Salt water intrusion in the coastal aquifer of the southern Po Plain, Italy, Hydrogeol. J., 16, 1541–1556, 2008.

Arnell, N. W.: Climate change and global water resources, Global Environ. Change, 9, 31–49, 1999.

Arnell, N. W.: Climate change and global water resources: SRES emissions and socio-economic scenarios, Global Environ. Change, 14, 31–52, 2004.

Arthington, A. H., Bunn, S. E., Poff, N. L., and Naiman, R. J.: The challenge of providing environmental flow rules to sustain river ecosystems, Ecol. Appl., 16, 1311–1318, 2006.

Barella-Ortiz, A., Polcher, J., Tuzet, A., and Laval, K.: Potential evaporation estimation through an unstressed surface-energy balance and its sensitivity to climate change, Hydrol. Earth Syst. Sci., 17, 4625–4639, doi:10.5194/hess-17-4625-2013, 2013.

Barnston, A. G. and Schickedanz, P. T.: The effect of irrigation on warm season precipitation in the southern Great Plains, J. Clim. Appl. Meteorol., 23, 865–888, 1984.

Beddington, J.: Catalysing sustainable water security: role of science, innovation and partnerships, Philos. T. Roy. Soc. A, 371, 414, doi:10.1098/rsta.2012.0414, 2013.

Blanc, E., Strzepek, K., Schlosser, A., Jacoby, H. D., Gueneau, A., Fant, C., Rausch, S., and Reilly, J.: Analysis of U.S. water resources under climate change, MIT Joint Program on the Science and Policy of Global Change, Report No. 239, http://globalchange.mit.edu/files/document/MITJPSPGC_Rpt239.pdf (last access: 6 May 2014), 2013.

Blyth, E. and Jacobs, C.: Including climate feedbacks in regional water resource assessments, WATCH Water and global change, Report No. 38, http://www.eu-watch.org/publications/technical-reports (last access: 6 May 2014), 2011.

Bondeau, A., Smith, P. C., Zaehle, S., Schaphoff, S., Lucht, W., Cramer, W., Gerten, D., Lotze-Campen, H., Müller, C., Reichstein, M., and Smith, B.: Modelling the role of agriculture for the 20th century global terrestrial carbon balance, Global Change Biol., 13, 679–706, doi:10.1111/j.1365-2486.2006.01305.x, 2007.

Bormann, H., Holländer, H. M., Blume, T., Buytaert, W., Chirico, G. B., Exbrayat, J.-F., Gustafsson, D., Hölzel, H., Kraft, P., Krauße, T., Nazemi, A., Stamm, C., Stoll, S., Blöschl, G., and Flühler, H.: Comparative discharge prediction from a small artificial catchment without model calibration: Representation of initial hydrological catchment development, Bodenkultur, 62, 23–29, 2011.

Boucher, O., Myhre, G., and Myhre, A.: Direct human influence of irrigation on atmospheric water vapour and climate, Clim. Dynam., 22), 597–603, 2004.

Bowman, D. M. J. S., Balch, J. K., Artaxo, P., Bond, W. J., Carlson, J. M., Cochrane, M. A., D'Antonio, C. M., DeFries, R. S., Doyle, J. C., Harrison, S. P., Johnston, F. H., Keeley, J. E., Krawchuk, M. A., Kull1, C. A., Marston, J. B., Moritz, M. A., Prentice, I. C., Roos, C. I., Scott, A. C., Swetnam, T. W., van der Werf, G. R., and Pyne, S. J.: Fire in the Earth system, Science, 324, 481–484, 2009.

Brovkin, V., Claussen, M., Driesschaert, E., Fichefet, T., Kicklighter, D., Loutre, M.-F., Matthews, H. D., Ramankutty, N., Schaeffer, M., and Sokolov, A.: Biogeophysical effects of historical land cover changes simulated by six Earth system models of intermediate complexity, Clim. Dynam., 26, 587–600, 2006.

Calvin, K., Wise, M., Clarke, L., Edmonds, J., Kyle, P., Luckow, P., and Thomson, A.: Implications of simultaneously mitigating and adapting to climate change: initial experiments using GCAM, Climatic Change, 117, 545–560, 2013.

Cayan, D. R., Das, T., Pierce, D. W., Barnett, T. P., Tyree, M., and Gershunov, A.: Future dryness in the southwest US and the hydrology of the early 21st century drought, P. Natl. Acad. Sci., 107, 21271–21276, 2010.

Chaturvedi, V., Hejazi, M., Edmonds, J., Clarke, L., Kyle, P., Davies, E., and Wise, M.: Climate mitigation policy implications for global irrigation water demand, Mitig. Adapt. Strat. Global Change, 18, 1–19, 2013a.

Chaturvedi, V., Hejazi, M., Edmonds, J., Clarke, L., Kyle, P., Davies, E., Wise, M., and Calvin, K. V.: Climate Policy Implications for Agricultural 5 Water Demand, Pacific Northwest National Laboratory, Richland, WA, available at: http://www.globalchange.umd.edu/wp-content/uploads/projects/PNNL-22356.pdf (last access: 6 May 2014), 2013b.

Chen, F. and Dudhia, J.: Coupling an advanced land surface-hydrology model with the Penn State-NCAR MM5 modeling system Part I: Model implementation and sensitivity, Mon. Weather Rev., 129, 569–585, 2001a.

Chen, F. and Dudhia, J.: Coupling an advanced land surface-hydrology model with the Penn State-NCAR MM5 modeling system Part II: Preliminary model validation, Mon. Weather Rev., 129, 587–604, 2001b.

Chenoweth, J., Hadjikakou, M., and Zoumides, C.: Quantifying the human impact on water resources: a critical review of the water footprint concept, Hydrol. Earth Syst. Sci., 18, 2325–2342, doi:10.5194/hess-18-2325-2014, 2014.

CIA: CIA World Factbook [CD-ROM], Washington, D.C., https://www.cia.gov/library/publications/the-world-factbook (last access: 6 May 2014), 2001.

CIAT: Gridded Population of the World, Version 3 (GPWv3): Population Density Grid, NASA Socioeconomic Data and Applications Center (SEDAC), http://sedac.ciesin.columbia.edu/data/set/gpw-v3-population-density (last access: 6 May 2014), 2005.

Claussen, M.: Earth system models, in: Understanding the Earth System: Compartments, Processes and Interactions, edited by: Ehlers, E., and Krafft, T., Springer-Verlag, Heidelberg, 145–162, 2001.

Cole, M. A.: Economic growth and water use, Appl. Econ. Lett., 11, 1–4, 2004.

Collins, W. D., Rasch, P. J., Boville, B. A., Hack, J. J., McCaa, J. R., Williamson, D. L., Briegleb, B., Bitz, C., Lin, S.-J., Zhang, M., and Dai, Y.: Description of the NCAR community atmosphere model (CAM 3.0), NCAR Tech. Note NCAR/TN-464+STR, 226, available at: http://hanson.geog.udel.edu/hanson/hanson/

CLD_GCMExperimentS11_files/description.pdf (last access: 6 May 2014), 2004.

Collins, W. D., Rasch, P. J., Boville, B. A., Hack, J. J., McCaa, J. R., Williamson, D. L., Briegleb, B., Bitz, C., Lin, S.-J., and Zhang, M.: The formulation and atmospheric simulation of the Community Atmosphere Model version 3 (CAM3), J. Climate, 19, 2144–2161, 2006.

Compton, E. and Best, M.: Impact of spatial and temporal resolution on modelled terrestrial hydrological cyce components, WATCH Water and global change. Report No. 44, http://www.eu-watch.org/publications/technical-reports (last access: 6 May 2014), 2011.

Cosgrove, B. A., Lohmann, D., Mitchell, K. E., Houser, P. R., Wood, E. F., Schaake J. C., Robock, A., Marshall, C., Sheffield, J., Duan, Q., Luo, L., Wayne Higgins, R., Pinker R. T., Dan Tarpley, J., and Meng, J.: Real-time and retrospective forcing in the North American Land Data Assimilation System (NLDAS) project, J. Geophys. Res., 108, 8842, doi:10.1029/2002JD003118, 2003.

Crutzen, P. J.: The "anthropocene", in: Earth System Science in the Anthropocene, edited by: Ehlers, E., Krafft, T., and Moss, C., Springer, Berlin, Heidelberg, 13–18, 2006.

Crutzen, P. J. and Steffen, W.: How long have we been in the Anthropocene era?, Climatic Change, 61, 251–257, 2003.

Dadson, S., Acreman, M., and Harding, R.: Water security, global change and land–atmosphere feedbacks, Philos. T. Roy. Soc. A, 371, 2002, doi:10.1098/rsta.2012.0412, 2013.

Davies, E. G., Kyle, P., and Edmonds, J. A.: An integrated assessment of global and regional water demands for electricity generation to 2095, Adv. Water Resour., 52, 296–313, 2013.

DeAngelis, A., Dominguez, F., Fan, Y., Robock, A., Kustu, M. D., and Robinson, D.: Evidence of enhanced precipitation due to irrigation over the Great Plains of the United States, J. Geophys. Res., 115, D15115, doi:10.1029/2010JD013892, 2010.

Deardorff, J. W.: Efficient prediction of ground surface temperature and moisture, with inclusion of a layer of vegetation, J. Geophys. Res.-Oceans, 83, 1889–1903, 1978.

Dee, D. P., Uppala, S. M., Simmons, A. J., Berrisford, P., Poli, P., Kobayashi, S., Andrae, U., Balmaseda, M. A., Balsamo, G., Bauer, P., Bechtold, P., Beljaars, A. C. M., van de Berg, L., Bidlot, J., Bormann, N., Delsol, C., Dragani, R., Fuentes, M., Geer, A. J., Haimberger, L., Healy, S. B., Hersbach, H., Hólm, E. V., Isaksen, L., Kållberg, P., Köhler, M., Matricardi, M., McNally, A. P., Monge-Sanz, B. M., Morcrette, J.-J., Park, B.-K., Peubey, C., de Rosnay, P., Tavolato, C., Thépaut, J.-N., and Vitart, F.: The ERA-interim reanalysis: configuration and performance of the data assimilation system, Q. J. Roy. Meteorol. Soc., 137, 553–597, doi:10.1002/qj.828, 2011.

de Rosnay, P., Polcher, J., Laval, K., and Sabre, M.: Integrated parameterization of irrigation in the land surface model ORCHIDEE: Validation over Indian Peninsula, Geophys. Res. Lett., 30, 1986, doi:10.1029/2003GL018024, 2003.

Destouni, G., Asokan, S. M., and Jarsjö, J.: Inland hydro-climatic interaction: Effects of human water use on regional climate, Geophys. Res. Lett., 37, L18402, doi:10.1029/2010GL044153, 2010.

Dickinson, R. E.: Land surface processes and climate-surface albedos and energy balance, Adv. Geophys., 25, 305–353, 1983.

Dickinson, R. E.: Modeling evapotranspiration for three-dimensional global climate models, Geophys. Monogr. Ser., 29, 58–72, 1984.

Döll, P.: Vulnerability to the impact of climate change on renewable groundwater resources: a global-scale assessment, Environ. Res. Lett., 4, 035006, doi:10.1088/1748-9326/4/3/035006, 2009.

Döll, P. and Siebert, S.: A digital global map of irrigated areas, ICID J., 49, 55–66, 2000.

Döll, P. and Siebert, S.: Global modeling of irrigation water requirements, Water Resour. Res., 38, 8-1–8-10, doi:10.1029/2001WR000355, 2002.

Döll, P., Fiedler, K., and Zhang, J.: Global-scale analysis of river flow alterations due to water withdrawals and reservoirs, Hydrol. Earth Syst. Sci., 13, 2413–2432, doi:10.5194/hess-13-2413-2009, 2009.

Döll, P., Müller Schmied, H., Schuh, C., Portmann, F. T., and Eicker, A.: Global-scale assessment of groundwater depletion and related groundwater abstractions: Combining hydrological modeling with information from well observations and GRACE satellites. Water Resour. Res., 50, 5698–5720, doi:10.1002/2014WR015595, 2014.

Douglas, E. M., Beltrán-Przekurat, A., Niyogi, D., Pielke Sr., R. A., and Vörösmarty, C. J.: The impact of agricultural intensification and irrigation on land–atmosphere interactions and Indian monsoon precipitation – A mesoscale modeling perspective, Global Planet. Change, 67, 117–128, 2009.

Ducoudré, N. I., Laval, K., and Perrier, A.: SECHIBA, a new set of parameterizations of the hydrologic exchanges at the land-atmosphere interface within the LMD atmospheric general circulation model, J. Climate, 6, 248–273, 1993.

Ek, M. B., Mitchell, K. E., Lin, Y., Rogers, E., Grunmann, P., Koren, V., Gayno, G., and Tarpley, J. D.: Implementation of Noah land surface model advances in the National Centers for Environmental Prediction operational mesoscale Eta model, J. Geophys. Res., 108, 8851, doi:10.1029/2002JD003296, 2003.

Eltahir, E. A.: A soil moisture–rainfall feedback mechanism: 1. Theory and observations, Water Resour. Res., 34, 765–776, 1998.

Entekhabi, D. and Eagleson, P. S.: Land surface hydrology parameterization for atmospheric general circulation models including subgrid scale spatial variability, J. Climate, 2, 816–831, 1989.

Evans, J. P. and Zaitchik, B. F.: Modeling the large-scale water balance impact of different irrigation systems, Water Resour. Res., 44, W08448, doi:10.1029/2007WR006671, 2008.

Falkenmark, M.: Growing water scarcity in agriculture: future challenge to global water security, Philos. T. Roy. Soc. A, 371, 2002, doi:10.1098/rsta.2012.0410, 2013.

Farmer, W., Strzepek, K., Schlosser, C. A., Droogers, P., and Gao, X.: A Method for Calculating Reference Evapotranspiration on Daily Time Scales, MIT Joint Program on the Science and Policy of Global Change, Report number 195, http://18.7.29.232/handle/1721.1/61773 (last access: 6 May 2014), 2011.

Fischer, G., Tubiello, F. N., Van Velthuizen, H., and Wiberg, D. A.: Climate change impacts on irrigation water requirements: effects of mitigation, 1990–2080, Technol. Forecast. Social Change, 74, 1083–1107, 2007.

Flörke, M. and Alcamo, J.: European outlook on water use, Final Report, EEA/RNC/03/007, Center for Environmental Systems Research, University of Kassel, http://www.improve.novozymes.

com/Documents/European_OutlookonWaterUse.pdf, (last access: 6 May 2014), 2004.

Flörke, M., Kynast, E., Bärlund, I., Eisner, S., Wimmer, F., and Alcamo, J.: Domestic and industrial water uses of the past 60 years as a mirror of socio-economic development: A global simulation study, Global Environ. Change, 23, 144–156, 2013.

Förster, H. and Lilliestam, J.: Modeling thermoelectric power generation in view of climate change, Reg. Environ. Change, 10, 327–338, 2010.

Friedl, M. A., McIver, D. K., Hodges, J. C., Zhanga, X. Y., Muchoneyb, D., Strahlera, A. H., Woodcocka, C. E., Gopala, S., Schneidera, A., Coopera, A., Baccinia, A., Gaoa, F., and Schaafa, C.: Global land cover mapping from MODIS: algorithms and early results, Remote Sens. Environ., 83, 287–302, 2002.

Gao, H., Birkett, C., and Lettenmaier, D. P.: Global monitoring of large reservoir storage from satellite remote sensing, Water Resour. Res., 48, W09504, doi:10.1029/2012WR012063, 2012.

Gaybullaev, B., Chen, S. C., and Kuo, Y. M.: Large-scale desiccation of the Aral Sea due to over exploitation after 1960, J. Mount. Sci., 9, 538–546, 2012.

Gerten, D.: A vital link: water and vegetation in the Anthropocene, Hydrol. Earth Syst. Sci., 17, 3841–3852, doi:10.5194/hess-17-3841-2013, 2013.

Gerten, D. and Rost, S.: Climate change impacts on agricultural water stress and impact mitigation potential, World Bank, Washington, D.C., USA, http://siteresources.worldbank.org/INTWDR2010/Resources/5287678-1255547194560/WDR2010_BGNoteGerten.pdf (last access: 3 November 2014), 2010.

Gerten, D., Schaphoff, S., Haberlandt, U., Lucht, W., and Sitch, S.: Terrestrial vegetation and water balance – hydrological evaluation of a dynamic global vegetation model, J. Hydrol., 286, 249–270, 2004.

Gerten, D., Hagemann, S., Biemans, H., Saeed, F., and Konzmann, M.: Climate Change and Irrigation: Global Impacts and Regional Feedbacks, WATCH Technical Report Number 47, http://www.eu-watch.org/publications/technical-reports (last access: 6 May 2014), 2011.

GEWEX: GEWEX plans for 2013 and beyond - GEWEX science questions (version 1), GEWEX document series No. 2012-2, http://www.gewex.org/pdfs/GEWEXScience_Questionsfinal.pdf, (last access: 6 May 2014), 2012.

Giordano, M.: Global groundwater? Issues and solutions, Annu. Rev. Environ. Resour., 34, 153–178, 2009.

Gleeson, T., Wada Y., Bierkens, M. F., and van Beek, L. P.: Water balance of global aquifers revealed by groundwater footprint, Nature, 488, 197–200, 2012.

Gleick, P. H.: Basic water requirements for human activities: Meeting basic needs, Water Int., 21, 83–92, 1996.

Gleick, P. H.: Water use, Annual Rev. Environ. Resour., 28, 275–314, 2003.

Gleick, P. H., Cooley, H., Famiglietti, J. S., Lettenmaier, D. P., Oki, T., Vörösmarty, C. J., and Wood, E. F.: Improving Understanding of the Global Hydrologic Cycle, in: Climate Science for Serving Society, edited by: Asrar, G. R. and Hurrell, J. W., Springer Netherlands, 151–184, 2013.

Gober, P. and Wheater, H. S.: Socio-hydrology and the science–policy interface: a case study of the Saskatchewan River basin, Hydrol. Earth Syst. Sci., 18, 1413–1422, doi:10.5194/hess-18-1413-2014, 2014.

Gosling, S. N., Taylor, R. G., Arnell, N. W., and Todd, M. C.: A comparative analysis of projected impacts of climate change on river runoff from global and catchment-scale hydrological models, Hydrol. Earth Syst. Sci., 15, 279–294, doi:10.5194/hess-15-279-2011, 2011.

Grey, D., Garrick, D., Blackmore, D., Kelman, J., Muller, M., and Sadoff, C.: Water security in one blue planet: twenty-first century policy challenges for science, Philos. T. Roy. Soc. A, 371, 2002, doi:10.1098/rsta.2012.0406, 2013.

Gueneau, A., Schlosser, C. A., Strzepek, K. M., Gao, X., and Monier, E.: CLM-AG: An Agriculture Module for the Community Land Model version 3.5, MIT Joint Program on the Science and Policy of Global Change, http://dspace.mit.edu/handle/1721.1/73007 (last access: 6 May 2014), 2012.

Guimberteau, M., Laval, K., Perrier, A., and Polcher, J.: Global effect of irrigation and its impact on the onset of the Indian summer monsoon, Clim. Dynam., 39, 1329–1348, doi:10.1007/s00382-011-1252-5, 2012.

Haddeland, I., Lettenmaier, D. P., and Skaugen T.: Effects of irrigation on the water and energy balances of the Colorado and Mekong river basins, J. Hydrol., 324, 210–223, 2006.

Haddeland, I., Skaugen, T., and Lettenmaier, D. P.: Hydrologic effects of land and water management in North America and Asia: 1700–1992, Hydrol. Earth Syst. Sci., 11, 1035–1045, doi:10.5194/hess-11-1035-2007, 2007.

Haddeland, I., Clark, D. B., Franssen, W., Ludwig, F., Voß, F., Arnell, N. W., Bertrand, N., Best, M., Folwell, S., Gerten, D., Gomes, S., Gosling, S. N., Hagemann, S., Hanasaki, N., Harding, R., Heinke, J., Kabat, P., Koirala, S., Oki, T., Polcher, J., Stacke, T., Viterbo, P., Weedon, G. P., and Yeh, P.: Multimodel estimate of the global terrestrial water balance: setup and first results, J. Hydrometeorol., 12, 869–884, doi:10.1175/2011JHM1324.1, 2011.

Haddeland, I., Biemans, H., Eisner, S., Flörke, M., Hanasaki, N., Konzmann, M., Ludwig, F., Masaki, Y., Schewe, J., Stacke, T., Tessler, Z. D., Wada, Y., and Wisser, D.: Global water resources affected by human interventions and climate change, P. Natl. Acad. Sci., 111, 3251–3256, doi:10.1073/pnas.1222475110, 2014.

Hagemann, S. and Dümenil, L.: A parameterization of the lateral waterflow for the global scale, Clim. Dynam., 14, 17–31, doi:10.1007/s003820050205, 1997.

Hanasaki, N., Kanae, S., and Oki, T.: A reservoir operation scheme for global river routing models, J. Hydrol., 327, 22–41, 2006.

Hanasaki, N., Kanae, S., Oki, T., Masuda, K., Motoya, K., Shirakawa, N., Shen, Y., and Tanaka, K.: An integrated model for the assessment of global water resources – Part 1: Model description and input meteorological forcing, Hydrol. Earth Syst. Sci., 12, 1007–1025, doi:10.5194/hess-12-1007-2008, 2008a.

Hanasaki, N., Kanae, S., Oki, T., Masuda, K., Motoya, K., Shirakawa, N., Shen, Y., and Tanaka, K.: An integrated model for the assessment of global water resources – Part 2: Applications and assessments, Hydrol. Earth Syst. Sci., 12, 1027–1037, doi:10.5194/hess-12-1027-2008, 2008b.

Hanasaki, N., Inuzuka, T., Kanae, S., and Oki, T.: An estimation of global virtual water flow and sources of water withdrawal for major crops and livestock products using a global hydrological model, J. Hydrol., 384, 232–244, 2010.

Hanasaki, N., Fujimori, S., Yamamoto, T., Yoshikawa, S., Masaki, Y., Hijioka, Y., Kainuma, M., Kanamori, Y., Masui, T., Takahashi, K., and Kanae, S.: A global water scarcity assessment under Shared Socio-economic Pathways – Part 1: Water use, Hydrol. Earth Syst. Sci., 17, 2375–2391, doi:10.5194/hess-17-2375-2013, 2013a.

Hanasaki, N., Fujimori, S., Yamamoto, T., Yoshikawa, S., Masaki, Y., Hijioka, Y., Kainuma, M., Kanamori, Y., Masui, T., Takahashi, K., and Kanae, S.: A global water scarcity assessment under Shared Socio-economic Pathways – Part 2: Water availability and scarcity, Hydrol. Earth Syst. Sci., 17, 2393–2413, doi:10.5194/hess-17-2393-2013, 2013b.

Harding, K. J. and Snyder, P. K.: Modeling the Atmospheric Response to Irrigation in the Great Plains, Part I: General Impacts on Precipitation and the Energy Budget, J. Hydrometeorol., 13, 1667–1686, 2012a.

Harding, K. J. and Snyder, P. K.: Modeling the atmospheric response to irrigation in the Great Plains, Part II: The precipitation of irrigated water and changes in precipitation recycling, J. Hydrometeorol., 13, 1687–1703, 2012b.

Hejazi, M. I., Edmonds, J., Chaturvedi, V., Davies, E., and Eom, J.: Scenarios of global municipal water-use demand projections over the 21st century, Hydrolog. Sci. J., 58, 519–538, 2013a.

Hejazi, M. I., Edmonds, J., Clarke, L., Kyle, P., Davies, E., Chaturvedi, V., Wise, M., Patel, P., Eom, J., and Calvin, K.: Integrated assessment of global water scarcity over the 21st century – Part 1: Global water supply and demand under extreme radiative forcing, Hydrol. Earth Syst. Sci. Discuss., 10, 3327–3381, doi:10.5194/hessd-10-3327-2013, 2013b.

Hejazi, M. I., Edmonds, J., Clarke, L., Kyle, P., Davies, E., Chaturvedi, V., Eom, J., Wise, M., Patel, P., and Calvin, K.: Integrated assessment of global water scarcity over the 21st century – Part 2: Climate change mitigation policies, Hydrol. Earth Syst. Sci. Discuss., 10, 3383–3425, doi:10.5194/hessd-10-3383-2013, 2013c.

Hejazi M. I., Edmonds, J. A., Clarke, L. A., Kyle, G. P., Davies, E., Chaturvedi, V., Wise, M. A., Patel, P. L., Eom, J., Calvin, K. V., Moss, R. H., and Kim, S. H.: Long-term global water projections using six socioeconomic scenarios in an integrated assessment modeling framework, Technol. Forecast. Soc., 81, 205–226, 2013d.

Hobbins, M. T., Dai, A., Roderick, M. L., and Farquhar, G. D.: Revisiting the parameterization of potential evaporation as a driver of long-term water balance trends, Geophys. Res. Lett., 35, L12403, doi:10.1029/2008GL033840, 2008.

Hossain, F., Degu, A. M., Yigzaw, W., Burian, S., Niyogi, D., Shepherd, J. M., and Pielke Sr., R.: Climate Feedback–Based Provisions for Dam Design, Operations, and Water Management in the 21st Century, J. Hydrol. Eng., 17, 837–850, 2012.

Hourdin, F., Musat, I., Bony, S., Braconnot, P., Codron, F., Dufresne, J.-L., Fairhead, L., Filiberti, M.-A., Friedlingstein, P., Grandpeix, J.-Y., Krinner, G., LeVan, P., Li, Z.-X.: The LMDZ4 general circulation model: climate performance and sensitivity to parametrized physics with emphasis on tropical convection, Clim. Dynam., 27, 787–813, doi:10.1007/s00382-006-0158-0, 2006.

Hughes, G., Chinowsky, P., and Strzepek, K.: The costs of adaptation to climate change for water infrastructure in OECD countries, Utilities Policy, 18, 142–153, 2010.

IIASA/FAO: Global Agro-ecological Zones (GAEZ v3.0), IIASA, Laxenburg, Austria and FAO, Rome, Italy, available at: http://webarchive.iiasa.ac.at/Research/LUC/GAEZv3.0/docs/GAEZModelDocumentation.pdf (last access: 15 July 2014), 2012.

IPCC: The IPCC Special Report on Emissions Scenarios (SRES), IPCC, Geneva, http://www.ipcc.ch/pdf/special-reports/spm/sres-en.pdf (last access: 6 May 2014), 2000.

Kalnay, E., Kanamitsu, M., Kistler, R., Collins, W., Deaven, D., Gandin, L., Iredell, M., Saha, S., White, G., Woollen, J., Zhu, Y., Leetmaa, A., Reynolds, R., Chelliah, M., Ebisuzaki, W., Higgins, W., Janowiak, J., Mo, K. C., Ropelewski, C., Wang, J., Jenne, R., and Joseph, D.: The NCEP/NCAR 40-year reanalysis project, B. Am. Meteorol. Soc., 77, 437–471, 1996.

Kanamitsu, M., Ebisuzaki, W., Woollen, J., Yang, S.-K., Hnilo, J. J., Fiorino, M., and Potter, G. L.: NCEP-DOE AMIP-II Reanalysis (R-2), B. Am. Meteorol. Soc., 83, 1631–1644, 2002.

Karl, T. R. and Trenberth K. E.: Modern global climate change, Science, 302, 1719–1723, 2003.

Kim, H., Yeh, P. J.-F., Oki, T., and Kanae, S.: Role of rivers in the seasonal variations of terrestrial water storage over global basins, Geophys. Res. Lett., 36, L17402, doi:10.1029/2009GL039006, 2009.

Konar, M., Hussein, Z., Hanasaki, N., Mauzerall, D. L., and Rodriguez-Iturbe, I.: Virtual water trade flows and savings under climate change, Hydrol. Earth Syst. Sci., 17, 3219–3234, doi:10.5194/hess-17-3219-2013, 2013.

Konzmann, M., Gerten, D., and Heinke, J.: Climate impacts on global irrigation requirements under 19 GCMs, simulated with a vegetation and hydrology model, Hydrolog. Sci. J., 58, 88–105, 2013.

Koster, R. D., Dirmeyer, P. A., Guo, Z., Bonan, G., Chan, E., Cox, P., Gordon, C. T., Kanae, S., Kowalczyk, E., Lawrence, D., Liu, P., Lu, C.-H., Malyshev, S., McAvaney, B., Mitchell, K., Mocko, D., Oki, T., Oleson, K., Pitman, A., Sud, Y. C., Taylor, C. M., Verseghy, D., Vasic, R., Xue, Y., and Yamada, T.: Regions of strong coupling between soil moisture and precipitation, Science, 305, 1138–1140, 2004.

Kriegler, E., O'Neill, B. C., Hallegatte, S., Kram, T., Lempert, R. J., Moss, R. H., and Wilbanks, T.: The need for and use of socio-economic scenarios for climate change analysis: a new approach based on shared socio-economic pathways, Global Environ. Change, 22, 807–822, 2012.

Krysanova, V., Müller-Wohlfeil, D. I., and Becker, A.: Development and test of a spatially distributed hydrological/water quality model for mesoscale watersheds, Ecol. Model., 106, 261–289, 1998.

Kump, L. R., Kasting, J. F., and Crane, R. G.: The earth system, Prentice Hall, San Francisco, 2010.

Kyle, P., Davies, E. G., Dooley, J. J., Smith, S. J., Clarke, L. E., Edmonds, J. A., and Hejazi, M.: Influence of climate change mitigation technology on global demands of water for electricity generation, Int. J. Greenh. Gas Con., 13, 112–123, 2013.

Lai, X., Jiang, J., Yang, G., and Lu, X. X.: Should the Three Gorges Dam be blamed for the extremely low water levels in the middle–lower Yangtze River?, Hydrol. Process., 28, 150–160, doi:10.1002/hyp.10077, 2014.

Lawrence, D. M., Oleson, K. W., Flanner, M. G., Thornton, P. E., Swenson, S. C., Lawrence, P. J., Zeng, X., Yang, Z.-L., Levis, S., Sakaguchi, K., Bonan, G. B., and Slater, A. G.: Parameterization improvements and functional and structural advances in Version 4 of the Community Land Model, J. Adv. Model. Earth Syst., 3, M03001, doi:10.1029/2011MS00045, 2011.

Lawrence, D., Maxwell, R., Swenson, S., Lopez, S., and Famiglietti, J.: Challenges of Representing and Predicting Multi-Scale Human–Water Cycle Interactions in Terrestrial Systems, http://climatemodeling.science.energy.gov/sites/default/files/Topic_3_final.pdf (last access: 6 May 2014), 2012.

Leng, G., Huang, M., Tang, Q., Sacks, W. J., Lei, H., and Leung, L. R.: Modeling the effects of irrigation on land surface fluxes and states over the conterminous United States: Sensitivity to input data and model parameters, J. Geophys. Res.-Atmos., 118, 9789–9803, doi:10.1002/jgrd.50792, 2013.

Lenton, T. M.: Land and ocean carbon cycle feedback effects on global warming in a simple Earth system model, Tellus B, 52, 1159–1188, 2000.

Levis, S. and Sacks W.: Technical descriptions of the interactive crop management (CLM4CNcrop) and interactive irrigation models in version 4 of the Community Land Model, http://www.cesm.ucar.edu/models/cesm1.1/clm/CLMcropANDirrigTechDescriptions.pdf (last access: 6 May 2014), 2011.

Levis, S., Bonan, G. B., Kluzek, E., Thornton, P. E., Jones, A., Sacks, W. J., and Kucharik, C. J.: Interactive Crop Management in the Community Earth System Model (CESM1): Seasonal Influences on Land-Atmosphere Fluxes, J. Climate, 25, 4839–4859, 2012.

Li, H., Wigmosta, M. S., Wu, H., Huang, M., Ke, Y., Coleman, A. M., and Leung, L. R.: A physically based runoff routing model for landsurface and earth system models, J. Hydrometeorol., 14, 808–828, 2013.

Liang, X., Lettenmaier, D. P., Wood, E. F., and Burges, S. J.: A simple hydrologically based model of land surface water and energy fluxes for general circulation models, J. Geophys. Res.-Atmos., 99, 14415–14428, 1994.

Lissner, T. K., Sullivan, C. A., Reusser, D. E., and Kropp, J. P.: Water stress and livelihoods: A review of data and knowledge on water needs, use and availability, in: 4th EGU Leonardo Conference: Hydrology and Society – Connections between Hydrology and Population dynamics, Policymaking and Power generation, 14–16 November, Torino, Italy, 2012.

Liu, J., Zhang, Z., Xu, X., Kuang, W., Zhou, W., Zhang, S., Li, R., Yan, C., Yu, D., Wu, S., and Jiang N.: Spatial patterns and driving forces of land use change in China during the early 21st century, J. Geogr. Sci., 20, 483–494, 2010.

Livneh, B., Restrepo, P. J., and Lettenmaier, D. P.: Development of a Unified Land Model for prediction of surface hydrology and land-atmosphere interactions, J. Hydrometeorol., 12, 1299–1320, 2011.

Lo, M.-H. and Famiglietti, J. S.: Irrigation in California's Central Valley strengthens the southwestern U.S. water cycle, Geophys. Res. Lett., 40, 301–306, doi:10.1002/grl.50108, 2013.

Lobell, D. B., Bala, G., and Duffy, P. B.: Biogeophysical impacts of cropland management changes on climate, Geophys. Res. Lett., 33, L06708, doi:10.1029/2005GL025492, 2006.

Lorenz, C. and Kunstmann, H.: The Hydrological Cycle in Three State-of-the-Art Reanalyses: Intercomparison and Performance Analysis, J. Hydrometeorol., 13, 1397–1420, 2012.

Lu, Y.: Development and application of WRF3.3-CLM4crop to study of agriculture-climate interaction, PhD Thesis, University of California, Merced, http://escholarship.org/uc/item/12b6p87z (last access: 6 May 2014), 2013.

Lucas-Picher, P., Christensen, J. H., Saeed, F., Kumar, P., Asharaf, S., Ahrens, B., Wiltshire, A. J., Jacob, D., and Hagemann, S.: Can regional climate models represent the Indian monsoon?, J. Hydrometeorol., 12, 849–868, 2011.

Macknick, J., Newmark, R., Heath, G., and Hallett, K. C.: A review of operational water consumption and withdrawal factors for electricity generating technologies, Technical Report NREL/TP-6A20-5090, http://www.cwatershedalliance.com/pdf/SolarDoc01.pdf (last access: 6 May 2014), 2011.

Manabe, S.: Climate and the ocean circulation part I. The atmospheric circulation and the hydrology of the earth's surface, Mon. Weather Rev., 97, 739–774, 1969.

Maurer, E. P., Wood, A. W., Adam, J. C., Lettenmaier, D. P., and Nijssen, B.: A long-term hydrologically based dataset of land surface fluxes and states for the conterminous United States, J. Climate, 15, 3237–3251, 2002.

McKenney, M. S. and Rosenberg, N. J.: Sensitivity of some potential evapotranspiration estimation methods to climate change, Agr. Forest Meteorol., 64, 81–110, 1993.

McNeill, J. R.: Something New Under the Sun: An Environmental History of the Twentieth-Century World, WW Norton & Company, New York, 2000.

Mehta, V. K., Haden V. R., Joyce, B. A., Purkey, D. R., and Jackson, L. E.: Irrigation demand and supply, given projections of climate and land-use change in Yolo County, California, Agr. Water Manage., 117, 70–82, 2013.

Meinshausen, M., Smith, S. J., Calvin, K., Daniel, J. S., Kainuma, M. L. T., Lamarque, J.-F., Matsumoto, K., Montzka, S. A., Raper, S. C. B., Riahi, K., Thomson, A., Velders, G. J. M., and van Vuuren, D. P. P.: The RCP greenhouse gas concentrations and their extensions from 1765 to 2300, Climatic Change, 109, 213–241, 2011.

Meybeck, M.: Global analysis of river systems: from Earth system controls to Anthropocene syndromes, Philos. T. Roy. Soc. Lond. B, 358, 1935–1955, 2003.

Milano, M., Ruelland, D., Fernandez, S., Dezetter, A., Fabre, J., Servat, E., Fritsch, J.-M., Ardoin-Bardin, S., and Thivet, G.: Current state of Mediterranean water resources and future trends under climatic and anthropogenic changes, Hydrolog. Sci. J., 58, 498–518, 2013.

Miller, J. R., Russell, G. L., and Caliri, G.: Continental-scale river flow in climate models, J. Climate, 7, 914–928, doi:10.1175/1520-0442(1994)007<0914:CSRFIC>2.0.CO;2, 1994.

Milly, P. C. D.: Potential evaporation and soil moisture in general circulation models, J. Climate, 5, 209–226, 1992.

Mitchell, T. D. and Jones, P. D.: An improved method of constructing a database of monthly climate observations and associated high-resolution grids, Int. J. Climatol., 25, 693–712, doi:10.1002/joc.1181, 2005.

Moore, N. and Rojstaczer, S.: Irrigation-induced rainfall and the Great Plains, J. Appl. Meteorol., 40, 1297–1309, 2001.

Moss, R. H., Edmonds, J. A., Hibbard, K. A., Manning, M. R., Rose, S. K., van Vuuren, D. P., Carter, T. R., Emori, S., Kainuma, M., Kram, T., Meehl, G. A., Mitchell, J. F. B., Nakicenovic, N., Riahi, K., Smith, S. J., Stouffer, R. J., Thomson, A. M., Weyant, J. P., and Wilbanks, T. J.: The next generation of scenarios for climate change research and assessment, Nature, 463, 747–756, 2010.

Music, B. and Caya, D.: Evaluation of the hydrological cycle over the Mississippi River basin as simulated by the Canadian Regional Climate Model (CRCM), J. Hydrometeorol., 8, 969–988, 2007.

Nakayama, T.: Simulation of the effect of irrigation on the hydrologic cycle in the highly cultivated Yellow River Basin, Agr. Forest Meteorol., 151, 314–327, 2011.

Nakayama, T. and Shankman D.: Evaluation of uneven water resource and relation between anthropogenic water withdrawal and ecosystem degradation in Changjiang and Yellow River basins, Hydrol. Process., 27, 3350–3362, doi:10.1002/hyp.9835, 2013.

Nassopoulos, H., Dumas, P., and Hallegatte, S.: Climate change, precipitation and water management infrastructures, presented at: Water in Africa: Hydro-Pessimism or Hydro-Optimism, 2–3 October 2008, Porto, Portugal, available at:http://www.slideshare.net/water.in.africa/hypatia-nassopoulos-ppt-presentation (last access: 15 October 2014), 2008.

Nassopoulos, H., Dumas, P., and Hallegatte, S.: Adaptation to an uncertain climate change: cost benefit analysis and robust decision making for dam dimensioning, Climatic Change, 114, 497–508, doi:10.1007/s10584-012-0423-7, 2012.

Nazemi, A. and Wheater, H. S.: Assessing the vulnerability of water supply to changing streamflow conditions, Eos Trans. Am. Geophys. Un., 95, 288, doi:10.1002/2014EOS320007, 2014a.

Nazemi, A. and Wheater, H. S.: How can the uncertainty in the natural inflow regime propagate into the assessment of water resource systems?, Adv. Water Resour., 63, 131–142, doi:10.1016/j.advwatres.2013.11.009, 2014b.

Nazemi, A. and H. S. Wheater: On inclusion of water resource management in Earth System models – Part 2: Representation of water supply and allocation and opportunities for improved modeling, Hydrol. Earth Syst. Sci., 19, 63–90, doi:10.5194/hess-19-63-2015, 2015.

Nazemi, A., Akbarzadeh-T, M. R., and Hosseini, S. M.: Fuzzy-stochastic linear programming in water resources engineering, in: Proceedings of IEEE Annual Meeting of Fuzzy Information, IEEE, 227–232, doi:10.1109/NAFIPS.2002.1018060, 2002.

Nazemi, A., Wheater, H. S., Chun, K. P., and Elshorbagy, A.: A stochastic reconstruction framework for analysis of water resource system vulnerability to climate-induced changes in river flow regime, Water Resour. Res., 49, 291–305, doi:10.1029/2012WR012755, 2013.

New, M., Hulme, M., and Jones, P.: Representing Twentieth-Century Space–Time Climate Variability Part I: Development of a 1961–90 Mean Monthly Terrestrial Climatology, J.f Climate, 12, 829–857, 1999.

New, M., Hulme, M., and Jones, P. D.: Representing twentieth century space-time climate variability, part II Development of 1901–96 monthly grids of terrestrial surface climate, J. Climate, 13, 2217–2238, 2000.

Ngo-Duc, T., Laval, K., Polcher, J., Lombard, A., and Cazenave, A.: Effects of land water storage on global mean sea level over the past half century, Geophys. Res. Lett., 32, L09704, doi:10.1029/2005GL022719, 2005a.

Ngo-Duc, T., Polcher, J., and Laval, K.: A 53-year forcing data set for land surface models, J. Geophys. Res., 110, D06116, doi:10.1029/2004JD005434, 2005b.

Ngo-Duc, T., Laval, K., Ramillien, G., Polcher, J., and Cazenave, A.: Validation of the land water storage simulated by Organising Carbon and Hydrology in Dynamic Ecosystems (ORCHIDEE) with Gravity Recovery and Climate Experiment (GRACE) data, Water Resour. Res., 43, W04427, doi:10.1029/2006WR004941, 2007.

Nicholson, S. E.: Land surface atmosphere interaction, Prog. Phys. Geogr., 12, 36–65, 1988.

Nilsson, C., Reidy, C. A., Dynesius, M., and Revenga, C.: Fragmentation and flow regulation of the world's large river systems, Science, 308, 405–408, 2005.

Noilhan, J. and Planton, S.: A simple parameterization of land surface processes for meteorological models, Mon. Weather Rev., 117, 536–549, 1989.

Oki, T. and Kanae, S.: Global hydrological cycles and world water resources, Science, 313, 1068–1072, 2006.

Oki, T. and Sud, Y. C.: Design of Total Runoff Integrating Pathways (TRIP) – A global river channel network, Earth Interact., 2, 1–37, 1998.

Oki, T., Blyth, E. M., Berbery, E. H., and Alcaraz-Segura, D.: Land Use and Land Cover Changes and Their Impacts on Hydroclimate, Ecosystems and Society, in: Climate Science for Serving Society, edited by: Asrar, G. R. and Hurrell, J. W., Springer Netherlands, 185–203, 2013.

Oleson, K. W., Dai, Y., Bonan, G. B., Bosilovichm, M., Dickinson, R., Dirmeyer, P., Hoffman, F., Houser, P., Levis, S., Niu, G.-Y., Thornton, P., Vertenstein, M., Yang, Z., and Zeng, X.: Technical description of the community land model (CLM), NCAR Tech. Note NCAR/TN-461+STR, 173 pp., doi:10.5065/D6N877R0, http://www.cesm.ucar.edu/models/cesm1.0/clm/CLM4_Tech_Note.pdf (last access: 28 December 2014), 2004.

Oleson, K. W., Niu, G. Y., Yang, Z. L., Lawrence, D. M., Thornton, P. E., Lawrence, P. J., Stöckli, R., Dickinson, R. E., Bonan, G. B., Levis, S., Dai, A., and Qian, T.: Improvements to the Community Land Model and their impact on the hydrological cycle, J. Geophys. Res.-Biogeo., 113, G01021, doi:10.1029/2007JG000563, 2008.

Ozdogan, M. and Gutman, G.: A new methodology to map irrigated areas using multi-temporal MODIS and ancillary data: An application example in the continental US, Remote Sens. Environ., 112, 3520–3537, 2008.

Ozdogan, M., Rodell, M., Beaudoing, H. K., and Toll, D. L.: Simulating the Effects of Irrigation over the United States in a Land Surface Model Based on Satellite-Derived Agricultural Data, J. Hydrometeorol., 11, 171–184, 2010.

Pielke, R. A., Cotton, W. R., Walko, R. L., Tremback, C. J., Lyons, W. A., Grasso, L. D., Nicholls, M. E., Moran, M. D., Wesley, D. A., Lee, T. J., and Copeland, J. H.: A comprehensive meteorological modeling system – RAMS, Meteorol. Atmos. Phys., 49, 69–91, 1992.

Pietroniro, A., Fortin, V., Kouwen, N., Neal, C., Turcotte, R., Davison, B., Verseghy, D., Soulis, E. D., Caldwell, R., Evora, N., and Pellerin, P.: Development of the MESH modelling system for hydrological ensemble forecasting of the Laurentian Great Lakes at the regional scale, Hydrol. Earth Syst. Sci., 11, 1279–1294, doi:10.5194/hess-11-1279-2007, 2007.

Pitman, A. J.: The evolution of, and revolution in, land surface schemes designed for climate models, Int. J. Climatol., 23, 479–510, 2003.

Pitman, A. J., Henderson-Sellers, A., and Yang, Z. L.: Sensitivity of regional climates to localized precipitation in global models, Nature, 346, 734–737, 1990.

Pitman, A. J., de Noblet-Ducoudré, N., Cruz, F. T., Davin, E. L., Bonan, G. B., Brovkin, V., Claussen, M., Delire, C., Ganzeveld, L., Gayler, V., van den Hurk, B. J. J. M., Lawrence, P. J., van der Molen, M. K., Müller, C., Reick, C. H., Seneviratne, S. I., Strengers, B. J., and Voldoire, A.: Uncertainties in climate responses to past land cover change: first results from the LUCID intercomparison study, Geophys. Res. Lett., 36, L14814, doi:10.1029/2009GL039076, 2009.

Pokhrel, Y., Hanasaki, N., Koirala, S., Cho, J., Yeh, P. J.-F., Kim, H., Kanae, S., and Oki, T.: Incorporating anthropogenic water regulation modules into a land surface model, J. Hydrometeorol., 13, 255–269, 2012.

Polcher, J.: Interactive comment on "On inclusion of water resource management in Earth System models – Part 1: Problem definition and representation of water demand" by A. Nazemi and H. S. Wheater, Hydrol. Earth Syst. Sci. Discuss., 11, C3403–C3410, 2014.

Polcher, J., Bertrand, N., Biemans, H., Clark, D. B., Floerke, M., Gedney, N., Gerten, D., Stacke, T., van Vliet, M., and Voss, F.: Improvements in hydrological processes in general hydrological models and land surface models within WATCH, WATCH Technical Report Number 34, available at: http://www.eu-watch.org/publications/technical-reports (last access: 6 May 2014), 2011.

Poff, N. L., Richter, B. D., Arthington, A. H., Bunn, S. E., Naiman, R. J., Kendy, E., Acreman, M., Apse, C., Bledsoe, B. P., Freeman, M. C., Henriksen, J., Jacobson, R. B., Kennen, J. G., Merritt, D. M., O'Keeffe, J. H., Olden, J. D., Rogers, K., Tharme, R. E., and Warner, A.: The ecological limits of hydrologic alteration (ELOHA): a new framework for developing regional environmental flow standards, Freshwater Biol., 55, 1365–2427, doi:10.1111/j.1365-2427.2009.02204.x, 2009.

Portmann, F. T., Siebert, S., and Döll, P.: MIRCA2000 – Global monthly irrigated and rainfed crop areas around the year 2000: A new high-resolution data set for agricultural and hydrological modeling, Global Biogeochem. Cycl., 24, GB1011, doi:10.1029/2008GB003435, 2010.

Postel, S. L., Daily, G. C., and Ehrlich, P. R.: Human appropriation of renewable fresh water, Science, 271, 785–788, 1996.

Precoda, N.: Requiem for the Aral Sea, Ambio, 20, 109-114, 1991.

Priestley, C. H. B. and Taylor, R. J.: On the assessment of surface heat flux and evaporation using large-scale parameters, Mon. Weather Rev., 100, 81–92, 1972.

Prudhomme, C., Giuntoli, I., Robinson, E. L., Clark, D. B., Arnell, N. W., Dankers, R., Fekete, B. M., Franssen, W., Gerten, D., Gosling, S. N., Hagemann, S., Hannah, D. M., Kim, H., Masaki, Y., Satoh, Y., Stacke, T., Wada, Y., and Wisser, D.: Hydrological droughts in the 21st century, hotspots and uncertainties from a global multimodel ensemble experiment, P. Natl. Acad. Sci. USA, 111, 3262–3267, doi:10.1073/pnas.1222473110, 2014.

Qian, Y., Huang, M., Yang, B., and Berg, L. K.: A Modeling Study of Irrigation Effects on Surface Fluxes and Land-Air-Cloud Interactions in the Southern Great Plains, J. Hydrometeorol., 14, 700–721, 2013.

Rausch, S. and Mowers, M.: Distributional and efficiency impacts of clean and renewable energy standards for electricity, Resour. Energy Econ., 36, 556–585, 2013.

Rodell, M., Velicogna, I., and Famiglietti, J. S.: Satellite-based estimates of groundwater depletion in India, Nature, 460, 999–1002, 2009.

Rohling, E. J. and Bryden, H. L.: Man-induced salinity and temperature increases in western Mediterranean deep water, J. Geophys. Res.-Oceans (1978–2012), 97, 11191–11198, 1992.

Rosenzweig, C., Elliott, J., Deryng, D., Ruane, A. C., Müller, C., Arneth, A., Boote, K. J., Folberth, C., Glotter, M., Khabarov, N., Neumann, K., Piontek, F., Pugh, T. A. M., Schmid, E., Stehfest, E., Yang, H., and Jones, J. W.: Assessing agricultural risks of climate change in the 21st century in a global gridded crop model intercomparison, P. Natl. Acad. Sci. USA, 111, 3268–3273, doi:10.1073/pnas.1222463110, 2014.

Rost, S., Gerten, D., Bondeau, A., Luncht, W., Rohwer, J., and Schaphoff, S.: Agricultural green and blue water consumption and its influence on the global water system, Water Resour. Res., 44, W09405, doi:10.1029/2007WR006331, 2008.

Rost, S., Gerten, D., Hoff, H., Lucht, W., Falkenmark, M., and Rockström, J.: Global potential to increase crop production through water management in rainfed agriculture, Environ.l Res. Lett., 4, 044002, doi:10.1088/1748-9326/4/4/044002, 2009.

Rudolf, B., Beck, C., Grieser, J., and Schneider, U.: Global precipitation analysis products of the GPCC, Climate Monitoring – Tornadoklimatologie – Aktuelle Ergebnisse des Klimamonitorings, available at: http://www.juergen-grieser.de/publications/publications_pdf/GPCC-intro-products-2005.pdf, last access: 16 July 2014, 163–170, 2005.

Sacks, W. J., Cook, B. I., Buenning, N., Levis, S., and Helkowski, J. H.: Effects of global irrigation on the near-surface climate, Clim. Dynam., 33, 159–175, 2009.

Saeed, F., Hagemann, S., and Jacob, D.: Impact of irrigation on the South Asian summer monsoon, Geophys. Res. Lett., 36, L20711, doi:10.1029/2009GL040625, 2009.

Schellnhuber, H. J.: Discourse: Earth System Analysis – The Scope of the Challenge, in: Earth System Analysis – Integrating science for sustainability, edited by: Schellnhuber, H. J. and Wenzel, V., Springer, Heidelberg, 1998.

Schellnhuber, H. J.: Earth system analysis and the second Copernican revolution, Nature, 402, C19–C23, 1999.

Schewe, J., Heinke, J., Gerten, D., Haddeland, I., Arnell, N. W., Clark, D. B., Dankers, R., Eisner, S., Fekete, B. M., Colón-González, F. J., Gosling, S. N., Kim, H., Liu, X., Masaki, Y., Portmann, F. T., Satoh, Y., Stacke, T., Tang, Q., Wada, Y., Wisser, D., Albrecht, T., Frieler, K., Piontek, F., Warszawski, L., and Kabat, P.: Multimodel assessment of water scarcity under climate change, P. Natl. Acad. Sci. USA, 111, 3245–3250, doi:10.1073/pnas.1222460110, 2014.

Schiermeier, Q.: Water risk as world warms, Nature, 505, 7481, doi:10.1038/505010a, 2014.

Schlosser, C. A., Kicklighter, D., and Sokolov, A.: A global land system framework for integrated climate-change assessments, MIT Joint Program on the Science and Policy of Global Change, Report No. 147, http://dspace.mit.edu/handle/1721.1/38461 (last access: 6 May 2014), 2007.

Sellers, P. J.: Biophysical models of land surface processes, in: Climate system modeling, edited by: Trenberth, K. E., Cambridge University Press, Cambridge, UK, 451–490, 1992.

Sellers, P. J., Mintz, Y. C. S. Y., Sud, Y. E. A., and Dalcher, A.: A simple biosphere model (SiB) for use within general circulation models, J. Atmos. Sci., 43, 505–531, 1986.

Sellers, P. J., Tucker, C. J., Collatz, G. J., Los, S. O., Justice, C. O., Dazlich, D. A., and Randall, D. A.: A global 1 by 1 NDVI data set for climate studies – Part 2: The generation of global fields of terrestrial biophysical parameters from the NDVI, Int. J. Remote Sens., 15, 3519–3545, 1994.

Sellers, P. J., Randall, D. A., Collatz, G. J., Berry, J. A., Field, C. B., Dazlich, D. A., Zhang, C., Collelo, G. D., and Bounoua, L.: A revised land surface parameterization (SiB2) for atmospheric GCMs – Part I: Model formulation, J. Climate, 9, 676–705, 1996a.

Sellers, P. J., Meeson, B. W., Closs, J., Collatz, J., Corprew, F., Dazlich, D., Hall, F. G., Kerr, Y., Koster, R., Los, S., Mitchell, K., McManus, J., Myers, D., Sun, K.-J., and Try, P.: The ISLSCP Initiative I global datasets: surface boundary conditions and atmospheric forcings for land–atmosphere studies, B. Am. Meteorol. Soc., 77, 1987–2005, 1996b.

Shiklomanov, I. A.: World water resources, UNESCO, 1998, Paris, http://www.ce.utexas.edu/prof/mckinney/ce385d/Papers/Shiklomanov.pdf (last access: 6 May 2014), 1993.

Shiklomanov, I. A.: Assessment of Water Resources and Water Availability in the World, Comprehensive Assessment of the Freshwater Resources of the World, WMO and SEI, Geneva, 1997.

Shiklomanov, I. A.: World water resources and water use: Present assessment and outlook for 2025, in: World water scenarios, edited by: Rijsberman, F. R., Earthscan, London, 160–203, 2000.

Short, W., Blair, N., Sullivan, P., and Mai, T.: ReEDS model documentation: base case data and model description, Golden, CO: National Renewable Energy Laboratory, http://www.nrel.gov/analysis/reeds/documentation.html (last access: 6 May 2014), 2009.

Siebert, S. and Döll, P.: The Global Crop Water Model (GCWM): Documentation and first results for irrigated crops, https://www2.uni-frankfurt.de/45217788/FHP07_SiebertandDoell__2008.pdf (last access: 6 May 2014), 2008.

Siebert, S. and Döll, P.: Quantifying blue and green virtual water contents in global crop production as well as potential production losses without irrigation, J. Hydrol., 384, 198–217, 2010.

Siebert, S., Döll, P., Hoogeveen, J., Faures, J.-M., Frenken, K., and Feick, S.: Development and validation of the global map of irrigation areas, Hydrol. Earth Syst. Sci., 9, 535–547, doi:10.5194/hess-9-535-2005, 2005.

Siebert, S., Döll, P., Feick, S., Hoogeveen, J., and Frenken, K.: Global map of irrigation areas version 4.0.1, Food and Agriculture Organization of the United Nations, Rome, Italy, https://www2.uni-frankfurt.de/45218039/Global_IrrigationMap (last access: 6 May 2014), 2007.

Siebert, S., Burke, J., Faures, J. M., Frenken, K., Hoogeveen, J., Döll, P., and Portmann, F. T.: Groundwater use for irrigation – a global inventory, Hydrol. Earth Syst. Sci., 14, 1863–1880, doi:10.5194/hess-14-1863-2010, 2010.

Sitch, S., Smith, B., Prentice, I. C., Arneth, A., Bondeau, A., Cramer, W., Kaplan, J. O., Levis, S., Lucht, W., Sykes, M. T., Thonicke, K., and Venevsky, S.: Evaluation of ecosystem dynamics, plant geography and terrestrial carbon cycling in the LPJ dynamic global vegetation model, Global Change Biol., 9, 161–185, 2003.

Sivapalan, M., Savenije, H. H., and Blöschl, G.: Socio-hydrology: A new science of people and water, Hydrol. Process., 26, 1270–1276, 2012.

Skamarock, W. C., Klemp, J. B., Dudhia, J., Gill, D. O., Barker, D. M., Wang, W., and Powers, J. G.: A description of the advanced research WRF version 2 (No. NCAR/TN468+STR), available at: http://oai.dtic.mil/oai/oai?verb=getRecord&metadataPrefix=html&identifier=ADA487419 (last access: 6 May 2014), 2005.

Skliris, N. and Lascaratos, A.: Impacts of the Nile River damming on the thermohaline circulation and water mass characteristics of the Mediterranean Sea, J. Mar. Syst., 52, 121–143, doi:10.1016/j.jmarsys.2004.02.005, 2004.

Smakhtin, V., Revenga, C., and Döll, P.: A pilot global assessment of environmental water requirements and scarcity, Water Int., 29, 307–317, 2004.

Small, I., Van der Meer, J., and Upshur, R. E.: Acting on an environmental health disaster: the case of the Aral Sea, Environ. Health Perspect., 109, 547–549, 2001.

Smith, M.: CROPWAT – A computer program for irrigation planning and management, Irrigation and Drainage, Pap. 46, Food and Agric. Org. of the UN, Rome, http://www.fao.org/nr/water/infores_databasescropwat.html (last access: 6 May 2014), 1992.

Solomon, S., Plattner, G. K., Knutti, R., and Friedlingstein, P.: Irreversible climate change due to carbon dioxide emissions, P. Natl. Acad. Sci., 106, 1704–1709, 2009.

Sophocleous, M.: Interactions between groundwater and surface water: the state of the science, Hydrogeol. J., 10, 52–67, 2002.

Sorooshian, S., Li, J., Hsu, K.-L., and Gao, X.: How significant is the impact of irrigation on the local hydroclimate in California's Central Valley? Comparison of model results with ground and remote-sensing data, J. Geophys. Res., 116, D06102, doi:10.1029/2010JD014775, 2011.

Soulis, E. D., Snelgrove, K. R., Kouwen, N., Seglenieks, F., and Verseghy, D. L.: Towards closing the vertical water balance in Canadian atmospheric models: coupling of the land surface scheme CLASS with the distributed hydrological model WATFLOOD, Atmos.-Ocean, 38, 251–269, 2000.

Steffen, W., Crutzen, P. J., and McNeill, J. R.: The Anthropocene: are humans now overwhelming the great forces of nature, Ambio, 36, 614–621, 2007.

Steffen, W., Grinevald, J., Crutzen, P., and McNeill, J.: The Anthropocene: conceptual and historical perspectives, Philos. T. Roy. Soc. A, 369, 842–867, 2011.

Steinfeld, H., Gerber, P., Wassenaar, T., Castel, V., Rosales, M., and de Haan, C.: Livestock's long shadow: Environmental issues and options, Food and Agriculture Organization – LEAD,

Rome, Italy, http://www.fao.org/docrep/010/a0701e/a0701e00.HTM (last access: 6 May 2014), 2006.

Strzepek, K., Schlosser, A., Farmer, W., Awadalla, S., Baker, J., Rosegrant M., and Gao, X.: Modeling the global water resource system in an integrated assessment modeling framework: IGSM-WRS, MIT Joint Program on the Science and Policy of Global Change, Report No. 189, available at: http://dspace.mit.edu/handle/1721.1/61767 (last access: 6 May 2014), 2010.

Strzepek, K., Baker, J., Farmer, W., and Schlosser, C. A.: Modeling water withdrawal and consumption for electricity generation in the United States, MIT Joint Program on the Science and Policy of Global Change, Report No. 222, http://dspace.mit.edu/handle/1721.1/71168 (last access: 6 May 2014), 2012a.

Strzepek, K., Schlosser, A., Gueneau, A. Gao, X., Blanc, É., Fant, C., Rasheed, B., and Jacoby, H. D.: Modeling water resource system under climate change: IGSM-WRS, MIT Joint Program on the Science and Policy of Global Change, Report No. 236, http://dspace.mit.edu/handle/1721.1/75774 (last access: 6 May 2014), 2012b.

Sulis, M., Paniconi, C., Rivard, C., Harvey, R., and Chaumont, D.: Assessment of climate change impacts at the catchment scale with a detailed hydrological model of surface-subsurface interactions and comparison with a land surface model, Water Resour. Res., 47, W01513, doi:10.1029/2010WR009167, 2011.

Takata, K., Emori, S., and Watanabe, T.: Development of the minimal advanced treatments of surface interaction and runoff, Global Planet. Change, 38, 209–222, 2003.

Tang, Q., Gao, H., Yeh, P., Oki, T., Su, F., and Lettenmaier, D. P.: Dynamics of Terrestrial Water Storage Change from Satellite and Surface Observations and Modeling, J. Hydrometeorol., 11, 156–170, 2010.

Tao, F., Yokozawa, M., Hayashi, Y., and Lin, E.: Terrestrial water cycle and the impact of climate change, Ambio, 32, 295–301, 2003.

Taylor, C. M.: Feedbacks on convection from an African wetland, Geophys. Res. Lett., 37, L05406, doi:10.1029/2009GL041652, 2009.

Taylor, C. M., de Jeu, R. A., Guichard, F., Harris, P. P., and Dorigo, W. A.: Afternoon rain more likely over drier soils, Nature, 489, 423–426, doi:10.1038/nature11377, 2012.

Taylor, K. E., Stouffer, R. J., and Meehl, G. A.: An Overview of CMIP5 and the Experiment Design, B. Am. Meteorol. Soc., 93, 485–498, 2012.

Taylor, R. G., Scanlon, B., Döll, P., Rodell, M., van Beek, R., Wada, Y., Longuevergne, L., Leblanc, M., Famiglietti, J. S., Edmunds, M., Konikow, L., Green, T. R., Chen, J., Taniguchi, M., Bierkens, M. F. P., MacDonald, A., Fan, Y., Maxwell, R. M., Yechieli, Y., Gurdak, J. J., Allen, D. M., Shamsudduha, M., Hiscock, K., Yeh, P. J.-F., Holman, I., and Treidel, H.: Ground water and climate change, Nat. Clim. Change, 3, 322–329, 2013.

Tesfa, T. K., Li, H.-Y., Leung, L. R., Huang, M., Ke, Y., Sun, Y., and Liu, Y.: A subbasin-based framework to represent land surface processes in an Earth system model, Geosci. Model Dev., 7, 947–963, doi:10.5194/gmd-7-947-2014, 2014.

Tharme, R. E.: A global perspective on environmental flow assessment: emerging trends in the development and application of environmental flow methodologies for rivers, River Res. Appl., 19, 397–441, doi:10.1002/rra.736, 2003.

Thenkabail, P. S., Biradar, C. M., Noojipady, P., Dheeravath, V., Li, Y., Velpuri, M., Gumma, M., Gangalakunta, O. R. P., Turral, H., Cai, X., Vithanage, J., Schull, M. A., and Dutta, R.: Global irrigated area map (GIAM), derived from remote sensing, for the end of the last millennium, Int. J. Remote Sens., 30, 3679–3733, 2009.

Trenberth, K. E. (Ed.): Climate Systems Modeling, Cambridge University Press, Cambridge, UK, 1992.

Trenberth, K. E. and Asrar, G. R.: Challenges and opportunities in water cycle research: WCRP contributions, Surv. Geophys., 35, 515–532, 2012.

Trenberth, K. E. and Dai, A.: Effects of Mount Pinatubo volcanic eruption on the hydrological cycle as an analog of geoengineering, Geophys. Res. Lett., 34, L15702, doi:10.1029/2007GL030524, 2007.

Tuinenburg, O. A., Hutjes, R. W. A., Jacobs, C. M. J., and Kabat, P.: Diagnosis of Local Land–Atmosphere Feedbacks in India, J.f Climate, 24, 251–266, 2011.

UN: Statistical Yearbook, Stat. Div., New York, 1997.

USDA: 2002 census of agriculture, National Agricultural Statistics Service, http://www.agcensus.usda.gov/Publications/2002/ (last access: 6 May 2014), 2002.

USDA: 2007 census of agriculture, Farm and Ranch Irrigation Survey, Volume 3, Special studies, part 1, http://www.agcensus.usda.gov/Publications/2007/Online_Highlights/FarmandRanch_IrrigationSurvey/fris08.pdf (last access: 6 May 2014), 2008.

van Beek, L. P. H., Wada, Y., and Bierkens, M. F. P.: Global monthly water stress: 1. Water balance and water availability, Water Resour. Res., 47, W07517, doi:10.1029/2010WR009791, 2011.

van Woerden, J., Diedericks, J., and Klein-Goldewjik, K.: Data management in support of integrated environmental assessment and modelling at RIVM – including the 1995 RIVM Catalogue of International Data Sets, RIVM Report no. 402001006, National Institute of Public Health and the Environment, Bilthoven, the Netherlands, 1995.

Vargas-Yáñez, M., Moya, F., García-Martínez, M. C., Tel, E., Zunino, P., Plaza, F., Salat, J., Pascual, J., López-Jurado, J. L., and Serra, M.: Climate change in the Western Mediterranean sea 1900–2008, J. Mar. Syst., 82, 171–176, doi:10.1016/j.jmarsys.2010.04.013, 2010.

Vassolo, S. and Döll, P.: Global-scale gridded estimates of thermoelectric power and manufacturing water use, Water Resour. Res., 41, W04010, doi:10.1029/2004WR003360, 2005.

Verseghy, D. L.: CLASS – A Canadian land surface scheme for GCMs I. Soil model, Int. J. Climatol., 11, 111–133, 1991.

Verseghy, D. L.: The Canadian land surface scheme (CLASS): Its history and future, Atmos.-Ocean, 38, 1–13, 2000.

Verseghy, D. L., McFarlane, N. A., and Lazare, M.: CLASS – A Canadian land surface scheme for GCMs II. Vegetation model and coupled runs, Int. J. Climatol., 13, 347–370, 1993.

Vitousek, P. M., Mooney, H. A., Lubchenco, J., and Melillo, J. M.: Human domination of Earth's ecosystems, Science, 277, 494–499, 1997.

Voisin, N., Liu, L., Hejazi, M., Tesfa, T., Li, H., Huang, M., Liu, Y., and Leung, L. R.: One-way coupling of an integrated assessment model and a water resources model: evaluation and implications of future changes over the US Midwest, Hydrol. Earth Syst. Sci., 17, 4555–4575, doi:10.5194/hess-17-4555-2013, 2013.

Vörösmarty, C. J. and Sahagian, D.: Anthropogenic disturbance of the terrestrial water cycle, BioScience, 50, 753–765, 2000.

Vörösmarty, C. J., Sharma, K. P., Fekete, B. M., Copeland, A. H., Holden, J., Marble, J., and Lough, J. A.: The storage and aging of continental runoff in large reservoir systems of the world, Ambio, 26, 210–219, 1997.

Vörösmarty, C. J., Federer, C. A., and Schloss A. L.: Potential evaporation functions compared on US watersheds: Possible implications for global-scale water balance and terrestrial ecosystem modeling, J. Hydrol., 207, 147–169, 1998.

Vörösmarty, C. J., Leveque, C., and Revenga, C.: Millennium Ecosystem Assessment Volume 1: Conditions and Trends, chap. 7: Freshwater ecosystems, Island Press, Washington, D.C., USA, 165–207, 2005.

Wada, Y., van Beek, L. P. H., van Kempen, C. M., Reckman, J. W. T. M., Vasak, S., and Bierkens, M. F. P.: Global depletion of groundwater resources, Geophys. Res. Lett., 37, L20402, doi:10.1029/2010GL044571, 2010.

Wada, Y., van Beek, L. P. H., Viviroli, D., Dürr, H. H., Weingartner, R., and Bierkens, M. F. P.: Global monthly water stress: 2. Water demand and severity of water stress, Water Resour. Res., 47, W07518, doi:10.1029/2010WR009792, 2011.

Wada, Y., van Beek, L. P. H., and Bierkens, M. F. P.: Nonsustainable groundwater sustaining irrigation: A global assessment, Water Resour. Res., 48, W00L06, doi:10.1029/2011WR010562, 2012.

Wada, Y., Wisser, D., Eisner, S., Flörke, M., Gerten, D., Haddeland, I., Hanasaki, N., Masaki, Y., Portmann, F. T., Stacke, T., Tessler, Z., Schewe, J.: Multimodel projections and uncertainties of irrigation water demand under climate change, Geophys. Res. Lett., 40, 4626–4632, 2013.

Wada, Y., Wisser, D., and Bierkens, M. F. P.: Global modeling of withdrawal, allocation and consumptive use of surface water and groundwater resources, Earth Syst. Dynam., 5, 15–40, doi:10.5194/esd-5-15-2014, 2014.

Walko, R. L., Band, L. E., Baron, J., Kittel, T. G. F., Lammers, R., Lee, T. J., Ojima, D., Pielke, R. A., Taylor, C., Tague, C., Tremback, C. J., and Vidale, P. L.: Coupled atmosphere-biophysics-hydrology models for environmental modeling, J. Appl. Meteorol., 39, 931–944, 2000.

Wei, J., Dirmeyer, P. A., Wisser, D., Bosilovich, M. G., and Mocko, D. M.: Where does the irrigation water go? An estimate of the contribution of irrigation to precipitation using MERRA, J. Hydrometeorol., 14, 275–289, 2013.

Wise, M. and Calvin, K.: GCAM 3.0 agriculture and land use: technical description of modeling approach, Pacific Northwest National Laboratory, Richland, WA, https://wiki.umd.edu/gcam/images/8/87/GCAM3AGTechDescript12511.pdf (last access: 6 May 2014), 2011.

Wise, M., Calvin, K., Thomson, A., Clarke, L., Bond-Lamberty, B., Sands, R., Smith, S. J., Janetos, A., and Edmonds, J.: The implications of limiting CO_2 concentrations for agriculture, land-use change emissions and bioenergy, Technical report PNNL-17943, available at: http://www.usitc.gov/researchand_analysis/economics_seminars/2009/200902_co2landuse.pdf (last access: 6 May 2014), 2009a.

Wise, M., Calvin, K., Thomson, A., Clarke, L., Bond-Lamberty, B., Sands, R., Smith, S. J., Janetos, A., and Edmonds, J.: Implications of limiting CO_2 concentrations for land use and energy, Science, 324, 1183–1186, 2009b.

Wisser, D., Frolking, S., Douglas, E. M., Fekete, B. M., Vörösmarty, C. J., and Schumann, A. H.: Global irrigation water demand: Variability and uncertainties arising from agricultural and climate data sets, Geophys. Res. Lett., 35, L24408, doi:10.1029/2008GL035296, 2008.

Wisser, D., Fekete, B. M., Vörösmarty, C. J., and Schumann, A. H.: Reconstructing 20th century global hydrography: a contribution to the Global Terrestrial Network-Hydrology (GTN-H), Hydrol. Earth Syst. Sci., 14, 1–24, doi:10.5194/hess-14-1-2010, 2010.

Wood, E. F., Roundy, J. K., Troy, T. J., van Beek, L. P. H., Bierkens, M. F. P., Blyth, E., de Roo, A., Döll, P., Ek, M., Famiglietti, J., Gochis, D., van de Giesen, N., Houser, P., Jaffé, P. R., Kollet, S., Lehner, B., Lettenmaier, D. P., Peters-Lidard, C., Sivapalan, M., Sheffield, J., Wade, A., and Whitehead, P.: Hyperresolution global land surface modeling: meeting a grand challenge for monitoring Earth's terrestrial water, Water Resour. Res., 47, W05301, doi:10.1029/2010WR010090, 2011.

WRI: World Resources 1998–99, Oxford Press, New York, USA, 1998.

WRI: World Resources 2000–01, Oxford Press, New York, USA, 2000.

Yoshikawa, S., Cho, J., Yamada, H. G., Hanasaki, N., and Kanae, S.: An assessment of global net irrigation water requirements from various water supply sources to sustain irrigation: rivers and reservoirs (1960–2050), Hydrol. Earth Syst. Sci., 18, 4289–4310, doi:10.5194/hess-18-4289-2014, 2014.

Zhao, M. and Dirmeyer, P. A.: Production and analysis of GSWP-2 near-surface meteorology data sets (Vol. 159), Center for Ocean-Land-Atmosphere Studies, Calverton, http://ww.w.monsoondata.org/gswp/gswp2data.pdf (last access: 6 May 2014), 2003.

Zhao, M., Pitman, A. J., and Chase, T.: The impact of land cover change on the atmospheric circulation, Clim. Dynam., 17, 467–477, 2001.

2

On inclusion of water resource management in Earth system models – Part 2: Representation of water supply and allocation and opportunities for improved modeling

A. Nazemi and H. S. Wheater

Global Institute for Water Security, University of Saskatchewan, 11 Innovation Boulevard, Saskatoon, SK S7N 3H5, Canada

Correspondence to: A. Nazemi (ali.nazemi@usask.ca)

Abstract. Human water use has significantly increased during the recent past. Water withdrawals from surface and groundwater sources have altered terrestrial discharge and storage, with large variability in time and space. These withdrawals are driven by sectoral demands for water, but are commonly subject to supply constraints, which determine water allocation. Water supply and allocation, therefore, should be considered together with water demand and appropriately included in Earth system models to address various large-scale effects with or without considering possible climate interactions. In a companion paper, we review the modeling of demand in large-scale models. Here, we review the algorithms developed to represent the elements of water supply and allocation in land surface and global hydrologic models. We note that some potentially important online implications, such as the effects of large reservoirs on land–atmospheric feedbacks, have not yet been fully investigated. Regarding offline implications, we find that there are important elements, such as groundwater availability and withdrawals, and the representation of large reservoirs, which should be improved. We identify major sources of uncertainty in current simulations due to limitations in data support, water allocation algorithms, host large-scale models as well as propagation of various biases across the integrated modeling system. Considering these findings with those highlighted in our companion paper, we note that advancements in computation and coupling techniques as well as improvements in natural and anthropogenic process representation and parameterization in host large-scale models, in conjunction with remote sensing and data assimilation can facilitate inclusion of water resource management at larger scales. Nonetheless, various modeling options should be carefully considered, diagnosed and intercompared. We propose a modular framework to develop integrated models based on multiple hypotheses for data support, water resource management algorithms and host models in a unified uncertainty assessment framework. A key to this development is the availability of regional-scale data for model development, diagnosis and validation. We argue that the time is right for a global initiative, based on regional case studies, to move this agenda forward.

1 Introduction

The water cycle is fundamental to the functioning of the Earth system and underpins the most basic needs of human society. However, as noted in our companion paper (hereafter referred to as Nazemi and Wheater, 2015), the current scale of human activities significantly perturbs the terrestrial water cycle, with local, regional and global implications. Such disturbances affect both hydrological functioning and land–atmospheric interactions, and therefore should be explicitly represented in large-scale models. We consider both land surface models (LSMs) and global hydrologic models (GHMs). LSMs generally represent water, energy and carbon cycles, and can be coupled with climate models (i.e., online simulations) for integrated Earth system modeling, or uncoupled from climate models (i.e., offline simulations) for large-scale impact assessment. GHMs are also run in uncoupled mode for impact assessment; however, they focus exclusively on the water cycle. In this survey, we consider the representation of water resources management in these large-scale models and focus on water quantity rather than water quality. We note that while historically the effects of water management have largely been neglected in LSMs and GHMs, there has been increasing interest in recent years in their inclusion and a common first step is to estimate the demand for water, in particular associated with irrigation (see Nazemi and Wheater, 2015). However, in practice water resource systems are often complex, and associated infrastructure may have competing functional requirements and constraints (e.g., flood protection, water supply, environmental flows), exacerbated during drought. In this paper, we turn to the issues around water supply and allocation and associated representations in large-scale models.

Major implications are associated with water allocation from surface-water and groundwater sources. For instance, large dams and reservoirs can significantly modify downstream streamflow characteristics (e.g., Vörösmarty et al., 1997, 2003; Oki and Kanae, 2006; Wisser et al., 2010; Tang et al., 2010; Tebakari et al., 2012; Lehner and Grill, 2013; Lai et al., 2014; Haddeland et al., 2014), with large regional variability (see, e.g., Pokhrel et al., 2012a). Considering that almost all major river systems in the Northern Hemisphere (except for the arctic and subarctic regions) are dammed (e.g., Meybeck, 2003; Nilsson et al., 2005), it can be argued that accurate simulation of continental and global runoff is impossible without considering the effects of reservoirs. Such hydrologic impacts and associated environmental consequences can be studied through offline LSMs or GHMs. There are, however, important land-surface implications associated with reservoir operation that require online simulations. For instance, it has also been argued that large dams can have important footprints on surface energy (Hossain et al., 2012), with associated effects on land-surface boundary conditions and potential interactions with local and regional climate (MacKay et al., 2009). To understand these effects, online LSMs, i.e., coupled with climate models, are required to provide quantitative knowledge of the extent of such impacts in time and space.

Groundwater resources have also been extensively perturbed during the "Anthropocene". Every year, a large amount of groundwater is pumped to the land surface for both irrigative and non-irrigative purposes (e.g., Zektser and Lorne, 2004; Siebert et al., 2010). Such extraction has already caused large groundwater depletion in some areas (Rodell et al., 2007, 2009; Gleeson et al., 2010, 2012) and changed the surface water balance due to return flows from demand locations to river systems and ultimately to oceans (e.g., Lettenmaier and Milly, 2009; Wada et al., 2010; Pokhrel et al., 2012b; Döll et al., 2014). In parallel, a considerable proportion of the surface water diverted into irrigated areas may recharge groundwater (Döll et al., 2012). From a broader perspective, groundwater aquifers (particularly shallow groundwater) can also be an important control on soil moisture and wetlands, and thus influence atmospheric surface boundary conditions (e.g., Maxwell et al., 2007, 2011; Fan and Miguez-Macho, 2011; Dadson et al., 2013). These online effects are widely unquantified at the global scale, as the sub-surface processes below the root zone have been generally assumed to be disconnected from the atmosphere (see Taylor et al., 2013).

In addition, representing water allocation practice in large-scale models is urgently required to address various emerging water security concerns including (but not limited to) human water supply (e.g., Postel et al., 1996), ecosystem health (e.g., Vörösmarty et al., 2010), sedimentation (e.g., Syvitsky et al., 2005) and water quality (e.g., Skliris and Lascaratos, 2004). These latter areas are beyond the scope of this paper, but highlight the need to represent human water allocation in large-scale models for regional and global impact assessments. For instance, the most densely populated parts of the globe suffer from extremely fragile water supply conditions (e.g., Grey et al., 2013; Falkenmark, 2013; Nazemi and Wheater, 2015) and this will be amplified under future climate change and population growth (e.g., Arnell, 2004; Wada et al., 2013; Rosenzweig et al., 2014; Schiermeier, 2014; Haddeland et al., 2014). While population growth directly affects water demand, indirect effects include changing land and water management, with associated impacts on the aquatic environment. Similarly, climate change is expected to perturb both water demand and supply, as it also results in greater seasonal and inter-annual variability with increase in the risk of extreme conditions (e.g., Dankers et al., 2014; Prudhomme et al., 2014). Looking to the future, Yoshikawa et al. (2014) argued that current sources can only account for 74 % of the global net irrigation requirements of the 2050s and supply–demand imbalance will cause a major increase in global water scarcity (Alcamo et al., 2007; Hanasaki et al., 2008a, b, 2013a, b; Schewe et al., 2014). In water-scarce conditions, competition for water resources becomes increasingly important and the details of water allocation practice play a key role in the spatial and temporal distribution of water stress. These issues necessitate adaptation strategies to mitigate the effects of water stress and extreme conditions; large-scale models are, therefore, required to assess the effects of various global changes and to examine the impact of alternative management strategies.

Representation of water allocation practice introduces a set of issues associated with management and societal preferences, local and regional differences in decision making, complexity of water resources systems (particularly at larger scales), as well as lack of data support. At local and basin scales, water allocation practice is mainly defined as an optimization problem, in which the aim is to minimize the adverse effects of water shortage and/or to maximize the economic benefits of the water resource system. The advent of search algorithms such as linear programming (Dantzig, 1965), dynamic programming (Bellman, 1952) and genetic algorithms (Goldberg, 1989) has resulted in a wide variety of operational models for water resource management at small basin scale (e.g., Rani and Moreira, 2010; Hossain and El-shafie, 2013; see Revelle et al. (1969) for the early developments). These small-scale water allocation models, however, typically do not include processes related to water supply and demand and receive these variables as prescribed inputs. Moreover, small-scale operational models often require detailed information about policy constraints and operational management. This information is not generally available over larger regions and at the global scale. Even if all related information were to be available, the level of complexity within small-scale operational models cannot be supported globally due to high dimensionality in decision variables and computational burdens. These restrictions have resulted in the progressive development of macro-scale algorithms to represent water allocation practice and competition among demands at regional and global scales.

The main objective of this paper is to overview the current literature and to identify the state of available methods and applications for large-scale representations of water supply and allocation in LSMs and GHMs, with relevance to both Earth system and large-scale hydrological modeling. Section 2 addresses the representation of surface-water and groundwater sources. Section 3 discusses the linkage between available sources and prescribed demands (see Nazemi and Wheater, 2015) through macro-scale allocation algorithms. Section 4 reviews current large-scale modeling applications and discusses the quality of available simulations. Section 5 merges the findings of Nazemi and Wheater (2015) with those obtained in Sects. 2 to 4, and highlights current gaps and opportunities from an integrated water resources, hydrology and land-surface modeling perspective. This is finalized by suggesting a systematic framework for model development and uncertainty assessment to guide future efforts in inclusion of water resource management in large-scale models. Section 6 closes our survey and provides some concluding remarks.

2 Available representations of water sources in large-scale models

2.1 Lakes and reservoirs

Natural lakes and man-made reservoirs cover more than 2 % of the global land surface area except for Antarctica and glaciated Greenland (Lehner and Döll, 2004). Lakes and reservoirs are important water sources due to their ability to store and release surface water for human demand. While natural lakes have been historically an important water source for human civilization, man-made reservoirs have been mainly constructed over the last 50 years. Currently, there are more than 16 million reservoirs worldwide (Lehner et al., 2011), retaining around 20 % of the annual runoff and 10 % of the total volume of the world's freshwater lakes (Gleick, 2000; Meybeck, 2003; Wood et al., 2011). This makes an important global water resource: Yoshikawa et al. (2014) estimated that reservoirs allocated 500 km^3 just for irrigation during the year 2000, worldwide.

From the large-scale modeling perspective, lakes and reservoirs introduce heterogeneity into land-surface parameterizations, with both offline and online implications. To represent these open water bodies, first they should be identified at the grid and sub-grid scales. The availability of basic data for larger lakes and reservoirs is relatively good (see Lehner and Döll, 2004, for a comprehensive list of data sources). For instance, the Global Lakes and Wetlands Database (GLWD; http://www.worldwildlife.org/pages/global-lakes-and-wetlands-database) includes more than 250 000 lakes globally. In addition, the International Commission of Large Dams (ICOLD; http://www.icold-cigb.net/) and Global Reservoir and Dam (GRanD; http://www.gwsp.org/products/grand-database.html) databases contain information about the location, purpose and capacity of 33 000 and 7000 large dams, worldwide. However, to estimate evaporation, as well as storage and release, more specific physical characteristics, such as storage-area–depth relationships, are required. These data are generally not available and parametric relationships have been used to approximate these properties based on various assumptions (e.g., Takeuchi, 1997; Liebe et al., 2005). Nonetheless, at this stage of model development, reservoir simulations cannot be directly verified, due to the lack of observations of reservoir level and storage (Gao et al., 2012). These data limitations may be largely solved in the relatively near future by upcoming satellite missions – see the discussion of Sect. 5.3.

Depending on their size, lakes and reservoirs can be represented either within channel or sub-grid routing components of host large-scale models. While larger lakes and reservoirs are normally represented within the river routing component and regulate the channel streamflow, smaller bodies are mainly considered within sub-grid parameterizations as an additional pond (e.g., Döll et al., 2003; Wisser et al., 2010). Ideally, natural lakes and reservoirs should differ in their representation due to human management. If human management is neglected, reservoir releases can be represented similar to natural lakes using simple parametric equations that link the reservoir release to reservoir storage (or level) – see e.g., Meigh et al. (1999), Döll et al. (2003), Pietroniro et al. (2007) and Rost et al. (2008). Lake algorithms, however, have had limited success in highly regulated basins. This is rather intuitive: for natural lakes, the dynamics of lake storage (and hence discharge) are regulated by climate and inflow variability, whereas the dynamics of reservoir discharge (and hence storage) are mainly controlled by pressures of downstream demands and management decisions. Moreover, reservoirs are often multi-functional and deal with competing demands with varying priority in time; therefore, simple lake routing algorithms are unable to fully describe reservoir functionality. Alternatively, macro-scale algorithms for reservoir operation have been suggested, which attempt to link reservoir releases to inflows, storage and prescribed human demands considering water allocation objectives – see Sect. 3.3.

Considering online implications, the effects of dams on near-surface energy and moisture conditions and hence land–atmospheric feedbacks can be important for large reservoirs (Hossain et al., 2012). Addressing this issue using coupled LSMs is currently a major gap in the literature and presents a challenging problem at the grid scale, since the impact of dams on the local climate can be masked by regional climate variability and surrounding land cover (e.g., Zhao and Shepherd, 2012).

2.2 Streamflow diversions and inter-basin water transfers

Streamflow diversions of any magnitude require dams or barrages. At smaller scales, these include within-basin water transfers from local streams to nearby demands. In-basin diversions are often represented in large-scale models by instantaneous abstractions (e.g., Hanaski et al., 2008a, 2010; Döll et al., 2009). Hydrologic routing can be alternatively considered for improved representation (e.g., Wisser et al., 2010). It should be noted that a proportion of the diverted flow normally returns to the river systems. Heuristic algorithms have been advised to mimic the mechanism of diversion based on returning the excess water to the river with some lag. Biemans et al. (2011) for instance represented the dynamics of diverted/return flows for irrigated areas by making water available for consumption for 5 days; if unused, it is released back to the river. This can have important implications for differentiating between the actual use and total withdrawals, in the case where water is over-allocated.

Inter-basin water transfers normally involve major infrastructure and can significantly perturb the regional streamflow regime. For instance, proposed south to north water transfer schemes in China (see Liu and Zheng, 2002; Liu and Yang, 2012) would divert 44.8 billion m^3 of water annually (http://www.internationalrivers.org/). The associated hydrological impacts are estimated to be as or more significant than land-use and/or land-cover changes (J. Liu et al., 2013). Inter-basin water transfer can be adequately represented by hydrologic routing. Examples are available for some regional applications (e.g., Nakayama and Shankman, 2013a, b; Ye et al., 2013); however, efforts to represent long-distance diversions at the global scale are limited. This is mainly due to data issues regarding the location and specification of diversion channels globally. This could be largely resolved in future due to improvements in remote sensing observations – see the discussion of Sect. 5.3.

2.3 Groundwater

Even large-scale models with detailed water resource management schemes have limited representation of groundwater availability (see Table 1), largely due to the limitations in data related to groundwater storage, withdrawals and subsurface properties as well as computational difficulties. There have been some efforts to include groundwater in LSMs to describe the aquifer dynamics, land–atmospheric feedbacks and watershed responses, mainly at basin and small regional scales (e.g., Maxwell and Miller, 2005; Maxwell et al., 2007, 2011; Kollet and Maxwell, 2008; Ferguson and Maxwell, 2010). These studies consider a physically based groundwater store, which can be updated at each modeling time step using a 3-D representation of groundwater movement, and linked to land-surface calculations through soil moisture dynamics. Such representations are computationally expensive and limited at the global scale, since temporal and spatial domains should be finely gridded for accurate representations of groundwater movement and soil-moisture interactions, particularly in online studies. To the best of our knowledge, no online study characterizing the feedback effects between groundwater management and climate is available at the global scale. Offline representation of groundwater management has mainly been performed in the context of GHMs and involves estimation of available groundwater storage, sub-grid groundwater recharge and groundwater withdrawals. In this section, we focus on groundwater availability and recharge and leave the discussion related to groundwater withdrawals to Sect. 3.2.

In current representations, often groundwater availability in general, or the nonrenewable and nonlocal blue water (NNBW) in particular, is assumed as an unlimited local source (e.g., Rost et al., 2008; Biemans et al., 2011; Pokhrel et al., 2012a, b). NNBW is a technical term defined as an "imaginary" source that implicitly accounts for nonrenewable fossil groundwater or other water sources that are not explicitly represented in the model. This can cause major uncertainties in estimation of actual withdrawals (see Sect. 3.2). Efforts have been made to improve this assumption. For instance, Strzepek et al. (2012) bounded groundwater availability by considering a threshold for groundwater allocation. Wada et al. (2014) proposed a conceptual linear groundwater reservoir, parameterized globally based on lithology and topography, to estimate the groundwater availability at the grid scale using the baseflow as a proxy. Although this conceptual representation provides an efficient scheme for global simulations, it ignores the baseflow reduction due to groundwater depletion. In a more recent attempt, Döll et al. (2014) continuously simulated the daily groundwater storage using the difference between groundwater recharge and the sum of baseflow and net groundwater abstraction, with base flow declining with decreasing groundwater storage. Both algorithms, however, do not consider inter-grid lateral groundwater movement, which can have an important impact on water availability across various scales. Although lateral groundwater movement is widely studied in aquifer studies at smaller basin and regional scales (e.g., Ye et al., 2013),

Table 1. Examples of available representations of water supply and allocation in large-scale models.

Reference	Water supply				Water allocation		
	Diversions	Reservoirs	Groundwater store	Desalination and reuse	Supply–demand dependency	Priorities in demands	Operational objectives
Haddeland et al. (2006b)	In- and inter-grid abstraction	Macro-scale operation[1]	N/A	N/A	Reservoir can supply up to 5 grids downstream[2]	Irrigation, flood control, hydropower, others	Minimize deficit, maximize hydropower
Hanasaki et al. (2008a)	In- and inter-grid abstraction	Macro-scale operation	N/A	N/A	Reservoir can supply up to 10 grids downstream	Municipal, industrial, irrigation	Allocate available water
Rost et al. (2008)	Local abstraction	Lake routing	NNBW assumed unlimited[3]		Local grid	Irrigation only	Meet demand using available water
Döll et al. (2009)	In- and inter-grid abstraction	Macro-scale operation	N/A	N/A	Reservoir can supply up to 5 grids downstream	Irrigation, non-irrigation	Meet total demand[4]
Hanasaki et al. (2010)	Local abstraction	Macro-scale operation/local abstraction	NNBW assumed unlimited		Local grid	Irrigation and livestock only	Meet total demand using unlimited NNBW
Strzepek et al. (2010)	Local abstraction	Macro-scale operation[5]	Countrywide estimates	N/A	Local basin	Domestic, industry, livestock, irrigation	Maximize profitability
Wisser et al. (2010)	In-grid hydrologic routing	Macro-scale operation	Unlimited local source[6]	N/A	Local grid	Irrigation only	Meet total demand using unlimited groundwater
Biemans et al. (2011)	Local abstraction, heuristic routing	Macro-scale operation	NNBW assumed unlimited[7]		Reservoir can supply up to 5 grids downstream	Irrigation only	Proportional allocation of available water
Wada et al. (2011)	In- and inter-grid abstraction	Macro-scale operation	Countrywide estimates	Countrywide estimates	Reservoir can supply up to 600 km downstream	Irrigation, flood control, hydropower, others	Minimize deficit, maximize hydropower
Pokhrel et al. (2012a)	Local abstraction	Macro-scale operation	NNBW assumed unlimited		Local grid	Irrigation, non-irrigation	Meet total demand using unlimited NNBW
Strzepek et al. (2012)	Local abstraction	Macro-scale operation[5]	Basin-scale threshold	Function of capacity	Local basin	Non-agricultural, agricultural	Minimize groundwater use and spill
Blanc et al. (2013)	Local abstraction, heuristic routing	Macro-scale operation[5]	Basin-scale threshold	N/A	Local basin agricultural	Non-agricultural, spill	Minimize groundwater use and
Hanasaki et al. (2013b)	Local abstraction	Macro-scale operation	N/A	N/A	Local grid	Municipal, industrial, irrigation	Allocate available water
Voisin et al. (2013a, b)	In- and inter-grid abstraction	Macro-scale operation	N/A	N/A	Reservoir can supply up to 200 km downstream	Irrigation, flood control, hydropower and others	Allocate available water
Wada et al. (2014)	In- and inter-grid abstraction	Macro-scale operation	Conceptual reservoir	Countrywide estimates	Reservoir can supply up to 600 km downstream	Irrigation, non-irrigation	Allocate available water

1 Simultaneous operation of multiple dams in a river basin was not considered.
2 See Haddeland et al. (2006a).
3 Simulations without assuming unlimited groundwater store were also performed.
4 Demand that cannot be allocated in any given day is allocated later in the year or in the next year, when water is available.
5 A virtual reservoir is considered for each basin.
6 Shallow groundwater is represented as a runoff retention pool, which delays runoff before it enters streams.
7 Simulations with considering only surface water availability were also performed.

it is currently a key missing process representation at larger regional and global scales (Taylor et al., 2013).

Groundwater recharge includes the movement of water from the unsaturated soil zone to a saturated groundwater body. There are a number of approaches to represent the vertical water movement in large-scale models, including heuristic methods (e.g., Döll et al., 2003), conceptual "leaky buckets" (e.g., Wada et al., 2010), or numerical solutions of the physically based Richards' equation (Best et al., 2011; D. B. Clark et al., 2011). These approaches are based on various assumptions and are subject to large uncertainties. Heuristic schemes relate the recharge rate to surface runoff, using a set of parameters based on catchment, soil and aquifer characteristics. These representations are often simplistic and may result in large estimation errors, particularly in arid and semi-arid regions (Polcher et al., 2011). Conceptual approaches widely assume a steady-state condition and use the unsaturated hydraulic conductivity to represent groundwater recharge with or without considering capillary rise (van Beek and Bierkens, 2008; Wada et al., 2010, 2014; van Beek et al., 2011; Ye et al., 2013). In a global study, Wada et al. (2012) used this approach to account for additional recharge from irrigated lands based on the unsaturated hydraulic conductivity at field capacity. This can be important for representing the excess water diverted from both surface and groundwater sources. Although conceptual representations are efficient for large-scale studies, still limitations remain in these schemes due to large heterogeneities in soil characteristics, a common assumption of steady-state recharge rate, as well as the inherent uncertainty associated with soil hydraulic properties. The physically based approaches remove the steady-state assumption; nonetheless as discussed above, they require a detailed numerical scheme for solving a highly non-linear partial differential equation. This is subject to various computational difficulties at larger scales, and invariably there is a gap between the scale for which Richards' equation was developed and the scale at which it is implemented in large-scale groundwater and hydrologic models (Beven, 2006a; Gentine et al., 2012).

2.4 Desalination and water reuse

Water reuse and desalination are currently minor water resources at the global scale and have been widely ignored in large-scale models. Nonetheless, it should be noted that these water sources have local relevance and are important in several water-limited regions (Wade Miller, 2006; Pokhrel et al., 2012a). Wada et al. (2011) estimated that annual desalinated water use is around 15 km^3 globally, of which Kazakhstan uses 10 % of the total volume. Desalinated water availability can be estimated using a bottom-up approach based on the information available about treatment and water reuse capacity at the grid scale (Strzepek et al., 2012). These data, however, are limited and uncertain globally. Alternatively, top-down approaches try to downscale countrywide data. Wada et al. (2011, 2014), for instance, downscaled the countrywide data on water reuse and desalination using a gridded population map. Considering that water reuse and desalination will likely be more important in future due to increased water scarcity at the global scale, we suggest more effort in representing these sources, including data collection to support future algorithm developments – see Sect. 5.3.

3 Available representations of water allocation in large-scale models

Water allocation distributes the available water sources among competing demands and should typically include a set of management decisions to systematically (1) link the prescribed demands to available sources of water; (2) determine allocation objectives as well as priorities in case of water shortage; and (3) withdraw the available water based on allocation objectives and management constraints. At this stage of model development, there are limited examples for representation of water allocation at larger scales. These studies are offline and have multiple sources of uncertainty. Table 1 summarizes some examples from the recent literature. In this section, we briefly discuss the main requirements and available algorithms for representing water allocation in large-scale models.

3.1 Main requirements

The first basic requirement is to identify which sources are available to supply the water demands within each computational grid. The majority of current allocation schemes assume that grid-based demands can be supplied from the sources available within the local grid. This assumption is intuitive and easy to implement; however, it naturally ignores long-distance water transfers. Various modifications have been proposed to overcome this limitation. Relative elevation and travel time of water from source to demand have been used to condition demands to available sources upstream. For example, Hanaskai et al. (2006) assumed that large reservoirs can potentially supply downstream demands that are located within 1100 km (based on a travel time of 1 month). Similarly, Wada et al. (2011) considered a criterion of approximately 600 km and Biemans et al. (2011) 250 km. These rules are evidently simplistic but can be easily implemented. They also generally assume steady-state conditions, so that the allocated water can be simply abstracted from the source and added at the demand location at the same time step. Alternatively, routing schemes can provide a more accurate basis for representing the water delivery and avoid this limitation – see the discussion of Sect. 5.5.

The second important issue is to determine objectives of and priorities for water allocation, particularly during shortage. In the absence of access to local operating rules, this requires defining a set of generic rules to assign the relative preference of each demand and to define the purpose of water allocation. Both irrigative (e.g., Rost et al., 2008; Döll et al., 2009; Wada et al., 2014) and non-irrigative demands (e.g., Hanasaki et al., 2008a; Strzepek et al., 2010, 2012; Blanc et al., 2013) have been given the highest priority. In cases where multiple demands with the same priority are derived from a unique source of water, the deficit is typically shared proportionately to the demands (e.g., Biemans et al., 2011). Based on priorities and assumptions made regarding water availability, several allocation objectives have been used (see Table 1). It should be noted that water resource management is commonly multi-purpose, and allocation objectives and priorities can change within a typical operational year. For example, many reservoirs are designed for two conflicting objectives, i.e., irrigation supply and flood control. To account for this, Voisin et al. (2013a) developed flood control storage targets to complement the irrigation release targets. They used rule curves to drop the reservoir storages before snowmelt starts while maintaining the storage in the reservoir to provide releases for irrigation, water supply and hydropower in the remaining part of the year. They showed that this modification can improve the simulation of regulated flow and maintain the spatiotemporal consistency of reservoir levels – see Voisin et al. (2013a) for more details.

Finally, allocation algorithms are required to estimate groundwater abstractions and reservoir releases at each simulation time step based on allocation objectives and priorities. Groundwater abstraction algorithms are generally limited, due to significant gaps in information about groundwater availability and actual groundwater withdrawals at the global scale. Although current data availability for lakes and reservoirs storages is also poor, runoff data are relatively available regionally and globally, which can be used for algorithm development and performance assessment through comparison of simulated and observed discharges downstream of reservoirs. Apart from local or national data, data of the Global Runoff Data Centre (GRDC; http://www.bafg.de/GRDC/) have been widely used for validation of macro-scale reservoir operation algorithms.

3.2 Grid-based groundwater abstractions

Groundwater abstractions include both sustainable (renewable) and unsustainable (non-renewable) water uses. While sustainability of groundwater withdrawals is a complex issue, in particular related to environmental impacts of abstraction, the distinction between these for large-scale applications is generally based on the grid-based groundwater recharge, as any abstraction exceeding recharge rate results in groundwater depletion and, therefore, can be considered as unsustainable. So far, groundwater withdrawals have been estimated through either bottom-up or top-down algorithms, both subject to large uncertainty.

In bottom-up procedures, the groundwater abstraction is identified using grid-based estimates of surface and groundwater availability as well as the water demand. If the groundwater and/or NNBW is considered as an infinite sources (Rost et al., 2008; Hanasaki et al., 2010; Wisser et al., 2010; Pokhrel et al., 2012a, b), then the groundwater or NNBW abstraction is equal to estimated demand minus estimated surface water availability at the grid scale. In this case, priorities are not inherently considered; however NNBW has the advantage that it explicitly accounts for the water that should come to the system from outside the modeled domain. If the groundwater availability is bounded at the grid or basin scale, then the maximum groundwater withdrawal cannot exceed the local groundwater availability (e.g., Strzepek et al., 2012; Wada et al., 2014); however, errors in estimations of surface water availability and water demands can still directly propagate into estimation of groundwater withdrawals.

Top-down approaches are based on using recorded regional groundwater withdrawals or downscaling national groundwater abstractions data to finer spatial scales. Siebert et al. (2010) created a global data set for irrigation water supply from groundwater abstractions based on FAO-AQUASTAT (http://www.fao.org/nr/water/aquastat/main/index.stm) and other census and sub-national data. In an another effort, Wada et al. (2010, 2012) used the data of the International Groundwater Resources Assessment Center (IGRAC; http://www.igrac.net) to estimate the countrywide groundwater use for year 2000. These estimates were further downscaled to $0.5° \times 0.5°$ grids, based on a global map of yearly total water demand. In a countywide study, Blanc et al. (2013) used the groundwater withdrawal data of the USGS for the year 2005 (USGS, 2011) and repeated the data for every year of simulation. These approaches are also limited by the fact that the actual groundwater pumping might be considerably more than the recorded data (e.g., Foster and Loucks, 2006; Wada et al., 2012) and groundwater withdrawals can have considerable inter-annual variability. Current and upcoming remote sensing technologies can address some of the issues around groundwater data availability – see Sect. 5.3.

3.3 Macro-scale reservoir operation

Current macro-scale reservoir operation algorithms are designed for offline applications and included in large-scale models for characterizing the impacts of reservoirs on terrestrial water storage, runoff and water supply. These algorithms can be roughly divided into two general categories based on either simulating the reservoir release using a set of prescribed operational rules or using search algorithms to find optimal reservoir release. In brief, simulation-based schemes are based on a set of functional rules that use initial storage as well as inflows and demand pressure during a typical operational period to simulate releases during the operational period. In contrast, optimization-based algorithms search for optimal releases at each time step given an ideal storage at the end of the operational year, storage at the beginning of the year and expected inflows and demands during the year. Naturally, optimization-based algorithms are more computationally expensive; nonetheless, they are more suitable for evaluating competition among water demands and effects of policy change, due to the ability to explicitly include multiple allocation objectives to guide the search for optimal releases. In contrast, simulation-based algorithms are more efficient and can be potentially modified to support online simulations – see Sect. 5.4. Table 2 summarizes some representative examples from the current literature.

3.3.1 Available simulation-based algorithms

Current simulation-based algorithms are heavily influenced by the work of Hanasaki et al. (2006), which was initially proposed for global routing models but extended to GHMs (Hanasaki et al., 2008a, 2010) and LSMs (Pokhrel et al., 2012a, b). The algorithm distinguishes between operational rules for irrigation and non-irrigation purposes. The algorithm also accounts for both inter-annual variability and seasonality in reservoir releases. In simple terms, the total release in a typical operational year is first determined based on the reservoir capacity, initial storage and the annual mean natural inflow to the reservoir. Second, the monthly fluctuations in the reservoir release are parameterized based on annual mean natural inflow, mean annual demand and the prescribed monthly demand. Note that demands are considered as total water withdrawals rather than consumptive uses. Finally, monthly fluctuations are corrected based on inter-annual variability in total reservoir releases (estimated during the first step) to provide actual monthly reservoir releases. The correction, depending on the purpose and size of reservoir, is based on the ratio of initial reservoir storage to total capacity, the ratio of reservoir capacity to annual mean inflow, and/or the monthly mean natural inflows to the reservoir – see Hanasaki et al. (2006) for related formulations.

The Hanasaki et al. algorithm has been widely used in the recent literature as it provides a generic and flexible framework to represent reservoir operation. Döll et al. (2009) implemented this algorithm to represent operation of large reservoirs within the framework of WaterGAP (Alcamo et al., 2003). They considered some modifications to accommodate losses from the reservoir and to characterize the dynamics of demand pressure on reservoirs based on consumptive uses rather than total water withdrawals. Biemans et al. (2011) modified the algorithm of Hanasaki et al. (2006) by extracting the reservoir releases using annual and monthly mean regulated inflows (rather than corresponding natural flows), limiting the demand pressure only to irrigation and changing the release rules during high-demand periods. These modifications were further added to the Joint UK Land Environment Simulator (JULES; Best et al., 2011; D. B. Clark et al., 2011) for offline simulations (Polcher et al., 2011). Voisin et al. (2013a) made a regional intercomparison between various simulation-based algorithms for the Columbia River basin and concluded that deriving releases based on withdrawals rather than consumptive uses results in improved simulations of downstream flows. They also indicated that the choice of natural or regulated inflows depends on the severity of the demand pressure and water allocation: if the overall water demand is high with respect to mean annual inflow, it would be better to drive the algorithm with mean monthly regulated inflow; otherwise it is better to use the natural flow, due to large uncertainties associated with water demand estimates and, therefore, regulated flows. Although this study is limited to one region, it provided an assessment of uncertainties in estimating the reservoir releases due to uncertainties in estimating both inflows and water demand – see the discussion of Sect. 4.

Existing simulation-based schemes are not limited to the above algorithms. Efforts have been made to simulate the reservoir releases using parametric functions, in which the parameters can be calibrated using observed downstream flows. For example, Wisser et al. (2010) advised a set of functional rules to parameterize the release from large reservoirs using the actual inflow and the long-term mean inflow to the reservoirs. More recently, Wu and Chen (2012) proposed a new algorithm by explicit consideration of operational rule curves, locally specified for each reservoir. In brief, rule curves are a set of pre-defined reservoir levels that divide the total reservoir capacity into different storage zones. These storage zones can be further associated with demands conditioned on the reservoir using various assumptions. The algorithm considers the reservoir operation at a given day as a deviation from mean releases at that day and represents this by a weighted sum of individual variations as the result of allocation for each individual water demand. Demand-specific allocations can be therefore characterized based on rule curves, the available storage, total capacity as well as the history of inflow to the reservoir. Accordingly the total release at any given day can be defined as a parametric function, in which the parameters can be tuned using observed downstream flows. Although they noted that the operational parameters are inherently time-varying, as the purpose of dam can change with time, a systematic scheme for dealing with non-stationary parametric estimation has not been provided. This remains for future efforts – see Sect. 5.4.

3.3.2 Available optimization-based algorithms

Optimization-based schemes were initially proposed by Haddeland et al. (2006a) and implemented further in Haddeland et al. (2006b, 2007). These algorithms are heavily inspired by small-scale reservoir operation algorithms within the engineering literature, particularly dynamic programming (see Voisin et al., 2013a), and strongly rely on estimates of expected inflow and demand. Therefore, they are not suitable for online simulations; however, they can be valuable for in-

Table 2. Representative examples of available macro-scale reservoir operation algorithms implemented in large-scale models.

Reference	Host model	Routing algorithm	Type of operation	Reservoir data	Validation discharge data
Hanasaki et al. (2006)	N/A	TRIP (Oki and Sud, 1998)	Simulation-based	WRD98 (ICOLD)	GSWP (Dirmeyer et al., 1999; Oki et al., 2001)
Haddeland et al. (2006a, b, 2007)	VIC (Liang et al., 1994)	Linearized Saint-Venant (Lohmann et al., 1996, 1998)	Optimization-based	ICOLD; Vörösmarty et al. (1997, 2003)	USGS (http://waterdata.usgs.gov) USBR (http://www.usbr.gov) GRDC (http://www.bafg.de/GRDC/)
Adam et al. (2007)	VIC (Liang et al., 1994)	Unit hydrograph and Linearized Saint-Venant (Lohmann et al., 1996, 1998)	Optimization-based	ICOLD; Vörösmarty et al. (1997, 2003)	Adam and Lettenmaier (2008)
Hanasaki et al. (2008a)	H08 (Hanasaki et al., 2008a, b)	TRIP (Oki and Sud, 1998)	Simulation-based	WRD98 (ICOLD)	GRDC (http://www.bafg.de/GRDC/)
Döll et al. (2009)	WaterGAP (Alcamo et al., 2003)	HBV (Bergström and Singh, 1995)	Simulation-based	GRanD (Lehner et al., 2008)	GRDC (http://www.bafg.de/GRDC/)
Wisser et al. (2010)	WBMplus (Vörösmarty et al., 1998)	Muskingum-Cunge (Ponce and Changanti, 1994)	Simulation-based	ICOLD	UNH-GRDC (Fekete et al., 1999, 2002)
Biemans et al. (2011)	LPJmL (Gerten et al., 2004; Rost et al., 2008)	Linear reservoir model (Huggins and Burney, 1982)	Optimization-based	GRanD (Lehner et al., 2011)	GRDC (http://www.bafg.de/GRDC/)
Van Beek et al. (2011)	PCR-GLOBWB (van Beek and Bierkens, 2009)	Kinematic Saint-Venant (Chow et al., 1998)	Optimization-based	GLWD1 (Lehner and Döll, 2004)	GRDC (http://www.bafg.de/GRDC/)
Wu and Chen (2012)	SWAT (Arnold et al., 1998)	SWAT (Arnold et al., 1998)	Simulation-based	Wu et al. (2007)	Chen and Wu (2008)*
Pokhrel et al. (2012a)	MASTIRO (Takata et al., 2003)	TRIP (Oki et al., 2001)	Simulation-based	WRD98 (ICOLD)	GRDC (http://www.bafg.de/GRDC/)
Voisin et al. (2013a)	VIC (Liang et al., 1994)	MOSART (Li et al., 2013; Tesfa et al., 2014)	Simulation-based	GRanD (Lehner et al., 2011)	USGS (http://waterdata.usgs.gov) USBR (http://www.usbr.gov) GRDC (http://www.bafg.de/GRDC/)
Voisin et al. (2013b)	SCLM (Li et al., 2011; Lawrence et al., 2011)	MOSART (Li et al., 2013; Tesfa et al. 2014)	Simulation-based	GRanD (Lehner et al., 2011)	USGS (http://waterdata.usgs.gov) USBR (http://www.usbr.gov) GRDC (http://www.bafg.de/GRDC/)

* Discharge data used for calibration as well.

tegrated impact assessment over large grids and/or assessment regions in offline mode (see, e.g., Strzepek et al., 2010, 2012; Blanc et al., 2013). In brief, the calculation starts by targeting the reservoir storage at the end of a typical operational year based on expected demands. Then, the minimum release at each daily time step is defined based on the expected streamflow at the dam's location to maintain a minimum flow requirement downstream of the reservoir. Accordingly, the maximum allowable daily release is determined based on simulated daily inflow, minimum release, reservoir storage at the beginning of the operational year and the targeted storage at the end of the year. Minimum and maximum releases introduce a feasible release range, where a search algorithm can be used to find the optimal monthly releases that provide the minimum deficit during the year and the least violation from the target storage at the end of the year. Adam et al. (2007) slightly changed this algorithm by considering new thresholds for allowable release and storage and used maximization of hydropower revenue as the objective function for reservoir operation.

There are two main issues with the proposed scheme. First, feasible reservoir releases are determined based on forecasted (or expected – Haddeland, 2014; personal communication) flow at dam location; and uncertainties in flow estimates can largely affect the search for optimal releases. Second, a high-dimensional search (e.g., 12 releases in the case of a monthly release simulation) must be performed for each operational year, which is computationally demanding. These issues were noted by van Beek et al. (2011). They modified the Haddeland et al. (2006a) algorithm to decrease the complexity and uncertainty associated with the algorithm. First, they defined the expected inflow for each month prospectively as a function of the flow in the same month of the previous years; therefore, they omitted using prognostic flow forecasts. In order to reduce the dimensionality of search, they considered reservoir release as a harmonic function; therefore, only release at beginnings of the release and the discharge periods needed to be determined. As the actual inflow values become available, the release can be consequently updated so that the final storage at the end of release period can meet the predefined target storage. With respect to determining the reservoir inflow based on naturalized or regulated flows, van Beek et al. (2011) noted that either setups can be used, depending on how the observed discharge is simulated at the large scale. This is due to large uncertainties in simulating the regulated runoff.

4 Current large-scale modeling applications

Water supply and allocation schemes reviewed in Sects. 2 and 3 have been used in a wide range of offline applications for estimation of human impacts on the terrestrial water cycle. Despite disagreements between different simulation results, the current literature agrees that the effects of water allocation are more pronounced at finer spatial and temporal scales. Earlier, Haddeland et al. (2007) studied the impacts of reservoir operation coupled with irrigation on continental runoff and argued that water allocation has resulted in 2.5 and 6 % increase in annual runoff volume in North America and Asia, respectively. This is almost canceled out by increased evaporation due to irrigation. Nonetheless, as the analysis moves from global and continental to regional and large catchment scales, the effects of water allocation become more profound. For instance, while the mean annual runoff decreased in the western US by around 9 % during a historical control period, the rate of decrement is around 37 % in the Colorado River during the same period (Haddeland et al., 2006b). The results of the most recent global multi-model intercomparison showed that direct impacts of the water resource management in some regions, e.g., parts of Asia and in the western US, are similar or even more than the climate change effects (see Haddeland et al., 2014). Similarly, the effects of water allocation are more significant at finer timescales. For instance, Adam et al. (2007) noted that reservoirs have a minor effect on annual flows in Eurasian watersheds but have significant seasonal effects by changing the flow timing and seasonal amplitudes (see also Döll et al., 2009; van Beek et al., 2011; Biemans et al., 2011).

These simulations, however, are highly uncertain (see, e.g., Haddeland et al., 2011, 2014) due to major limitations in algorithms reviewed above, host large-scale models and data support. The efficiency of available water allocation algorithms can be diagnosed by comparing the streamflow obtained from simulations with observations. Currently, macro-scale water allocation schemes cannot fully describe the dynamics of regulated streamflows and there can be major disagreements between the regulated discharges obtained from different reservoir algorithms (Voisin et al., 2013a). It has been shown that calibration can improve the quality of reservoir operation algorithms (e.g., Wu and Chen, 2012); however, calibration is also associated with uncertainty and can potentially hinder model applications for future projections due to possible temporal and spatial variations in optimal parameters. Hanasaki et al. (2006) as well as Döll et al. (2009) showed that simulation-based algorithms can generally provide improved discharge simulations compared to lake routing algorithms. However, it should be noted that simulations still remain substantially biased in highly regulated catchments (e.g., San Francisco River, US; Syr Darya, central Asia) and in cold regions (e.g., Saskatchewan and Churchill rivers in Canada), particularly during high flows (e.g., Hanasaki et al., 2008a; Biemans et al., 2011; Pokhrel et al., 2012a). The simulation algorithm of Wu and Chen (2012) was found to be more accurate in simulating both storage and release compared to simple multi-linear regression and the target-release scheme embedded in SWAT (Arnold et al., 1998); however, it was tested only at the local scale and it is not clear how the algorithm can perform in other regions with different climate, level of regulation and allocation objectives. Very similar conclusions were obtained for optimization-based algorithms. Discharge simulations are generally improved compared to the no reservoir condition (e.g., Haddeland et al., 2006a); however, there are still significant deficiencies in simulating highly regulated flows, particularly in mountainous and cold regions such as Colorado River in the US as well as Yukon and Mackenzie rivers in Canada (e.g., Haddeland et al., 2006b; Adam et al., 2007). This relates in particular to prognostic reservoir inflows, which remain highly uncertain in these environments; this uncertainty contributes to the uncertainty in assigning optimal reservoir releases, often in dynamic and complex manners (Nazemi and Wheater, 2014b; Müller Schmied et al., 2014).

From a broader perspective, the current performance of reservoir operation and water allocation algorithms must be seen in the context of the hydrological performance of the host large-scale models, including how well the water demand has been represented (see Nazemi and Wheater, 2015). Currently, there are large biases in modeling hydrological processes across various scales, and runoff estimates remain widely divergent (e.g., Wisser et al., 2010; Haddeland et al., 2011; Gudmundsson et al., 2012; Hejazi et al., 2014). In particular, it has been shown that current simulations systematically underestimate streamflow in the arctic and subarctic regions and overestimate the observations in dry catchments; and reservoir operation algorithms mainly improve the timing of the flow, but not the volume (e.g., van Beek et al., 2011). While there are many potential reasons for this, one key source of this limitation is the quality of gridded precipitation products (Biemans et al., 2009, 2011). Rost et al. (2008) used different precipitation products to simulate the regulated river discharge and found substantial variations in simulated discharge due to the choice of precipitation data. Moreover, they showed that sometimes the total precipitation estimate could be less than the total observed discharge after abstraction and regulation. Upcoming satellite missions can address some of the issues regarding historical forcing (see the discussion of Sect. 5.3); however, uncertainty in future precipitation (and other climate variables) should be dealt systematically using multiple climate forcing options based on various combinations of concentration pathways, climate models and downscaling procedures.

Turning from surface water to groundwater issues, almost all available global studies agree on a significant increasing trend in groundwater withdrawal from the late 20th century onward. As an example, Wada et al. (2014) argued that, from 1990 to 2010, the rate of global groundwater withdrawal increased by around 3 % a year. These results are in relatively good agreement with major observed depletions in some regional aquifers (see Gleeson et al., 2012). However, various quantified assessments and further conclusions such as regarding groundwater-induced sea-level rise remain highly uncertain and show major disagreements due to crude representation of groundwater availability, recharge and withdrawal, as discussed in Sects. 2.3 and 3.2 (see, e.g., Wada et al., 2010; Pokhrel et al., 2012b; Döll et al., 2014). This highlights an urgent necessity for improving the representation of human–groundwater interactions at larger scales.

5 Towards an improved representation of water resource management in large-scale models

5.1 Ideal representation and remaining gaps

Throughout our survey, we highlighted the importance of including water supply and allocation in conjunction with water demand (see Nazemi and Wheater, 2015) in models that are relevant to Earth system modeling and/or required for understanding large-scale hydrological responses, with both online and offline implications. From an integrated water resource management and land-surface modeling perspective, water demands can be considered as functions of climate, vegetation and soil moisture as well as socio-economic and policy variables (see Nazemi and Wheater, 2015). As shown in this paper, water supply is driven by water demands but controlled by natural surface-water and groundwater availability, which determine the maximum possible water allocation. Therefore, water demand and water supply should be systematically linked through a feedback loop, represented by water allocation. This integrated water resource system should be then linked to natural land-surface processes at the grid scale. This is rather intuitive: when considered in a typical grid, water allocation perturbs hydrological and land-surface variables within the grid. In parallel, the combined effects of land-surface and hydrological processes govern the variations in surface-water and groundwater availability, which consequently determine water demand (and accordingly water allocation) in the next simulation step. Figure 1 shows a simplified schematic for this integrated modeling framework, in which grid-based calculations of natural and anthropogenic land-surface are further coupled with climate through grid-based land–atmospheric feedbacks.

Major gaps remain in representing water resource management in LSMs in the way defined above. First, as also discussed in Nazemi and Wheater (2015), the key consideration in Earth system modeling is the conservation of mass, energy and water; however, this is widely violated in current models that include elements of water resource management (see Polcher, 2014). For instance, considering groundwater or NNBW as unlimited water sources necessitates bringing water to the system from outside the modeling domain, breaking the assumption that the Earth system is a closed system. This has particular importance when understanding the effects of human–water interactions on the climate and sea-level rise is sought.

Second, water resource management often takes place at the sub-grid resolution of current LSMs used for simulations over large regional and global scales (i.e., 50 km and more).

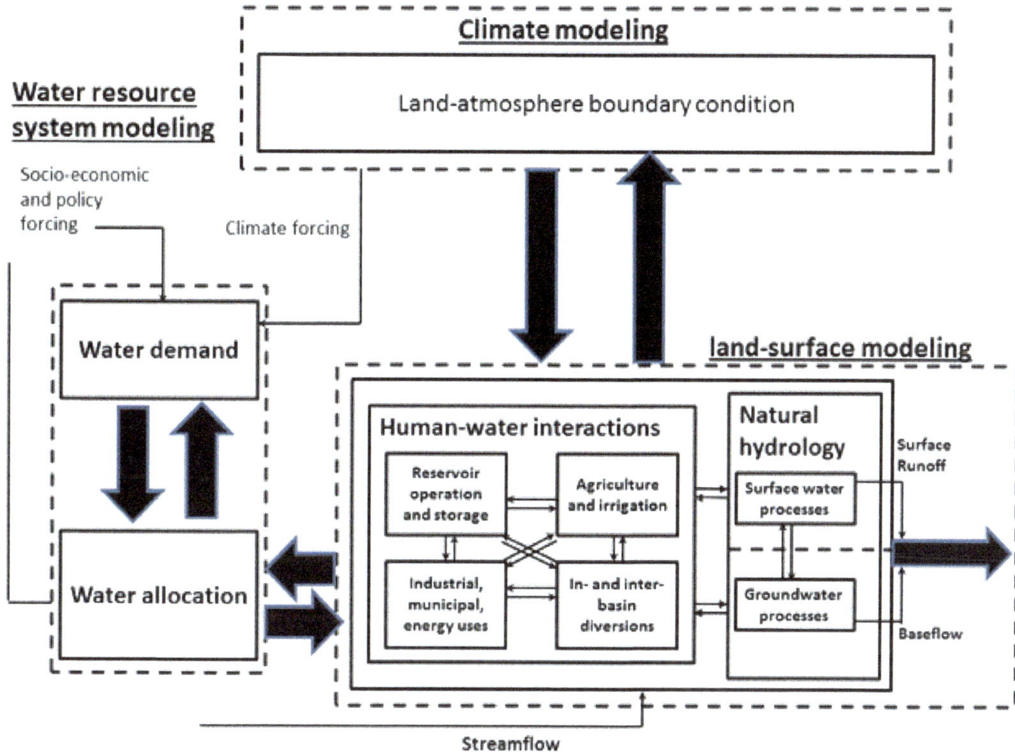

Figure 1. A fully coupled framework for inclusion of water resources management in a typical LSM grid.

Including the elements of water resource management therefore requires moving towards a "hyperresolution" scale (a few kilometers or less) for explicit representation (see Wood et al., 2011) and/or adding new sub-grid parameterizations related to human–water interactions, as illustrated in Fig. 1. However, as the resolutions become finer or more sub-grid parameterizations are added, modeling complexity, computational burdens and data requirements increase significantly, particularly in online simulation in which finer modeling resolution and better discretization of soil and vegetation is generally required to capture land–atmospheric feedbacks and possible climate responses (see Sorooshian et al., 2011a).

Third, we have noted that all currently available efforts in including water supply and allocation in large-scale models are offline and have been made mainly in the context of GHMs. GHMs provide an efficient platform for algorithm development and testing given the relative lack of computational constraints. However, self-evidently understanding online effects of large reservoir storage and large-scale groundwater pumping needs online simulations using coupled LSMs. At this stage of model development, however, many algorithms originally designed for offline applications might not be suitable for online implementations. An important example is reservoir operation, as both optimization- and simulation-based algorithms have some levels of prognosis that hinder their application in coupled simulations.

Fourth, online applications are associated with complexity in representing various feedbacks and time-scaling mismatch among different LSM component and water resource management (see Wang et al., 2004). In addition, current performance of online simulations is limited due to significant biases across different components and propagation of these biases throughout the fully coupled system.

Fifth, we have highlighted major limitations even in offline representation of water resource management at larger scales due to various sources of uncertainty. These uncertainties are due to (1) data support, particularly with respect to precipitation, actual water use and land-surface characteristics; (2) water demand, supply and allocation algorithms, particularly with respect to irrigation demand estimation, reservoir operation and groundwater withdrawals; as well as (3) host large-scale models, particularly with respect to those calculations that determine surface-water and groundwater availability. It should be noted that here we only focus on epistemic sources of uncertainty, which needs to be addressed, quantified, communicated and possibly reduced (see Beven and Alcock, 2012). Table 3 summarizes various aspects of uncertainty related to data support, algorithmic procedures and host models, identified for estimation of water demand (see Nazemi and Wheater, 2015) as well as water supply and allocation (see Sects. 2 to 4) in offline mode. It is often quite difficult to identify the exact source of uncertainty due to complex interconnections between various elements;

Table 3. Uncertainties in current offline representations of water resource management in large-scale models.

Component	Type of activity	Specification	Data uncertainty	Algorithm uncertainty	Host model uncertainty[1]
Water demand (Nazemi and Wheater, 2015)	Irrigative demands	Irrigation	Climate forcing; soil, crop, land-use and land management including sub-grid heterogeneities; actual diversions; socio-economy and technological variables; agricultural management	Characterizing the potential evapotranspiration and crop water demand; representing the sub-grid crop diversity, irrigation expansion, crop change, return flows	Estimation of actual evapotranspiration, soil water movement, runoff and canopy losses; considering CO_2 effects
	Non-irrigative demands	Industrial uses	Location, diversity and capacity of uses; actual diversions; downscaling proxies; socio-economy and technological variables	Seasonal variations in industrial water needs; structural and parametric uncertainty in estimation and projection of industrial demand.	N/A
		Energy-related uses	Location, diversity and capacity of uses; actual diversions; downscaling proxies; socio-economy and technological variables	Seasonal variations in energy-related water needs; structural and parametric uncertainty in estimation and projection of industrial demand.	N/A
		Municipal uses	Population; diversity in uses; actual diversions and uses; downscaling proxies; socio-economy and technological variables	Seasonal variations in municipal water needs, structural and parametric uncertainty in estimation and projection of municipal demand	N/A
		Livestock uses	Heads; socio-economy	Seasonal variations in livestock water need; return flows	N/A
		Environmental flows	Habitat and ecosystem needs in time and space	Over-simplicity of demand calculation	Hydrological processes upstream
Water allocation (see Sects. 2 to 4)	Water supply	River diversion	Location of diversion; capacity, slope and other properties of diversion networks	Diversion losses, return flows	Channel routing
		Lake and reservoir storages[2]	Precipitation; reservoir location and characteristics; actual storage; small dams	Crude representation of reservoir releases using representations of natural lake, losses from reservoir	Hydrological processes upstream of dams, channel routing
		Inter-basin transfer	Location of diversion; capacity, slope and other water transfer properties; management policies; actual water transfer.	Diversion losses, simplicity of heuristic algorithms	Channel routing, calculation of demands
		Reused water	Location, capacity and actual desalinated water supply	Limited representations	N/A
		Groundwater storage	Soil properties, groundwater movement	Crude representation of groundwater availability, ignoring inter-cell lateral groundwater movements	Estimation of groundwater storage, recharge and discharge, calculation of demand.
	Water allocation practice	Operational objectives	Management policies and local constraints	Limitations of common objective functions; temporal and spatial variations in operational objective	Estimation of water demand and supply
		Demand–supply dependency	Management policies and local constraints, topography, diversion channels	Steady-state assumption	Estimation of water demand and supply
		Priorities	Management policies and local constraints	Temporal and spatial variations in priorities	Estimation of water demand and supply
		Reservoir operations	Management policies and local constraints	Simplicity of operational rules in simulation-based approaches, complexity of optimization-based algorithms, prognosis of both approaches	Operational objectives, inflow to reservoirs, water demand
		Groundwater withdrawal	Well locations, groundwater management, actual pumping capacities	Crude representation of groundwater withdrawals based on both top-down and bottom-up algorithms	Groundwater storage, surface water availability, grid-based water demands

[1] Uncertainties from host model also include the uncertainties that can extend from other algorithms, related to water resource management, embedded in host models (see Fig. 1).
[2] See also reservoir operations.

currently, a formal framework to test and validate the water resource management components in the face of various sources of uncertainty is not available (see also Beven and Cloke, 2012). In following sections, we briefly focus on these gaps and highlight the opportunities to address them and move towards the integrated representation proposed in Fig. 1.

5.2 Outstanding challenges – closing the water balance and online simulations

At this stage of research, issues around closing the water balance and online simulations are the most fundamental challenges in representing water resource management in Earth system models. Closing the water balance requires considering all sources of human water supply and withdrawals in the system and integrating them into the host large-scale models. One major gap in representing the water sources is groundwater, which is ignored or crudely represented in most current models. In parallel, as noted above, performing online simulations requires moving towards finer spatial and temporal scales and handling various sources of bias within the integrated system. Although providing an extensive discussion on issues around integrating groundwater models with LSMs as well as online Earth system modeling remains beyond the scope of this paper, here we attempt to briefly point to the main challenges and highlight a few opportunities for future developments.

Technically, the issues around coupling LSMs with groundwater and/or climate models are rather similar. In principle, (1) both require couplers to build an integrated model from independent models; (2) both require refining temporal and spatial resolutions; (3) both substantially increase the complexity of calculations; (4) both need research in terms of improving and adding new algorithms for process representations; and finally (5) both require handling various sources of uncertainty. Research on coupling individual models in an integrated Earth system modeling framework is ongoing, and currently there are various coupling strategies available (e.g., Dunlap et al., 2014). One challenge in coupling the elements of water resource management with climate is the mismatch between temporal scales of water resource management and natural cycles in the Earth system (Wang et al., 2004; Michetti and Zampieri, 2014). For instance, capturing the online effects of evaporation from reservoirs requires running the climate model with fine temporal resolution; although the reservoir evaporation is mainly a function of reservoir temperature and area, which vary slowly. Research, therefore, should be done to compare and optimize existing coupling strategies to handle such inconsistencies in time scaling.

One major need for representing groundwater and for online simulations is the necessity for moving towards finer spatial resolutions. This can result in various challenges. First, even if the spatial resolution increases, several sources of heterogeneity would still be ignored, as current LSMs do not consider them. For instance, LSMs usually define plant species based on plant functional types (PFTs), within which all parameters are identical. However, current LSMs recognize only limited PFTs and hence they typically ignore much of the biodiversity (Sato et al., 2014). Improvement in LSMs in terms of adding more detail into land-surface parameterization can provide opportunities to represent such sources of heterogeneity. Second, going toward finer modeling resolutions requires improved data support at finer scales. Although fine-resolution data are becoming more and more available (e.g., for soil properties – see Sato et al., 2014), such data sets are normally obtained from multiple independent sources, which differ in terms of their quality (see S. Liu et al., 2013). More efforts towards producing standardized and accurate data sources can support future fine-grid Earth system modeling. Finally, moving towards finer scales requires a new set of process representations and parameterizations (Hurrell et al., 2013). There are new developments along scale-aware parameterizations (e.g., Hurrell et al., 2009) that can help refine parameterizations for finer spatial scales.

One important issue with online simulations and groundwater modeling is the computational complexities compared to offline surface water simulations (e.g., Hill et al., 2004; Kollet et al., 2010; Wood et al., 2011). Wehner et al. (2008) suggested opportunities to address computational burdens, including hardware design (i.e., building enhanced computer processors for a specific application) and use of distributed and grid systems. A wide range of applications exists for grid and cloud computing systems (see Schwiegelshohn et al., 2010; Lecca et al., 2011; Fernández-Quiruelas et al., 2011). Improved computational power can also provide a basis to explore various model resolutions to identify critical scales for process representations (see Gentine et al., 2012) and to support computationally expensive offline calculations, such as groundwater processes, dynamic crop growth, river routing and model calibration (e.g., von Bloh et al., 2010; Rouholahnejad et al., 2012; Wu et al., 2013).

Understanding and handling various sources of uncertainty requires activities towards evaluating model performance against observations, which includes new diagnostics for systematic assessments of the modeling system. One key challenge is the fact that LSMs are run over large grids, whereas validation data for land-surface variables and groundwater can be only obtained at local scales. There are several attempts to overcome this issue. For instance, FLUXNET (http://www.daac.ornl.gov/FLUXNET/fluxnet.shtml) coordinates regional and global analyses of observations from micrometeorological tower sites to fill validation gap for online LSMs. As Sato et al. (2014) indicated, such observation networks can facilitate diagnosing the LSM efficiency and sources of errors over large geographical scales. Moreover, a large number of combinations of model configurations should be tested to ensure reliability and performance of individual components and characterize the bias propagation from one component to others (Hurrell et al., 2013). For that purpose, it should be noted that increased modeling complexity does not necessarily result in an improved precision (see Sato et al., 2014; Smith et al., 2014); therefore, a systematic approach is required to test, intercompare and falsify modeling options in the light of validation data available. This will be discussed in more detail in Sect. 5.6.

5.3 Data support

As noted through our survey, major data limitations exist in representing various aspects of water resource management, which are related to forcing, parameterization, calibration and validation of water demand, supply and allocation algorithms (see also Table 3). At this stage of research, major gaps are noted in spatial and temporal quality and coverage of the data related to climate, hydrology, socio-economy, policy and water resource management that are required to drive or to support large-scale models (see Wood et al., 2011; Gleick et al., 2013; Oki et al., 2013).

One important opportunity to improve data support is the use of remote sensing technology, which can provide a synoptic view of the state of land-surface and atmospheric variables (see Sorooshian et al., 2011b; Asrar et al., 2013) and a reliable data support for dynamic forcing, parameter estimation as well as evaluation of large-scale models (see van Dijk and Renzullo, 2011; Trenberth and Asrar, 2012). For instance, Landsat missions (http://landsat.gsfc.nasa.gov; see Williams et al., 2006) have captured long-term variations in global land cover with a temporal resolution of 16 days and spatial resolution of up to 30 m, which can help to parameterize anthropogenic activities such as crop growth and reservoir area. More recently, passive MODerate Resolution Imaging Spectroradiometer (MODIS; http://modis.gsfc.nasa.gov; see Savtchenko et al., 2004) has provided a wide range of land-surface information and has already been applied for various large-scale modeling studies, including validation of online models (Sorooshian et al., 2011a), high-resolution parameterization (Ke et al., 2012) and monitoring storage in large reservoirs (Gao et al., 2012). Assimilation of MODIS land measurements with meteorological data and the Penman–Monteith equation has also provided 8-day, monthly and annual evapotranspiration estimates at 1 km resolution globally (Mu et al., 2007, 2011). This can provide a basis to evaluate simulated evapotranspiration over land surface (see, e.g., Yilmaz et al., 2014). Another important product is the Gravity Recovery and Climate Experiment (GRACE; http://www.csr.utexas.edu/grace/; see Tapley et al., 2004), measuring changes in the total terrestrial water storage at rather coarser resolutions. GRACE data have already been used in studies related to regional groundwater depletion (e.g., Rodell et al., 2007, 2009), model calibration (e.g., Sun et al., 2012) and validation of large-scale simulations (e.g., Pokhrel et al., 2012a, b; Döll et al., 2014).

Upcoming satellite missions can further support representation of water resources management. For instance, precipitation is a key limitation in hydrological modeling in general, but is also important for irrigation demand and scheduling. The upcoming Global Precipitation Measurement mission (GPM; http://gpm.nasa.gov) will collect data at 10 km resolution, every 3 h, globally. The upcoming Soil Moisture Active Passive mission (SMAP; see Entekhabi et al., 2010) will provide improved global soil moisture measurements every 24 h without sensitivity to cloud cover. This can be considered as an important data support for irrigation demand algorithms. Another upcoming remote sensing mission is the Surface Water and Ocean Topography mission (SWOT; see Fu et al., 2009; Biancamaria et al., 2010; Durand et al., 2010), which will provide fine-scale measurements of various surface water stores, including reservoirs as well as natural and man-made channels. Such information at the global scale has the potential to revolutionize representation, calibration and validation of algorithms related to estimation of inflow to reservoirs, reservoir releases and inter-basin water transfers.

There are also important improvements in sharing ground-based data and simulation results, including some inspiring grass-root data collection efforts. For example, the International Groundwater Resources Assessment Centre (IGRAC; http://www.un-igrac.org) assigns an associate expert to each 1-degree grid cell to submit monthly groundwater levels. Such data can be a critical source for testing groundwater withdrawal algorithms. Similar grass-root efforts could be made to record other water resource management data, particularly with respect to actual (rather than licensed) water uses, local management policies and water technologies. We also note that sharing of gridded climate forcing and simulation results is important and provides a basis for consistent model intercomparison efforts. One example is the recently finished EU-WATCH program (http://www.eu-watch.org/), which provides forcing and simulation results of WATCH's Model Intercomparison Project (WaterMIP; http://www.eu-watch.org/watermip).

5.4 Water resource management algorithms

Computational algorithms for representing the elements of water resource management have various sources of uncertainty (see Table 3), and improving the related representations and reducing the modeling uncertainty can be considered as an important avenue for future developments. Some important opportunities include enhancing the simulation-based reservoir operation algorithms for online applications and various applications of calibration, data assimilation and system identification techniques.

- One crucial limitation of current reservoir operation algorithms, as noted above, is in online applications. Simulation-based schemes provide a basis to move forward; however, modifications are required to relax prognostic inputs and to represent the thermal and evaporative functions of reservoirs for online applications. Modeling schemes have already been developed for representing energy balance of natural lakes at sub-grid scale (e.g., MacKay, 2011; MacKay and Seglenieks, 2013) and can be merged with improved simulation-based reservoir operation algorithms to simultaneously characterize reservoir release, storage and evaporation as well as land–atmospheric feedbacks. However, an important question remains in how to address substantial biases in estimation of reservoir release due to the uncertainty in estimation of reservoir inflows, particularly in online simulations. This issue can be partially handled using data assimilation frameworks; but substantial uncertainty remains in future simulation, where assimilation is not possible. Therefore, efforts should be made to represent reservoirs in a robust manner that can handle the inflow biases.

- Calibration using observed, simulated or assimilated system behavior can be used to implicitly represent management and sub-grid heterogeneity. One example would be to address diversity in irrigation demand by finding "representative parameters" that match the assimilated evaporation over a typical irrigated grid. Calibration with ability to identify time-varying parameters could also be used to improve the performance of reservoir operation algorithms and provide a basis to account for variations in water allocation practice in time and potentially in space by considering functioning of multiple reservoirs.

- Another opportunity is to improve functional mappings of system response and demand through system identification techniques. These techniques can range from statistical regression models to more sophisticated machine-learning techniques such as artificial neural networks (e.g., Nazemi et al., 2006a) and genetic symbolic regression (e.g., Hassanzadeh et al., 2014). One example would be building functional relationships for estimation of irrigative or non-irrigative water demands and/or uses. Another would be to represent reservoir operations through transfer functions and enhanced rule-based models as well as building different decision support systems for handling operations taking place at different timescales (i.e., hydropower with a 5 min market, floods with sub-hourly to hourly time step, and monthly seasonal water supply). This can provide an interesting prospect to extract operational rules from observed data and to incorporate soft variables such as social values and expert insights into modeling water resource management (e.g., Nazemi et al., 2002). This can provide various opportunities, for instance for describing the operation of multiple reservoirs at the basin scale, which is widely ignored in the current large-scale reservoir operation schemes.

5.5 Host models

Limitations in host models can introduce a wide range of uncertainties (see Table 3). This is due to the fact that water resource management algorithms are fully embedded within the host models and interact with calculations related to land-surface processes at the grid scale (see Fig. 1). For instance, estimation of antecedent soil moisture affects estimation of irrigation demand. Similarly, estimates of inflows to reservoirs govern the calculations related to reservoir releases and storage. Currently, there are major limitations in representing soil moisture, snow cover, permafrost, evapotranspiration, deep percolation and runoff in large-scale models and they cannot be represented without large uncertainty (Lawrence et al., 2012; Trenberth and Asrar, 2012; Oki et al., 2013). Moreover, host models often contain missing processes. For instance, current host models often ignore the effects of increased CO_2 concentration on irrigation demand. This may result in large uncertainties under climate change effects (see Wada et al., 2013).

While an extensive review of these issues goes beyond the scope of this paper, we note that substantial efforts continue to be made to include missing processes and to improve current parameterizations of natural and anthropogenic processes in large-scale models, particularly in the context of LSMs. For instance, the Community Land Model (CLM; Oleson et al., 2004; 2008; Lawrence et al., 2011) has been recently improved by new algorithms for representing permafrost (Swenson et al., 2012), agriculture (Drewniak et al., 2013) and irrigation (Levis and Sacks, 2011; Levis et al., 2012). Another important development is the vector-based river routing algorithms (e.g., Li et al., 2013; Tesfa et al., 2014) that can improve the representation of natural and anthropogenic channel processes such as reservoir stores, streamflow diversions and inter-basin water transfers (see Lehner and Grill, 2013). Another key opportunity is the application of data assimilation and/or calibration techniques to reduce parametric uncertainty and to improve prediction capability. Some systematic frameworks for calibration and parameterization of land-surface processes are suggested (Rosolem et al., 2012, 2013). We expect improvements in process representations and parameterizations related to LSMs will increase in near future due to the need that has been already recognized (e.g., Wood et al., 2011; Lawrence et al., 2012; Trenberth and Asrar, 2012; Gleick et al., 2013; Oki et al., 2013; Dadson et al., 2013).

5.6 A framework to move forward

Several improvements need to be made in order to appropriately represent the elements of water resource management in Earth system models. We noted that moving towards including the elements of water resource management in a way described in Fig. 1 requires continuous developments in water resource management algorithms, host LSMs, on-

Table 4. Required developments to include the elements of water resource management in Earth system models (see also Table 3).

Water resource management component	Required algorithmic improvements	Targeted spatial scale	Targeted temporal scale	Data support for parameterization and validation
Irrigation demands	Improving the calculation of crop-specific water demand considering the effect of CO_2, considering soil-water movement and other losses	Hyperresolution and sub-grid scale	Sub-daily/sub-hourly (for online simulations)	Crop and soil diversity, measured or assimilated evaporation over irrigated lands
Non-irrigative human demands	Improving the mapping relationship, representing the diversity of non-irrigative demands	Large grids with the ability to be downscaled into finer resolutions using socio-economic and climate proxies	Yearly and monthly with the ability to be downscaled into finer scales using socio-economic and climate proxies	Water use data, gridded climate and regional socio-economic data
Environmental flow needs	Improving the demand approximation considering the diversity in the aquatic life	Catchment scale	Monthly and less	Aquatic biodiversity and water use, climate information, water temperature, water quality
Lakes and reservoirs	Improving the representation of release and storage, linking hydrologic representation with energy-balance components	Grid and sub-grid	Daily	Reservoir storage and water level, release downstream of reservoirs, storage-area–elevation relationships, operational objectives
Water diversions	Representing in-grid and inter-grid water diversions including losses	Grid and inter-grid	Daily	Water distribution specifications, location of abstractions
Groundwater	Improving the representation of groundwater storage and recharge	Grid	Daily (shorter in online simulations)	Soil properties, well locations, pumping capacities
Water reuse and desalination	Improving the representation of water reuse and desalination and the annual dynamics of water supply from each facility	Grid	Yearly with the ability to be downscaled into finer timescales using climate and socio-economic proxies	Location and capacity of facilities, gridded climate, regional socio-economic data

line land–atmospheric coupling and data support. We pointed to the main gaps and provided a brief overview on the opportunities for overcoming these limitations. As far as the algorithms related to representing water resource management are concerned, Table 4 summarizes improvements that need to be made before we can properly represent human–water interaction in Earth system models, along with targeted temporal and spatial resolutions. Modeling resolutions can vary across various elements of water resource managements due to the difference in how different elements affect water and energy balance at the land surface. For instance, irrigation and crop growth directly affect both energy and water balance at the sub-grid scale, with substantial difference between crop function during a day. Therefore, irrigation should be represented at a fine temporal and spatial resolution to capture potential climate responses. Reservoirs also affect water and energy balance; however, as noted above reservoir area and surface temperature vary slowly and therefore there is no need to approach a finer timescale than the scale needed for representing the water balance and downstream releases.

As noted throughout our survey, a variety of modeling options for representing key elements of water resource management at larger scales is currently available and new details about natural and anthropogenic processes are continually being added to Earth system models. Nonetheless, major limitations exist in current data, algorithms and host models, which induce major biases within components and complicate uncertainty quantification and model tractability. At this juncture, a primary task for model development should be to test and compare different data and modeling alternatives in an integrated system. This requires considering model hierarchy and the links between different components and exploring individual and integrated model space with respect to accuracy, identifiability and capability for generalization. This, in turn, can direct where future attempts should be focused to reduce uncertainties further (see also Smith et al., 2014; Michetti and Zampieri, 2014). Guidelines are available for (1) considering multiple working hypotheses for supporting and representing relevant sub-processes and modeling component; (2) constructing different simulations based on various combinations of the considered options and (3) rejecting them if they fail to describe new data, violate their underlying assumptions and/or can be equally described by simpler models (M. P. Clark et al., 2011; see Popper, 1959). Modular systems, such as the recently released WRF-Hydro (Gochis et al., 2013), are particularly suitable for building such a framework as they provide a tool for constructing/falsifying different hypotheses for process representations, parameterizations and data support in a unified computational platform.

To address this and to move towards the integrated representation of water resource management in LSMs, suggested in Fig. 1, we propose a systematic framework for improving the incorporation of water resource management through building, testing and falsifying various modeling options. Figure 2 shows this framework based on the links between different modeling components. In brief, Fig. 2 divides the model development into six components, related to (a) modeling setup and data configuration, (b) climate modeling, (c) land-surface modeling, (d) water resource management representation, (e) calibration and parametric identification, as well as (f) testing and falsification. The framework starts with prior knowledge (a), coming from the modeling purpose, current modeling capabilities and limitations and the knowledge obtained from previous modeling attempts. According to the prior knowledge and emerging advancements, a range of modeling scales can be selected and multiple working hypotheses can be configured to represent the data and modeling options in (b) to (e). Depending on the mode and period of simulation, climate data or more generally climate models (b) are required to force or to be coupled with land-surface processes. The land-surface component (c) includes relevant sub-modules related to natural processes, water supply and allocation and irrigative and non-irrigative withdrawals. The anthropogenic activities are controlled by the water resource management component (d), which requires inputs from land-surface and climate components to determine water availability and to estimate various demands with the aid of these and/or other proxies (priori knowledge). Rules for prioritizing, partitioning and allocating water demands are reflected in a management decision sub-module that further drives water allocation in the land-surface modeling component. Sub-modules within (c) and (d) often contain unknown parameters that need to be identified through prior knowledge or calibration. As a result, calibration and parameter identification algorithms (e) with capability for further uncertainty assessment are a key requirement. Population-based optimization algorithms are particularly suitable for parameter identification as they provide a range of behavioral parameters, which can be analyzed through advanced visualization schemes and provide valuable insights into modeling uncertainty, identifiability and multiple performance measures (e.g., Nazemi et al., 2006b, 2008; Pryke et al., 2007). Moreover, population-based algorithms can provide methodological linkage to uncertainty assessment through various diagnostic tests. Guidelines are provided to test and falsify models through various evaluation criteria such as parametric identifiability (e.g., Beven, 2006b), Pareto optimality (Gupta et al., 1998), predictive uncertainty (Wagener et al., 2004) and limits of acceptability (Beven and Alcock, 2012).

Due to the current stage of model development, there is a need to approach the framework suggested in Fig. 2 with a sequential workflow, as certain improvements should be made first before we can improve others. Figure 3 divides the suggested framework into four sequential working packages. First, various options for data support, water resource management (WRM) algorithms and host models should be benchmarked, tested and intercompared individually to highlight their relative suitability in further offline simulation. This would naturally result in falsifying some of the working hypotheses. The selected options then should be mixed and matched in an offline mode. The offline simulation efficiency should be then explored and intercompared between various integrated settings to assess the biases propagated across the system and examine the robustness of the individual components in an integrated offline simulation. The non-falsified options in this stage can be further improved and configured for online simulation, which can be then coupled with climate models in a way described in Fig. 2.

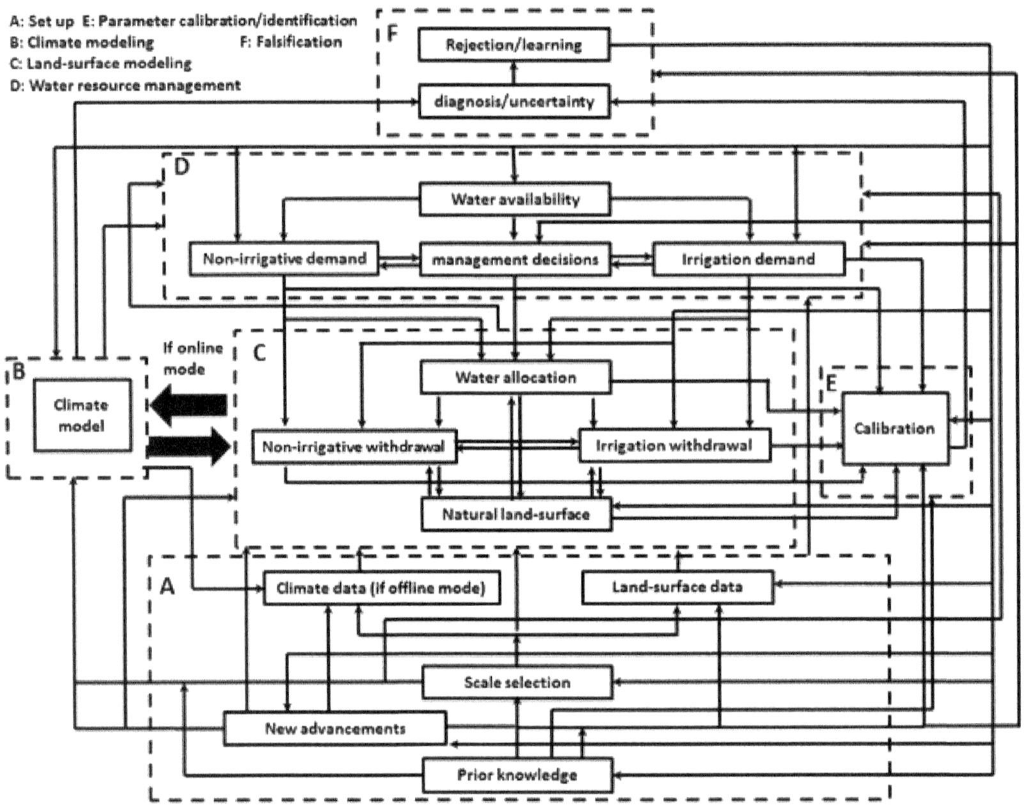

Figure 2. A modular framework for improving the inclusion of water resource management in LSMs through building, testing and falsifying multiple working hypotheses.

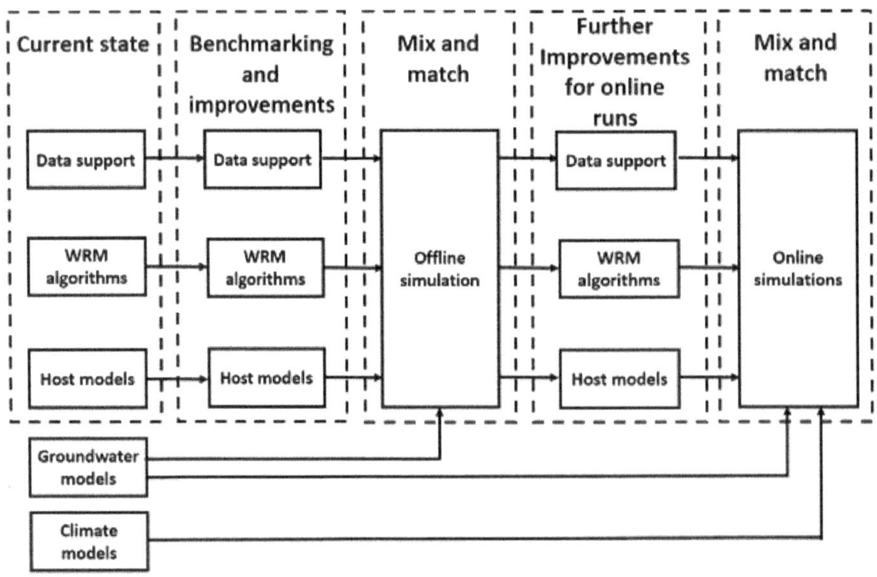

Figure 3. A sequential workflow for benchmarking, improving and including the elements of water resource management into offline and online Earth system simulations.

A key requirement for implementing the suggested framework is the availability of suitable data, at an appropriate scale, for algorithm development and intercomparison. Although global studies are important to improve our knowledge of the Earth system and global water supply, our ability to conduct a comprehensive global study as proposed in Fig. 2 is currently limited due to methodological, computational and funding barriers. We argue that a network of regional case studies, however, could provide access to local data, and a sample of comparative examples to support algorithm intercomparison and further development. We note, for example, the success of model intercomparison projects such as MOPEX (Duan et al., 2006) for hydrological modeling, and suggest that the time is right to develop a similar initiative for the incorporation of anthropogenic effects in hydrological models. One possibility is to draw on the resources of the set of Regional Hydroclimate Projects (RHPs) supported by the Global Energy and Water Exchanges (GEWEX) initiative of the World Climate Research Program (WCRP). As an example, our home river basin in western Canada, the 340 000 km^2 transboundary Saskatchewan River basin (SaskRB), is a GEWEX RHP, embodies a complex large-scale water resources system (Nazemi et al., 2013), and poses globally relevant science and management challenges (see Wheater and Gober, 2013). These require improved representation of water resource management at larger scales to diagnose the changes in the regional discharge, climate and water security as the result of current and future water resource management and climate change. Such RHPs could provide a basis for model development and intercomparison to support inclusion of water resource management in Earth system models for fully coupled global simulations. We have already started to explore various modeling options and the ways of improving individual algorithms (i.e., stage 1 of sequential model development protocol illustrated in Fig. 3) throughout the SaskRB. For instance, we have benchmarked several reservoir operation algorithms using observed inflows and assessed the possibility of improving simulation using calibration. We have realized that the efficiency of reservoir operation algorithms can be considerably improved if the assumption of fixed model parameterization is relaxed and the algorithm parameters are identified through calibration against observed reservoir level and discharge. We are about to finalize this study and will present our findings through a technical paper in near future.

6 Summary and concluding remarks

Human water supply and allocation have intensively perturbed the water cycle. We noted that the inclusion of these anthropogenic activities in Earth system models poses a new set of modeling challenges and progress has remained incomplete. Despite some major developments, we noted that current limitations significantly degrade the modeling capability at larger scales, particularly with respect to future conditions, and neglect potentially significant sources of change to land–atmospheric system. We highlighted important deficiencies related to representing groundwater stores and withdrawals as well as online implications of large reservoirs. We also noted that current water allocation algorithms have considerable limitations in representing streamflow in regulated catchments. We argued that these limitations are attributed to uncertainties in data support, water allocation algorithms and host large-scale models.

We identified four opportunities for improvements. These are advancements in (1) high-performance computing and coupling techniques; (2) remote sensing, data collection and data sharing; (3) calibration algorithms, system identification techniques and assimilation products; and (4) ongoing improvements in host models including both process representation and parameter identification. As there are several options available for data support, water resource management algorithms and host models, we proposed a modular framework for testing various modeling and data options, which can be configured by multiple working hypotheses and implemented in a unified and fully integrated modeling framework. The selected working hypotheses can be tested and falsified on the basis of available information, intercomparison and/or various model diagnosis frameworks. Similar to other recent commentaries (e.g., M. P. Clark et al., 2011; see also Beven et al., 2012), we believe that such a systematic framework in essential for improving current modeling capability in both offline and online modes and can be pursued using regional case studies, before aiming for fully coupled global simulations. WCRP RHPs are one source of suitable examples to move this agenda forward.

It should be noted that filling current gaps in the inclusion of water resource management in Earth system models requires substantial efforts across a wide range of disciplines, from social and policy sciences to economics and water management, from natural sciences to engineering and mathematical modeling, and from remote sensing to hardware technology and computer science. Interdisciplinary research efforts, therefore, are important. Moreover, for various reasons including funding limitations, the community needs to fully recognize the role of collaboration and explore various opportunities to share data and resources for efficient model developments and for consistent intercomparisons.

Finally, it should be indicated that our survey considered water resource management from a water quantity perspective. Water quality concerns are increasingly associated with growing human water demand and can also impact water supply and allocation. Coupling water quality and quantity in Earth system models is however very much in its infancy, and much future effort will be required to fill this gap. We hope that our survey will trigger more attention towards the necessity for improving current Earth system modeling capability to respond to the needs and challenges of the "Anthropocene".

Acknowledgements. The first author has attended NASA's Applied Remote Sensing Training free webinar series (http://water.gsfc.nasa.gov/) and would like to thank Amita Mehta, Evan Johnson and John Bolten for providing useful materials related to remote sensing technology. The authors gratefully acknowledge the constructive comments from the editor and two anonymous reviewers as well as Jan Polcher, Ingjerd Haddeland, Naota Hanasaki, Jason Evans and Bruce Davison, which have enabled us to make significant improvements to this paper. Financial support for this survey was provided by the Canada Excellence Research Chair in Water Security at the University of Saskatchewan..

Edited by: W. Buytaert

References

Adam, J. C. and Lettenmaier, D. P.: Application of new precipitation and reconstructed streamflow products to streamflow trend attribution in northern Eurasia, J. Climate, 21, 807–1828, 2008.

Adam, J. C., Haddeland, I., Su, F., and Lettenmaier, D. P.: Simulation of reservoir influences on annual and seasonal streamflow changes for the Lena, Yenisei and Ob'rivers, J. Geophys. Res.-Atmos., 112, D24114, doi:10.1029/2007JD008525, 2007.

Alcamo, J., Döll, P., Henrichs, T., Kaspar, F., Lehner, B., Rösch, T., and Siebert, S.: Development and testing of the WaterGAP 2 global model of water use and availability, Hydrolog. Sci. J., 48, 317–337, 2003.

Alcamo, J., Flörke, M., and Märker, M.: Future long-term changes in global water resources driven by socio-economic and climatic changes, Hydrolog. Sci. J., 52, 247–275, 2007.

Arnell, N. W.: Climate change and global water resources: SRES emissions and socio-economic scenarios, Global Environ. Change, 14, 31–52, 2004.

Arnold, J. G., Srinivasan, R., Muttiah, R. S., and Williams, J. R.: Large area hydrologic modeling and assessment part i: model development, J. Ame. Water Resour. Assoc., 34, 73–89, doi:10.1111/j.1752-1688.1998.tb05961.x, 1998.

Asrar, G. R., Hurrell, J. W., and Busalacchi, A. J.: A need for "actionable" climate science and information: summary of WCRP Open Science Conference, B. Am. Meteorol. Soc., 94, ES8–ES12, 2013.

Bellman, R.: On the theory of dynamic programming, P. Natl. Acad. Sci., 38, 716–719, 1952.

Bergström, S. and Singh, V. P.: The HBV model, in: Computer models of watershed hydrology, edited by: Singh, V. P., Water Resources Publications, Colorado, USA, 443-476, 1995.

Best, M. J., Pryor, M., Clark, D. B., Rooney, G. G., Essery, R .L. H., Ménard, C. B., Edwards, J. M., Hendry, M. A., Porson, A., Gedney, N., Mercado, L. M., Sitch, S., Blyth, E., Boucher, O., Cox, P. M., Grimmond, C. S. B., and Harding, R. J.: The Joint UK Land Environment Simulator (JULES), model description – Part 1: Energy and water fluxes, Geosci. Model Dev., 4, 677–699, doi:10.5194/gmd-4-677-2011, 2011.

Beven, K.: Searching for the Holy Grail of scientific hydrology: $Q_t = (S, R, \Delta t)A$ as closure, Hydrol. Earth Syst. Sci., 10, 609–618, doi:10.5194/hess-10-609-2006, 2006a.

Beven, K.: A manifesto for the equifinality thesis, J. Hydrol., 320, 18–36, 2006b.

Beven, K., Smith, P., Westerberg, I., and Freer, J.: Comment on "Pursuing the method of multiple working hypotheses for hydrological modeling" by P. Clark et al., Water Resour. Res., 48, W11801, doi:10.1029/2012WR012282, 2012.

Beven, K. J. and Alcock, R. E.: Modelling everything everywhere: a new approach to decision-making for water management under uncertainty, Freshwater Biol., 57, 124–132, 2012.

Beven, K. J. and Cloke, H. L.: Comment on "Hyperresolution global land surface modeling: Meeting a grand challenge for monitoring Earth's terrestrial water" by Eric F. Wood et al., Water Resour. Res., 48, W01801, doi:10.1029/2011WR010982, 2012.

Biancamaria, S., Andreadis, K. M., Durand, M., Clark, E. A., Rodriguez, E., Mognard, N. M., Alsdorf, D. E., Lettenmaier, D. P., and Oudin, Y.: Preliminary characterization of SWOT hydrology error budget and global capabilities, IEEE J. Sel. Top. Appl., 3, 6–19, 2010.

Biemans, H., Hutjes, R. W. A., Kabat, P., Strengers, B. J., Gerten, D., and Rost, S.: Effects of Precipitation Uncertainty on Discharge Calculations for Main River Basins, J. Hydrometeorol., 10, 1011–1025, 2009.

Biemans, H., Haddeland, I., Kabat, P., Ludwig, F., Hutjes, R. W. A., Heinke, J., von Bloh, W., and Gerten, D.: Impact of reservoirs on river discharge and irrigation water supply during the 20th century, Water Resour. Res., 47, W03509, doi:10.1029/2009WR008929, 2011.

Blanc, E., Strzepek, K., Schlosser, A., Jacoby, H. D., Gueneau, A., Fant, C., Rausch, S., and Reilly, J.: Analysis of U.S. water resources under climate change, MIT Joint Program on the Science and Policy of Global Change, Report No. 239, http://globalchange.mit.edu/files/document/MITJPSPGC_Rpt239.pdf (last access: 6 May 2014), 2013.

Chen, J. and Wu, Y.: Exploring hydrological process features of the East River (Dongjiang) basin in south China using VIC and SWAT, in: Proceedings of the International Association of Hydrological Sciences and the International Water Resources Association Conference, Guangzhou, China, IAHS Press, Wallingford, UK, 116–123, 2008.

Chow, V. T., Maidment, D. R., and Mays, L. W.: Applied Hydrology, McGraw-Hill Series in Water Resources and Environmental Engineering, McGraw-Hill, New York, 572 pp., 1998.

Clark, D. B., Mercado, L. M., Sitch, S., Jones, C. D., Gedney, N., Best, M. J., Pryor, M., Rooney, G. G., Essery, R. L. H., Blyth, E., Boucher, O., Harding, R. J., Huntingford, C., and Cox, P. M.: The Joint UK Land Environment Simulator (JULES), model description – Part 2: Carbon fluxes and vegetation dynamics, Geosci. Model Dev., 4, 701–722, doi:10.5194/gmd-4-701-2011, 2011.

Clark, M. P., Kavetski, D., and Fenicia, F.: Pursuing the method of multiple working hypotheses for hydrological modeling, Water Resour. Res., 47, W09301, doi:10.1029/2010WR009827, 2011.

Dadson, S., Acreman, M., and Harding, R.: Water security, global change and land–atmosphere feedbacks, Philos. T. Roy. Soc. A, 371, 2002, doi:10.1098/rsta.2012.0412, 2013.

Dankers, R., Arnell, N. W., Clark, D. B., Falloon, P. D., Fekete, B. M., Gosling, S. N., Heinke, J., Kim, H., Masaki, Y., Satoh, Y., Stacke, T., Wada, Y., and Wisser, D.: First look at changes in flood hazard in the Inter-Sectoral Impact Model Intercomparison Project ensemble, P. Natl. Acad. Sci. USA, 111, 3257–3261, doi:10.1073/pnas.1302078110, 2014.

Dantzig, G. B.: Linear Programming and Extensions, Princeton University Press, New Jersey, USA, 1965.

Dirmeyer, P. A., Dolman, A. J., and Sato, N.: The pilot phase of the global soil wetness project, B. Am. Meteorol. Soc., 80, 851–878, 1999.

Döll, P., Kaspar, F., and Lehner, B.: A global hydrological model for deriving water availability indicators: model tuning and validation, J. Hydrol., 270, 105–134, 2003.

Döll, P., Fiedler, K., and Zhang, J.: Global-scale analysis of river flow alterations due to water withdrawals and reservoirs, Hydrol. Earth Syst. Sci., 13, 2413–2432, doi:10.5194/hess-13-2413-2009, 2009.

Döll, P., Hoffmann-Dobrev, H., Portmann, F. T., Siebert, S., Eicker, A., Rodell, M., Strassberg, G., and Scanlon, B. R.: Impact of water withdrawals from groundwater and surface water on continental water storage variations, J. Geodyn., 59, 143–156, 2012.

Döll, P., Müller Schmied, H., Schuh, C., Portmann, F. T., and Eicker, A.: Global-scale assessment of groundwater depletion and related groundwater abstractions: Combining hydrological modeling with information from well observations and GRACE satellites, Water Resour. Res., 50, 5698–5720, doi:10.1002/2014WR015595, 2014.

Drewniak, B., Song, J., Prell, J., Kotamarthi, V. R., and Jacob, R.: Modeling agriculture in the Community Land Model, Geosci. Model Dev., 6, 495–515, doi:10.5194/gmd-6-495-2013, 2013.

Duan, Q., Schaake, J., Andreassian, V., Franks, S., Goteti, G., Gupta, H. V., Gusev, Y. M., Habets, F., Hall, A., Hay, L., Hogue, T., Huang, M., Leavesley, G., Liang, X., Nasonova, O. N., Noilhan, J., Oudin, L., Sorooshian, S., Wagener, T., and Wood, E. F.: Model Parameter Estimation Experiment (MOPEX): an overview of science strategy and major results from the second and third workshops, J. Hydrol., 320, 3–17, 2006.

Dunlap, R., Vertenstein, M., Valcke, S., and Craig, T.: Second Workshop on Coupling Technologies for Earth System Models, B. Am. Meteorol. Soc., 95, ES34–ES38, 2014.

Durand, M., Rodriguez, E., Alsdorf, D. E., and Trigg, M.: Estimating river depth from remote sensing swath interferometry measurements of river height, slope, and width, IEEE J. Select. Top. Appl. Earth Obs. Rem. S., 3, 20–31, 2010.

Entekhabi, D., Njoku, E. G., O'Neill, P. E., Kellogg, K. H., Crow, W. T., Edelstein, W. N., Entin, J. K., Goodman, S. D., Jackson, T. J., Johnson, J., Kimball, J., Piepmeier, J. R., Koster, R. D., Martin, N., McDonald, K. C., Moghaddam, M., Moran, S., Reichle, R., Shi, J.-C., Spencer, M. W., Thurman, S. W., Leung, T., and Van Zyl, J.: The Soil Moisture Active Passive (SMAP) mission, Proc. IEEE, 98, 704–716, 2010.

Falkenmark, M.: Growing water scarcity in agriculture: future challenge to global water security, Philos. T. Roy. Soc. A, 371, 2002, doi:10.1098/rsta.2012.0410, 2013.

Fan, Y. and Miguez-Macho, G.: A simple hydrologic framework for simulating wetlands in climate and earth system models, Clim. Dyn., 37, 253–278, 2011.

Fekete, B. M., Vörösmarty, C. J., and Grabs, W.: Global, composite runoff fields based on observed river discharge and simulated water balances, http://www.bafg.de/GRDC/EN/02_srvcs/24rprtsrs/report22.pdf?__blob=publicationFile (last access: 6 May 2014), 1999.

Fekete, B. M., Vöröosmarty, C. J., and Grabs, W.: High-resolution fields of global runoff combining observed river discharge and simulated water balances, Global Biogeochem. Cy., 16, 15-1–15-10, doi:10.1029/1999GB001254, 2002.

Ferguson, I. M. and Maxwell, R. M.: The role of groundwater in watershed response and land surface feedbacks under climate change, Water Resour. Res., 46, W00F02, doi:10.1029/1999GB001254, 2010.

Fernández-Quiruelas, V., Fernández, J., Cofiño, A. S., Fita, L., and Gutiérrez, J. M.: Benefits and requirements of grid computing for climate applications: An example with the community atmospheric model, Environ. Model. Softw., 26, 1057–1069, 2011.

Foster, S. and Loucks, D. P.: Non-renewable groundwater resources: A guidebook on Socially-sustainable Management for Water-policy Makers, UNESCO, http://unesdoc.unesco.org/images/0014/001469/146997e.pdf (last access: 6 May 2014), 2006.

Fu, L. L., Alsdorf, D., Rodriguez, E., Morrow, R., Mognard, N., Lambin, J., Vaze, P., and Lafon, T.: The SWOT (Surface Water and Ocean Topography) mission: spaceborne radar interferometry for oceanographic and hydrological applications, in: Proceedings of OCEANOBS'09 Conference, available at: http://bprc.osu.edu/water/publications/oceanobs09_swot.pdf (last access: 6 May 2014), 2009.

Gao, H., Birkett, C., and Lettenmaier, D. P.: Global monitoring of large reservoir storage from satellite remote sensing, Water Resour. Res., 48, W09504, doi:10.1029/2012WR012063, 2012.

Gentine, P., Troy, T. J., Lintner, B. R., and Findell, K. L.: Scaling in surface hydrology: progress and challenges, J. Contemp. Water Res. Educ., 147, 28–40, 2012.

Gerten, D., Schaphoff, S., Haberlandt, U., Lucht, W., and Sitch, S.: Terrestrial vegetation and water balance – hydrological evaluation of a dynamic global vegetation model, J. Hydrol., 286, 249–270, 2004.

Gleeson, T., VanderSteen, J., Sophocleous, M. A., Taniguchi, M., Alley, W. M., Allen, D. M., and Zhou, Y.: Groundwater sustainability strategies, Nat. Geosci., 3, 378–379, 2010.

Gleeson, T., Wada, Y., Bierkens, M. F., and van Beek, L. P.: Water balance of global aquifers revealed by groundwater footprint, Nature, 488, 197–200, 2012.

Gleick, P. H.: The world's water 2000–2001: the biennial report on freshwater resources, Island Press, Washington, D.C., 2000.

Gleick, P. H., Cooley, H., Famiglietti, J. S., Lettenmaier, D. P., Oki, T., Vörösmarty, C. J., and Wood, E. F.: Improving Understanding of the Global Hydrologic Cycle, in: Climate Science for Serving Society, edited by: Asrar, G. R. and Hurrell, J. W., Springer Netherlands, 151–184, 2013.

Gochis, D. J., Yu, W., and Yates, D. N.: The WRF-Hydro model technical description and user's guide, version 1.0, NCAR Technical Document, http://www.ral.ucar.edu/projects/wrfhydro/ (last access: 6 May 2014), 2013.

Goldberg, D. E.: Genetic algorithms in search, optimization, and machine learning, Addison-wesley, Reading, Menlo Park, 1989.

Grey, D., Garrick, D., Blackmore, D., Kelman, J., Muller, M., and Sadoff, C.: Water security in one blue planet: twenty-first century policy challenges for science, Philos. T. Roy. Soc. A, 371, 2002, doi:10.1098/rsta.2012.0406, 2013.

Gudmundsson, L., Tallaksen, L. M., Stahl, K., Clark, D. B., Dumont, E., Hagemann, S., Bertrand, N., Gerten, D., Heinke, J.,

Hanasaki, N., Voss, F., and Koirala, S.: Comparing large-scale hydrological model simulations to observed runoff percentiles in Europe, J. Hydrometeorol., 13, 604–620, 2012.

Gupta, H. V., Sorooshian, S., and Yapo, P. O.: Toward improved calibration of hydrologic models: Multiple and noncommensurable measures of information, Water Resour. Res., 34, 751–763, doi:10.1029/97WR03495, 1998.

Haddeland, I., Skaugen, T., and Lettenmaier, D. P.: Anthropogenic impacts on continental surface water fluxes, Geophys. Res. Lett., 33, L08406, doi:10.1029/2006GL026047, 2006a.

Haddeland, I., Lettenmaier, D. P., and Skaugen, T.: Effects of irrigation on the water and energy balances of the Colorado and Mekong river basins, J. Hydrol., 324, 210–223, 2006b.

Haddeland, I., Skaugen, T., and Lettenmaier, D. P.: Hydrologic effects of land and water management in North America and Asia: 1700–1992, Hydrol. Earth Syst. Sci., 11, 1035–1045, doi:10.5194/hess-11-1035-2007, 2007.

Haddeland, I., Clark, D. B., Franssen, W., Ludwig, F., Voß, F., Arnell, N. W., Bertrand, N., Best, M., Folwell, S., Gerten, D., Gomes, S., Gosling, S. N., Hagemann, S., Hanasaki, N., Harding, R., Heinke, J., Kabat, P., Koirala, S., Oki, T., Polcher, J., Stacke, T., Viterbo, P., Weedon, G. P., and Yeh, P.: Multimodel estimate of the global terrestrial water balance: setup and first results, J. Hydrometeorol., 12, 869–884, doi:10.1175/2011JHM1324.1, 2011.

Haddeland, I., Biemans, H., Eisner, S., Flörke, M., Hanasaki, N., Konzmann, M., Ludwig, F., Masaki, Y., Schewe, J., Stacke, T., Tessler, Z. D., Wada, Y., and Wisser, D.: Global water resources affected by human interventions and climate change, P. Natl. Acad. Sci., 111, 3251–3256, doi:10.1073/pnas.1222475110, 2014.

Hanasaki, N., Kanae, S., and Oki, T.: A reservoir operation scheme for global river routing models, J. Hydrol., 327, 22–41, 2006.

Hanasaki, N., Kanae, S., Oki, T., Masuda, K., Motoya, K., Shirakawa, N., Shen, Y., and Tanaka, K.: An integrated model for the assessment of global water resources – Part 1: Model description and input meteorological forcing, Hydrol. Earth Syst. Sci., 12, 1007–1025, doi:10.5194/hess-12-1007-2008, 2008a.

Hanasaki, N., Kanae, S., Oki, T., Masuda, K., Motoya, K., Shirakawa, N., Shen, Y., and Tanaka, K.: An integrated model for the assessment of global water resources – Part 2: Applications and assessments, Hydrol. Earth Syst. Sci., 12, 1027–1037, doi:10.5194/hess-12-1027-2008, 2008b.

Hanasaki, N., Inuzuka, T., Kanae, S., and Oki, T.: An estimation of global virtual water flow and sources of water withdrawal for major crops and livestock products using a global hydrological model, J. Hydrol., 384, 232–244, 2010.

Hanasaki, N., Fujimori, S., Yamamoto, T., Yoshikawa, S., Masaki, Y., Hijioka, Y., Kainuma, M., Kanamori, Y., Masui, T., Takahashi, K., and Kanae, S.: A global water scarcity assessment under Shared Socio-economic Pathways – Part 1: Water use, Hydrol. Earth Syst. Sci., 17, 2375–2391, doi:10.5194/hess-17-2375-2013, 2013a.

Hanasaki, N., Fujimori, S., Yamamoto, T., Yoshikawa, S., Masaki, Y., Hijioka, Y., Kainuma, M., Kanamori, Y., Masui, T., Takahashi, K., and Kanae, S.: A global water scarcity assessment under Shared Socio-economic Pathways – Part 2: Water availability and scarcity, Hydrol. Earth Syst. Sci., 17, 2393–2413, doi:10.5194/hess-17-2393-2013, 2013b.

Hassanzadeh, E., Nazemi, A., and Elshorbagy, A.: Quantile-Based Downscaling of Precipitation Using Genetic Programming: Application to IDF Curves in Saskatoon, J. Hydrol. Eng., 19, 943–955, 2014.

Hejazi, M. I., Edmonds, J., Clarke, L., Kyle, P., Davies, E., Chaturvedi, V., Wise, M., Patel, P., Eom, J., and Calvin, K.: Integrated assessment of global water scarcity over the 21st century under multiple climate change mitigation policies, Hydrol. Earth Syst. Sci., 18, 2859–2883, doi:10.5194/hess-18-2859-2014, 2014.

Hill, C., DeLuca, C., Suarez, M., and Da Silva, A.: The architecture of the Earth System Modeling framework, Comput. Sci. Eng., 6, 18–28, 2004.

Hossain, F., Degu, A. M., Yigzaw, W., Burian, S., Niyogi, D., Shepherd, J., and Pielke, R.: Climate feedback-based provisions for dam design, operations, and water management in the 21st century, J. Hydrol. Eng., 17, 837–850, 2012.

Hossain, M. S. and El-shafie, A.: Intelligent Systems in Optimizing Reservoir Operation Policy: A Review, Water Resour. Manage., 27, 3387–3407, 2013.

Huggins, L. F. and Burney, J. R.: Surface runoff, storage and routing, in: Hydrologic Modeling of Small Watersheds, edited by: Haan, C. T., Johnson, H. P., and Brakensiek, D. L., American Society of Agricultural Engineers, St. Joseph, Michigan, USA, 169–225, 1982.

Hurrell, J., Meehl, G. A., Bader, D., Delworth, T. L., Kirtman, B., and Wielicki, B.: A unified modeling approach to climate system prediction. B. Am. Meteorol. Soc., 90, 1819–1832, 2009.

Hurrell, J. W., Holland, M. M., Gent, P. R., Ghan, S., Kay J. E., Kushner, P. J., Lamarque, J.-F., Large, W. G., Lawrence, D., Lindsay, K., Lipscomb, W. H., Long, M. C., Mahowald, N., Marsh, D. R., Neale, R. B., Rasch, P., Vavrus, S., Vertenstein, M., Bader, D., Collins, W. D., Hack, J. J., Kiehl, J., and Marshall, S.: The community earth system model: a framework for collaborative research, B. Am. Meteorol. Soc., 94, 1339–1360, doi:10.1175/BAMS-D-12-00121.1, 2013.

Ke, Y., Leung, L. R., Huang, M., Coleman, A. M., Li, H., and Wigmosta, M. S.: Development of high resolution land surface parameters for the Community Land Model, Geosci. Model Dev., 5, 1341–1362, doi:10.5194/gmd-5-1341-2012, 2012.

Kollet, S. J. and Maxwell, R. M.: Capturing the influence of groundwater dynamics on land surface processes using an integrated, distributed watershed model, Water Resour. Res., 44, W02402, doi:10.1029/2007WR006004, 2008.

Kollet, S. J., Maxwell, R. M., Woodward, C. S., Smith, S., Vanderborght, J., Vereecken, H., and Simmer, C.: Proof of concept of regional scale hydrologic simulations at hydrologic resolution utilizing massively parallel computer resources, Water Resour. Res., 46, W04201, doi:10.1029/2009WR008730, 2010.

Lai, X., Jiang, J., Yang, G., and Lu, X. X.: Should the Three Gorges Dam be blamed for the extremely low water levels in the middle–lower Yangtze River?, Hydrol. Process., 28, 150–160, doi:10.1002/hyp.10077, 2014.

Lawrence, D. M., Oleson, K. W., Flanner, M. G., Thornton, P. E., Swenson, S. C., Lawrence, P. J., Zeng, X., Yang, Z.-L., Levis, S., Sakaguchi, K., Bonan, G. B., and Slater, A. G.: Parameterization improvements and functional and structural advances in Version 4 of the Community Land Model, J. Adv. Model. Earth Syst., 3, M03001, doi:10.1029/2011MS00045, 2011.

Lawrence, D. M., Maxwell, R., Swenson, S., Lopez, S., and Famiglietti, J.: Challenges of Representing and Predicting Multi-Scale Human–Water Cycle Interactions in Terrestrial Systems, available at: http://climatemodeling.science.energy.gov/sites/default/files/Topic_330final.pdf (last access: 6 May 2014), 2012.

Lecca, G., Petitdidier, M., Hluchy, L., Ivanovic, M., Kussul, N., Ray, N., and Thieron V.: Grid computing technology for hydrological applications, J. Hydrol., 403, 186–199, 2011.

Lehner, B. and Döll, P.: Development and validation of a global database of lakes, reservoirs and wetlands, J. Hydrol., 296, 1–22, 2004.

Lehner, B. and Grill, G.: Global river hydrography and network routing: baseline data and new approaches to study the world's large river systems, Hydrol. Process., 27, 2171–2186, doi:10.1002/hyp.9740, 2013.

Lehner, B., Verdin, K., and Jarvis, A.: New Global Hydrography Derived From Spaceborne Elevation Data, Eos Trans. AGU, 89, 93–94, doi:10.1029/2008EO100001, 2008.

Lehner, B., Liermann, C. R., Revenga, C., Vörösmarty, C., Fekete, B., Crouzet, P., Döll, P., Endejan, M., Frenken, K., Magome, J., Nilsson, C., Robertson, J. C., Rödel, R., Sindorf, N., and Wisser, D.: High-resolution mapping of the world's reservoirs and dams for sustainable river-flow management, Front. Ecol. Environ., 9, 494–502, 2011.

Lettenmaier, D. P. and Milly, P. C. D.: Land waters and sea level, Nat. Geosci., 2, 452–454, 2009.

Levis, S. and Sacks, W.: Technical descriptions of the interactive crop management (CLM4CNcrop) and interactive irrigation models in version 4 of the Community Land Model, http://www.cesm.ucar.edu/models/cesm1.1/clm/CLMcropANDirrigTechDescriptions.pdf (last access: 6 May 2014), 2011.

Levis, S., Bonan, G. B., Kluzek, E., Thornton, P. E., Jones, A., Sacks, W. J., and Kucharik, C. J.: Interactive Crop Management in the Community Earth System Model (CESM1): Seasonal Influences on Land-Atmosphere Fluxes, J. Climate, 25, 4839–4859, 2012.

Li, H., Huang, M., Wigmosta, M., Ke, Y., Coleman, A., Leung, L. R., Wang, A., and Ricciuto, D. M.: Evaluating runoff simulations from the Community Land Model 4.0 using observations from flux towers and a mountainous watershed, J. Geophys. Res., 116, D24120, doi:10.1029/2011JD016276, 2011.

Li, H., Wigmosta, M. S., Wu, H., Huang, M., Ke, Y., Coleman, A. M., and Leung, L. R.: A physically based runoff routing model for land surface and earth system models, J. Hydrometeorol., 14, 808–828, 2013.

Liang, X., Lettenmaier, D. P., Wood, E. F., and Burges, S. J.: A simple hydrologically based model of land surface water and energy fluxes for general circulation models, J. Geophys. Res.-Atmos., 99, 14415–14428, 1994.

Liebe, J., Van De Giesen, N., and Andreini, M.: Estimation of small reservoir storage capacities in a semi-arid environment: A case study in the Upper East Region of Ghana, Phys. Chem. Earth A/B/C, 30, 448–454, 2005.

Liu, C. and Zheng, H.: South-to-north water transfer schemes for China, Int. J. Water Resour. Develop., 18, 453–471, 2002.

Liu, J. and Yang, W.: Water sustainability for China and beyond, Science, 337, 649–650, 2012.

Liu, J., Zang, C., Tian, S., Liu, J., Yang, H., Jia, S., You, L., Liu, B., and Zhang, M.: Water conservancy projects in China: achievements, challenges and way forward, Global Environ. Change, 23, 633–643, 2013.

Liu, S., Wei, Y., Post, W. M., Cook, R. B., Schaefer, K., and Thornton, M. M.: The Unified North American Soil Map and its implication on the soil organic carbon stock in North America, Biogeosciences, 10, 2915–2930, doi:10.5194/bg-10-2915-2013, 2013.

Lohmann, D., Nolte-Holube, R., and Raschke, E.: A large-scale horizontal routing model to be coupled to land surface parametrization schemes, Tellus A, 48, 708–721, doi:10.1034/j.1600-0870.1996.t01-3-00009.x, 1996.

Lohmann, D., Raschke, E., Nijssen, B., and Lettenmaier, D. P.: Regional scale hydrology: I. Formulation of the VIC-2L model coupled to a routing model, Hydrolog. Sci. J., 43, 131–141, 1998.

MacKay, M. D.: A process oriented small lake dynamical scheme for coupled climate modeling applications, in: Vol. 1, AGU Fall Meeting Abstracts, San Francisco, p. 1359, 2011.

MacKay, M. D. and Seglenieks, F.: On the simulation of Laurentian Great Lakes water levels under projections of global climate change, Climatic Change, 117, 55–67, 2013.

MacKay, M. D., Neale, P. J., Arp, C. D., De Senerpont Domis, L. N., Fang, X., Gal, G., Jöhnk, K. D., Kirillin, G., Lenters, J. D., Litchman, E., MacIntyre, S., Marsh, P., Melack, J., Mooij, W. M., Peeters, F., Quesada, A., Schladow, S. G., Schmid, M., Spence, C., and Stokes, S. L.: Modeling lakes and reservoirs in the climate system, Limnol. Oceanogr., 54, 2315–2329, 2009.

Maxwell, R. M. and Miller, N. L.: Development of a coupled land surface and groundwater model, J. Hydrometeorol, 6, 233–247, 2005.

Maxwell, R. M., Chow, F. K., and Kollet, S. J.: The groundwater-land-surface-atmosphere connection: soil moisture effects on the atmospheric boundary layer in fully-coupled simulations, Adv. Water Resour., 30, 2447–2466, 2007.

Maxwell, R. M., Lundquist, J. K., Mirocha, J. D., Smith, S. G., Woodward, C. S., and Tompson, A. F. B.: Development of a coupled groundwater–atmosphere model, Mon. Weather Rev., 139, 96–116, 2011.

Meigh, J. R., McKenzie, A. A., and Sene, K. J.: A grid-based approach to water scarcity estimates for eastern and southern Africa, Water Resour. Manage., 13, 85–115, 1999.

Meybeck, M.: Global analysis of river systems: from Earth system controls to Anthropocene syndromes, Philos. T. Roy. Soc. Lond. B, 358, 1935–1955, 2003.

Michetti, M. and Zampieri, M.: Climate–Human–Land Interactions: A Review of Major Modelling Approaches. Land, 3, 793–833, doi:10.3390/land3030793, 2014.

Mu, Q., Zhao, M., and Running, S. W.: Development of a global evapotranspiration algorithm based on MODIS and global meteorology data, Remote Sens. Environ., 111, 519–536, 2007.

Mu, Q., Zhao, M., and Running, S. W.: Improvements to a MODIS global terrestrial evapotranspiration algorithm, Remote Sens. Environ., 115, 1781–1800, 2011.

Müller Schmied, H., Eisner, S., Franz, D., Wattenbach, M., Portmann, F. T., Flörke, M., and Döll, P.: Sensitivity of simulated global-scale freshwater fluxes and storages to input data, hydrological model structure, human water use and calibration, Hydrol. Earth Syst. Sci., 18, 3511–3538, doi:10.5194/hess-18-3511-2014, 2014.

Nakayama, T. and Shankman, D.: Impact of the Three-Gorges Dam and water transfer project on Changjiang floods, Global Planet. Change, 100, 38–50, 2013a.

Nakayama, T. and Shankman, D.: Evaluation of uneven water resource and relation between anthropogenic water withdrawal and ecosystem degradation in Changjiang and Yellow River basins, Hydrol. Process., 27, 3350–3362, doi:10.1002/hyp.9835, 2013b.

Nazemi, A. and Wheater, H. S.: Assessing the Vulnerability of Water Supply to Changing Streamflow Conditions, Eos Trans. AGU, 95, 288, doi:10.1002/2014EO320007, 2014a.

Nazemi, A. and Wheater, H. S.: How can the uncertainty in the natural inflow regime propagate into the assessment of water resource systems?, Adv. Water Resour., 63, 131–142, doi:10.1016/j.advwatres.2013.11.009, 2014b.

Nazemi, A. and Wheater, H. S.: On inclusion of water resource management in Earth System models – Part 1: Problem definition and representation of water demand, Hydrol. Earth Syst. Sci., 19, 33–61, doi:10.5194/hess-19-33-2015, 2015.

Nazemi, A., Akbarzadeh, M. R., and Hosseini, S. M.: Fuzzy-stochastic linear programming in water resources engineering, in: Proceeding of Fuzzy Information Processing Society, NAFIPS 2002, IEEE, New Jersey, USA, 227–232, doi:10.1109/NAFIPS.2002.1018060, 2002.

Nazemi, A., Hosseini, S. M., and Akbarzadeh-T, M. R.: Soft computing-based nonlinear fusion algorithms for describing non-Darcy flow in porous media, J. Hydraul. Res., 44, 269–282, 2006a.

Nazemi, A., Yao, X., and Chan, A. H.: Extracting a set of robust Pareto-optimal parameters for hydrologic models using NSGA-II and SCEM, in: Proceedings of IEEE Congress on Evolutionary Computation (CEC 2006), Vancouver, Canada, 1901–1908, doi:10.1109/CEC.2006.1688539, 2006b.

Nazemi, A., Chan, A. H., and Yao, X.: Selecting representative parameters of rainfall-runoff models using multi-objective calibration results and a fuzzy clustering algorithm, in: BHS 10th National Hydrology Symposium, Exeter, UK, 13–20, 2008.

Nazemi, A., Wheater, H. S., Chun, K. P., and Elshorbagy, A.: A stochastic reconstruction framework for analysis of water resource system vulnerability to climate-induced changes in river flow regime, Water Resour. Res., 49, 291–305, doi:10.1029/2012WR012755, 2013.

Nilsson, C., Reidy, C. A., Dynesius, M., and Revenga, C.: Fragmentation and flow regulation of the world's large river systems, Science, 308, 405–408, 2005.

Oki, T. and Kanae, S.: Global hydrological cycles and world water resources, Science, 313, 1068–1072, 2006.

Oki, T. and Sud, Y. C.: Design of Total Runoff Integrating Pathways (TRIP) – A global river channel network, Earth Interact., 2, 1–37, 1998.

Oki, T., Agata, Y., Kanae, S., Saruhashi, T., Yang, D., and Musiake, K.: Global assessment of current water resources using total runoff integrating pathways, Hydrolog. Sci. J., 46, 983–995, 2001.

Oki, T., Blyth, E. M., Berbery, E. H., and Alcaraz-Segura, D.: Land Use and Land Cover Changes and Their Impacts on Hydroclimate, Ecosystems and Society, in: Climate Science for Serving Society, edited by: Asrar, G. R. and Hurrell, J. W., Springer Netherlands, 185–203, 2013.

Oleson, K. W., Dai, Y., Bonan, G. B., Bosilovichm, M., Dickinson, R., Dirmeyer, P., Hoffman, F., Houser, P., Levis, S., Niu, G.-Y., Thornton, P., Vertenstein, M., Yang, Z., and Zeng, X.: Technical description of the community land model (CLM), NCAR Tech. Note NCAR/TN-461+STR, 173 pp., doi:10.5065/D6N877R0, http://www.cesm.ucar.edu/models/cesm1.0/clm/CLM4_Tech_Note.pdf (last access: 28 December 2014), 2004.

Oleson, K. W., Niu, G. Y., Yang, Z. L., Lawrence, D. M., Thornton, P. E., Lawrence, P. J., Stöckli, R., Dickinson, R. E., Bonan, G. B., Levis, S., Dai, A., and Qian, T.: Improvements to the Community Land Model and their impact on the hydrological cycle, J. Geophys. Res.-Biogeo., 113, G01021, doi:10.1029/2007JG000563, 2008.

Pietroniro, A., Fortin, V., Kouwen, N., Neal, C., Turcotte, R., Davison, B., Verseghy, D., Soulis, E. D., Caldwell, R., Evora, N., and Pellerin, P.: Development of the MESH modelling system for hydrological ensemble forecasting of the Laurentian Great Lakes at the regional scale, Hydrol. Earth Syst. Sci., 11, 1279–1294, doi:10.5194/hess-11-1279-2007, 2007.

Pokhrel, Y. N., Hanasaki, N., Koirala, S., Cho, J., Yeh, P. J.-F., Kim, H., Kanae, S., and Oki, T.: Incorporating anthropogenic water regulation modules into a land surface model, J. Hydrometeorol., 13, 255–269, 2012a.

Pokhrel, Y. N., Hanasaki, N., Yeh, P. J., Yamada, T. J., Kanae, S., and Oki, T.: Model estimates of sea-level change due to anthropogenic impacts on terrestrial water storage, Nat. Geosci., 5, 389–392, doi:10.1038/ngeo1476, 2012b.

Polcher, J.: Interactive comment on "On inclusion of water resource management in Earth System models – Part 1: Problem definition and representation of water demand" by A. Nazemi and H. S. Wheater, Hydrol. Earth Syst. Sci. Discuss., 11, C3403–C3410, 2014.

Polcher, J., Bertrand, N., Biemans, H., Clark, D. B., Floerke, M., Gedney, N., Gerten, D., Stacke, T., van Vliet, M., and Voss, F.: Improvements in hydrological processes in general hydrological models and land surface models within WATCH, WATCH Technical Report Number 34, available at: http://www.eu-watch.org/publications/technical-reports (last access: 6 May 2014), 2011.

Ponce, V. M. and Changanti, P. V.: Variable-parameter Muskingum-Cunge method revisited, J. Hydrol., 162, 433–439, 1994.

Popper, K.: The logic of scientific discovery, 1995 Edn., Routledge, London, 1959.

Postel, S. L., Daily, G. C., and Ehrlich, P. R.: Human appropriation of renewable fresh water, Science, 271, 785–788, 1996.

Prudhomme, C., Giuntoli, I., Robinson, E. L., Clark, D. B., Arnell, N. W., Dankers, R., Fekete, B. M., Franssen, W., Gerten, D., Gosling, S. N., Hagemann, S., Hannah, D. M., Kim, H., Masaki, Y., Satoh, Y., Stacke, T., Wada, Y., and Wisser, D.: Hydrological droughts in the 21st century, hotspots and uncertainties from a global multimodel ensemble experiment, P. Natl. Acad. Sci. USA, 111, 3262–3267, doi:10.1073/pnas.1222473110, 2014.

Pryke, A., Mostaghim, S., and Nazemi, A.: Heatmap visualization of population based multi objective algorithms, In Evolutionary multi-criterion optimization, Springer, Berlin, Heidelberg, 361–375, 2007.

Rani, D. and Moreira, M. M.: Simulation–optimization modeling: a survey and potential application in reservoir systems operation, Water Resour. Manage., 24, 1107–1138, 2010.

Revelle, C., Joeres, E., and Kirby, W.: The Linear Decision Rule in Reservoir Management and Design: 1. Development of the Stochastic Model, Water Resour. Res., 5, 767–777, doi:10.1029/WR005i004p00767, 1969.

Rodell, M., Chen, J., Kato, H., Famiglietti, J. S., Nigro, J., and Wilson, C. R.: Estimating groundwater storage changes in the Mississippi River basin (USA) using GRACE, Hydrogeol. J., 15, 159–166, 2007.

Rodell, M., Velicogna, I., and Famiglietti, J. S.: Satellite-based estimates of groundwater depletion in India, Nature, 460, 999–1002, 2009.

Rosenzweig, C., Elliott, J., Deryng, D., Ruane, A. C., Müller, C., Arneth, A., Boote, K. J., Folberth, C., Glotter, M., Khabarov, N., Neumann, K., Piontek, F., Pugh, T. A. M., Schmid, E., Stehfest, E., Yang, H., and Jones, J. W.: Assessing agricultural risks of climate change in the 21st century in a global gridded crop model intercomparison, P. Natl. Acad. Sci. USA, 111, 3268–3273, doi:10.1073/pnas.1222463110, 2014.

Rosolem, R., Gupta, H. V., Shuttleworth, W. J., Zeng, X., and de Gonçalves, L. G. G.: A fully multiple-criteria implementation of the Sobol' method for parameter sensitivity analysis, J. Geophys. Res., 117, D07103, doi:10.1029/2011JD016355, 2012.

Rosolem, R., Gupta, H. V., Shuttleworth, W. J., de Gonçalves, L. G. G., and Zeng, X.: Towards a comprehensive approach to parameter estimation in land surface parameterization schemes, Hydrol. Process., 27, 2075–2097, doi:10.1002/hyp.9362, 2013.

Rost, S., Gerten, D., Bondeau, A., Luncht, W., Rohwer, J., and Schaphoff, S.: Agricultural green and blue water consumption and its influence on the global water system, Water Resour. Res., 44, W09405, doi:10.1029/2007WR006331, 2008.

Rouholahnejad, E., Abbaspour, K. C., Vejdani, M., Srinivasan, R., Schulin, R., and Lehmann, A.: A parallelization framework for calibration of hydrological models, Environ. Model. Softw., 31, 28–36, 2012.

Sato, H., Ito, A., Ito, A., Ise, T., and Kato, E.: Current status and future of land surface models, Soil Sci. and Plant Nutr., doi:10.1080/00380768.2014.917593, in press, 2014.

Savtchenko, A., Ouzounov, D., Ahmad, S., Acker, J., Leptoukh, G., Koziana, J., and Nickless, D.: Terra and Aqua MODIS products available from NASA GES DAAC, Adv. Space Res., 34, 710–714, 2004.

Schewe, J., Heinke, J., Gerten, D., Haddeland, I., Arnell, N. W., Clark, D. B., Dankers, R., Eisner, S., Fekete, B. M., Colón-González, F. J., Gosling, S. N., Kim, H., Liu, X., Masaki, Y., Portmann, F. T., Satoh, Y., Stacke, T., Tang, Q., Wada, Y., Wisser, D., Albrecht, T., Frieler, K., Piontek, F., Warszawski, L., and Kabat, P.: Multimodel assessment of water scarcity under climate change, P. Natl. Acad. Sci. USA, 111, 3245–3250, doi:10.1073/pnas.1222460110, 2014.

Schiermeier, Q.: Water risk as world warms, Nature, 505, 7481, doi:10.1038/505010a, 2014.

Schwiegelshohn, U., Badia, R. M., Bubak, M., Danelutto, M., Dustdar, S., Gagliardi, F., Geiger, A., Hluchy, L., Kranzlmüller, D., Erwin Laure, E., Priol, T., Reinefeld, A., Resch, M., Reuter, A., Rienhoff, O., Rüter, T., Sloot, S., Talia, D., Ullmann, K., Yahyapour, R., and von Voigt, G.: Perspectives on grid computing, Future Gener. Comp. Sy., 26, 1104–1115, 2010.

Siebert, S., Burke, J., Faures, J. M., Frenken, K., Hoogeveen, J., Döll, P., and Portmann, F. T.: Groundwater use for irrigation – a global inventory, Hydrol. Earth Syst. Sci., 14, 1863–1880, doi:10.5194/hess-14-1863-2010, 2010.

Skliris, N. and Lascaratos, A.: Impacts of the Nile River damming on the thermohaline circulation and water mass characteristics of the Mediterranean Sea, J. Mar. Syst., 52, 121–143, doi:10.1016/j.jmarsys.2004.02.005, 2004.

Smith, M. J., Palmer, P. I., Purves, D. W., Vanderwel, M. C., Lyutsarev, V., Calderhead, B., Joppa, L. N., Bishop, C. M., and Emmott, S.: Changing How Earth System Modeling is Done to Provide More Useful Information for Decision Making, Science, and Society, B. Am. Meteorol. Soc., 95, 1453–1464, doi:10.1175/BAMS-D-13-00080.1, 2014.

Sorooshian, S., Li, J., Hsu, K.-L., and Gao, X.: How significant is the impact of irrigation on the local hydroclimate in California's Central Valley? Comparison of model results with ground and remote-sensing data, J. Geophys. Res., 116, D06102, doi:10.1029/2010JD014775, 2011a.

Sorooshian, S., AghaKouchak, A., Arkin, P., Eylander, J., Foufoula-Georgiou, E., Harmon, R., Hendrickx, J. M. H., Imam, B., Kuligowski, R., Skahill, B., and Skofronick-Jackson, G.: Advanced concepts on remote sensing of precipitation at multiple scales, B. Am. Meteorol. Soc., 92, 1353–1357, 2011b.

Strzepek, K., Schlosser, A., Farmer, W., Awadalla, S., Baker, J., Rosegrant, M., and Gao, X.: Modeling the global water resource system in an integrated assessment modeling framework: IGSM-WRS, MIT Joint Program on the Science and Policy of Global Change, Report No. 189, available at: http://dspace.mit.edu/handle/1721.1/61767 (last access: 6 May 2014), 2010.

Strzepek, K., Schlosser, A., Gueneau, A., Gao, X., Blanc, É., Fant, C., Rasheed, B., and Jacoby, H. D.: Modeling water resource system under climate change: IGSM-WRS, MIT Joint Program on the Science and Policy of Global Change, Report No. 236, http://dspace.mit.edu/handle/1721.1/75774 (last access: 6 May 2014), 2012.

Sun, A. Y., Green, R., Swenson, S., and Rodell, M.: Toward calibration of regional groundwater models using GRACE data, J. Hydrol., 422, 1–9, 2012.

Swenson, S. C., Lawrence, D. M., and Lee, H.: Improved simulation of the terrestrial hydrological cycle in permafrost regions by the Community Land Model, J. Adv. Model. Earth Syst., 4, M08002, doi:10.1029/2012MS000165, 2012.

Syvitski, J. P. M., Vorosmarty, C. J., Kettner, A. J., and Green, P.: Impact of humans on the flux of terrestrial sediment to the global coastal ocean, Science, 308, 376–380, 2005.

Takata, K., Emori, S., and Watanabe, T.: Development of the minimal advanced treatments of surface interaction and runoff, Global Planet. Change, 38, 209–222, 2003.

Takeuchi, K.: Least marginal environmental impact rule for reservoir development, Hydrolog. Sci. J., 42, 583–597, 1997.

Tang, Q., Gao, H., Yeh, P., Oki, T., Su, F., and Lettenmaier, D. P.: Dynamics of Terrestrial Water Storage Change from Satellite and Surface Observations and Modeling, J. Hydrometeorol., 11, 156–170, 2010.

Tapley, B. D., Bettadpur, S., Ries, J. C., Thompson, P. F., and Watkins, M. M.: GRACE measurements of mass variability in the Earth system, Science, 305, 503–505, 2004.

Taylor, R. G., Scanlon, B., Döll, P., Rodell, M., van Beek, R., Wada, Y., Longuevergne, L., Leblanc, M., Famiglietti, J. S., Edmunds, M., Konikow, L., Green, T. R., Chen, J., Taniguchi, M., Bierkens, M. F. P., MacDonald, A., Fan, Y., Maxwell, R. M., Yechieli, Y., Gurdak, J. J., Allen, D. M., Shamsudduha, M., Hiscock, K., Yeh, P. J.-F., Holman, I., and Treidel, H.: Ground water and climate change, Nat. Clim. Change, 3, 322–329, 2013.

Tebakari, T., Yoshitani, J., and Suvanpimol, P.: Impact of large-scale reservoir operation on flow regime in the Chao Phraya River basin, Thailand, Hydrol. Process., 26, 2411–2420, 2012.

Tesfa, T. K., Li, H.-Y., Leung, L. R., Huang, M., Ke, Y., Sun, Y., and Liu, Y.: A subbasin-based framework to represent land surface processes in an Earth system model, Geosci. Model Dev., 7, 947–963, doi:10.5194/gmd-7-947-2014, 2014.

Trenberth, K. E. and Asrar, G. R.: Challenges and opportunities in water cycle research: WCRP contributions, Surv. Geophys., 35, 515–532, 2012.

USGS: Water Use in the United States, http://water.usgs.gov/watuse/data/2005/index.html (last access: 6 May 2014), 2011.

van Beek, L. P. H. and Bierkens, M. F. P.: The Global Hydrological Model PCR-GLOBWB: Conceptualization, Parameterization and Verification, Report Department of Physical Geography, Utrecht University, Utrecht, the Netherlands, http://vanbeek.geo.uu.nl/suppinfo/vanbeekbierkens2009.pdf (last access: 6 May 2014), 2009.

van Beek, L. P. H., Wada, Y., and Bierkens, M. F. P.: Global monthly water stress: 1. Water balance and water availability, Water Resour. Res., 47, W07517, doi:10.1029/2010WR009791, 2011.

van Dijk, A. I. J. M. and Renzullo, L. J.: Water resource monitoring systems and the role of satellite observations, Hydrol. Earth Syst. Sci., 15, 39–55, doi:10.5194/hess-15-39-2011, 2011.

Voisin, N., Li, H., Ward, D., Huang, M., Wigmosta, M., and Leung, L. R.: On an improved sub-regional water resources management representation for integration into earth system models, Hydrol. Earth Syst. Sci., 17, 3605–3622, doi:10.5194/hess-17-3605-2013, 2013a.

Voisin, N., Liu, L., Hejazi, M., Tesfa, T., Li, H., Huang, M., Liu, Y., and Leung, L. R.: One-way coupling of an integrated assessment model and a water resources model: evaluation and implications of future changes over the US Midwest, Hydrol. Earth Syst. Sci., 17, 4555–4575, doi:10.5194/hess-17-4555-2013, 2013b.

Von Bloh, W., Rost, S., Gerten, D., and Lucht, W.: Efficient parallelization of a dynamic global vegetation model with river routing, Environ. Model. Softw., 25, 685–690, 2010.

Vörösmarty, C. J., Sharma, K. P., Fekete, B. M., Copeland, A. H., Holden, J., Marble, J., and Lough, J. A.: The storage and aging of continental runoff in large reservoir systems of the world, Ambio, 26, 210–219, 1997.

Vörösmarty, C. J., Federer, C. A., and Schloss A. L.: Potential evaporation functions compared on US watersheds: Possible implications for global-scale water balance and terrestrial ecosystem modeling, J. Hydrol., 207, 147–169, 1998.

Vörösmarty, C. J., Meybeck, M., Fekete, B., Sharma, K., Green, P., and Syvitski, J. P.: Anthropogenic sediment retention: major global impact from registered river impoundments, Global Planet. Change, 39, 169–190, 2003.

Vörösmarty, C. J., McIntyre, P. B., Gessner, M. O., Dudgeon, D., Prusevich, A., Green, P., Glidden, S., Bunn, S. E., Sullivan, C. A., Reidy Liermann, C., and Davies, P. M.: Global threats to human water security and river biodiversity, Nature, 467, 555–561, 2010.

Wada, Y., van Beek, L. P. H., van Kempen, C. M., Reckman, J. W. T. M., Vasak, S., and Bierkens, M. F. P.: Global depletion of groundwater resources, Geophys. Res. Lett., 37, L20402, doi:10.1029/2010GL044571, 2010.

Wada, Y., van Beek, L. P. H., Viviroli, D., Dürr, H. H., Weingartner, R., and Bierkens, M. F. P.: Global monthly water stress: 2. Water demand and severity of water stress, Water Resour. Res., 47, W07518, doi:10.1029/2010WR009792, 2011.

Wada, Y., van Beek, L. P. H., and Bierkens, M. F. P.: Nonsustainable groundwater sustaining irrigation: A global assessment, Water Resour. Res., 48, W00L06, doi:10.1029/2011WR010562, 2012.

Wada, Y., Wisser, D., Eisner, S., Flörke, M., Gerten, D., Haddeland, I., Hanasaki, N., Masaki, Y., Portmann, F. T., Stacke, T., Tessler, Z., and Schewe, J.: Multimodel projections and uncertainties of irrigation water demand under climate change, Geophys. Res. Lett., 40, 4626–4632, 2013.

Wada, Y., Wisser, D., and Bierkens, M. F. P.: Global modeling of withdrawal, allocation and consumptive use of surface water and groundwater resources, Earth Syst. Dynam., 5, 15–40, doi:10.5194/esd-5-15-2014, 2014.

Wade Miller, G.: Integrated concepts in water reuse: managing global water needs, Desalination, 187, 65–75, 2006.

Wagener, T., Wheater, H. S., and Gupta, H. V.: Rainfall-Runoff Modelling in Gauged and Ungauged Catchments, Imperial College Press, London, UK, 2004.

Wang, Y., Leung, L. R., McGregor, J. L., Lee, D. K., Wang, W. C., Ding, Y., and Kimura, F.: Regional climate modeling: progress, challenges, and prospects, J. Meteorol. Soc. Jpn., 82, 1599–1628, 2004.

Wehner, M., Oliker, L., and Shalf, J.: Towards ultra-high resolution models of climate and weather, Int. J. High Perform. Comput. Appl., 22, 149–165, 2008.

Wheater, H. and Gober, P.: Water security in the Canadian Prairies: science and management challenges, Philos. T. Roy. Soc. A, 371, 20120409, doi:10.1098/rsta.2012.0409, 2013.

Williams, D. L., Goward, S., and Arvidson, T.: Landsat: Yesterday, today, and tomorrow, Photogramm. Eng. Remote Sens., 72, 1171–1178, 2006.

Wisser, D., Fekete, B. M., Vörösmarty, C. J., and Schumann, A. H.: Reconstructing 20th century global hydrography: a contribution to the Global Terrestrial Network-Hydrology (GTN-H), Hydrol. Earth Syst. Sci., 14, 1–24, doi:10.5194/hess-14-1-2010, 2010.

Wood, E. F., Roundy, J. K., Troy, T. J., van Beek, L. P. H., Bierkens, M. F. P., Blyth, E., de Roo, A., Döll, P., Ek, M., Famiglietti, J., Gochis, D., van de Giesen, N., Houser, P., Jaffé, P. R., Kollet, S., Lehner, B., Lettenmaier, D. P., Peters-Lidard, C., Sivapalan, M., Sheffield, J., Wade, A., and Whitehead, P.: Hyperresolution global land surface modeling: meeting a grand challenge for monitoring Earth's terrestrial water, Water Resour. Res., 47, W05301, doi:10.1029/2010WR010090, 2011.

Wu, Y. and Chen, J.: An Operation-Based Scheme for a Multiyear and Multipurpose Reservoir to Enhance Macroscale Hydrologic Models, J. Hydrometeorol., 13, 270–283, 2012.

Wu, Y., Chen, J., and Sivakumar, B.: Numerical Modeling of Operation and Hydrologic Effects of Xinfengjiang Reservoir in Southern China, in: Proc. MODSIM 2007 International Congress on Modelling and Simulation, http://mssanz.org.au/MODSIM07/papers/24_s17/NumericalModelings17_Wu.pdf, last access: 6 May 2014, 1561–1567, 2007.
Wu, Y., Li, T., Sun, L., and Chen, J.: Parallelization of a hydrological model using the message passing interface, Environ. Model. Softw., 43, 124–132, 2013.
Ye, A., Duan, Q., Chu, W., Xu, J., and Mao, Y.: The impact of the South–North Water Transfer Project (CTP)'s central route on groundwater table in the Hai River basin, North China, Hydrol. Process., 28, 5755–5768, doi::10.1002/hyp.10081, 2013.
Yilmaz, M. T., Anderson, M. C., Zaitchik, B., Hain, C. R., Crow, W. T., Ozdogan, M., Chun, J. A., and Evans, J.: Comparison of prognostic and diagnostic surface flux modeling approaches over the Nile River basin, Water Resour. Res., 50, 386–408, doi:10.1002/2013WR014194, 2014.
Yoshikawa, S., Cho, J., Yamada, H. G., Hanasaki, N., and Kanae, S.: An assessment of global net irrigation water requirements from various water supply sources to sustain irrigation: rivers and reservoirs (1960–2050), Hydrol. Earth Syst. Sci., 18, 4289–4310, doi:10.5194/hess-18-4289-2014, 2014.
Zektser, I. S. and Lorne, E.: Groundwater resources of the world: and their use, http://unesdoc.unesco.org/images/0013/001344/134433e.pdf (last access: 6 May 2014), 2004.
Zhao, F. and Shepherd, M.: Precipitation Changes near Three Gorges Dam, China, Part I: A Spatiotemporal Validation Analysis, J. Hydrometeorol., 13, 735–745, 2012.

3

Estimating the water needed to end the drought or reduce the drought severity in the Carpathian region

T. Antofie, G. Naumann, J. Spinoni, and J. Vogt

European Commission, Joint Research Centre, Institute for Environment and Sustainability (IES), Climate Risk Management Unit, Ispra, Italy

Correspondence to: T. Antofie (tiberiuantofie@yahoo.com)

Abstract. A drought severity climatology for the Carpathian region has been produced using the self-calibrating Palmer Drought Severity Index (Sc-PDSI) for the period 1961–2010. Using the Sc-PDSI and the assumptions of the Palmer drought model (PDM) the precipitation required for drought termination (when Sc-PDSI reaches −0.5) and amelioration (when Sc-PDSI reaches −2.0) are computed for periods of 1, 3, and 6 months. We discuss the reduction of the uncertainty in the determination of the beginning and ending of drought conditions, and provide a quantitative measure of the probability that any drought could be ameliorated or terminated. We present how the spatial variability of the amount of water needed for drought recovery and the climatological probability of receiving that amount of water is determined by the local conditions against the general climate characteristics of a small area such as the Carpathian region. Regionally, the Pannonian Basin, the Transylvanian Plateau and the external Carpathian foothills and plains in the southern and eastern part of the region require the highest quantity of precipitation to recover from a drought while having the lowest climatological probabilities for such amounts of rainfall. High precipitation amounts over the northern and northwest part of the region result in higher soil moisture supplies and higher climatological probabilities to end a given drought event. Moreover, the succession and/or predominance of particular types of general atmospheric circulation patterns produce a seasonal cycle and inter-annual variability of precipitation that is quantitatively reflected in the excess of precipitation that is above normal required for drought recovery. Overall, the results of this study provide an overview on the chances of recovery from a drought period with moderate or severe drought and present information useful in decision making in water and drought management.

1 Introduction

Drought is one of the most far-reaching natural and socioeconomic disasters (WMO/UNCCD/FAO/UNW-DPC, 2013). Traditionally, the acknowledgement and attempts to manage droughts were mostly orientated towards crisis management, while little attention has been given to proactive drought risk management. More recently, European as well as international policies and initiatives have highlighted the need for a more proactive, risk-based management of droughts. Examples are the requirement for the set-up of River Basin Management Plans, including Drought Management Plans under the European Water Framework Directive (WFD), the High-Level Meeting on National Drought Policies (HMNDP, http://www.hmndp.org), or the Integrated Drought Management Programme (IDMP, http://www.droughtmanagement.info) established by the World Meteorological Organization (WMO) and the Global Water Partnership (GWP) in 2013 .

An essential element in risk management is the reduction of drought impacts (i.e., mitigation) based on an assessment of the cost of damages associated with droughts as compared to the costs for efficient early warning and preparedness, including the adaptation to climate change. Drought as a natural hazard has been the subject of a great number of studies, focusing on the definition of drought and the development of drought indicators (e.g., Palmer, 1965; McKee et al., 1993; Wells et al., 2004; Vicente-Serrano et al., 2010) as well as on drought assessment and monitoring (e.g., Briffa et al., 1994; Guttman, 1998; Lloyd-Hughes and Saunders 2002; Dai et al., 2004; van der Schrier et al. 2006; Dai, 2011). However, little attention was given to the analysis of probabilities that a given drought (and its impacts) could be ameliorated or terminated through adequate rainfalls. The number of studies addressing the drought recovery topic is few (Karl et al., 1986, 1987) and articles focused on drought as a natural hazard (Wilhite, 2000; Wilhite and Buchanan, 2005) as well as reports on drought management and monitoring (e.g., WMO, 2006; IPCC, 2007; ISDR, 2007), address the subject only in a general manner.

This paper provides a quantitative measure of the probability that any drought could be ameliorated or terminated over some defined period of time using the Palmer drought model (PDM) (Karl et al., 1987) with its assumptions and limitations. The study was partially implemented in the framework of the CARPATCLIM project (http://www.carpatclim-eu.org). Within this project, a consortium of meteorological services and environmental institutes of nine countries of the region joined forces with the purpose of improving the availability and accessibility of quality-controlled meteorological and climatological data. Based on the CARPATCLIM daily and monthly gridded data (0.1° × 0.1° resolution for the 1961–2010 period), a series of indicators (self-calibrating Palmer Drought Severity Index – Sc-PDSI, Standardized Precipitation-Evapotranspiration Index – SPEI, Standardized Precipitation Index – SPI, Reconnaissance Drought Indicator – RDI, and Palfai Drought Index – PADI) were computed with the purpose of defining the climate characteristics of the region. Among them the Sc-PDSI was selected due to its use in measuring the intensity and severity of drought events in Europe (van der Schrier et al., 2006, 2007). Also, it was selected due to its ability to quantify the impact of droughts on a wide range of economic sectors (it serves as a meteorological, hydrological and agricultural drought index, Karl, 1983; Karl and Knight, 1985), using a physical based model build on a complex soil water budget system. As opposed to the PDSI, the Sc-PDSI is more spatially comparable across regions using fixed parameters related to the soil/surface characteristics at each location. In addition, it can be used (following the assumptions of the Palmer drought model) to assess the chances of drought recovery. Despite its importance, quantifying drought recovery has not been examined in the Carpathian region yet. Moreover, agriculture is a major economic sector in the Carpathian region (UNEP/DEWA, 2007). The main crops in the region are winter wheat, maize and potatoes (UNEP/DEWA, 2007), which are highly vulnerable to droughts throughout the whole year. Therefore information on ending or ameliorating droughts such as climatological probability that the droughts could be recovered and the seasonal analysis of drought occurrence could be useful in decisions concerning water and agricultural resources management.

The Sc-PDSI is a drought indicator based on the principles of balance between moisture supply and demand. A series of articles have pointed out the assumptions, strengths and weaknesses of the Palmer drought model along with details on calculation procedures (Alley, 1984; Karl, 1987, 1986a, b; Wells et al., 2004; van der Schrier et al., 2006). The PDSI or modified versions of PDSI have been used to quantify drought as a recurrent extreme climate event both at continental (Europe, North America) and global level (Dai et al., 2004; Dai, 1998, 2011; Wells et al., 2004; van der Schrier et al., 2006, 2007). By changing the standardization used by Palmer (1965), which was based on data from the US, Wells et al. (2004) proposed the Sc-PDSI and it was recognized as an improvement of the original PDSI (Dai, 2011). PDSI was developed with the intention of measuring the departure of soil moisture from the normal conditions, using a hydrological accounting system. Different from PDSI other drought indicators are based on past statistics of certain climate variables which often include only precipitation (Dai, 2011) and assumes that droughts are directly controlled by the temporal variability of the precipitation. Recent studies have confirmed the importance of the effect of other variables, such as temperature, on drought conditions. These studies (Williams et al., 2011; Martínez-Villalta et al., 2008; McGuire et al., 2010; Linares and Camarero, 2011) have shown that temperature rise affects the severity of the droughts and mainly the drought stress induced by heat waves on net primary production and tree mortality. For examples, the heat waves in Europe in 2003 and 2010 played extreme roles in influencing drought severity, which increased evapotranspiration and aggravated the drought severity (Rebetez et al., 2006). As a result – enhanced by precipitation shortage (Adams et al., 2009) – major decreasing in net primary production (Ciais et al., 2005) and high forest mortality occurred. This shows how drought stress – through increased evapotranspiration – is induced, to a large degree, by the availability of soil moisture. Therefore, the use of drought indices which is based on soil water balance model, such as PDSI or modified versions as Sc-PDSI, is required in order to calculate current soil moisture conditions. Moreover, the statistical based drought indicators are normalized measures with respect to location and period, which makes the frequency of their severity classes climatologically consistent for any site (Heinrich and Gobiet, 2012), not being able to identify regions that are more "drought-prone" than others (Hayes et al., 1999). Therefore, PDSI has been used as it allows for comparison of drought frequency within different severity classes on different locations and it is suitable to account the drought under global warming conditions. Various aspects of the hydrological model, on which the PDSI is based on, are directly used in the calculation procedure of the precipitation required to recover from drought, which not only confers homogeneity but also offers means of validation of the obtained results.

Based on these considerations and using the assumptions of the Palmer drought model (PDM), the precipitation

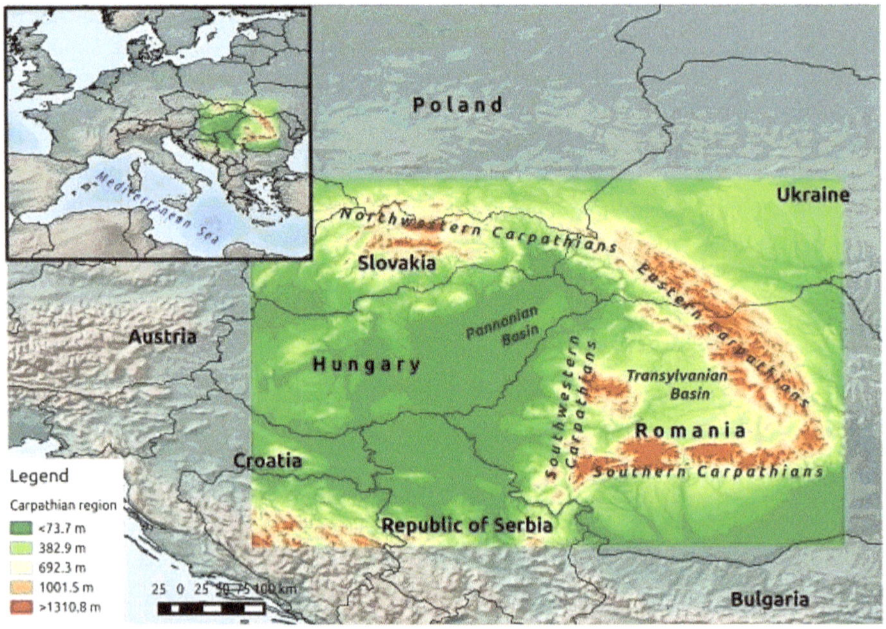

Figure 1. Carpathian region – geographical units.

needed to end or ameliorate a drought (in a 1-, 3-, or 6-month period) for different levels of severity (moderate when Sc-PDSI ≤ -2, severe when Sc-PDSI ≤ -3, extreme when Sc-PDSI ≤ -4), and their climatological probability have been computed. A spatial and temporal analysis of these results is presented, including information on the deviation (%) of the required precipitation from the normal annual rainfall cycle and an analysis of the months of the year with the highest/lowest probability for terminating a drought at different levels of severity.

2 Data and methodology

2.1 Data

The region covered by this study, depicted in Fig. 1, is centered on the Carpathian Mountains and the surrounding lowlands (17–27° E, 44–50° N). Stretching across Central and Eastern Europe, the Carpathian Mountains spans over seven countries (in the studied region), starting from the Czech Republic, Slovakia and Poland in the northwest, then continuing east and southwards through Ukraine, Hungary, Romania and Serbia. The region also spans over parts of Croatia, Bosnia-Herzegovina, Bulgaria and Republic of Moldova. The Carpathian Mountains represent a prolongation of the Alps to the east and northeast, but their structure is less compact, and they are split up into a number of mountain blocks (with heights reaching over 2000 m in altitude) separated by basins (such as Pannonian and Transylvanian) and surrounded by lowlands. The Carpathian region receives polar-continental air masses arriving from the east and northeast in the winter, while during other seasons, oceanic air masses from the west and also Mediterranean in the southern part (UNEP/DEWA, 2007). The calculation procedure for the water needed to recover from drought events is elaborated upon the Palmer drought model used to compute the Sc-PDSI. The computation of the Sc-PDSI (Wells et al., 2004) is carried out on monthly temporal scale and is based on the moisture demand and supply (water-balance model) and takes into account precipitation, evapotranspiration and soil moisture conditions. The basic input data are the following:

- Gridded monthly precipitation (from the CARPATCLIM project at $0.1° \times 0.1°$ spatial resolution for the 1961–2010 period),
- Gridded monthly mean surface air temperature (from the CARPATCLIM project $0.1° \times 0.1°$ resolution for the 1961–2010 period) used to compute Thornthwaite's potential evapotranspiration – PET, (Thornthwaite, 1948).

Temperature and precipitation gridded data have been interpolated within the CARPATCLIM project from quality-checked, completed, homogenized and harmonized daily station data. Please see Spinoni et al. (2013) for a more detailed description.

- The available water holding capacity (AWC) of the soil is computed from the soil texture classes and soil profile depths in the European Soil Database (http://eusoils.jrc.ec.europa.eu/) and the Soil Geographical Database of Eurasia (Toth and Weynants, 2012). The AWC value for each grid cell, shown in Fig. 2, is assumed to be constant

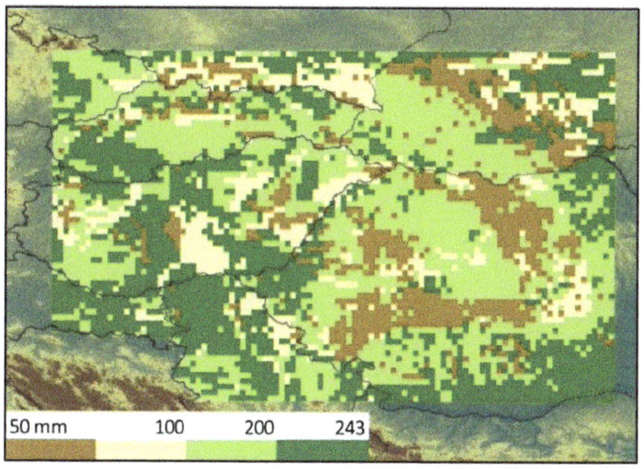

Figure 2. Available water holding capacity (AWC) of the soil (mm) in the Carpathian region.

over the considered period and calculated using the van Genuchten equation for which the parameters are obtained from the Hydraulic Properties of European Soils (HYPRES) pedotransfer class functions (based on the texture classes) (Wösten et al., 1999).

For the computation of the precipitation needed to recover from drought gridded data sets of CAFEC (climatically appropriate for existing conditions) precipitation (precipitation needed to maintain a normal soil moisture level), climate characteristic coefficient and the moisture anomaly index from the Palmer drought model for 1961–2010 period were used. A complete description of these variables is presented in Sect. 2.2.2

2.2 Computation methodologies

2.2.1 Sc-PDSI computation

Sc-PDSI is based on the Palmer Drought Severity Index (PDSI), first introduced by Palmer (1965), and modified by Wells et al. (2004) in order to allow a more accurate comparison of the index at different locations. Sc-PDSI measures the cumulative departure of moisture supply and demand from the normal conditions and is computed on monthly time scale. The supply in this model is the precipitation, the water demand is the potential evapotranspiration and the outputs are the actual evapotranspiration and runoff. Often discussed in other studies (e.g., Alley 1984; Karl, 1986a; Guttman et al., 1992; Weber and Nkemdirim, 1998; Wells et al., 2004; Dai et al., 2004; Vicente-Serrano et al., 2010) the strengths, weakness and differences of these two drought indicators will not be examined in this study. The major difference comes from the reduced frequency of extreme events of Sc-PDSI when compared with PDSI. This aspect is the result of the overall effect of the calibration based on the actual climatic characteristics of a given location that makes Sc-PDSI more consistent over diverse climatological regions.

The Sc-PDSI calculation procedure starts with the calculation of the monthly hydrological parameters of a rather complex soil water balance system: evapotranspiration, recharge, runoff, water loss from the soil and their potential values. The hydrological system is confined by the assumptions that the soil is split in two layers (with the upper soil layer holding 25.4 mm of water) and the saturation level (of both soil layers) is conditioned by the top layer, both on supply and demand.

The potential evapotranspiration (PET) is estimated following Thornthwaite (1948), while the other potential parameters are defined as follows: the potential recharge (PR) is the amount of moisture required to bring the soil moisture up to filed capacity (AWC less the total amount of moisture stored in both soil layers), the potential loss (PL) is the moisture that could be lost from the soil if precipitation is zero for the month and the potential runoff (PRO) is defined as total AWC less potential recharge (PR). The climatically appropriate for existing conditions (CAFEC) precipitation (or the precipitation needed to maintain a normal soil moisture level) is obtained by summing the monthly mean potential values which are previously scaled by their ratio with the monthly mean actual values.

The difference between monthly precipitation and CAFEC-precipitation, weighted by a local climate characteristic coefficient (an empirical derived normalization factor) results in Palmer moisture anomaly index (Palmer's Z-Index). A description of the modifications made to obtain Sc-PDSI is presented in Appendix A.

2.2.2 Ending and ameliorating the drought

The Sc-PDSI$_i$ values and the assumptions of the Palmer drought model (PDM) were used for setting the theoretical basis of the calculation of precipitations needed to recover from the drought. Gridded data sets of CAFEC precipitation ($\hat{P}$), the climate characteristic coefficient (K_i) and the moisture anomaly index (Z_i) (from the Palmer drought model) for the 1961–2010 period are employed. The calculation starts by rewriting PDM's equation used to compute the moisture anomaly index (Z_i), as follows:

$$P_i = \left(\frac{Z_i}{K_i}\right) + \hat{P}_i, \tag{1}$$

where i denotes the months of the year, P_i = precipitation needed to end or ameliorate the drought, $\hat{P}_i$ = CAFEC precipitation and K_i = the coefficient of climate characteristic.

However, before being able to compute P_i, Z_i has to be adapted to recovering drought conditions (end or ameliorate) and $\hat{P}_i$ has to be related with the Sc-PDSI$_{i-1}$ (of the previous month) as CAFEC precipitation (with its soil water balance variables) cannot be computed until the end of the month.

The first step

This represents the transformation of the moisture anomaly index (Z_i) from the self-calibrated drought severity formula in Eq. (2) into the moisture anomaly index needed to end the drought (Z_e) and the moisture anomaly index needed to ameliorate the drought (Z_a).

$$\mathrm{Sc\_PDSI}_i = p\mathrm{Sc\_PDSI}_{i-1} + qZ_i, \tag{2}$$

where q and p are weighted factors.

From the PDSI severity classes (Palmer, 1965), adopted also for the Sc-PDSI (Table 1), it can be stated that a drought event ends when the Sc-PDSI increases above −0.5. Therefore, when the $\mathrm{Sc\text{-}PDSI}_i$ in Eq. (2) is set to −0.5 and solving for Z_i – which now should be mentioned as the moisture anomaly index needed to end the drought (Z_e) – the new formula becomes the following:

$$Z_e = \left(\frac{-0.5}{q}\right) - \left(\frac{p}{q}\mathrm{Sc\_PDS}I_{i-1}\right). \tag{3}$$

Considering the same severity classes, it can be assumed that a drought is ameliorated when the Sc-PDSI reaches a value of −2.0. Applying the same hypothetical basis when trying to calculate the moisture anomaly index needed to ameliorate the drought (Z_a), the $\mathrm{Sc\text{-}PDSI}_i$ in Eq. (2) is set to −2.0 and the formula becomes the following:

$$Z_a = \left(\frac{-2.0}{q}\right) - \left(\frac{p}{q}\mathrm{Sc\_PDSI}_{i-1}\right). \tag{4}$$

The weighted factors q and p are computed at all the locations (grid points) and they are specific for the dry spells. They are site-dependent which make the Z_a and Z_e unique for every grid point. Moreover, these two formulas can be computed not only for different values of $\mathrm{Sc\text{-}PDSI}_{i-1}$ but also for periods of time longer than a month. Once these simultaneous equations are solved, moisture anomaly indexes needed to end (Z_e) or ameliorate (Z_a) a drought are computed for different $\mathrm{Sc\text{-}PDSI}_i$ intensities and different time periods (1, 3, and 6 months in our study).

The second step

The next step is assigning values to the CAFEC precipitation ($\hat{P}_i$) in Eq. (1) since the balance of the demand and supply at the level of soil moisture is solved only at the end of the month. Once this balance reaches a deficit of water, the anomaly is reproduced at the level of the drought indicator in the next month. So, in order to supply the model with precipitations needed to recover the drought at the time when this anomaly happens, the values of CAFEC precipitations ($\hat{P}_i$) were regressed against $\mathrm{Sc\text{-}PDSI}_{i-1}$ for each month during a drought. In order to solve this relation for different time periods (1-, 3-, and 6-month periods) $\hat{P}_i$ is linearly regressed against $\mathrm{Sc\text{-}PDSI}_i$ at time $i-1$, $i-3$ and $i-6$. The new $\hat{P}_i$ can be called the CAFEC precipitation regressed, matching the time (month) when the drought indicator registers the drought event.

Table 1. Cumulative frequency, severity classes, and SC-PDSI values in the Carpathian region.

Cumulative frequency (%)	Severity classes	Sc-PDSI value
2.4	Extremely wet	4 or more
4.1	Severe wet	3.00–3.99
7.4	Moderately wet	2.00–2.99
11.6	Slightly wet	1.00–1.99
7.2	Incipient wet spell	0.50–0.99
17.3	Near normal	0.49–−0.49
9.1	Incipient dry spell	−0.50–−0.99
16.7	Slightly dry	−1.00–−1.99
12.5	Moderately dry	−2.00–−2.99
7.6	Severely dry	−3.00–−3.99
4.0	Extremely dry	−4 or less

The third step

In the third step, the precipitation needed to end or ameliorate the drought is computed as in Eq. (1), using both the moisture anomaly index needed to end (Z_e) or ameliorate (Z_a) the drought, as well as the regressed CAFEC.

2.2.3 Probability calculation

The climatological probability of receiving the amount of precipitation needed to end and ameliorate the drought was calculated using the gamma distribution. Gamma distribution has been frequently used in literature to represent precipitation (Thom, 1966; Wilks, 1990, 1995; Oeztuerk, 1981) due to the advantage that it excludes negative values, being bounded on the left at zero (Thom, 1966; Wilks, 1995). Analysis of rainfall data strongly depends on its distribution pattern (Sharma et al., 2010). This is especially important as gamma distribution is positively skewed and represents an advantage as it mimics the actual rainfall distributions for many geographical areas (Ananthakrishnan et al., 1989). Also it provides a flexible representation of a variety of rainfall regimes while utilizing only two parameters, the shape and the scale (Wilks, 1990).

The calculations were performed separately for each month and each location (grid point) on the basis of the entire 50 years of available data (1961–2010). Input data were the computed precipitation needed to end or ameliorate the drought ($0.1^\circ \times 0.1^\circ$ resolution) in the next 1, 3, and 6 months and the actual gridded monthly precipitation ($0.1^\circ \times 0.1^\circ$ resolution) accumulated for the same time periods. The probability statistics should not be considered as a forecast. They represent a quantitative measure of the proba-

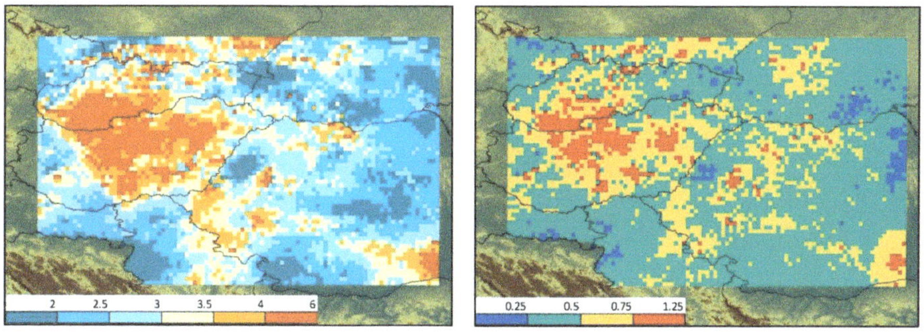

Figure 3. Average number of months per year with moderate drought (Sc-PDSI ≤ -2.0; left) and extreme drought (Sc-PDSI ≤ -4.0; right) in the Carpathian region (1961–2010).

bility computed on the basis of past actual precipitation data. Practically, the probability density function (PDF) of the actual precipitation data is used to find the cumulative probability (CDF) of the precipitation needed to recover from the drought for the required month and temporal scale.

All the procedures followed in the calculation of the climatological probability of recovering from a drought are based on the processes used by Oeztuerk (1981) to compute the probability distribution for precipitation. In the first step the actual precipitation data on "moving windows" of 1, 3, and 6 months are matched with the precipitation needed to recover from the drought in the next 1, 3, and 6 months. In the second step the cumulative probability (CDF) of the computed precipitation needed to end or ameliorate the drought is derived.

3 Results

Previous studies of drought in the Carpathian region were based on the analysis of intensity, duration and spatial extent, either at national level (e.g., Palfai, 1990; Szinell et al., 1998; Szalai, 2000; Popova et al., 2006; Trnka et al., 2009; Cheval, 2013) or at inter-regional level (e.g., Bartholy et al., 2013; Spinoni et al., 2013). Our results show that the incidence of drought in this region is rather high. During the period 1961–2010, every part of the region experienced on average between 0.5 and 6 drought months per year, (Sc-PDSI ≤ -2, Fig. 3 left). Moreover, the incidence of extreme drought (Sc-PDSI ≤ -4) has an occurrence of 5–45 days per year for the same time interval as shown in Fig. 3, right. When compared with other drought indicators Sc-PDSI shows good correlation with indices of long accumulation periods. The correlation between Sc-PDSI and SPI or SPEI over each grid point and for entire Carpathian region and time period (1961–2010) shows high values with the SPI_9 (0.85) and SPEI_9 (0.82) detecting the drought events on comparable spatial and temporal resolution and lower values with SPI_1 (0.33) and SPEI_1 (0.35) (Antofie et al., 2013).

The spatial and temporal analysis of the results for precipitations needed to recover from drought and their climatological probability is related to three levels of severity: moderate drought when $-3 <$ Sc-PDSI ≤ -2, severe drought when $-4 <$ Sc-PDSI ≤ -3, and extreme drought when Sc-PDSI ≤ -4, which are evaluated on a temporal window of 1, 3, and 6 months. In order to ease the interpretation of the results with their temporal and spatial variability a review of the climatological conditions of the area related with the 0 characteristics of the Palmer drought model will be presented.

PDSI originally was designed to measure the soil moisture departure as a difference between a climatological moisture supply which in our case is the actual precipitation and the precipitation needed to maintain a normal soil moisture level (CAFEC precipitation; Palmer, 1965). In this study, other means of moisture supply such as precipitation in form of snow water equivalent are not considered. Since the regional spatial variation of precipitation in this region is mainly determined by the mountain orography and the large scale atmospheric processes (UNEP/DEWA, 2007), it is expected (in a temperate-continental climate) that moisture supply is more significant in the high altitudes while the moisture demand is higher in the low altitudes (higher rate of evapotranspiration due to higher temperatures). With increasing continental conditions from west to east and temperature decreasing from north to south, a higher moisture demand in the south and southwest, and higher moisture supplies in the northern, western and southwestern parts of the region are expected.

Based on these general climatological characteristics, the PDM will produce the highest Z values (soil moisture) in the areas and for the period of the year with highest precipitation amount. The model will also indicate the most favorable period of the year for recovering from drought. The most favorable periods are the months that have the highest frequency of excess of precipitation compared to the normal. This is not necessarily the wettest month of the year but the month with the largest positive skew as the PDM is based on departures from the normal.

For both the required precipitation and the probabilities of recovery from drought a spatial pattern linked with the atmospheric circulation patterns responsible for the climate variability in the Carpathian region can be noticed. The southern

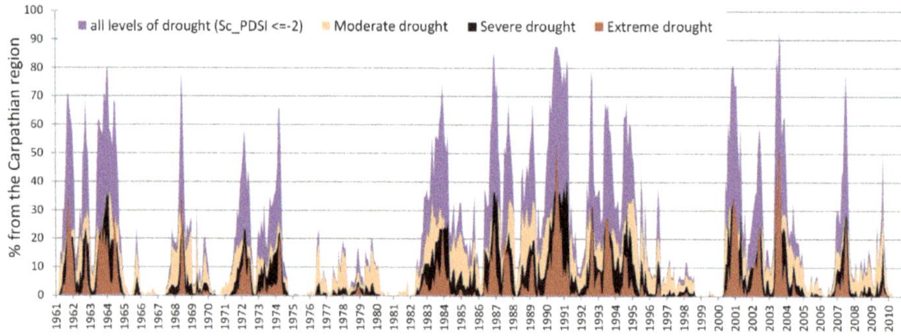

Figure 4. Incidence (% surface of the region) of different severity levels of drought per month.

and southwestern Carpathians and the western Carpathians act like a barrier for the main sources of moisture (Mediterranean and North Atlantic air masses; Busuioc and von Storch, 1996; Busuioc, 2001). These systems are causing first, high precipitation amounts and a pronounced annual precipitation cycle, as it is the case of North Atlantic circulation in the western, northern and northwestern part of the Carpathian region. Secondly, is causing a highly variable precipitation intensity and a relatively constant distributed precipitation regime through the year (by creating a second precipitation peak in autumn), as it is the case of Mediterranean cyclones in the southwestern and southern part of the Carpathian region. The cyclonic presence and trajectories have been the subject of extensive climatological research (e.g., van Bebber, 1891; Radinovic, 1987; Katsoulis, 1998; Flocas, 1998; Maheras, 2001). Often these studies establish a connection between the advance of the cyclones from the Mediterranean area and intense precipitation events. High amounts of precipitation with genesis in the Mediterranean space (Gulf of Genoa) are produced on the cyclonal trajectory V (from the Tyrrhenian Sea to Ukraine). Most important for the Carpathian region are the trajectory Vc that crosses from west to east (the south of the Carpathian region) in spring and very rarely in summer and trajectory Vb (important for the western part of the Carpathian region, passing over the Pannonian Plain) towards Poland. For both trajectories, the cyclones circulate especially in autumn, winter and spring with the largest probability of occurrence in April and a secondary maximum in early autumn. The cyclone circulation diminishes and migrates southwards in December-January, due to the intensification of both the Azores and Siberian anticyclones (Maheras, 2001).

The annual cycles of the moisture supply and demand follow a continental pattern in the Carpathian region (imposed by the North Atlantic circulation) with a maximum of supply and demand at the beginning of the summer (May-June-July) and end of summer (July-August) respectively a minimum in the winter months (December-January-February). Regardless of the annual cycles of precipitation, the months with the higher probability of substantial excess of precipitation from the normal (April-May in spring and October-November in autumn) will be related with the cyclonic presence from the Mediterranean area. Nevertheless the joined influence of the circulations moving either from the Atlantic or the Mediterranean Sea is a common characteristic of the Carpathian region (Busuioc and von Storch, 1996; Busuioc, 2001).

3.1 Drought recovery and its temporal variability

As shown in Fig. 4 the incidence of drought events (Sc-PDSI ≤ -2) is most pronounced during the early years of the 1960's, 1970's and 2000's, as well as during almost the entire decade of the 1980's and 1990's and more isolated in the years 1968, 2007 and 2009. The recorded drought occurrence in the region presented through country reports at UNCCD's 1st Regional Workshop on Capacity Development to Support National Drought Management Policies for Eastern European Countries (9–11 July 2013, Bucharest) confirms the drought-prone characteristic of the region. The years with the highest drought incidence mentioned in the region are 2000, 2003, 2007 and 2012 (Holjevac et al. 2013), beginning of the 1990's (Gregorič et al., 2013), the time intervals from 1961–1965, 1973–1974 and also 1980's since when an increasing in the incidence of droughts it is noticed (Mateescu et al., 2013; Gregorič et al., 2013).

One of the characteristics of these drought events is the strong prevalence (% from the area) of extreme droughts (Sc-PDSI ≤ -4) as compared to other severity levels. This can be seen especially in the years with the highest general drought incidence over the region: 1961, 1964, 1968, 1987, 1990, 1992, 1993, 2001–2003, 2007. For these cases most of the drought events happened either in the summer period (June–August) or in the winter months (December–February); for a few cases, drought occurred in October and March–April.

As shown in Table 2 for selected drought events between 200 % to more than 480 % of the normal 1-monthly precipitation would have been required for recovery (i.e., bringing Sc-PDSI to a level of -0.5). For a 3-month period, the percentage is reduced from 100 % to almost 230 % of the 3-monthly precipitation, and for a 6-month period still up to 50 % above normal 6-monthly precipitation would have been required. To ameliorate a drought (i.e., reaching Sc-PDSI of

Table 2. Percentage above the normal precipitation needed to end a drought – Pp (%) – in the next 1, 3, and 6 months for the drought events with the highest incidence (% surface from the region)

Years	2003	1990	1990	2003	1986	2003	1990	1990	1991	1990
Month	8	8	7	9	12	6	9	6	3	10
Incidence (%)	93.3	92.6	88.5	88.1	85.8	85.6	85.5	84.7	83.9	83.9
1-month Pp (%)	414.5	376.9	625.1	204.1	368	228.2	415.7	374.3	462.5	482.3
3-month Pp (%)	195.7	230.8	160.9	172.5	110.7	129.7	360.3	118.7	99	233.3
6-month Pp (%)	25.2	57.9	52	21	36.3	34.3	63.1	49.2	48.2	62.3

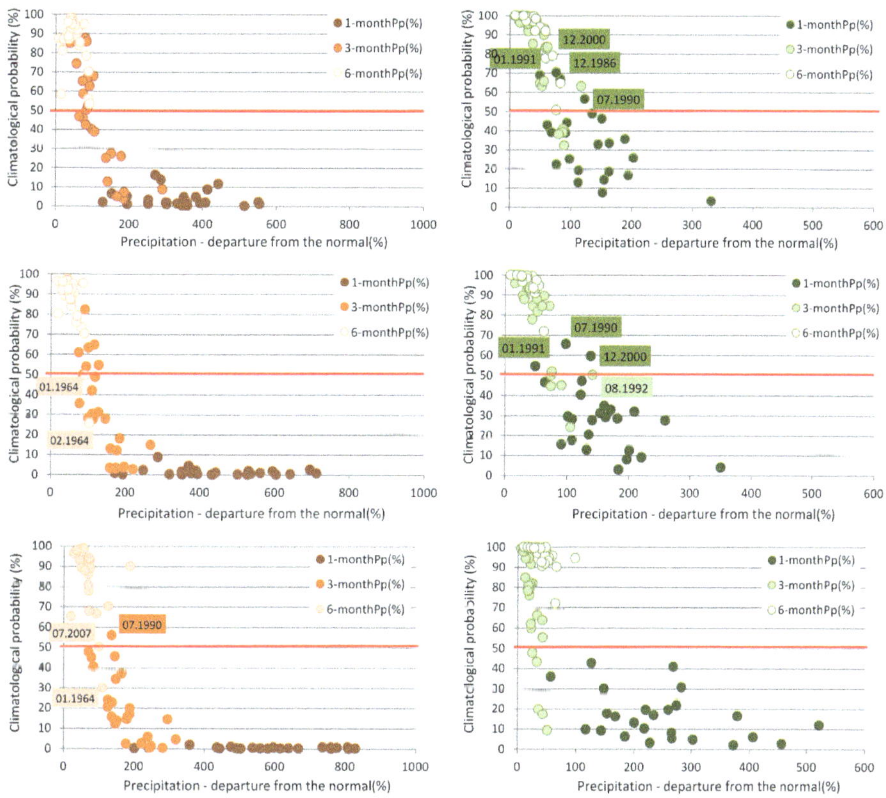

Figure 5. Probability (%) of ending (left) or ameliorating (right) moderate (top), severe (centre) and extreme drought (bottom) events with the highest incidence in the Carpathian region in the next 1, 3, 6 months.

≥ -2) smaller amounts of precipitation would be sufficient: 70–100 % above normal precipitation in a 1-month period, 30–60 % in a 3-month period, and less than 20 % in a 6-month period.

In order to get a better idea of the climatological probabilities to recover from such droughts, we analyzed the first 25 most significant events (droughts occurring on > 75 % of the area) for different drought intensity levels. Figure 5 shows the required precipitation in percent of the climatologically expected precipitation and the associated probabilities for different drought intensities and precipitation accumulation periods. It can be seen that moderate severity droughts ($-3 <$ Sc-PDSI ≤ -2) required between 110 and 550 % of the normal 1-monthly precipitation to recover from drought (top left), while for a 3-month period they ranged between 50 and 200 %. For a 6-month period the required precipitation is well within the climatologically expected. For the same drought cases, during the peak intensity of the drought (Sc-PDSI ≤ -4) the quantity of precipitation required increases up to approximately 8 times above normal 1-monthly precipitation, while for a 3-month period the values reach up to 300 %, only for the 6-month period the precipitation required to recover from drought is close to the climatologically expected (bottom left). Severe droughts ($-4 <$ Sc-PDSI ≤ -3) would have been ended with rainfall ranging between 2 to 7 times the 1-monthly normal precipitation and approximately 100 % of the 6-monthly normal precipitation (center left).

Most of these values indicate the improbability of ending or ameliorating the drought in a short period of time, as their climatological probability is (extremely) low. If we

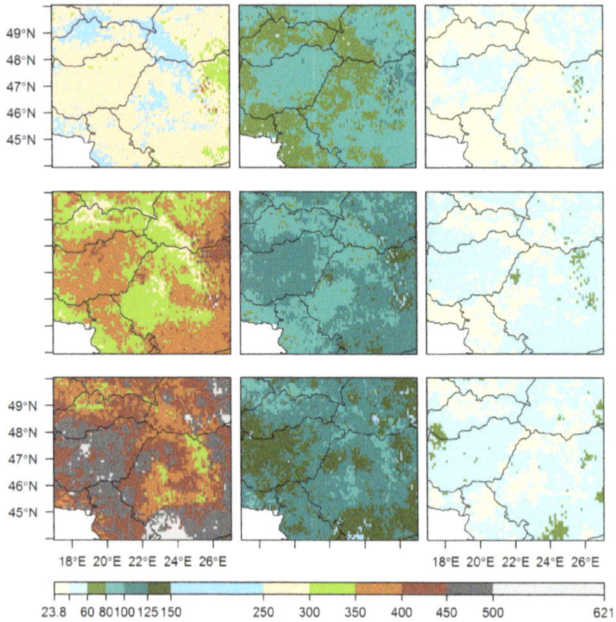

Figure 6. Percentage (%) above the normal precipitation needed to end a (top) moderate ($-3 < \text{Sc-PDSI} \leq -2$), (center) severe ($-4 < \text{Sc-PDSI} \leq -3$) and (bottom) extreme drought ($\text{SC-PDSI} \leq -4$) in the next month (left), next 3 months (center) and next 6 months (right) (1961–2010).

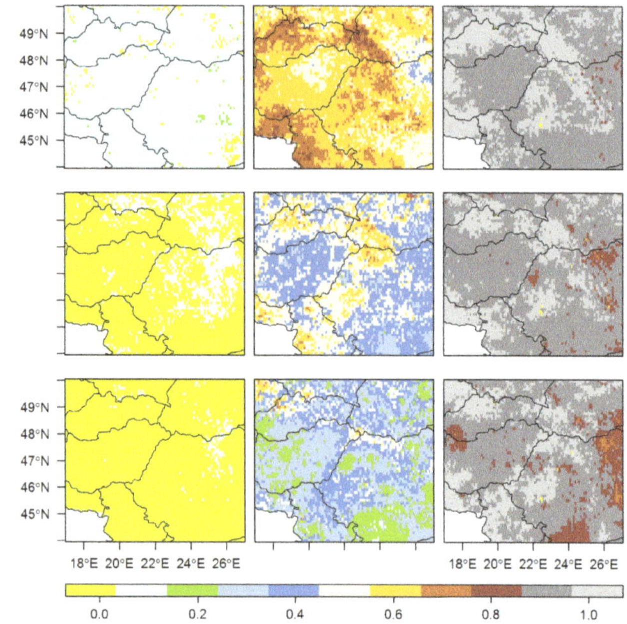

Figure 7. Climatological probability (%) of receiving the precipitation needed to end a (top) moderate ($-3 < \text{Sc-PDSI} \leq -2$), (center) severe ($-4 < \text{Sc-PDSI} \leq -3$) and (bottom) extreme drought ($\text{SC-PDSI} \leq -4$) in the next month (left), next 3 months (center) and next 6 months (right) (1961–2010).

settle a limit of 50 % probability, above which the amount of precipitation could be considered more likely than not (IPCC, 2007), none of the drought events could have been ended in the next month. However, a few of the moderate droughts (July 1990, December 1986, December 2000, January 1991) and of the severe droughts (July 1990, December 2000, January 1991) could have been ameliorated with 48 to 140 % above the normal precipitation in 1 month (top right). On the other hand, in a 6-month period almost all drought events considered could most probably have been ended with amounts 10 to 80 % above the normal precipitation (188 % for the extreme drought of April 1991). Only the severe drought from January and February 1964 and the extreme drought from July 2007 could not have been ended even in a 6-month period, making them the most excessive droughts of the studied period in the Carpathian region. Nevertheless, they could have been ameliorated with 45 to 65 % above the normal 6-monthly precipitation. In a 3-month period, only one drought event of extreme intensity (July 1990, requiring 136 % above the normal precipitation), 12 events of the moderate and 6 events of severe drought could have been ended with high probability. All the other events could only have been ameliorated with a range of 15 to 140 % (August 1992) above the normal 3-monthly precipitation.

3.2 Drought recovery and its spatial variability

Figure 6 presents the averaged positive deviations (in percent) from the normal precipitation needed to recover from a drought computed for the period 1961–2010. The Pannonian Basin, the Transylvanian Plateau and the external Carpathian foothills and plains in the southern and eastern part of the region require the highest relative amount of precipitation to recover from drought. In these regions, moderate droughts and extreme droughts needed between 250 and 300 % (sometimes up to 600 %) above the normal precipitation to end a drought in the next month, a decrease which is noticed with increasing altitude. The topographic pattern is lost when the moisture supply is required for a larger time window. This is due to the general climate characteristics that overwrite the variability introduced by the local physical conditions. Also, the longer time intervals require less relative amounts of precipitation to recover from drought (i.e., from 20 up to 40–60 % for all the drought intensities).

Figure 7 shows the corresponding probabilities. The probability of ending or ameliorating an extreme drought ($\text{Sc-PDSI} \leq -4$) or a severe drought ($-4 < \text{Sc-PDSI} \leq -3$) in a 1-month period is low (< 8 %), showing the improbability of recovering the high intensity droughts in such a short time interval. The probability remains below 20 % even for the moderate droughts. For a 3-month period, the probability of ending a drought increases from below 10 up to 40 %(for the extreme droughts), but is still unlikely (< 33 %) or about as likely as not ending a drought (33–66 %). More likely, with a probability of 60–80 % a moderate drought could be ended over almost the entire region in the 3-month time interval. Once we advance to the 6-month interval, all droughts, indifferent of their intensity level, move from likely (> 66 %) to virtually certain (> 99 %) to be ended.

As presented at the beginning of this section, the succession, intensity and the predominance of the cyclonic circulation may lead to a seasonal variability of the precipitation needed to recover from drought and climatological probability. The soil moisture supply and demand follow the annual cycle of precipitation and temperature (imposed by the general atmospheric circulation) but they are reflected differently at the level of the month with the highest and lowest probabilities of recovering from drought. More likely to recover from drought are the months with higher probability of substantial excess of precipitation from the normal and especially for the regions with a constant precipitation regime throughout the year. The more the precipitation regime presents a pronounced peak the more the preferred recovery moths are variable.

Corresponding to these characteristics, in almost the entire Carpathian region, the preferred months for ending a drought event are the months of April and May as in Fig. 8, corresponding to the months with the largest probability of receiving high precipitation amounts compared with the normal and a maximum activity of Mediterranean cyclones. The least preferred months for ending a drought event are the months of January and February for southern, southwestern, and northwestern regions, and October, November, and December for the northeastern part. This corresponds to the months with the largest probability of receiving low precipitation amounts compared with the normal and a minimum activity of the Mediterranean cyclones. This situation can be observed in Fig. 8 where we present the months with the highest and lowest probability for ending droughts at different intensities during the next month in Fig. 8a, the next 3 months in Fig. 8b and the next 6 months in Fig. 8c.

Moderate drought events in April appear to have the highest probability for being ended in the next month. Also, severe and extreme drought events in April and May (for northern and northeastern regions) are characterized by highest probabilities of being ended in following month. The late summer (July-August) and early autumn (September-October) drought events are ended with highest probability in the southern, western and northwestern parts of the Carpathian region as seen in the Fig. 8a, top.

In Fig. 8b top, we show that the drought events with the highest probability of being ended in 3 months are the droughts from the end of winter (January-February) in the western, southern and northwestern regions for the moderate droughts and spring droughts (April-May) in northern and northeastern regions, especially for the extreme drought events. The late autumn drought events (October-November) present the highest probability of being ended in the next 6-month period as seen in Fig. 8c, top.

Concerning the lowest probabilities for ending a drought event, the worst months for ending the droughts are the winter months, corresponding to the driest period of the annual precipitation cycle and the minimum activity of the Mediterranean cyclones in the Carpathian region. This makes drought events between October (in the northern, northeastern area of the Carpathian region) and February (in the southern and eastern part of the region and Pannonian Basin) the least probable to be ended in the next month as seen in the Fig. 8a, bottom.

In Fig. 8b bottom, we show that the drought events with the lowest probability of being ended in the next 3-month period are the droughts from the December in the northern and northeastern regions and autumn droughts in the other regions.

The least probable to be ended in the next 6 months are the summer droughts (June-July-August), especially in the southern and southwestern regions, while the winter droughts are the least probable to be ended in the north and northeastern part of the region as seen in Fig. 8c.

As shown, in Carpathian region, the water deficit occurs throughout the whole year. As the agriculture is an important economic sector in the Carpathian region, the drought impact could be essential. Most crops may experience water stress (deficit) at various stages in their growth cycle. The sequences of vegetative growth with their key physiological phases (i.e., crop phenology) and their sensitivity to water deficit can be used to highlight the importance of seasonal analysis of drought occurrence. Winter crops (i.e., winter wheat) are planted in Carpathian region in September through October and harvested July through August of the next year, while the spring crops (i.e., corn, spring wheat, sunflower, and potatoes) are planted April through May and harvested August through September or even October (potatoes) of the same year (UNEP/DEWA, 2007).

Early drought in the growing season (the end of autumn in October and November for winter crops and the end of spring in late April and May for spring crops) are affecting wheat germination and crop establishment (Bouaziz and Hicks, 1990). The water stress during the vegetative stages (the months of April and May for winter crops and late May and June for spring crops) may affect the leaf index development (Rickman et al., 1983). Soil water deficit increased towards harvesting (early summer for winter crops and late July or beginning of autumn in August for spring crops) is likely to produce a severe reduction in grain growth and quality which eventually cause reduction in final yields. On the other hand, it has been noted that water deficit in the maturity (anthesis) and harvesting period accelerates development (Simane et al., 1993) and significantly contribute to grain yield (Palta et al., 1994).

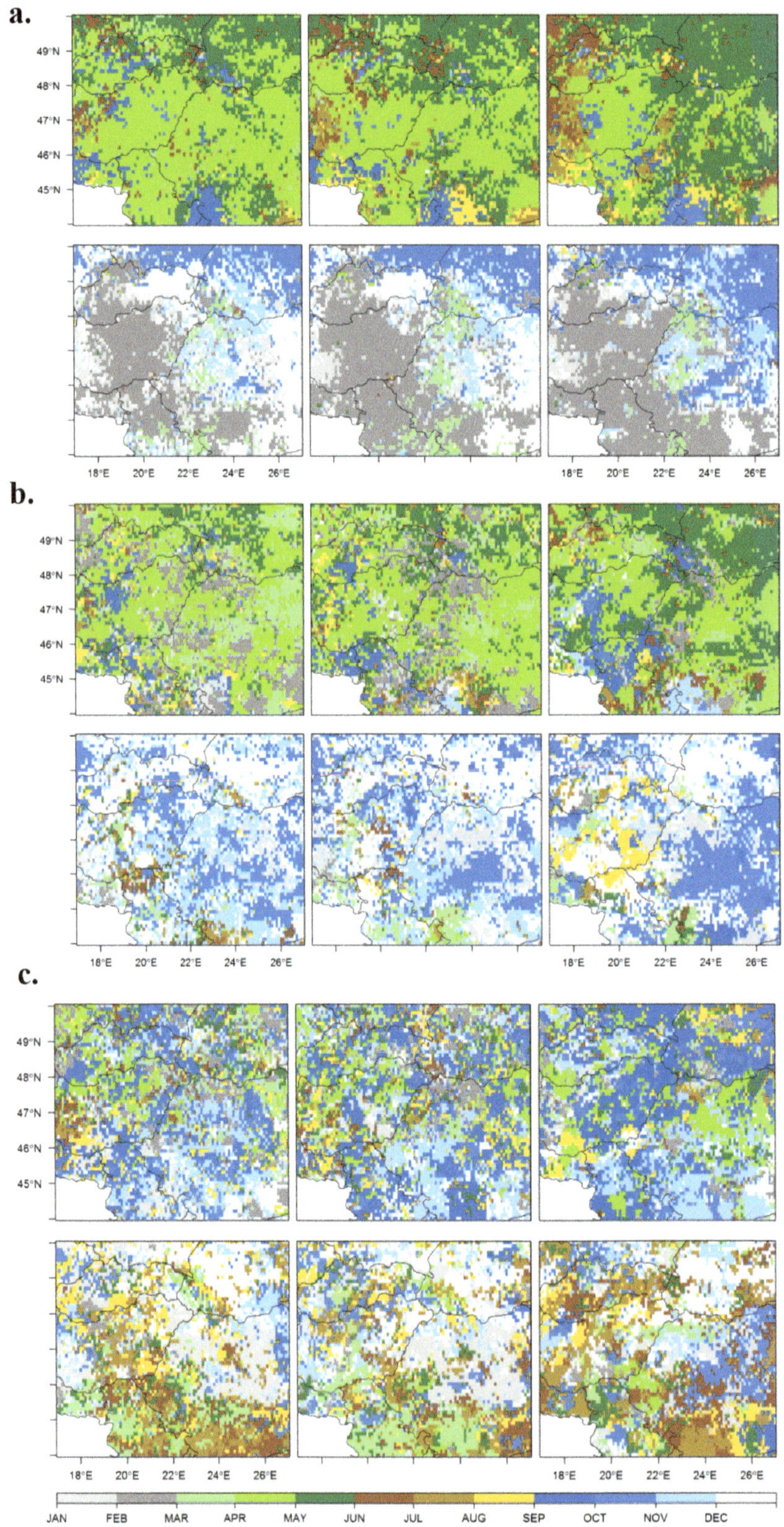

Figure 8. The months with the highest (top), and lowest (bottom) probability of having (left) a moderate drought ($-3 < \text{Sc-PDSI} \leq -2$), (center) severe ($-4 < \text{Sc-PDSI} \leq -3$) and (right) extreme drought ($-4 \leq \text{Sc-PDSI}$) terminated in **(a)** the next month, **(b)** the next 3 months and **(c)** the next 6 months.

4 Conclusions

The main characteristics of the spatial and temporal variability of the precipitation needed to end or ameliorate a drought in the Carpathian region are presented in this study. Sc-PDSI was used as a drought indicator, and the Palmer drought model was considered to calculate moisture supply and demand. The incidences of drought in the region is considerable. During the study period (1961–2010), the region experienced, on average, drought events from at least 0.5 months to 4–6 months per year for moderate droughts and less than a month per year for extreme droughts.

The amount of the precipitation needed to end a drought in the next month, reached, on average, between 200 and 480 % above the normal 1-monthly and up to 50 % above the 6-monthly total of the normal precipitation. It was also shown that most of the drought events, irrespective of their intensity, are extremely unlikely (< 5 %) to be ended in the next month.

Regionally, the Pannonian Basin, the Transylvanian Plateau and the external Carpathian foothills and plains in the southern and eastern part of the region require the highest quantity of precipitation to recover from drought and present the lowest climatological probabilities. In almost the entire Carpathian region, the preferred months for ending a drought event are the months of April and May corresponding to months with the largest probability of receiving high precipitation amounts. Often during this period of the year a connection between the cyclones from the Mediterranean area and intense precipitation events is established. The worst months for ending the droughts in the Carpathian region are the late autumn and winter months, corresponding to the driest period of the annual precipitation cycle and the minimum activity of the Mediterranean cyclones in the Carpathian region.

High precipitation amounts over the northern and northwestern part of the region are causing higher moisture supply and higher climatological probabilities to recover from drought. On the other hand for the eastern and northeastern area of the Carpathian region in October and for the southern and eastern part of the region and Pannonian Basin in February, the smallest climatological probabilities occur to recover from drought.

The early drought events for winter crops (in October and November) cause a high stress effect especially on germination and early crop establishment. On the other hand the water stress is much lower for the spring crops from April and May and for the same physiological phase due to a high cyclonic activity.

The summer droughts have a low probability to be ended, especially in the southern and southwestern regions, if in the maturity (anthesis) and harvesting periods of the crops, the water deficit can cause less damage in crop development, thereby producing even an increase in the grain yield.

Appendix A: Sc-PDSI calculation

The computation of the self-calibrating PDSI was done in four steps: (a) computation of the soil water budget (Thornthwaite, 1948), (b) normalization with respect to demand, (c) normalization with respect to location, and (d) computation of the drought severity.

A1 Computation of the soil water budget

The computation of the soil water budget was done considering the following assumptions: the soil is divided in two layers, the AWC value is site dependent – representative of the soils type, the top layer contains 25.4 mm of available moisture at field capacity, the moisture stored in the soil layers changes according to the priority conditions imposed by the top layer on supply and demand. Rainfall surplus is first added to the top layer until this layer is saturated and only then it passes to the second layer while on the other hand moisture is withdrawn from the top layer first, before removing from the second soil layer.

Following these rules eight hydrological parameters of the water balance are computed: the actual evapotranspiration (ET), the soil water recharge (R), the runoff (RO), the water loss from the soil (L) and their potential values used in the calculation of Palmer's constants to define the Climatically Appropriate for Existing Conditions (CAFEC) precipitation. The potential evapotranspiration was computed using the Thorntwaite formula while the other potential parameters are computed as follows (Weber and Nkemdirim, 1998): the potential recharge (PR) is the amount of moisture required to bring the soil moisture up to filed capacity (AWC minus the total amount of moisture stored in both soil layers), the potential loss (PL) is the moisture that could be lost from the soil if precipitation is zero for the month and the potential runoff (PRO) is defined as total AWC minus the potential recharge (PR). Dividing the mean actual quantity by the mean potential quantity, coefficients defining the usual climate for a specific location were obtained (for evapotranspiration – α, recharge – β, runoff – γ, and loss – δ) as in Eq. (A2). The four coefficients are determined for each of the 12 months. The mean of the actual and potential quantities were computed over a baseline equal to the data period available (1961–2010).

A2 Normalization with respect to demand

Normalization with respect to demand (or moisture departure for the month – D) was calculated by subtracting from the normal precipitation the amount of the precipitation needed to maintain a normal soil moisture level (CAFEC precipitation – $\hat{P}$, computed from the potential values of the water balance and their coefficients):

$$D = P - \hat{P} = P - (\alpha_i \mathrm{PE} + \beta_i \mathrm{PR} + \gamma_i \mathrm{PRO}\ Z\delta_i \mathrm{PL}), \quad \text{(A1)}$$

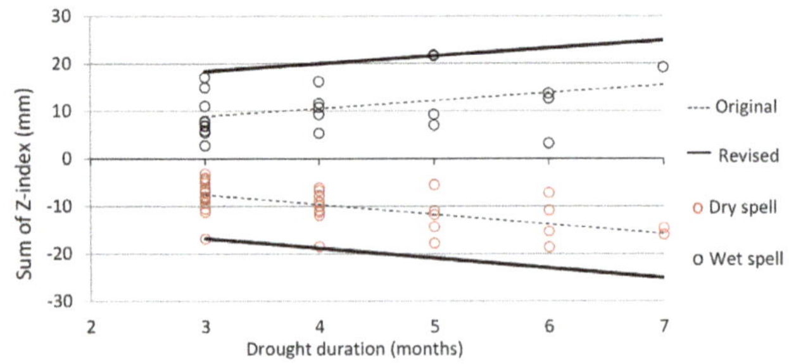

Figure A1. Accumulated z-index (mm) versus duration (months) with the intercept of the most extreme drought/wet spell.

where D is the moisture departure for the month, P the actual precipitation, $\hat{P}$ = CAFEC precipitation, and $\alpha, \beta, \gamma, \delta$ = water-balance coefficients computed as follows:

$$\alpha = \overline{\mathrm{ET}}_i / \overline{\mathrm{PE}}_i, \beta = \overline{R}_i / \overline{\mathrm{PR}}_i$$
$$\gamma = \overline{\mathrm{RO}}_i / \overline{\mathrm{PRO}}_i, \delta = \overline{L}_i / \overline{\mathrm{PL}}_i, \quad \text{(A2)}$$

where ET, R, RO, L are the evapotranspiration, recharge, runoff, and soil moisture loss. $\overline{\mathrm{PE}}, \overline{\mathrm{PR}}, \overline{\mathrm{PRO}}, \overline{\mathrm{PL}}$ are their potential values, the bars indicate the average value and i ranges over the months of the year.

A3 Normalization with respect to location

This was done by converting the moisture departure for the month (D) into an indicator of moisture anomaly (Z_i) by multiplying the moisture departure with a climatic characteristic coefficient (K). This is the point where the Sc-PDSI becomes different from the PDSI. The purpose of the climatic characteristic, K, is to adjust the value of PDSI according to the tails of its distribution in order to allow for an accurate comparison of PDSI values over time and space. Practically, the values of every location (pixel in this case) and each value of PDSI_i were weighted according to the 2nd and 98th percentile of the PDSI and compared with the expected −4.0 and +4.0 calibration:

$$K = \begin{bmatrix} \dfrac{-4.0}{\mathrm{PDSI}_{2\mathrm{nd}}} \acute{K}, & \text{if } D < 0 \\ \dfrac{4.0}{\mathrm{PDSI}_{98\mathrm{th}}} \acute{K}, & \text{if } D \geq 0 \end{bmatrix}, \quad \text{(A3)}$$

where $\mathrm{PDSI}_{2\mathrm{nd}}$ and $\mathrm{PDSI}_{98\mathrm{th}}$ are the 2nd and 98th percentile of the PDSI distribution computed using K':

$$K'_i = 1.5 \log_{10}\left[\left(\frac{\overline{PE} + \overline{R} + \overline{RO}}{\overline{P} + \overline{L}} + 2.8\right)\overline{D}^{-1}\right] + 0.5, \quad \text{(A4)}$$

where $\overline{D}$ = average absolute value of the moisture departure and $\overline{\mathrm{PE}}, \overline{R}, \overline{\mathrm{RO}}, \overline{P}, \overline{L}$ are the parameters of water balance

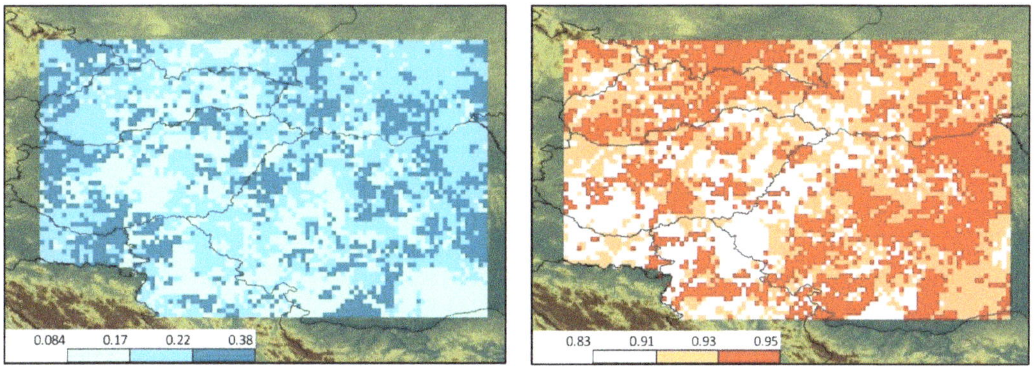

Figure A2. Duration factors p (left) and q (right) for dry cases in the Carpathian region (1961–2010).

values of evapotranspiration, recharge, runoff, precipitation and loss.

Using the climate characteristic coefficient (K) and the moisture departure (D) for the month i, the moisture anomaly index is computed as follows:

$$Z_i = D_i \cdot K_i. \tag{A5}$$

A4 Computation of drought severity

Once Z is computed for the month i, the computation of the drought severity begins by relating the previous month's PDSI_{i-1} with the current moisture anomaly Z_i. The weights assigned to these two components are given by the duration factors (p and q):

$$\mathrm{PDSI}_i = p\mathrm{PDSI}_{i-1} + qZ_i. \tag{A6}$$

Differently from the original computation (the original PDSI is computed using the duration factors $p = 0.897$ for PDSI_{i-1} and $q = 1/3$ for Z_i) the Sc-PDSI duration factors for wet and dry spells are computed separately, as it is assumed that different locations have different sensitivities to precipitation events. These duration factors (p and q) were computed using the least squares method by fitting straight lines to the lowest (highest) Z_i values accumulated over different lengths of time, aiming at representing most extreme dry/wet periods of various lengths. Practically the accumulated Z_i was regressed against its duration (months) taking into account the most extreme dry/wet periods of various lengths as shown in Fig. A1.

The most extreme wet/dry period is defined, in this study, as events with duration greater or equal to 3 consecutive months and with the highest intensity of Z_i (less/higher than 0.05/0.95 percentiles of accumulated negative/positive Z_i values are omitted). Once the intercepts of the most extreme wet/dry periods were computed, two sets of p and q (for dry/wet spells) were calculated as follows:

$$p = (1 - m/(m+b)), \tag{A7}$$

$$q = C/(m+b), \tag{A8}$$

where m is the slope, b the intercept of the extreme wet/dry period and C is a calibration factor, in this study −4 and +4 were assigned for dry and wet spells. Finally, $PDSI_{i-1}$ and Z_i from Eq. (A6) were added to compute the Sc-PDSI_i, using the p and q as weighting factors. The obtained values shown in Fig. A2 vary between 0.85 and 0.95 for p and 0.08 and 0.38 for q of a dry spell. These values are very important as they are to be used in the calculation of the moisture anomaly index needed to end (Z_e) and ameliorate (Z_a) a drought.

Acknowledgements. This study was supported, in part, by the results of the CARPATCLIM project (www.carpatclim-eu.org). We would like to acknowledge the work of (alphabetical order): Igor Antolovic, Ingeborg Auer, Oliver Bochnicek, Zita Bihari, Sorin Cheval, Natalia Gnatiuk, Johann Hiebl, Peter Kajaba, Piotr Kilar, Gabriela Ivanakova, Danuta Limanowka, Monika Lakatos, Monica Matei, Janja Milkovic, Dragan Mihic, Yurii Nabyvanets, Pavol Nejedlik, Predrag Petrovic, Robert Pyrc, Tatjana Savic, Oleg Skrynyk, Sandor Szalai, Tamás Szentimrey, Pavel Štastný, Petr Štěpánek, Radim Tolasz, and Pavel Zahradníček.

Edited by: S. Maskey

References

Adams, H. D., Maite, G. C., Greg, A. B. G., Juan, C. V., David, D. B., Chris, B. Z., Peter, A. T., and Travis, E. H.: Temperature sensitivity of drought-induced tree mortality portends increased regional die-off under global-change-type drought, P. Natl. Acad. Sci. USA, 106, 7063–7066, 2009.

Alley, W. M.: The Palmer drought severity index: Limitations and applications, J. Appl. Meteor., 23, 1100–1109, 1984.

Ananthakrishnan, R. and Soman, M. K.: Statistical distribution of daily rainfall and its association with the coefficient of variation of rainfall series, Int. J. Climatol., 9, 485–500, 1989.

Antofie T., Naumann G., Spinoni J., Weynants M., Szalai S., Szentimrey T., Bihari Z., and Vogt J.: A drought severity climatology for the Carpathian Region using Sc-PDSI, Geophysical Research Abstracts, EGU General Assembly, Vienna, Austria, 7–12 April 2013, 15, 2013.

Bartholy, J., Pongracz, R., and Sabitz, J.: Analysis of drought index trends for the Carpathian Basin using regional climate model simulations, Geophysical Research Abstracts, EGU General Assembly, Vienna, Austria, 7–12 April 2013, 15, 2013.

Briffa, K. R., Jones, P. D., and Hulme, M.: Summer moisture variability across Europe, 1892–1991: an analysis based on the Palmer drought severity index, Int. J. Climatol., 14, 475–506, 1994.

Bouaziz, A. and Hicks, D. R.: Consumption of wheat seed reserves during and during and early growth as affected by soil water potential, Plant Soil, 128, 161–165, 1990.

Busuioc, A.: Large-scale mechanism, influencing the winter Romanian climate variability, in: Detecting and Modelling Regional Climate Change and Associated Impacts, edited by: Brunet, M. and Lopez, D., Springer-Verlag, Berlin, 333–343, 2001.

Busuioc, A. and von Storch, H.: Changes in the winter precipitation in Romania and its relation to the large-scale circulation, Tellus A, 48, 538–552, doi:10.1034/j.1600-0870.1996.t01-3-00004.x, 1996.

Cheval S., Busuioc, A., Dumitrescu, A., Birsan, M. V.: Spatiotemporal Variability of the Meteorological Drought in Romania using the Standardized Precipitation Index, Geophysical Research Abstracts, EGU General Assembly, Vienna, Austria, 7–12 April 2013, 15, 2013.

Ciais, Ph., Reichstein, M., Viovy, N., Granier, A., Ogée, J., Allard, V., Aubinet, M., Buchmann, N., Bernhofer, C., Carrara, A., Chevallier, F., De Noblet, N., Friend, A. D., Friedlingstein, P., Grünwald, T., Heinesch, B., Keronen, P., Knohl, A., Krinner, G., Loustau, D., Manca, G., Matteucci, G., Miglietta, F., Ourcival, J. M., Papale, D., Pilegaard, K., Rambal, S., Seufert, G., Soussana, J. F., Sanz, M. J., Schulze, E. D., Vesala, T., and Valentini, R.: Europe-wide reduction in primary productivity caused by the heat and drought in 2003, Nature, 437, 529–533, 2005.

Dai, A., Trenberth, K. E., and Karl, T. R.: Global variations in droughts and wet spells: 1900–1995, Geophys. Res. Lett., 25, 3367–3370, doi:10.1029/98GL52511, 1998.

Dai, A.: Characteristics and trends in various forms of the Palmer Drought Severity Index (PDSI) during 1900–2008, J. Geophys. Res., 116, D12115, doi:10.1029/2010JD015541, 2011.

Dai, A., Fung, I. Y., Del Genio, A. D., and Qian, T.: A global dataset of Palmer Drought Severity Index for 1870–2002: Relationship with soil moisture and effects of surface warming, J. Hydrometeorol., 5, 1117–1130, 2004.

Gregorič, G. and Sušnik, A.,: Drought conditions and management strategies in Romania, Country Report, 1st Regional Workshop on Capacity Development to Support National Drought Management Policies for Eastern European Countries, Initiative on "Capacity Development to support National Drought Management Policy"(WMO, UNCCD, FAO and UNW-DPC), Bucharest, Romania, 9–11 July 2013.

Flocas, A. A.: Frontal depressions over the Mediterranean Sea and central southern Europe, Meìditerraneìe, 4, 43–52, 1998.

Guttman, N. B.: Comparing the Palmer Drought Index and the Standardized Precipitation Index, J. Am. Water Resour. As., 34, 113–121, 1998.

Guttman, N. B., Wallis, J. R., and Hosking, J. R. M.: Spatial comparability of the Palmer drought severity index, Water Resour. Bull., 28, 1111–1119, 1992.

Hayes, M. J., Svoboda, M. D., Wilhite, D. A., Vanyarkho, O. V.: Monitoring the 1996 drought using the standardized precipitation index, B. Am. Meteorol. Soc., 80, 429–438, 1999.

Heinrich, G. and Gobiet, A.: The future of dry and wet spells in Europe: a comprehensive study based on the ENSEMBLES regional climate models, Int. J. Climatol., 32, 1951–1970. doi:10.1002/joc.2421, 2012.

Holjevac, M. C., Pavlovic, D., and Pandzic, K.: Drought conditions and management strategies in Croatia, Country Report, 1st Regional Workshop on Capacity Development to Support National Drought Management Policies for Eastern European Countries, Initiative on "Capacity Development to support National Drought Management Policy"(WMO, UNCCD, FAO and UNW-DPC), Bucharest, Romania, 9–11 July 2013.

IPCC: Contribution of Working Group I to the Fourth Assessment Report of the Intergovernmental Panel on Climate Change, Summary for Policymakers Climate Change 2007: The Phisical Science Basis, available at: www.ipcc.ch (last access: 19 December 2014), 2007.

ISDR, International Strategy for Disaster Reduction: Drought Risk Reduction Framework and Practices: Contributing to the Implementation of the Hyogo Framework for Action, United Nations Secretariat of the International Strategy for Disaster Reduction (UN/ISDR), Geneva, Switzerland, 197 pp., 2007.

Karl, T. R.: Some Spatial Characteristics of Drought Duration in the United States, J. Clim. Appl. Meteor., 22, 1356–1366, 1983.

Karl, T. R.: The Sensitivity of the Palmer Drought Severity Index and Palmer's Z-Index to their Calibration Coefficients Including

Potential Evapotranspiration, J. Climate Appl. Meteor., 25, 77–86, 1986a.

Karl, T. R.: The relationship of soil moisture parameterizations to subsequent seasonal and monthly mean temperature in the United States, Mon. Weather Rev., 114, 675–686, 1986b.

Karl, T. R. and Knight, R. W.: Atlas of monthly Palmer hydrological drought indices (1931–1983) for the contiguous United States, Historical Climatology Series 3–7, Asheville: National Oceanic and Atmospheric Administration, National Climatic Data Center, 217 pp., 1985.

Karl, T. R., Knight, R. W., Ezell, D. S., and Quinlan, F. T.: Probabilities and precipitation required to end/ameliorate droughts, Historical Climatology Series 3–16, Asheville: National Oceanic and Atmospheric Administration, National Climatic Data Center, 315 pp., 1986.

Karl, T., Quinlan, F., and Ezel, D. S.: Drought termination and amelioration: its climatological probability, J. Clim. Appl. Meteor., 26, 1198–1209, 1987.

Katsoulis, B. D., Makrogiannis, T. D., and Goutsidou, Y. A.: Monthly anticyclonicity in southern Europe and the Mediterranean region, Theor. Appl. Climatol., 59, 51–59, 1998.

Linares, J. C. and Camarero, J. J.: From pattern to process: linking intrinsic water-use efficiency to drought-induced forest decline, Glob. Change Biol., 18, 1000–1015, 2011.

Lloyd-Hughes, B. and Saunders, M. A.: A drought climatology for Europe, Int. J. Climatol., 22, 1571–1592, 2002.

Maheras, P., Flocas, H. A., Patrikas, I., and Anagnostopoulou, C.: A 40 year objective analysis of surface cyclones in the Mediterranean region: Spatial and temporal distribution, Int. J. Climatol., 21, 359–367, 2001.

Martínez-Villalta, J., López, B. C., Adell, N., Badiella, L., and Ninyerola M.: Twentieth century increase of Scots pine radial growth in NE Spain shows strong climate interactions, Glob. Change Biol., 14, 2868–2881, 2008.

Mateescu, E., Smarandache, M., Jeler, N., and Apostol, V.: Drought conditions and management strategies in Romania, Country Report, 1st Regional Workshop on Capacity Development to Support National Drought Management Policies for Eastern European Countries, Initiative on "Capacity Development to support National Drought Management Policy"(WMO, UNCCD, FAO and UNW-DPC), Bucharest, Romania, 9–11 July 2013.

McGuire, A. D., Ruess, R. W., Lloyd, A., Yarie, J., Clein, J. C., and Juday, G. P.: Vulnerability of white spruce tree growth in interior Alaska in response to climate variability: dendro-chronological, demographic, and experimental perspectives, Can. J. Forest Res., 40, 1197–1209, 2010.

McKee, T. B., Doesken, N. J., and Kleist, J.: The relationship of drought frequency and duration of time scales, 8th Conference on Applied Climatology, Am. Meteorol. Soc., Anaheim, CA, USA, 17–23 January 1993, 179–186, 1993.

Oeztuerk, A.: On the Study of a Probability Distribution for Precipitation Totals, J. Appl. Meteorol., 20, 1499–1505, 1981.

Palfai, I.: Description and forecasting of droughts in Hungary, Proc. of 14th Congress on Irrigation and Drainage (ICID), Rio de Janario, Brasil, 30 April–4 May 1990, 151–158, 1990.

Palta, J. A., Kobata, T., Turner, N. C., and Fillery, I. R.: Remobilization of carbon and nitrogen in wheat as influenced by postanthesis water deficits, Crop Sci., 34, 118–124, 1994.

Palmer, W. C.: Meteorological drought, US Department of Commerce, Weather Bureau Research Paper, 45, 58 pp., 1965.

Popova, Z., Kercheva, M., and Pereira L. S.: Validation of the FAO methodology for computing ETo with limited data. Application to south Bulgaria, Irrig. Drain., 55, 201–215, 2006.

Radinovic, D.: Mediterranean Cyclones and their Influence on the Weather and Climate, WMO Programme on Short and Medium Range Weather Prediction Research (PSMP), WMO Sofia, 24, 131 pp., 1987.

Rebetez, M., Dupont, O., and Giroud, M.: Heat and drought 2003 in Europe: A climate synthesis, Ann. Forest Sci., 63, 569–577, 2006.

Rickman, R. W., Klepper, B. L., and Peterson, C. M.: Time distribution for describing appearance of specific culms of winter wheat, Agron. J., 75, 551–556, 1983.

Sharma, M. A. and Singh, J. B.: Use of Probability Distribution in Rainfall Analysis, New York Sci. J., 23, 40–49, 2010.

Spinoni, J., Antofie, T., Barbosa, P., Bihari, Z., Lakatos, M., Szalai, S., Szentimrey, T., and Vogt, J.: An overview of drought events in the Carpathian Region in 1961–2010, Adv. Sci. Res., 10, 21–32, doi:10.5194/asr-10-21-2013, 2013.

Szalai, S., Szinell, C. S., and Zoboki, J.: Drought monitoring in Hungary, in: Early warning systems for drought preparedness and drought management, World Meteorological Organization, Lisboa, 182–199, 2000.

Szinell, C., Bussay, A., and Szentimrey, T.: Drought Tendencies in Hungary, Int. J. Climat., 18, 1479–1491, 1998.

Simane, B., Peacock, J. M., and Struik, P. C.: Differences in development and growth rate among drought-resistant and susceptible cultivars of durum wheat (Triticum turgidum L. var. durum), Plant Soil, 157, 155–166, 1993.

Thom, H. C. S.: Some Methods of Climatological Analysis, WMO Technical note 81, Secretariat of the WMO, Geneva, Switzerland, 53 pp., 1966.

Thornthwaite, C. W.: An approach toward a rational classification of climate, Geogr. Rev., 38, 55–94, 1948.

Toth, G. and Weynants, M.: Multiscale thematic soil water database, Deliverable 4.4 of MyWater project (call: FP7-SPA.2010.1.1-04; contract No: 263188), 2012.

Trnka, M., Dubrovský, M., Svoboda, M., Semerádová, D., Hayes, M., Žalud, Z., and Wilhite, D.: Developing a regional drought climatology for the Czech Republic, Int. J. Climatol., 29, 863–883, doi:10.1002/joc.1745, 2009.

UNEP/DEWA: Carpathian Environment Outlook (KEO) Report 2007, Genève, 2007.

van Bebber, W.: Die Zugstrassen der barometrischen Minima, Meteorol. Z., 8, 361–366, 1891 (in German).

van der Schrier, G., Briffa, K. R., Jones, P. D., and Osborn, T. J.: Summer moisture variability across Europe, J. Clim., 19, 2818–2834, doi:10.1175/JCLI3734.1, 2006.

van der Schrier, G., Efthymiadis, D., Briffa, K. R., and Jones, P. D.: European alpine moisture variability 1800–2003, Int. J. Climatol., 27, 415–427, 2007.

Vicente-Serrano, S. M., Beguería, S., López-Moreno, J. I., Angulo, M., and El Kenawy, A.: A new global 0.5° gridded dataset (1901–2006) of a multiscalar drought index: Comparison with current drought index datasets based on the Palmer Drought Severity Index, J. Hydrometeorol., 11, 1033–1043, doi:10.1175/2010JHM1224.1, 2010.

Weber, L. and Nkemdirim, L.: Palmer's drought indices revisited, Geogr. Ann. A, 80, 153–172, 1998.

Wells N., Goddard, S., and Hayes, M. J.: A self-calibrating Palmer Drought Severity Index, J. Climate, 17, 2335–2351, 2004.

Wilks, D. S.: On the Combination of Forecast Probabilities for Consecutive Precipitation Periods, Weather Forecast., 5, 640–650, 1990.

Wilks, D. S.: Forecast verification, Statistical Methods in the Atmospheric Sciences, New York, Academic Press, p. 467, 1995.

Wilhite, D. A.: Drought as a natural hazard: Concepts and definitions, in: Drought: A Global Assessment, edited by: Wilhite, D. A., Routledge, New York, 1, 1–18, 2000.

Wilhite, D. A. and Buchanan, M.: Drought as hazard: Understanding the natural and social context, in: Drought and Water Crisis: Science, Technology, and Management Issues, edited by: Wilhite, D. A., CRC Press (Taylor and Francis), New York, 3–29, 2005.

Williams A. P., Xu, Ch., and McDowell, N. G.: Who is the new sheriff in town regulating boreal forest growth?, Environ. Res. Lett., 6, 041004, doi:10.1088/1748-9326/6/4/041004, 2011.

Wösten, J. H. M., Lilly, A., Nemes, A. and Le Bas, C.: Development and use of a database of hydraulic properties of European soils, Geoderma, 90, 169–185, 1999.

WMO, World Meteorological Organization: Drought monitoring and early warning: Concepts, progress and future challenges, WMO-No. 1006, 2006.

WMO/UNCCD/FAO/UNW-DPC: Capacity Development to Support National Drought Management Policies initiative, available at: http://www.ais.unwater.org/ais/course/view.php?id=37, last access: 29 November 2013.

4

High-resolution global topographic index values for use in large-scale hydrological modelling

T. R. Marthews[1], **S. J. Dadson**[1], **B. Lehner**[2], **S. Abele**[1], **and N. Gedney**[3]

[1]School of Geography and the Environment, University of Oxford, Oxford, OX1 3QY, UK
[2]Department of Geography, McGill University, Montreal, H3A 0B9, Quebec, Canada
[3]Met. Office, Hadley Centre for Climate Prediction and Research, (JCHMR), Wallingford, OX10 8BB, UK

Correspondence to: T. R. Marthews (toby.marthews@ouce.ox.ac.uk)

Abstract. Modelling land surface water flow is of critical importance for simulating land surface fluxes, predicting runoff and water table dynamics and for many other applications of Land Surface Models. Many approaches are based on the popular hydrology model TOPMODEL (TOPography-based hydrological MODEL), and the most important parameter of this model is the well-known *topographic index*. Here we present new, high-resolution parameter maps of the topographic index for all ice free land pixels calculated from hydrologically conditioned HydroSHEDS (Hydrological data and maps based on SHuttle Elevation Derivatives at multiple Scales) data using the GA2 algorithm (GRIDATB 2). At 15 arcsec resolution, these layers are 4 times finer than the resolution of the previously best-available topographic index layers, the compound topographic index of HYDRO1k (CTI). For the largest river catchments occurring on each continent we found that, in comparison with CTI our revised values were up to 20 % lower in, e.g. the Amazon. We found the highest catchment means were for the Murray–Darling and Nelson–Saskatchewan rather than for the Amazon and St. Lawrence as found from the CTI. For the majority of large catchments, however, the spread of our new GA2 index values is very similar to those of CTI, yet with more spatial variability apparent at fine scale. We believe these new index layers represent greatly improved global-scale topographic index values and hope that they will be widely used in land surface modelling applications in the future.

1 Introduction

Land surface models (LSMs) are widely used for predicting the effects of global climate change on vegetation development, runoff and inundation, evapotranspiration rates and land surface temperature (Gerten et al., 2004; Prentice et al., 2007; Clark and Gedney, 2008; Dadson and Bell, 2010; Dadson et al., 2010, 2011; Wainwright and Mulligan, 2013; IPCC, 2013). However, the simulation of hydrological dynamics within LSMs remains relatively simplified because these models are usually run at coarse spatial resolution (up to 300 km grid boxes) and the physics they follow is based predominantly on approximations of processes that occur at much finer spatial scales (Ducharne, 2009; Wainwright and Mulligan, 2013). Correctly characterising hydrology is very important because landscape-scale water movements ($\sim$ 10–100 km scale) and changes in the water cycle control many effects ranging from local energy and carbon fluxes to land–atmosphere feedbacks to the climate system to potentially catastrophic changes in vegetation distributions.

When coupled with atmospheric models, most LSMs can simulate a wide variety of natural and human-modified processes from soil moisture feedbacks on precipitation (Seneviratne et al., 2006, 2010; Coe et al., 2009) and river flow (Gedney et al., 2004, 2006; Clark and Gedney, 2008; Milly et al., 2008; Falloon and Betts, 2010; Sanderson et al., 2012) through to vegetation development and carbon productivity (Prentice et al., 2007; Marthews et al., 2012; IPCC, 2013). Although usually applied at landscape-scale resolutions (grid-cell sizes of 10–100 km; e.g. Harding and Warnaars, 2011), LSMs are increasingly finding applicability at finer resolutions approaching 1–10 km, at which the physics they encapsulate begins to approach the more detailed scales (0.1–1.0 km) typically required in process-based hydrological models or used in catchment-based water resources assessments (e.g. Ke et al., 2012; Choi, 2013; cf. discussions in Wood et al., 2011, 2012; Beven and Cloke, 2012). A growing body of work has lately emerged using LSMs to produce large-area projections of current and future water resources for use in applications related to climate change impact assessments (Gedney and Cox, 2003; Gerten et al., 2004; Falloon and Betts, 2010; Wood et al., 2012; Zulkafli et al., 2013; Harding et al., 2013).

Land Surface Models require a representation of surface and subsurface runoff. Models of runoff production used in regional and continental applications typically contain parameterised physics based on statistical representations of processes known to operate at finer scales (Ward and Robinson, 2000; Clark and Gedney, 2008), which can lead to inaccurate predictions in data-sparse regions and generally high uncertainty. The large quantity of detailed topographic information now widely available at sublandscape-scale resolutions offers an opportunity to improve the fidelity of large-area simulations of the hydrological cycle, for the benefit of both climate and hydrological models (Dharssi et al., 2009; Wainwright and Mulligan, 2013).

Currently, the most common approach to inundation prediction is to use a runoff production scheme such as TOPMODEL (TOPography-based hydrological MODEL), which partitions runoff from the soil column into surface and subsurface components (Beven and Kirkby, 1979; Quinn et al., 1991, 1995; Beven, 1997, 2012; e.g. MacKellar et al., 2013). One of the most important configurational parameters for TOPMODEL is the well-known *topographic index* (defined in Appendix A), which is widely used in hydrology and terrain-related applications (Ward and Robinson, 2000; Wilson and Gallant, 2000).

The HYDRO1k global values for the compound topographic index (CTI) were released by USGS in 2000 (USGS, 2000) and they have since become the most commonly used global ancillary files for topographic index values. HYDRO1k was a great step forward in the development of global hydrological modelling applications: it allowed spatially explicit hydrological routines to be incorporated into LSMs for the first time and large-scale applications of the TOPMODEL hydrological model to become standard (Beven, 2012). The recent availability of topographic maps at even higher spatial resolution with globally consistent coverage builds on this and means that further improvements can now be made.

The limitations of HYDRO1k CTI values become most apparent when considering wetland ares. Wetlands are critical nodes in the Earth system where land–atmosphere fluxes are strongly dependent on seasonal and interannual hydrological variability (Coe, 1998; Baker et al., 2009; O'Connor et al., 2010; Dadson et al., 2010). In wetlands, the availability of water introduces important feedbacks on climate via surface fluxes of energy and water and these areas form a key link between the hydrological and carbon cycles (Ward and Robinson, 2000; Gedney et al., 2004; Seneviratne et al., 2006, 2010; Coe et al., 2009; Dadson et al., 2010). Some analyses based on CTI values have consistently overestimated the extent and duration of tropical wetlands of various types. Notably, simulations using the Earth system model HadGEM2, which is parameterised using CTI (Collins et al., 2011), predict much larger and more persistent Amazonian wetlands than actually exist according to current surveys (e.g. Lehner and Döll, 2004; Prigent et al., 2007; Junk et al., 2011), which may at least partly be caused by the limited resolution and quality of the HYDRO1k CTI.

In the context of LSMs, the need for high-resolution topographical data across wide spatial domains has recently been highlighted (Lehner et al., 2008; Wood et al., 2011; Lehner and Grill, 2013). With the advent of satellite-based global mapping, notably the Shuttle Radar Topography Mission (SRTM), there has been a significant improvement in the availability of high-resolution data with continental coverage, such as in the high-resolution global HydroSHEDS (Hydrological data and maps based on SHuttle Elevation Derivatives at multiple Scales) data layers (Lehner et al., 2008), but such data have generally not yet been utilised to support large-scale hydrological modelling studies (Wood et al., 2011, 2012).

In this study, we respond to the need for higher-resolution data for use in LSMs. We have three main aims: (1) to calculate the topographic index using the GA2 algorithm based on high-resolution global HydroSHEDS data; (2) to compare our values to the current standard for values of this index (the CTI of HYDRO1k); and (3) to discuss current developments in large-scale hydrological modelling and how models can benefit from higher-resolution parameter maps such as these.

2 Methods

2.1 Topographic index

The topographic index is a parameter of the TOPMODEL hydrological model (Beven and Kirkby, 1979; Quinn et al., 1991, 1995; Beven, 1997, 2012). The algorithm required for calculating this index is relatively simple (Appendix A), but it has not previously been applied to generate a global data layer at very high spatial resolution because (1) the index must be calculated from harmonised topographic information, which only became available in the 2000s and (2) LSMs have only recently become sophisticated enough to make use of such a high-quality layer (Prentice et al., 2007).

The HydroSHEDS "hydrologically conditioned" layers

Grid-based topographic index calculations require a digital elevation model (DEM) and we have used the HydroSHEDS DEM (Lehner et al., 2008; http://www.hydrosheds.org/). The

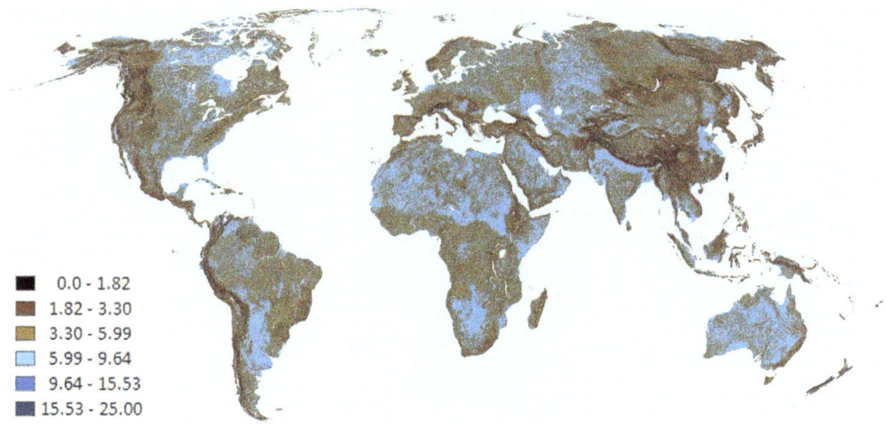

Figure 1. Global topographic index values based on GA2 applied to HydroSHEDS base data (Appendix A). Blue shades indicate pixels with index values above the global mean (5.99) and brown shades indicate below-average values.

HydroSHEDS data layers were derived from raw SRTM data at 3 arcsec pixel resolution (approximately 90 m at the Equator) through the application of hydrological conditioning in a sequence of correction steps (Lehner, 2013) which resulted in a globally consistent suite of grid layers which were subsequently upscaled to a resolution of 15 arcsec (approximately 450 m at the Equator). We acquired the HydroSHEDS DEM and also a layer of pre-calculated contributing upstream catchment areas for each 15 arcsec pixel (UPLAND, in m^2; Lehner, 2013). As of April 2014, HydroSHEDS data have only been produced at the highest quality for all land areas south of 60° N (the limit of SRTM); so for areas at higher latitude we substituted the HYDRO1k DEM to provide seamless global grids (more specifically, the GTOPO30 DEM underlying HYDRO1k disaggregated to 15 arcsec resolution by tiling the larger pixels and applying a 3 × 3 kernel average filter to smooth the resulting surface).

2.2 Generating the ancillary files

Our calculations had to be carried out over domains composed of complete watersheds, so we mosaicked both the DEM and UPLAND tiles into a global data layer using ArcGIS 10.1 (Esri Inc., Redlands, California). These two input layers were then converted to NetCDF format using GDAL (Geospatial Data Abstraction Library; OSGF, 2011).

Topographic index values were calculated using the GA2 algorithm, which is the widely used GRIDATB algorithm with some modifications for use with HydroSHEDS data (see Appendix A for details). Resulting index values for the global land surface were then filtered to remove areas for which topographic index values are invalid or meaningless, including lakes and reservoirs (masked out using the Global Lakes and Wetlands Database; Lehner and Döll, 2004), mountain glaciers and ice caps (using the Randolph Glacier Inventory; Pfeffer et al., 2014) and the Greenland ice sheet (using Lewis, 2009).

Because of the large layer size (1.2×10^9 land pixels), GA2 was run on the ARCUS server for all continental-scale calculations, a 1344-core computer cluster at the Oxford e-Research Centre (OeRC). Zonal histograms were plotted using ArcGIS 10.1 and subsequent statistics calculated using R (R Development Core Team, 2013).

3 Results

We produced a layer of topographic index (TI) values following the GA2 algorithm for all ice-free land pixels worldwide (Fig. 1). TI values calculated this way are not just relative measures but consistent and comparable between catchments (Appendix A), so we may compare global values: as expected, TI values are low at ridge-tops (minimal catchment area) and high in valleys (along drainage paths and in zones of water concentration in the landscape; Wilson and Gallant, 2000), yielding a global range of 0.00–25.00 and average of 5.99 (Fig. 2).

Wetter areas of the globe generated generally higher TI values (Fig. 1), although there are many exceptions to this (e.g. in desert areas where high TI values do not correlate with high flow accumulation, at least in the present climate). Zonal statistics calculated for the various lake and wetland types of the world (as defined by the Global Lakes and Wetlands Database, see Table 1) show that pixels representing rivers had the highest TI values (global mean 8.81 over $0.42 \times 10^6\,km^2$), but also the highest variance with some river pixels scoring below the global mean for ice-free land outside lakes, reservoirs, wetlands and wetland complexes (global mean 5.88 over $128.99 \times 10^6\,km^2$). In terms of TI, wetland complexes in Asia (mostly occurring in India and Tibetan China; Table 1) and mires (mostly occurring in boreal Canada and the Russian Federation) were indistinguishable from dry land (Fig. 3), indicating that wetlands in both

Table 1. Topographic index values from the GA2 algorithm applied to HydroSHEDS data (Appendix A) compared to CTI values from HYDRO1k for all global river basins larger than 10^6 km^2 (SD: standard deviation). Note that some sources quote much higher index values, but these are often not scale-corrected values and are therefore not directly comparable (e.g. Yang et al., 2007).

River catchments	Area[a] (million km^2)	CTI of HYDRO1k				GA2 based on HydroSHEDS				Percentage increase moving from CTI to GA2
		Mean	SD	Min	Max	Mean	SD	Min	Max	
Africa										
Congo	3.70 (0.04)	7.0	2.2	0.9	23.2	6.0	2.4	1.0	24.5	−14.5
Nile	3.40 (0.07)	6.7	2.4	0.6	23.3	6.7	2.5	0.5	24.2	+0.7
Niger	2.12 (0.00)	7.3	2.2	0.9	22.8	6.7	2.4	0.8	24.0	−7.6
Zambezi	1.39 (0.03)	6.3	2.2	0.5	21.3	6.4	2.4	0.6	23.4	+1.1
Asia										
Ob-Irtysh	2.97 (0.02)	6.6	2.1	0.6	22.2	6.7	2.4	0.1	24.6	+1.5
Yenisei	2.58 (0.07)	4.9	2.1	0.6	23.0	5.2	2.5	0.0	24.2	+5.5
Lena	2.50 (0.00)	4.9	2.0	0.7	22.6	5.2	2.5	0.0	24.3	+5.9
Amur/Heilong Jiang	1.86 (0.02)	5.2	2.1	0.8	21.8	5.5	2.7	0.7	24.2	+4.4
Yangtze/Chang Jiang	1.81 (0.01)	4.6	2.3	0.0	21.6	4.6	2.8	0.0	23.9	−0.2
Indus	1.17 (0.03)	5.6	2.9	0.0	21.6	5.7	3.0	0.0	23.0	+2.9
Ganges	1.08 (0.01)	6.4	2.6	0.1	21.8	6.4	2.8	0.0	23.2	+0.8
Mekong	0.80 (0.00)	5.3	2.5	0.5	22.0	5.4	2.9	0.5	23.0	+1.9
Yellow/Huang He	0.75 (0.00)	5.3	2.3	0.7	21.7	5.4	2.7	0.5	23.0	+3.2
Australasia										
Murray–Darling	1.06 (0.00)	–[b]	–[b]	–[b]	–[b]	6.9	2.4	1.0	22.7	–[b]
Europe										
Volga	1.38 (0.03)	6.4	2.1	1.3	22.6	6.2	2.3	1.1	23.7	−2.2
Danube	0.82 (0.00)	5.4	2.5	0.5	21.4	5.4	2.8	0.2	23.2	+0.0
Rhine	0.17 (0.00)	5.5	2.3	0.5	19.8	5.4	2.6	0.1	21.6	−2.9
North America										
Mississippi–Missouri	2.98 (0.02)	6.2	2.0	0.8	22.6	6.2	2.5	0.5	24.5	−0.8
Mackenzie	1.81 (0.17)	6.1	2.5	0.6	22.6	6.1	2.6	0.0	24.2	+0.3
St. Lawrence	1.34 (0.30)	7.3	2.7	1.3	21.5	6.1	2.4	0.9	23.5	−16.8
Nelson–Saskatchewan	0.89 (0.09)	7.2	2.1	0.7	21.4	6.8	2.3	0.1	23.5	−5.6
South America										
Amazon	7.05 (0.01)	7.7	2.4	0.0	24.0	6.1	2.5	0.4	25.0	−20.3
Paraná (excl. Río de la Plata)	2.58 (0.02)	7.1	2.3	0.6	23.3	6.4	2.6	0.5	24.3	−9.7
Orinoco	0.88 (0.00)	6.7	2.3	0.3	22.2	6.3	2.7	0.4	23.2	−6.8

[a] Area of lakes, reservoirs, glaciers and ice sheets within the basin given in parentheses (the topographic index is not evaluated at these pixels by GA2, whereas the HYDRO1k CTI calculation assigns values to lakes as if they are flat plains, Appendix A). [b] HYDRO1k did not include mainland Australia therefore no CTI values are available for the Murray–Darling (USGS, 2000).

these areas are maintained by factors other than topography (e.g. rainfall and evapotranspiration).

In comparison to HYDRO1k, the new TI values from GA2 based on HydroSHEDS were higher for river pixels and slightly higher for intermittent wetlands and lakes. TI values were lower at pixels in tropical swamp forests and inundated forests and also slightly lower in coastal wetlands.

The new TI values from GA2/HydroSHEDS were in line with HYDRO1k values for CTI (USGS, 2000) at most global pixels, but in certain areas there were significant divergences. Considering all river catchments larger than 10^6 km^2 in particular (Table 2), values from GA2 were lower for many basins, most notably the Amazon, Congo, Paraná, Niger and St. Lawrence rivers; in the case of the Amazon as

Table 2. Topographic index values from the GA2 algorithm applied to HydroSHEDS data (Appendix A). For a map of the extent of these wetland types (see Lehner and Döll, 2004).

Wetland type[a]	Area[b] (million km^2)	Topographic index (dimensionless) Mean value	SD	Min	Max
Ice-free land outside wetlands and wetland complexes	128.99	5.88	2.56	0.00	24.69
Intermittent wetlands/lakes (mostly in drylands)	0.66	8.07	2.89	0.59	24.03
Pans, brackish/saline wetlands (mostly temperate and subtropical)	0.40	7.91	2.59	0.82	23.09
Freshwater marsh, floodplains	2.72	7.38	2.45	0.56	24.89
Mires (e.g. bogs, fens) (mostly boreal)	1.23	5.97	2.56	0.00	24.06
Swamp forests, inundated forests (mostly S. America and Congo)	0.94	6.92	2.48	0.86	25.00
Coastal wetlands (e.g. mangroves, estuaries, deltas, lagoons)	0.45	7.03	2.22	0.58	24.54
River pixels	0.42	8.81	4.80	0.16	25.00
Wetland complex (0–25 % wetland) (Asia only, mostly India and Tibetan China)	0.83	5.61	2.52	0.30	22.28
25–50 % wetland (USA and Canada only)	4.01	6.47	2.30	0.09	24.33
50–100 % wetland (USA and Canada only)	2.76	6.84	2.38	0.00	24.45

[a] Following the Global Lakes and Wetlands Database (GLWD; Lehner and Döll, 2004). [b] These areas sum to 143.43 × 10^6 km^2 which is the global extent of land not covered by lakes, reservoirs, glaciers or ice sheets that lies outside Antarctica and other islands excluded from HydroSHEDS (viz. Antarctica, Polynesia east of the 180° meridian line, the Azores, St Helena, Ascension Is., Tristan da Cunha, South Georgia, the South Sandwich Is., the Kerguelen Archipelago and some smaller oceanic islands; Lehner et al., 2008). Permanent lakes and reservoirs cover 1.23 × 10^6 km^2 globally (Lehner and Döll, 2004), the Greenland ice sheet covers 1.99 × 10^6 km^2 (Lewis, 2009) and all glaciers cover 0.80 × 10^6 km^2 (Pfeffer et al., 2014).

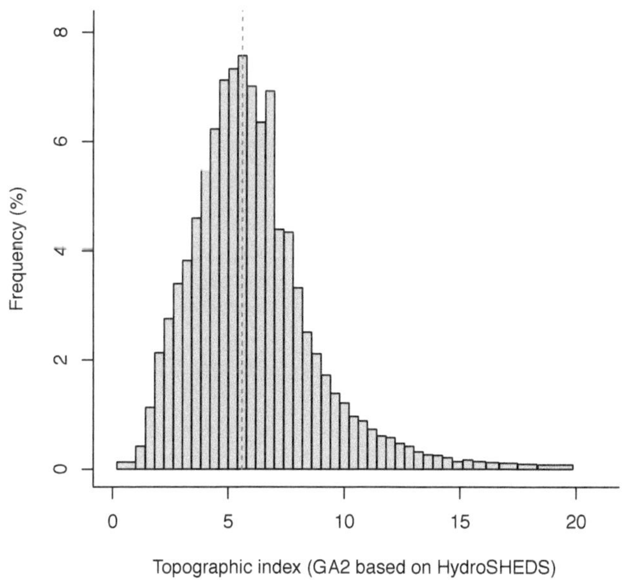

Figure 2. Histogram of global topographic index values (vertical line shows global mean of 5.99; global maximum is 25.0044 at a pixel within a river island at the confluence of the Amazon and Xingú rivers in Brazil).

much as 20 % lower than the CTI values (Fig. 4). According to our calculations, the catchments with the highest spatially averaged TI values were the Murray–Darling, Nelson–Saskatchewan, Nile and Niger (compared to the order Amazon, St. Lawrence, Niger and Nelson under the CTI calculations, although n.b. HYDRO1k's CTI included no estimates for the Murray–Darling; Table 2). Although it might be expected that the size of the Amazon floodplain would be enough to ensure it scored the highest TI, note that (i) there is no globally consistent correlation between wetland area and TI (Fig. 3) and (ii) because these are spatial averages, the density of wetland within each catchment is more important than the absolute wetland size (and the Nelson–Saskatchewan, for example, is known for a high density of wetland terrain).

As expected, our new index values reflect the same pattern of below-average values in mountainous areas and above-average values in lowland areas as seen in HYDRO1k, however more variability is visible in the histograms for GA2 because the higher resolution means that more of the smaller river valleys within the mountain ranges become visible (leading to an increase in the mean and spread of index values e.g. in the Mackenzie Mountains; Fig. 5c). Also visible on the zoomed comparison maps of the Rocky Mountains (Fig. 6) is an example of differing qualities of HYDRO1k vs. HydroSHEDS data: on the eastern half of the maps, the CTI version shows a series of blue, lake-shaped objects with topographic indices in the range of 10 (also visible as a small rise in the corresponding histogram; Fig. 5a), while the GA2 version does not show these features. These objects represent valleys that are drained, in reality, through narrow gorges or river channels. The higher-resolution data of HydroSHEDS (and possibly manual corrections) are capable of resolving this issue correctly. Yet due to the coarser resolution of HYDRO1k, the valleys would appear as closed depressions in the DEM; the standard GIS solution to enforce continued drainage in such cases is to lift the topography until overflow

Figure 3. Comparison of topographic index calculations, divided by wetland type (following Lehner and Döll, 2004, excluding lakes and reservoirs); TI (dark shaded box): calculations of topographic index from this study (also shown as a horizontal solid line; precise figures given in Table 1); and H1k (light shaded box): the compound topographic index of HYDRO1k (USGS, 2000), both of which applied the scale correction of Ducharne (2009). Boxes show mean ± SD index values for the global distribution of that wetland type. For reference, the mean topographic index value for ice-free land outside wetlands is shown by a broken line on all panels (Table 1).

occurs (using a sink-filling algorithm); the resulting (erroneous) flat topography then leads to overestimated CTI values.

Index values at 15 arcsec resolution are available at http://doi.org/10/t7d in NetCDF format (a version in GeoTIFF format – translated using GDAL (OSGF, 2011) – is available on request).

4 Discussion

Modelling soil water flow and runoff generation is of critical importance for simulating land surface fluxes, predicting water table dynamics, wetland inundation and river routing and, at a regional scale, quantifying surface evaporation rates and the growth, transpiration and seasonality of vegetation (Ward and Robinson, 2000; Baker et al., 2009; Dadson et al., 2010; Marthews et al., 2014). Landscape-scale hydrological processes are therefore key elements in modelling land surface–atmosphere exchange processes and critical to the successful use of coupled LSMs to predict the effects of climate change at larger scales.

The hydrological routines of LSMs have undergone steady improvement in recent years (Wood et al., 2011; Zulkafli et al., 2013; Wainwright and Mulligan, 2013). However, these landscape-scale processes remain generally less well-modelled than processes operating at the finer local scale (e.g. photosynthesis models) or larger continental scale (e.g. general circulation models). Arguably, the development of landscape-scale processes has been relatively slow not just because of a lack of complete understanding of the processes involved, but also, more simply, by the limited availability of high-resolution parameter maps for the models concerned (Wood et al., 2011; Wainwright and Mulligan, 2013; Marthews et al., 2014). Because LSMs are now being applied at increasingly high spatial resolution in order to analyse the distribution and movement of water resources, model development is gaining momentum. Large-scale gridded simulations based on high-resolution drivers are now becoming routine, and this has led to an increasingly recognised need for the high-resolution data sets required to drive those simulations (e.g. Wood et al., 2011, 2012; Beven and Cloke, 2012; Castanho et al., 2013).

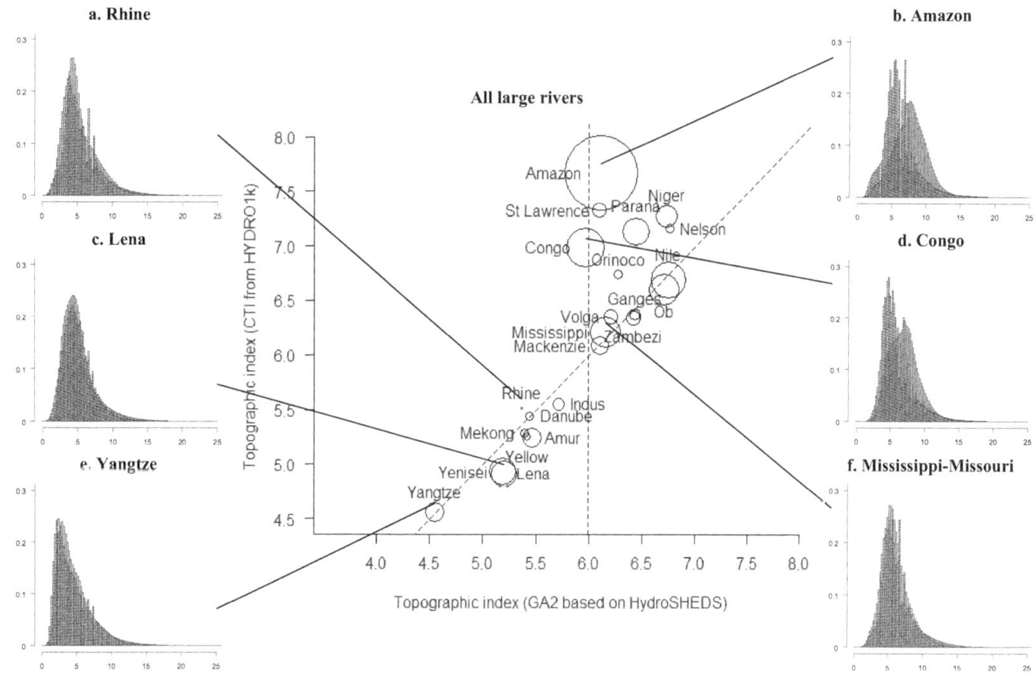

Figure 4. Comparison of the CTI and GA2 calculations of the topographic index (from Table 2), showing that CTI values are larger for some catchments, most notably the Amazon, Congo, Paraná, Niger and St. Lawrence. Circle areas are proportional to catchment area and a one–one line is shown for reference. The largest catchments tend to be closest to the global average index value of 5.99 (also shown for reference). Histograms are shown for six catchments: the Rhine, Amazon, Lena, Congo, Yangtze and Mississippi–Missouri (each grey histogram shows CTI values, hatched histogram shows GA2; axes on all histograms are omitted: all are topographic index – horizontal – and fraction of pixels – vertical): for catchments close to the one–one line, the corresponding histograms were closely similar.

4.1 High-resolution hydrological modelling

TOPMODEL was originally applied at the scale of small catchments, using pixels smaller than 50 m × 50 m in extent (Quinn et al., 1991, 1995; Ward and Robinson, 2000; Beven, 2012), with the index values understood to have relative significance only (i.e. similar values calculated in different catchments do not necessarily imply hydrological similarity; see Chappell et al., 2006). There have been many developments from this basic framework over the years (e.g. see Wolock, 1993; Wilson and Gallant, 2000; Hjerdt et al., 2004; Beven, 2012) and this study has likewise taken a novel approach. Notably, we have applied our calculations at continental scales with larger pixels (approximately 450 m × 450 m at the Equator), using the resolution correction of Ducharne (2009; also see Moore et al., 1993; Wolock and McCabe, 1995; Clark and Gedney, 2008). Additionally, because our calculations are carried out over complete continental land masses, the index values derived may be considered to be consistent and comparable between catchments.

Although we accept the arguments of Beven and Cloke (2012) that moving to higher-resolution data sets is not the only line of development that should be followed, ultimately we support the ideas of Wood et al. (2011, 2012) that increasing the resolution at which global hydrological simulations are carried out will have many benefits including the more realistic representation of processes currently at subgrid resolution and, ultimately, better weather and inundation prediction (Wood et al., 2011). Methane production in wetlands, for example, is critically dependent on the level of the water table (Gedney et al., 2004; O'Connor et al., 2010; Pangala et al., 2013), models of which are in turn dependent on accurate representation of the topography; therefore, higher-resolution simulations involving improved topographic index values should of necessity improve the representation of wetland fluxes of heat, water and trace gases to the atmosphere (Gedney et al., 2004) and overall estimates of methane release.

In this study we have refined the standard topographic index calculations and greatly improved their spatial resolution. We have presented our new maps of topographic index values both by wetland type (using the Global Lakes and Wetlands Database; Lehner and Döll, 2004) and also in terms of the largest river catchments occurring on each continent, finding that, in comparison to our revised values, HYDRO1k's CTI topographic index values were significantly higher in some catchments (Table 2). In most large catchments, however, the spread of our new GA2 index values was very similar to those of CTI, yet with more spatial variability apparent at fine scale (Figs. 4, 5).

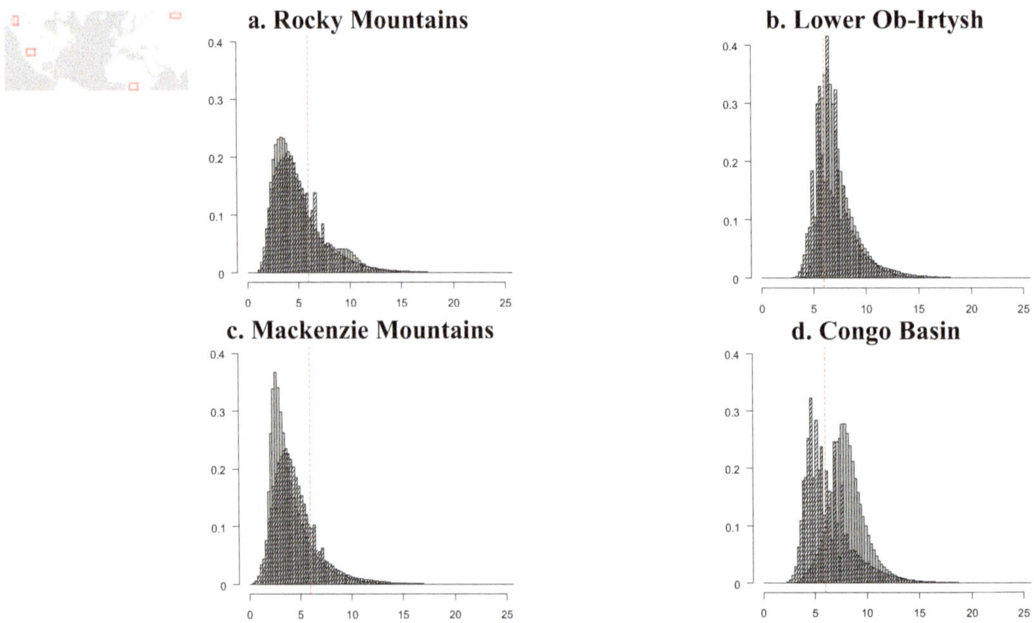

Figure 5. Comparison of the CTI and GA2 calculations of the topographic index for four example areas of 10^6 km^2 each: **(a)** an area of the Rocky Mountains (USA), **(b)** the Lower Ob-Irtysh (Russian Federation), **(c)** an area of the Mackenzie Mountains (Canada) and **(d)** the Congo Basin (Democratic Republic of the Congo, Republic of the Congo, Cameroon and the Central African Republic) (see inset). These examples were chosen so that two are mountainous, two lowland plains, two are north of 60° N and two south, to demonstrate that the new topographic index values are a refinement on the CTI values of HYDRO1k. On each histogram, grey bars show CTI values, hatched bars show GA2 and a red broken line shows the global average index value of 5.99 for reference. Axes on all histograms are omitted: all are topographic index (horizontal) and fraction of pixels (vertical).

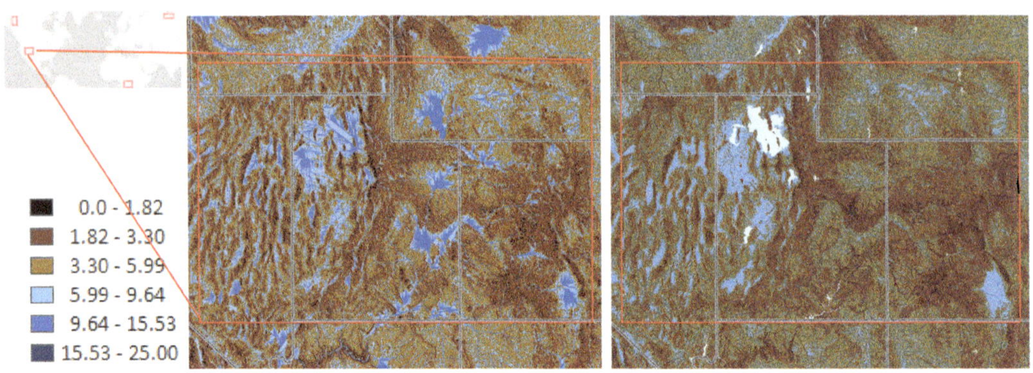

Figure 6. Comparison of the CTI and GA2 calculations of the topographic index for an area of the Rocky Mountains (USA). Maps of the CTI (left panel) and GA2 (right panel) values are shown (from which the histograms of Fig. 5a were calculated), with identical colour scale to Fig. 1. Note the 4400 km^2 Great Salt Lake, Utah, to the N of the area (which is masked out of the GA2 map (light blue) but included in CTI as if it were a flat plain) and the San Luis Valley, Colorado, to the SE, being the headwaters of the Rio Grande, USA.

4.2 Limitations of the GA2 algorithm

The topographic index is a measure of the relative propensity for soil to become saturated to the surface as a result of local topography (Beven, 2012). We have calculated it using a robust algorithm (GA2) based on the original implementation of these calculations (GRIDATB; Appendix A). Although topographic index values are comparable between different areas, it is important to remain careful when interpreting their meaning in different regions, such as arid vs. humid, or shallow vs. deep soils (i.e. when factors other than topography influence water accumulation in the landscape). In regions where saturation-excess overland flow (the component of runoff most affected by topography) is less than dominant as a runoff generation mechanism, uncertainties in inundation predictions based on TOPMODEL must be carefully calculated and predictions interpreted with care (see Beven, 2012).

A well-known limitation of topographic index values is that they are not absolute because the maximum value in any particular catchment is dependent on the catchment's area and slope profile. Therefore, we could not carry out more than an ad hoc comparison between TI and CTI (because of no independent baseline to refer to other than HYDRO1k itself). Histograms of TI and CTI values correspond closely (e.g. Fig. 5), though, and the consistency and rigour of the algorithm we have applied as well as the improved HydroSHEDS base data used for the calculation lead us to believe that our values are at least as robust as CTI at all spatial points.

A second limitation of our method is that we have used global base elevation data that is not on an equal-area projection. The HydroSHEDS data layers are projected using a geographic coordinate system (GCS, with WGS84 datum), i.e. a grid of unrotated cells that become increasingly distorted in the east–west direction as latitude increases. This implies that slopes will be underestimated in east–west directions at higher latitudes as true pixel distances are getting shorter (Appendix A). There is no appropriate method, however, to avoid uncertainty completely in the slope calculations as the underlying SRTM elevation measurements are already unequally spaced, and as there is no commonly agreed method to calculate slopes or drainage (flow) directions (see Appendix A). We assume that our calculations of the steepest gradients with average pixel distances provide a reasonable compromise to approximate the real slope of each pixel (see Appendix A).

Finally, as a related issue, we used HydroSHEDS UPLAND data which are ultimately based on the underlying D8 concept of deriving drainage directions from the steepest slopes. We acknowledge that recent advances in creating DEM-based drainage networks (e.g. D8-LTD or other options such as FD8 or MD8; Orlandini et al., 2014) provide avenues to alter and potentially improve the drainage-direction calculations and, in consequence, the topographic index values, but testing for the individual effects is beyond the scope of this project due to the multiscale complexity involved (see Appendix B for further explanations). We believe, however, that while these methods may have a significant effect on local drainage directions and channel routing, the cumulative calculation of "contributing upstream area" is less affected.

5 Conclusions

LSMs have now been applied over many years to the problem of explaining and predicting global climate change (Prentice et al., 2007; IPCC, 2013). Recent developments in land surface modelling and Earth observation have attempted to incorporate better hydrological understanding into these applications, with a particular focus on a better characterisation of the physical processes that control the water cycle (Coe, 1998; Gedney and Cox, 2003; Coe et al., 2009; Dadson and Bell, 2010; Dadson et al., 2010, 2011; Zulkafli et al., 2013). This study offers a new high-resolution, spatially consistent data layer of topographic index values for all ice-free land pixels worldwide based on the hydrologically conditioned HydroSHEDS data (Lehner et al., 2008). These data layers are at 4 times the resolution of the HYDRO1k compound topographic index layers (USGS, 2000) and we believe represent the most accurate global-scale calculation of topographic index values that exists to date.

Appendix A: Calculating the topographic index

The *topographic index* is a fundamental parameter of TOPMODEL (Kirkby, 1975; Beven and Kirkby, 1979; Quinn et al., 1991, 1995; Beven, 1997, 2012), alternatively known as the *topographic wetness index* (TWI; e.g. Wilson and Gallant, 2000) or the *compound topographic index* (CTI; e.g. USGS 2000, Evans 2003). The topographic index is essentially a means of grouping runoff-producing elements in the landscape (Kirkby, 1975; Beven and Kirkby, 1979). Different landscape pixels that have similar topographic index values should be observed to have similar hydrological dynamics (Wolock, 1993; Quinn et al., 1995), allowing for a great simplification of hydrology calculations (Beven, 1997, 2012).

The topographic index is a measure of the relative propensity for the soil to, at a point, become saturated to the surface, given the area that drains into it A and its local outflow slope β (Beven, 2012; increasing A will tend to increase the accumulation of water, but increasing β will tend to reduce it by increasing gravitational outflow; Quinn et al., 1991). The index is often calculated using an algorithm called GRIDATB, originally written in 1983 by K. Beven of the Hydrology Group, University of Lancaster (revised for distribution 1993–1995 by P. Quinn and J. Freer and described in Quinn et al., 1991, 1995; for alternative calculations see e.g. Wolock, 1993).

We calculated topographic index values for each pixel using the GA2 algorithm, which is a slightly modified version of GRIDATB version 95.01 that has been written specifically for this study based on the basic loop structure implemented in Buytaert (2011) with some modifications to allow for the use of HydroSHEDS data. GA2 calculates the outflow gradient of each pixel (Fig. A1) and uses precalculated UPLAND values from HydroSHEDS for the catchment area A of each pixel (corrected for latitudinal projection distortions, B. Lehner, 2013).

Because of the use of the HydroSHEDS DEM, we made three small modifications in GA2 to the standard GRIDATB calculations.

- We applied the correction for the DEM resolution suggested by Ducharne (2009) to allow calculations to be carried out at continental scales (see A1 below).

- GRIDATB used the multiple drainage-direction algorithm of Quinn et al. (1991, 1995), also known as the FD8 or MD8 routing model (Wolock and McCabe, 1995; Zhao et al., 2009; Lang et al., 2013). However, in GA2 we instead used a direction-of-steepest-descent model: the deterministic eight node (D8) routing model (Moore et al., 1993; Wolock and McCabe, 1995; Wilson and Gallant, 2000; Zhao et al., 2009; Orlandini et al., 2014). This was for consistency with the HydroSHEDS drainage direction approach used to derive UPLAND areas in this study, which were calculated using D8.

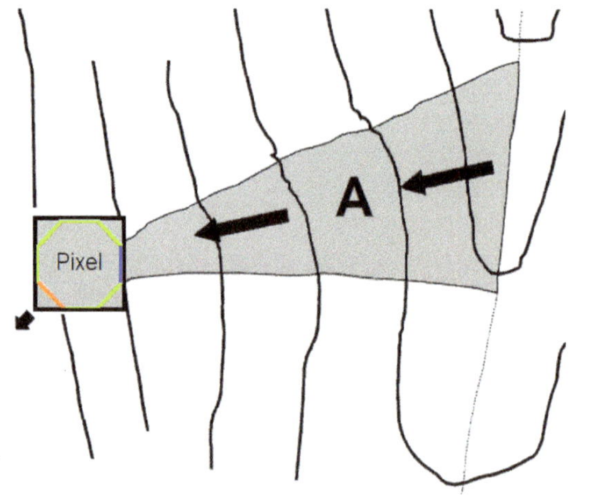

Figure A1. Illustration of the topographic index calculation of GA2 for one pixel of a DEM (the black square) downstream from a catchment area A (in m^2, defined to include the area of the pixel itself, which is usually negligible in comparison to A). The *inflow contour* of the pixel is shown in blue, the *outflow contour* in orange and the remaining perimeter of the octagon is shown green (q.v. the octagon of contour lengths shown in Quinn et al. (1991), Fig. 1). We calculate DX (pixel sidelength in m), $\tan(\beta)$ (mean slope across the outflow contour), $\tan(\beta')$ (mean slope across the non-outflow contour (blue + green)), clout (outflow contour length in m), a (specific catchment area in m) A/clout (n.b. called an "area" but units are $m^2\,m^{-1}$, i.e. m) and dfltsink $= \ln\left(\frac{A}{2\mathrm{DX}\tan(\beta')}\right)$ (this default value for *sinks*, i.e. pixels with no outflow, is described in Quinn et al. (1995), Fig. 14). The *topographic index value for this cell* is defined as $\ln\left(\frac{a}{\tan(\beta)}\right)$ if clout $\neq 0$, or dfltsink if clout $= 0$ (Quinn et al., 1991, 1995).

- The HydroSHEDS DEM does not have uniformly sized grid cells because of its native geographic projection (GCS WGS84) where pixel dimensions vary with latitude (i.e. the real width of a pixel gets increasingly shorter than the height towards the poles). Because slope directions are restricted to the eight cardinal and diagonal directions, we account for varying pixel dimensions in our slope calculations by taking an average distance between neighbouring pixels (rather than direction-dependent): we approximated DX as the square root of the area of each cell (with latitude-corrected pixel areas calculated using the Met. Office Unified Model routine realat1.f90 written by T. Oki in 1996; Dadson and Bell, 2010). When away from the Equator, this implies that slopes will be slightly overestimated in north–south directions and underestimated in east–west directions.

- Finally, because the value of dfltsink is undefined on plains (i.e. areas of no outflow and no inflow, which occur more often when vertical resolution is lower) we followed USGS (2000) and Evans (2003) in applying a minimum of 0.001 to $\tan(\beta')$.

Correcting for DEM resolution

A question arises when comparing catchments digitised at different resolutions (e.g. Chappell et al., 2006): how to compare topographic index values calculated from DEMs at different resolutions? Although not part of the original topographic index calculations, it has become accepted that topographic index values as calculated above should be reduced to the "equivalent" value for a 1 m resolution DEM by subtracting ln(DX) (and restricting the result to be ≥ 0). Applying this scale correction is becoming standard; e.g. see Ducharne (2009; also see Moore et al., 1993; Wolock and McCabe, 1995; Clark and Gedney, 2008).

Appendix B: Drainage direction and UPLAND calculations

Our calculations of topographic index values depend on the HydroSHEDS UPLAND layer containing the upstream catchment area draining into each point, and this layer in turn depends on the underlying drainage direction grid of HydroSHEDS. At its highest resolution of 3 arcsec, HydroSHEDS follows the D8 algorithm to determine drainage directions based on steepest slopes, which is considered the standard for use with large-scale routing models (e.g. TRIP, Grid-2-Grid, Dadson and Bell, 2010). However, in areas where turbulence or diffusional effects lead to significant hydrologic dispersion, flow lines may not coincide uniformly with slope lines (Rice et al., 2008; Orlandini et al., 2014). Deriving channel networks from terrain data has been an area of active recent research (e.g. Orlandini et al., 2014) and there is not yet universal agreement between the many different methods for calculating drainage/flow directions from DEM data (see discussions in Wilson and Gallant, 2000; Zhao et al., 2009; Orlandini et al., 2014).

At the upscaled 15 arcsec resolution of HydroSHEDS, the D8 concept is still valid in terms of providing one of eight possible neighbour pixels as the downstream direction; however, the direction values are not based on steepest slopes alone but also incorporate information from the 3 arcsec flow accumulation maps (Lehner, 2013). Additionally, a large number of manual corrections have been implemented over several years which modify the native DEM values ("hydrological conditioning"; Lehner 2013). As a consequence, our use of HydroSHEDS has unavoidably involved an acceptance of these algorithms and manipulations, and testing alternative settings to derive drainage directions or routing schemes is beyond the logistical limits of this study as it would require coordinated changes in slope, upscaling, and correction procedures at the multiple scales involved.

Acknowledgements. This study is part of the project Changing Land–Atmosphere Feedbacks in Tropical African Wetlands, coordinated by S. J. Dadson (http://www.geog.ox.ac.uk/research/landscape/projects/africanwetlands.html) and funded by the UK Natural Environment Research Council (NERC; NE/I01277X/2). N. Gedney is supported by the Joint UK DECC/Defra Met. Office Hadley Centre Climate Programme (GA01101). Thanks go to the University of Oxford Advanced Research Computing facility (http://www.arc.ox.ac.uk/) and the Oxford e-Research Centre (http://www.oerc.ox.ac.uk/) for their assistance and the use of the HAL and ARCUS servers for calculating the data layers presented here.

Edited by: R. Moussa

References

Baker, C., Thompson, J. R., and Simpson, M.: Hydrological Dynamics I: Surface Waters, Flood and Sediment Dynamics, in: The Wetlands Handbook, edited by: Maltby, E. and Barker, T., Wiley-Blackwell, Chichester, UK, 120–168, 2009.

Beven, K. J.: TOPMODEL: a critique, Hydrol. Process., 11, 1069–1085, 1997.

Beven, K. J.: Rainfall–Runoff Modelling The Primer, 2nd Edn., Wiley-Blackwell, Chichester, UK, 2012.

Beven, K. J. and Cloke, H. L.: Comment on "Hyperresolution global land surface modeling: Meeting a grand challenge for monitoring Earth's terrestrial water" by Eric F. Wood et al., Water Resour. Res., 48, W01801, doi:10.1029/2011WR010982, 2012.

Beven, K. J. and Kirkby, M. J.: A physically based, variable contributing area model of basin hydrology, Hydrological Sciences – Bulletin des Sciences Hydrologiques, 24, 43–69, 1979.

Buytaert, W.: topmodel: Implementation of the hydrological model TOPMODEL in R, version 0.7.2-2, R Package, available at: http://cran.r-project.org/web/packages/topmodel/index.html (last access: 6 June 2014), 2011.

Castanho, A. D. A., Coe, M. T., Costa, M. H., Malhi, Y., Galbraith, D., and Quesada, C. A.: Improving simulated Amazon forest biomass and productivity by including spatial variation in biophysical parameters, Biogeosciences, 10, 2255–2272, doi:10.5194/bg-10-2255-2013, 2013.

Chappell, N. A., Vongtanaboon, S., Jiang, Y., and Tangtham, N.: Return-flow prediction and buffer designation in two rainforest headwaters, Forest Ecol. Manage., 224, 131–146, 2006.

Choi, H. I.: Application of a Land Surface Model Using Remote Sensing Data for High Resolution Simulations of Terrestrial Processes, Remote Sensing, 5, 6838–6856, 2013.

Clark, D. B. and Gedney, N.: Representing the effects of subgrid variability of soil moisture on runoff generation in a land surface model, J Geophys. Res. D, 113, D10111, doi:10.1029/2007JD008940, 2008.

Coe, M. T.: A linked global model of terrestrial hydrologic processes: Simulation of modern rivers, lakes, and wetlands, J Geophys. Res.-Atmos., 103, 8885–8899, 1998.

Coe, M. T., Costa, M. H., and Soares-Filho, B. S.: The influence of historical and potential future deforestation on the stream flow of the Amazon River – Land surface processes and atmospheric feedbacks, J. Hydrol., 369, 165–174, 2009.

Collins, W. J., Bellouin, N., Doutriaux-Boucher, M., Gedney, N., Halloran, P., Hinton, T., Hughes, J., Jones, C. D., Joshi, M., Liddicoat, S., Martin, G., O'Connor, F., Rae, J., Senior, C., Sitch, S., Totterdell, I., Wiltshire, A., and Woodward, S.: Development and evaluation of an Earth-System model – HadGEM2, Geosci. Model Dev., 4, 1051–1075, doi:10.5194/gmd-4-1051-2011, 2011.

Dadson, S. J. and Bell, V. A.: Comparison of Grid-2-Grid and TRIP runoff routing schemes, Report, Centre for Ecology & Hydrology, Wallingford, UK, 2010.

Dadson, S. J., Ashpole, I., Harris, P., Davies, H. N., Clark, D. B., Blyth, E., and Taylor, C. M.: Wetland inundation dynamics in a model of land surface climate: Evaluation in the Niger inland delta region, J. Geophys. Res.-Atmos., 115, D23114, doi:10.1029/2010JD014474, 2010.

Dadson, S. J., Bell, V. A., and Jones, R. G.: Evaluation of a grid-based river flow model configured for use in a regional climate model, J. Hydrol., 411, 238–250, 2011.

Dharssi, I., Vidale, P. L., Verhoef, A., Macpherson, B., Jones, C., and Best, M.: New soil physical properties implemented in the Unified Model at PS18, Met Office Technical Report 528, The Met Office, Exeter, UK, 2009.

Ducharne, A.: Reducing scale dependence in TOPMODEL using a dimensionless topographic index, Hydrol. Earth Syst. Sci., 13, 2399–2412, doi:10.5194/hess-13-2399-2009, 2009.

Evans, J.: CTI.aml Compound Topographic Index AML script, available at: http://arcscripts.esri.com/details.asp?dbid=11863 (last access: 6 June 2014), 2003.

Falloon, P. and Betts, R.: Climate impacts on European agriculture and water management in the context of adaptation and mitigation – The importance of an integrated approach, Sci. Total Environ., 408, 5667–5687, 2010.

Gedney, N. and Cox, P. M.: The Sensitivity of Global Climate Model Simulations to the Representation of Soil Moisture Heterogeneity, J. Hydrometeorol., 4, 1265–1275, 2003.

Gedney, N., Cox, P. M., and Huntingford, C.: Climate feedback from wetland methane emissions, Geophys. Res. Lett., 31, L20503, doi:10.1029/2004GL020919, 2004.

Gedney, N., Cox, P. M., Betts, R. A., Boucher, O., Huntingford, C., and Stott, P. A.: Detection of a direct carbon dioxide effect in continental river runoff records, Nature, 439, 835–838, 2006.

Gerten, D., Schaphoff, S., Haberlandt, U., Lucht, W., and Sitch, S.: Terrestrial vegetation and water balance – hydrological evaluation of a dynamic global vegetation model, J. Hydrol., 286, 249–270, 2004.

Harding, R. J. and Warnaars, T. A.: Water and global change: The WATCH Project Outreach Report, Centre for Ecology and Hydrology, Wallingford, UK, 2011.

Harding, R. J., Blyth, E. M., Tuinenburg, O. A., and Wiltshire, A.: Land atmosphere feedbacks and their role in the water resources of the Ganges basin, Sci. Total Environ., 468, S85–S92, 2013.

Hjerdt, K. N., McDonnell, J. J., Seibert, J., and Rodhe, A.: A new topographic index to quantify downslope controls on local drainage, Water Resour. Res., 40, W05602, doi:10.1029/2004WR003130, 2004.

IPCC – Intergovernmental Panel on Climate Change: Climate Change 2013: The Physical Science Basis, CUP, Cambridge, UK, 2013.

Junk, W. J., Piedade, M. T. F., Schöngart, J., Cohn-Haft, M., Adeney, J. M., and Wittmann, F.: A Classification of Major Naturally-Occurring Amazonian Lowland Wetlands, Wetlands, 31, 623–640, 2011.

Ke, Y., Leung, L. R., Huang, M., Coleman, A. M., Li, H., and Wigmosta, M. S.: Development of high resolution land surface parameters for the Community Land Model, Geosci. Model Dev., 5, 1341–1362, doi:10.5194/gmd-5-1341-2012, 2012.

Kirkby, M.: Hydrograph Modelling Strategies., in: Processes in Physical and Human Geography, edited by: Peel, R., Chisholm, M., and Haggett, P., Heinemann, London, UK, 69–90, 1975.

Lang, M., McCarty, G., Oesterling, R., and Yeo, I.-Y.: Topographic Metrics for Improved Mapping of Forested Wetlands, Wetlands, 33, 141–155, 2013.

Lehner, B.: HydroSHEDS Technical Documentation (Version 1.2), World Wildlife Fund, Washington, D.C., available at: http://hydrosheds.org/page/development (last access: 6 June 2014), 2013.

Lehner, B. and Döll, P.: Development and validation of a global database of lakes, reservoirs and wetlands, J. Hydrol., 296, 1–22, 2004.

Lehner, B. and Grill, G.: Global river hydrography and network routing: baseline data and new approaches to study the world's large river systems, Hydrol. Process., 27, 2171–2186, 2013.

Lehner, B., Verdin, K., and Jarvis, A.: New Global Hydrography Derived From Spaceborne Elevation Data, Eos, 89, 93–94, 2008.

Lewis, S.: Hydrologic Sub-basins of Greenland. Dataset, National Snow and Ice Data Center (NSIDC), Boulder, Colorado, available at: http://nsidc.org/data/nsidc-0371.html (last access: 6 June 2014), 2009.

MacKellar, N. C., Dadson, S. J., New, M., and Wolski, P.: Evaluation of the JULES land surface model in simulating catchment hydrology in Southern Africa, Hydrol. Earth Syst. Sci. Discuss., 10, 11093–11128, doi:10.5194/hessd-10-11093-2013, 2013.

Marthews, T. R., Malhi, Y., Girardin, C. A. J., Silva-Espejo, J. E., Aragão, L. E. O. C., Metcalfe, D. B., Rapp, J. M., Mercado, L. M., Fisher, R. A., Galbraith, D. R., Fisher, J. B., Salinas-Revilla, N., Friend, A. D., and Restrepo-Coupe, N.: Simulating forest productivity along a neotropical elevational transect: temperature variation and carbon use efficiency, Global Change Biol., 18, 2882–2898, 2012.

Marthews, T. R., Quesada, C. A., Galbraith, D. R., Malhi, Y., Mullins, C. E., Hodnett, M. G., and Dharssi, I.: High-resolution hydraulic parameter maps for surface soils in tropical South America, Geosci. Model Dev., 7, 711–723, doi:10.5194/gmd-7-711-2014, 2014.

Milly, P. C. D., Betancourt, J., Falkenmark, M., Hirsch, R. M., Kundzewicz, Z. W., Lettenmaier, D. P., and Stouffer, R. J.: Stationarity Is Dead: Whither Water Management?, Science, 319, 573–574, 2008.

Moore, I. D., Lewis, A., and Gallant, J. C.: Terrain Attributes: Estimation Methods and Scale Effects, in: Modelling Change in Environmental Systems, edited by: Jakeman, A. J., Beck, M. B., and McAleer, M. J., John Wiley & Sons Ltd., Chichester, UK, 189–214, 1993.

O'Connor, F. M., Boucher, O., Gedney, N., Jones, C. D., Folberth, G. A., Coppell, R., Friedlingstein, P., Collins, W. J., Chappellaz, J., Ridley, J., and Johnson, C. E.: Possible role of wetlands, permafrost, and methane hydrates in the methane cycle under future climate change: A review, Rev. Geophys., 48, RG4005, doi:10.1029/2010RG000326, 2010.

Orlandini, S., Moretti, G., and Gavioli, A.: Analytical basis for determining slope lines in grid digital elevation models, Water Resour. Res., 50, 529–539, doi:10.1002/2013WR014606, 2014.

OSGF – Open Source Geospatial Foundation: Geospatial Data Abstraction Library (version 1.9.0), Translator library, available at: http://www.gdal.org/ (last access: 6 June 2014), 2011.

Pangala, S. R., Moore, S., Hornibrook, E. R. C., and Gauci, V.: Trees are major conduits for methane egress from tropical forested wetlands, New Phytol., 197, 524–531, 2013.

Pfeffer, W. T., Arendt, A. A., Bliss, A., Bolch, T., Cogley, J. G., Gardner, A. S., Hagen, J.-O., Hock, R., Kaser, G., Kienholz, C., Miles, E. S., Moholdt, G., Mölg, N., Paul, F., Radić, V., Rastner, P., Raup, B. H., Rich, J., Sharp, M. J., and The Randolph Consortium: The Randolph Glacier Inventory: a globally complete inventory of glaciers, J. Glaciol., 60, 537–552, 2014.

Prentice, I. C., Bondeau, A., Cramer, W., Harrison, S. P., Hickler, T., Lucht, W., Sitch, S., Smith, B., and Sykes, M. T.: Dynamic Global Vegetation Modeling: Quantifying Terrestrial Ecosystem Responses to Large-Scale Environmental Change, in: Terrestrial Ecosystems in a Changing World, edited by: Canadell, J. G., Pataki, D. E., and Pitelka, L. F., Springer, Berlin, Germany, 175–192, 2007.

Prigent, C., Papa, F., Aires, F., Rossow, W. B., and Matthews, E.: Global inundation dynamics inferred from multiple satellite observations, 1993–2000, J Geophys. Res., 112, D12107, doi:10.1029/2006JD007847, 2007.

Quinn, P., Beven, K., Chevallier, P., and Planchon, O.: The prediction of hillslope flow paths for distributed hydrological modelling using digital terrain models, Hydrol. Process., 5, 59–79, 1991.

Quinn, P. F., Beven, K. J., and Lamb, R.: The $\ln(a/\tan\beta)$ index: how to calculate it and how to use it within the TOPMODEL framework, Hydrol. Process., 9, 161–182, 1995.

R Development Core Team: R: A language and environment for statistical computing, version 3.0.2, R Foundation for Statistical Computing, Vienna, available at: http://www.R-project.org (last access: 6 June 2014), 2013.

Rice, S., Roy, A., and Rhoads, B.: River Confluences, Tributaries and the Fluvial Network. John Wiley & Sons, Chichester, UK, 2008.

Sanderson, M. G., Wiltshire, A. J., and Betts, R. A.: Projected changes in water availability in the United Kingdom, Water Resour. Res., 48, W08512, doi:10.1029/2012WR011881, 2012.

Seneviratne, S. I., Lüthi, D., Litschi, M., and Schär, C.: Land–atmosphere coupling and climate change in Europe, Nature, 443, 205–209, 2006.

Seneviratne, S. I., Corti, T., Davin, E. L., Hirschi, M., Jaeger, E. B., Lehner, I., Orlowsky, B., and Teuling, A.J.: Investigating soil moisture–climate interactions in a changing climate: A review, Earth-Sci. Rev., 99, 125–161, 2010.

USGS – US Geological Survey: HYDRO1k Elevation derivative database, US Geological Survey Earth Resources Observation and Science (EROS) Center, Sioux Falls, South Dakota, available at: https://lta.cr.usgs.gov/HYDRO1K (last access: 6 June 2014), 2000.

Wainwright, J. and Mulligan, M.: Environmental Modelling Finding Simplicity in Complexity, 2nd Edn., Wiley-Blackwell, Chichester, UK, 2013.

Ward, R. C. and Robinson, M.: Principles of Hydrology, 4th Edn., McGraw-Hill, Maidenhead, UK, 2000.

Wilson, J. P. and Gallant, J. C.: Terrain analysis Principles and Applications, John Wiley & Sons, New York, 2000.

Wolock, D. M.: Simulating the variable-source-area concept of streamflow generation with the watershed model TOPMODEL, United States Geological Survey Water-Resources Investigations Report 93-4124, US Geological Survey, Lawrence, Kansas, USA, 1993.

Wolock, D. M. and McCabe, G. J.: Comparison of single and multiple flow direction algorithms for computing topographic parameters in TOPMODEL, Water Resour. Res., 31, 1315–1324, 1995.

Wood, E. F., Roundy, J. K., Troy, T. J., van Beek, L. P. H., Bierkens, M. F. P., Blyth, E., de Roo, A., Döll, P., Ek, M., Famiglietti, J., Gochis, D., van de Giesen, N., Houser, P., Jaffé, P. R., Kollet, S., Lehner, B., Lettenmaier, D. P., Peters-Lidard, C., Sivapalan, M., Sheffield, J., Wade, A., and Whitehead, P.: Hyperresolution global land surface modeling: Meeting a grand challenge for monitoring Earth's terrestrial water, Water Resour. Res., 47, W05301, doi:10.1029/2010WR010090, 2011.

Wood, E. F., Roundy, J. K., Troy, T. J., van Beek, R., Bierkens, M., Blyth, E., de Roo, A., Döll, P., Ek, M., Famiglietti, J., Gochis, D., van de Giesen, N., Houser, P., Jaffe, P., Kollet, S., Lehner, B., Lettenmaier, D. P., Peters-Lidard, C. D., Sivapalan, M., Sheffield, J., Wade, A. J., and Whitehead, P.: Reply to comment by Keith J. Beven and Hannah L. Cloke on "Hyperresolution global land surface modeling: Meeting a grand challenge for monitoring Earth's terrestrial water", Water Resour. Res., 48, W01802, doi:10.1029/2011WR011202, 2012.

Yang, X., Chapman, G. A., Gray, J. M., and Young, M. A.: Delineating soil landscape facets from digital elevation models using compound topographic index in a geographic information system, Aust. J. Soil Res., 45, 569–576, 2007.

Zhao, G., Gao, J., Tian, P., and Tian, K.: Comparison of two different methods for determining flow direction in catchment hydrological modeling, Water Sci. Eng., 2, 1–15, doi:10.1029/2011WR011202, 2009.

Zulkafli, Z., Buytaert, W., Onof, C., Lavado, W., and Guyot, J. L.: A critical assessment of the JULES land surface model hydrology for humid tropical environments, Hydrol. Earth Syst. Sci., 17, 1113–1132, doi:10.5194/hess-17-1113-2013, 2013.

5

Climate change and non-stationary flood risk for the upper Truckee River basin

L. E. Condon[1,2], S. Gangopadhyay[1], and T. Pruitt[1]

[1]Bureau of Reclamation, Technical Service Center, Denver, Colorado, USA
[2]Hydrologic Science and Engineering Program and Department of Geology and Geological Engineering, Colorado School of Mines, Golden, Colorado, USA

Correspondence to: L. E. Condon (lcondon@mymail.mines.edu)

Abstract. Future flood frequency for the upper Truckee River basin (UTRB) is assessed using non-stationary extreme value models and design-life risk methodology. Historical floods are simulated at two UTRB gauge locations, Farad and Reno, using the Variable Infiltration Capacity (VIC) model and non-stationary Generalized Extreme Value (GEV) models. The non-stationary GEV models are fit to the cool season (November–April) monthly maximum flows using historical monthly precipitation totals and average temperature. Future cool season flood distributions are subsequently calculated using downscaled projections of precipitation and temperature from the Coupled Model Intercomparison Project Phase 5 (CMIP-5) archive. The resulting exceedance probabilities are combined to calculate the probability of a flood of a given magnitude occurring over a specific time period (referred to as flood risk) using recent developments in design-life risk methodologies. This paper provides the first end-to-end analysis using non-stationary GEV methods coupled with contemporary downscaled climate projections to demonstrate the evolution of a flood risk profile over typical design life periods of existing infrastructure that are vulnerable to flooding (e.g., dams, levees, bridges and sewers). Results show that flood risk increases significantly over the analysis period (from 1950 through 2099). This highlights the potential to underestimate flood risk using traditional methodologies that do not account for time-varying risk. Although model parameters for the non-stationary method are sensitive to small changes in input parameters, analysis shows that the changes in risk over time are robust. Overall, flood risk at both locations (Farad and Reno) is projected to increase 10–20 % between the historical period 1950 to 1999 and the future period 2000 to 2050 and 30–50 % between the same historical period and a future period of 2050 to 2099.

1 Introduction

"Stationarity is dead" (Milly et al., 2008), yet the standard practice for flood frequency analysis is predicated on this very assumption. This discrepancy has not gone unnoticed within the scientific community and there is a growing body of research investigating (1) trends in observed floods (e.g., Franks, 2002; Vogel et al., 2011), (2) ways to incorporate non-stationarity into frequency distributions (e.g., Katz et al., 2002; Raff et al., 2009) and (3) methodologies to interpret risk and approach design within a non-stationary framework (e.g., Mailhot and Duchesne, 2010; Rootzén and Katz, 2013; Salas and Obeysekara, 2014). Both the frequency and intensity of extreme events are particularly susceptible to change because small shifts in the center of a distribution can potentially have much larger impacts on the tails (Meehl et al., 2000). Regardless of climate change, naturally occurring long-term climate oscillations, such as the El Niño–Southern Oscillation (ENSO), have been linked to low frequency variability in flood frequency (e.g., Cayan et al., 1999; Jain and Lall, 2001). Anthropogenic climate change has the potential to amplify natural climatic variability throughout the interconnected climate and hydrologic systems.

Already trends in many hydrologic variables have been observed across the western United States (as well as around the world). For example, clear increases in temperature have been measured across the west (e.g., Cayan et al., 2001; Dettinger and Cayan, 1995). Precipitation trends are more variable. Regonda et al. (2005) found increased total winter precipitation (rain and snow) from 1950 to 1999 in many sites across the western United States, although springtime snow water equivalent (SWE) was shown to decline over the same period. Similarly, Mote et al. (2005) analyzed snowpack trends in western North America and reported widespread declines in springtime SWE over the period 1925–2000, especially since the middle of the 20th century. They attribute this decline predominantly to climatic factors such as ENSO, Pacific Decadal Oscillation (PDO) and positive trends in regional temperature. Easterling et al. (2000) summarized previous studies on precipitation trends. They note that trends vary from region to region but, in general, increases in precipitation have occurred disproportionately in the extremes. Several subsequent studies have observed increasing trends in extreme precipitation events, although the changes are relatively small (Gutowski et al., 2008; Kunkel, 2003; Madsen and Figdor, 2007).

Research has also demonstrated increasing trends in flood frequency in some regions. Walter and Vogel (2010) and Vogel et al. (2011) observed increasing flood magnitudes across the United States using stream gauge records, and Franks (2002) showed statistically significant increases in flood frequency since the 1940s. Still, non-stationary flood behavior has been historically difficult to quantify and there has been some debate about the significance of flood frequency trends. For example, Hirsch (2011) noted both increasing and decreasing trends in annual flood magnitudes in different regions of the US. Also, Douglas et al. (2000) found that, if one takes into account spatial correlation, many previous findings of flood trends are not statistically significant. Difficulty in diagnosing flood trends is not unique to the western US; a literature review of historical flood studies across Europe also found spatial variability in flood trends (Hall et al., 2014).

Even when significant trends are found, the complexity of flooding mechanisms, which depend on many variables that can vary regionally and seasonally, makes it difficult to attribute trends to specific causes. Illustrating the importance of seasonality, Small et al. (2006) showed that if a high-precipitation event occurs in the fall, as opposed to the spring, it will contribute to baseflow rather than inducing flooding. Also, urbanization can drastically increase the impervious area of a basin, thus amplifying floods by decreasing infiltration and speeding runoff. The largest flood magnitude increases observed by both Walter and Vogel (2010) and Vogel et al. (2011) were in basins with urban development. The influence of development trends on flood behavior can be difficult to separate from other variables. For example, Villarini et al. (2009) could not conclusively tie reduced stationarity (i.e., changes in mean and/or variance) in peak discharge records to climate change because of variability in the other factors that influence runoff.

Merz et al. (2012) note that attributing changes in flood hazard is complicated by the complex array of drivers that can include land cover change and infrastructure development as well as natural climate variability and change. Here we set aside the impacts of development and management practices and focus on the role of climate change. However, even with this simplification, future extremes can still be influenced by a number of interrelated variables such as changes in temperature, precipitation efficiency and vertical wind velocity (Mullet et al., 2011; O'Gorman and Schneider, 2009). Analyzing global circulation model (GCM) outputs, Pierce et al. (2012) found total changes in precipitation to be small relative to the existing variability, but noted larger seasonal changes in storm intensity and frequency. Despite uncertainty, many studies agree that warming will increase the potential for intense rainfall (Allan, 2011; Gutowski et al., 2008; Pall et al., 2011; Sun et al., 2007). Furthermore, Min et al. (2011) found that some GCM simulations may underestimate extreme precipitation events in the Northern Hemisphere, indicating that projections of extreme precipitation based on GCM outputs may be conservative.

Studies have also predicted increases in flood frequency and magnitude with a warmer climate, especially in snowmelt-dominated basins (e.g., Das et al., 2011). As with historical flooding trends, translating forecasted climate variables to flood frequency is a complex process and several methodologies have been used. Downscaled GCM climate forcings can be used to drive hydrologic models and simulate future floods directly (e.g., Das et al., 2011; Vogel et al., 2011; Raff et al., 2009). With this approach, traditional stationary flood frequency distributions can be fit to the simulated floods to calculate return periods of interest (e.g., Raff et al., 2009; Vogel et al., 2011). This allows for return periods and flood magnitudes that change over time, as with the flood magnification and recurrence reduction factors calculated by Walter and Vogel (2010) and Vogel et al. (2011). While these approaches do capture temporal changes between analysis periods, they still assume that flood mechanisms are stationary within each period of analysis.

This limitation can be overcome using non-stationary generalized extreme value (GEV) distributions where the model parameters, like mean (i.e., location) and spread (i.e., scale), are allowed to vary as a function of time (e.g., Gilroy and McCuen, 2012) or with relevant covariates (e.g., Griffis and Stedinger, 2007; Katz et al., 2002; Towler et al., 2010). This approach has been gaining popularity for flood frequency estimation. Using this technique, it is not necessary to simulate future floods directly by forcing a hydrologic model with projected hydroclimate fields (e.g., precipitation and temperature). The parameters of the GEV model, like mean and spread, change with time (i.e., non-stationary) based on a linear combination of covariates like precipitation and temperature. Historical relationships between extreme events and hydroclimate fields are used to identify the weighting of covariates. These weights are then used to estimate parameters for future time periods using precipitation and temperature outputs from hydroclimate projections. For example, Gilroy and McCuen (2012) used non-stationary GEV models of flood frequency that incorporated a linear trend in the location parameter. Similarly, Griffis and Stedinger (2007) and Towler et al. (2010) used climate variables as covariates for the distribution parameters.

While non-stationary flood forecasting methods provide flexibility in analyzing flood variability, they are also incongruent with many of the traditional metrics used in water resources planning. Historically, most infrastructures that are vulnerable to flooding (e.g., dams, levees, sewers and bridges) have been designed to withstand flooding of a specified return period (e.g., the 100-year flood). However, these calculations rely on a flood frequency distribution which is assumed to remain stationary with time, and hence the return period design metric is also assumed to be stationary. When non-stationary methods are used, the underlying flood frequency distributions, and associated return periods, vary with time. Thus, under a non-stationary climate, the notion of a static return period flood event (e.g., 100-year flood, 200-year flood) is no longer a valid concept.

To address this issue, Rootzén and Katz (2013) introduced the concept of design life level to calculate the risk of a given flood magnitude occurring over a specified time period. Salas and Obeysekera (2014) further demonstrated the relevance of this technique to the hydrologic community using flood frequency examples. However, this methodology has yet to receive widespread attention within the hydrologic community. Here, we present a non-stationary flood frequency assessment for the upper Truckee River basin (UTRB) using contemporary downscaled climate projections and the non-stationary design life level technique introduced by Rootzén and Katz (2013) to quantify flood risk. Note that, following the convention of Rootzén and Katz (2013), we use the term flood risk to refer to the probability of an extreme event occurring and not as a quantification of expected losses. While the methodology used for this analysis is previously established, this paper provides the first end-to-end demonstration of non-stationary GEV analysis coupled with contemporary downscaled climate projections (specifically, downscaled climate projections from the Coupled Model Intercomparison Project Phase 5, or CMIP-5) to quantify how the flood risk profiles may evolve in the upper Truckee River basin over the 21st century. The flood analysis presented here is part of a larger study on climate change impacts in the Truckee River basin (Reclamation, 2010). This project is supported by local water managers and conducted by the Bureau of Reclamation through the WaterSMART Basin Studies Program authorized under US Public Law 111-11, Subtitle F (SECURE Water Act). The intent of this work is (1) to investigate potential flood risk changes over time in the UTRB and (2) to demonstrate the applicability of non-stationary techniques in a regional flood analysis to make these tools more accessible to the hydrologic community.

The paper is organized as follows. Section 2 provides background on the study area along with the data sets and models used. The methodologies of using non-stationary spatial GEV analysis in conjunction with climate projections and time-evolving risk assessment are described in Sect. 3. Results and discussions of findings are given in Sect. 4. Summary and conclusions from the analysis are presented in Sect. 5.

2 Background

This section provides background on the study area (Sect. 2.1), streamflow data and simulations (Sect. 2.2) and climate data and models (Sect. 2.3).

2.1 Upper Truckee River basin

The Truckee River originates in the northern Sierra Nevada in California (above Lake Tahoe) and flows northeast to Nevada, where it ends in the Pyramid Lake (Fig. 1). The total basin area is roughly 7900 km^2; however, the area upstream of Reno (2763 km^2), henceforth referred to as the upper Truckee River, provides the majority of the basin's precipitation through snowpack. The focus of this analysis is on the Farad and Reno gauge locations shown in Fig. 1, henceforth referred to as Farad and Reno. The Farad gauge is located roughly 1.5 km downstream of the Farad hydropower plant and provides a cumulative measure of all of the upper basin tributaries (Stokes, 2002). Most of the available water supply is generated upstream of the Farad gauge (USACE, 2013a). The Reno gauge is located downstream of Farad in the heart of Reno and is a good reference point for analyzing urban flooding. The intervening area between the Farad and Reno gauges is small, roughly 350 km^2 [km], and there are only two small tributaries that enter the main stem between Farad and Reno.

Flooding in the upper Truckee generally takes one of three forms. Some of the most severe floods have resulted from heavy rain events covering most of the basin and lasting 1 to 6 days. These storms generally occur from November to April and may be linked to atmospheric rivers (Ralph and Dettinger, 2012). Snowmelt floods are also common from April to July. Although snowmelt floods transmit large volumes of water for longer durations, they generally do not cause damage because they are typically well predicted and can be regulated with upstream reservoirs. Finally, in late summer (July–August), local cloudbursts can generate high-intensity precipitation over small areas. These storms can cause local damage to tributaries but generally do not have a large impact on the main stem of the Truckee.

In the 20th century, nine major floods have been recorded on the Truckee River, all of which occurred from November to April (USACE, 2013b). The flood of record occurred in January of 1997 and was caused by warm rain falling on

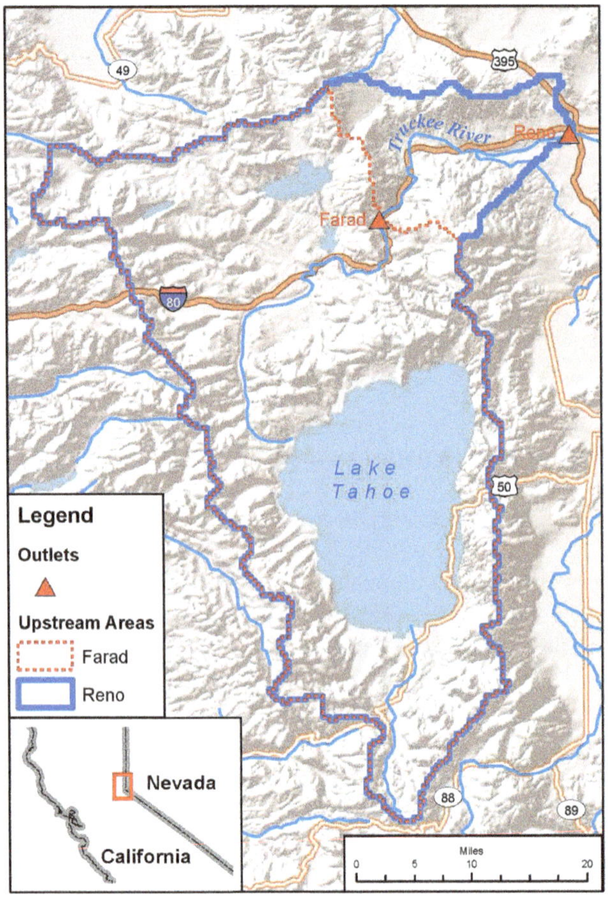

Figure 1. Map of model domain including the Farad and Reno gauges and their drainage areas.

a large snowpack (~ 180 % of normal) and melting nearly all of the snowpack below ~2100 m (USACE, 2013b). The floods of 1950, 1955 and 1963 were some of the most damaging due to the development of Reno along the river during this time period (USACE, 2013b). Subsequent flood damages have been, at least partially, mitigated by the implementation of flood infrastructure starting in the 1960s.

2.2 Streamflow data and simulations

Streamflow has been measured at both the Farad and Reno USGS gauges. However, gauge flows are not readily applicable to flood frequency analysis due to upstream developments of water supply and flood control structures. For example, upstream of Reno there are four dams with flood control capabilities (i.e., Martis Creek Dam, Prosser Creek Dam, Stampede Dam and Boca Dam); in addition to Tahoe, Donner and Independence lakes provide incidental flood regulation. Unregulated flow estimates were developed by the US Army Corps of Engineers (USACE) but are only available for historical flood events (USACE, 2013b). Therefore, we simulate unregulated flows from 1950 to 1999 using the three-layer variable infiltration capacity (VIC) model and validate results using the available unregulated flow estimates.

A brief summary of the VIC model is provided here, and for additional technical specifications the reader is referred to Liang et al. (1994, 1996) and Nijssen et al. (1997). VIC is a gridded hydrologic model designed to simulate macroscale (spatial resolution is greater than 1 km) water balances using parameterized sub-grid infiltration and vegetation processes. In the VIC model, surface water infiltrates to the subsurface based on soil properties, and soil moisture is distributed vertically through three model layers extending up to about 2 m below the land surface. At the surface, potential evapotranspiration (PET) is simulated using the Penman–Monteith PET model (Maidment, 1993). Surface flows are determined in a two-step process. First, the water balance for each grid cell is calculated independently to determine surface runoff and baseflow, and subsequently runoff from each cell is routed to river channels and outlets using a predefined routing network. Here we drive VIC with daily weather forcings including precipitation, maximum and minimum temperature, and wind speed. Additional climate variables such as short- and long-wave radiation, relative humidity and vapor pressure are calculated within the model using established empirical relationships. The VIC model is well documented and has already been used in a number of hydrologic and climate change studies (e.g., Christensen and Lettenmaier, 2007; Christensen et al., 2004; Gangopadhyay et al., 2011; Maurer et al., 2007; Payne et al., 2004; Reclamation, 2011; Van Rheenen et al., 2004). Recently VIC has also been applied for real-time flood estimation (Wu et al., 2014).

The VIC model used for this analysis was part of the Bureau of Reclamation (Reclamation) West Wide Climate Risk Assessment (WWCRA) effort and is described in Reclamation (2011). The WWCRA VIC model encompasses the western US. Simulated and observed streamflows were compared at 152 locations primarily from the USGS Hydro-climatic Data Network (Slack et al., 1993) and 43 additional locations of importance to Reclamation's water management activities. Among the evaluated locations are several in the Truckee basin including the Truckee River at Farad. For details on model calibration and development we refer the reader to Reclamation (2011) and Gangopadhyay et al. (2011). While we do not discuss model calibration further here, in the subsequent sections we provide additional model verification for flood simulation in the UTRB.

2.3 Climate data and models

As noted in the previous section, the VIC model requires daily climate inputs to drive water balance simulations. We use the national 1/8° (roughly 12 km) gridded data set from Maurer et al. (2002) for historical (i.e., 1950–1999) climate observations. Additionally, monthly total precipitation and average temperature were aggregated for the upstream area of each gauge for every month of the flood season (i.e., November through April). These values are used as covariates for fitting non-stationary GEV models discussed in Sect. 3.

Future gridded precipitation and temperature values from 2000 to 2099 were generated from GCM outputs. We analyzed 234 projections generated by 37 different climate models from the CMIP-5 archive (Taylor et al., 2012). In the absence of objective guidance in contemporary climate literature to limit the number of projections, we chose to include all of the available CMIP-5 projections of future climate in this study. However, it should be noted that other studies have demonstrated that a subset of projections could provide comparable results for specific study objective (e.g., water supply) (Pierce et al, 2009; Harding et al., 2012). Projections span four representative concentration pathways (RCPs) for greenhouse gas emissions. Each GCM projection includes monthly gridded precipitation and temperature from 1950 to 2099 at a coarse grid resolution ranging between ~65 and 250 km.

Reclamation in collaboration with other federal and non-federal partners has developed a monthly archive of downscaled CMIP-5 projections at the finer 1/8° resolution using the two-step bias correction and spatial disaggregation (BCSD) algorithm described by Wood et al. (2004). For this analysis we extended the existing hydrology archive to cover the UTRB domain for all 234 BCSD CMIP-5 climate projections following the steps detailed below. A subset of the CMIP-5 hydrology projections is publically accessible through the archive of the downscaled CMIP3 and CMIP5 climate and hydrology projections at http://gdo dcp.ucllnl.org/downscaled_cmipprojections/. Additional documentation on the archive and the methodology is provided in Reclamation (2014).

The downscaled climate variables include monthly total precipitation, monthly maximum and minimum temperatures and monthly average temperature. Before applying the BCSD algorithm, all 234 climate projections were first gridded from their respective native GCM scale to a common grid of 1° latitude by 1° longitude. Similarly, the observed 1/8° gridded data set (Maurer et al., 2002) was aggregated to the coarser 1° latitude by 1° longitude grid. Next, for a given climate variable, GCM and location (1° latitude by 1° longitude grid cell), the bias correction (BC) step uses quantile mapping between monthly cumulative distribution functions (CDFs) of historical simulated and historical observed values to identify biases over a common climatological period – in this case, 1950–1999. The projected future climate variables from the same GCM at the same location are then bias-corrected using the identified bias. The result of bias correction is an adjusted GCM data set (20th century and 21st century, linked together) that is statistically consistent with the observed data during the bias correction overlap period (i.e., 1950–1999 in this application). Note that the BC step happens at the coarse 1° latitude by 1° longitude grid. Next, multiplicative adjustment factors (ratio of bias-corrected GCM to observed) for precipitation and offset adjustment factors (bias-corrected GCM minus observed) for temperature are calculated for each of the 1° latitude by 1° longitude grid cells (Reclamation, 2013). These adjustments are then spatially disaggregated (SD) to a 1/8° latitude by 1/8° longitude grid. Finally, the adjustments are applied (multiplicative for precipitation, additive for temperature) to the finer resolution, 1/8° gridded observed precipitation and temperature fields (Maurer et al., 2002) to derive the 1/8° gridded BCSD climate projections.

3 Methodology

This section describes the methodology used for flood frequency analysis in the UTRB. Discussion is divided into two sections. First, we describe the process of extreme value modeling using non-stationary GEV distributions (Sect. 3.1). Second, the methodology for design-life level risk assessment is described (Sect. 3.2)

3.1 Extreme value modeling

Extreme values analysis (EVA) deals with the examination of the tail (i.e., extreme) values of a distribution (as opposed to standard approaches which are generally more concerned with the average system behavior). EVA methods are standard practice for flood frequency analysis because they are designed to capture the behavior of low-frequency high-impact events. Furthermore, Katz (2010) points out that in climate change studies traditional approaches are not sufficient and extreme value statistics are needed. For this analysis, we use the GEV, which is commonly applied to flood frequency analysis to model block maxima from streamflow time series (e.g., Katz et al., 2002; Towler et al., 2010). The cumulative distribution function (CDF) for the GEV, F, is as follows:

$$F(z;\boldsymbol{\theta}) = \exp\left\{-\left[1+\xi\left(\frac{z-\mu}{\sigma}\right)\right]^{-1/\xi}\right\}, \quad (1)$$

where z is the streamflow maxima value of interest and $\boldsymbol{\theta}$ is the parameter set (μ, σ, ξ) used to specify the distribution, such that the center is given by the location (μ), the spread by the scale (σ) and the behavior of the upper tail by the shape (ξ). Based on the shape parameter, the GEV can take one of three forms: Gumbel, or light tailed, when ξ is zero; Fréchet, or heavy tailed, if ξ is positive; and Weibull, or bounded, when ξ is negative. Following the methodology of Towler et al. (2010), GEV parameters (μ, σ, ξ) are fitted using the maximum likelihood estimation (MLE) technique.

In traditional stationary flood frequency analysis, it is assumed that observations are independent and identically distributed (IID), and therefore model parameters (μ, σ, ξ) derived from the observed flood record are assumed to remain constant across the period of record and into the future. Here, we introduce non-stationarity into the distribution by allowing location and scale parameters to change with relevant covariates, such that:

$$\mu(t) = \beta_{0,\mu} + \beta_{1,\mu}x_1 + \ldots + \beta_{n,\mu}x_n, \quad (2)$$

$$\sigma(t) = \beta_{0,\sigma} + \beta_{1,\sigma}x_1 + \ldots + \beta_{n,\sigma}x_n, \quad (3)$$

where the β variables represent the coefficients and the x variables are the covariates. In keeping with previous studies, the shape parameter, which is the most difficult to estimate, is assumed constant (e.g., Obeysekara and Salas, 2014; Salas and Obeysekera, 2013; Towler et al., 2010).

Some previous studies (e.g., Salas and Obeysekera, 2014; Stedinger and Griffis, 2011) have developed non-stationary location and scale parameters that are explicitly dependent on time. This approach requires first the derivation of temporal flooding trends and second the projection of this trend into the future. Here we derive location and scale parameters based on time-varying meteorological variables (i.e., temperature and precipitation). With the approach used here, temporal trends in flooding are introduced as a function of temporal variability in precipitation and temperature, but no explicit trend is specified a priori.

To determine the optimal set of covariates for a non-stationary model, additional statistical methods must be employed. The Akaike information criterion (AIC; Akaike, 1974), given in Eq. (4), weighs the goodness of the fit of a model with the level of complexity.

$$\text{AIC} = 2(\text{NLLH}) + 2K \quad (4)$$

Here, NLLH is the negative log-likelihood estimated for a model fitted with K parameters. In this formulation, higher-ranked models have lower AIC scores. For this analysis, the best model is selected using pairwise comparisons of NLLH scores following the methods of Salas and Obeysekera (2014) and others. Models are compared using the deviance statistic (D), which is equal to twice the difference in NLLH scores. The D is then tested for significance based on a chi-squared distribution with the degrees of freedom set equal to the difference in the number of parameters (K) between models. Finally, p values less than 0.05 indicate a statistically significant improvement in model performance at the 5 % significance level.

Following the methodology described above, GEV distributions are fitted to time series of maximum monthly historical (1950–1999) 1-day simulated streamflows (detailed in Sect. 2) for the cool season (November to April). Although there are some unregulated historical flow estimates, the available data set only covers six storms. Therefore to be consistent, we fit our model only to the simulated flows. The data set includes maximum daily streamflows for each month in the cool season defined by the block of months November through April, as opposed to the more traditional single value per year. This technique was also used by Towler et al. (2010), who noted that expanding the data set helps avoid the problems associated with using maximum likelihood estimate on small data sets. However, as noted by Towler et al. (2010), when multiple values are used per year the calculated probabilities must be adjusted appropriately to derive annual values. Floods during the cool season generally last between 1 and 4 days. Here we focus on the 1-day flood peak, as opposed to multi-day flood volumes, because this is a representative metric for flood damage. Additionally, using the 1-day flood maximum focuses the analysis on flood magnitude rather than duration.

Two covariates were considered, monthly total precipitation (P) and mean temperature (T), averaged over the upstream area for each gauge. As discussed in Sect. 2, precipitation is a relevant covariate because many of the floods in this season are rain-on-snow events or extreme rainfall events. Similarly, temperature drives snowmelt and is an important contributor to UTRB flood events (e.g., January 1997 event). Both stationary and non-stationary GEV models were evaluated using the extRemes package (Gilleland and Katz, 2011) in the R statistical computing environment (R Core Team, 2012).

3.2 Time-varying risk assessment

Traditional flood planning relies on the concept of return periods, which are usually calculated as the inverse of annual exceedance probability for a given flood magnitude, assuming a stationary distribution: for example, the log-Pearson Type III (LP3) distribution described by the Interagency Advisory Committee on Water Data bulletin 17B (IACWD, 1982). However, when non-stationary models are used, the distribution parameters, and hence the exceedance probabilities, vary with time. Table 1 compares various flood probability calculations between stationary and non-stationary approaches (Salas and Obeysekera, 2014). As shown here, when the flood distribution is stationary, the return period for a given flood magnitude is constant and relies only on the exceedance probability (Eq. 4a in Table 1). However, if distribution parameters are non-stationary, then the return period will vary based on the period of interest (Eq. 4b in Table 1). This concept is easily extended to flood risk (here defined as the probability of a flood of a given magnitude occurring, not expected losses). In traditional analyses, the risk of a flood occurring in a given period depends only on the length of the period (Eq. 5a), while in a non-stationary analysis, risk depends on both the length of time considered and the time period itself (Eq. 5b). This is the concept of design life level proposed by Rootzén and Katz (2013). Here we adopt the design-life level risk framework given by Eq. (5b) in Table 1 and calculate the risk of flood for a range of future periods and design life lengths.

Table 1. Flood calculations using stationary and non-stationary distributions (adapted from Salas and Obeysekera, 2014).

Eq. #	Description	(a) Stationary	(b) Non-stationary
1	Exceedance probability (probability of flood[1] occurring in year x)	p	p_x
2	Probability of the first flood occurring in year x[2]	$f(x)=(1-p)^{x-1}p$	$f(x)=p_x\prod_{t=1}^{x-1}1-p_t$
3	Probability of a flood occurring before year x[3]	$F(x)=\sum_{i=1}^{x}f(i)$ $F(x)=1-(1-p)^x$	$F(x)=1-\prod_{t=1}^{x}(1-p_t)$
4	Return period (expected waiting time between flood occurrences[4,5])	$E(X)=\sum_{x=1}^{\infty}x\cdot P(X=x)$ $E(X)=1/p$	$E(X)=1+\sum_{x=1}^{x_{\max}}\prod_{t=1}^{x}(1-P_t)$
5	Probability of a flood occurring before the design life n[6]	$R=P(X\leq n)=F(n)$ $R=1-(1-p)^n$	$R=1-\prod_{t=1}^{n}(1-p_t)$

[1] Flood is defined as a flow exceeding a predefined threshold; [2] $f(x)$ is the probability density function of X; [3] $F(x)$ is the cumulative distribution function of X; [4] X is a random variable denoting the waiting time for the first flood occurrence; [5] $x_{\max}$ is the time when p_x equals 1; [6] n is the length of the time period over which flood risk is calculated.

4 Results and discussion

Results are grouped into three sections. First we present the development of the non-stationary GEV models (Sect. 4.1). Next the models are verified by comparing simulated results to observations (Sect. 4.2). Finally we present future projections of flood frequency analysis (Sect. 4.3).

4.1 Extreme value model development

A suite of models were fit to the logarithms of block cool season (November–April) maxima flows simulated by the calibrated VIC model with different non-stationary parameter combinations. The model structures tested include stationary, non-stationary location, non-stationary scale and non-stationary location and scale. For all model structures, model fit was tested using one or both covariates (i.e., precipitation and temperature). Models were also tested using the block maxima flows directly; however, performance was improved considerably with the logarithmic transformation. Validation of the VIC-simulated flows as well as the GEV models is presented in the following section.

Table 2 summarizes NLLH and AIC scores for each model configuration. The D for pairwise comparisons of NLLH scores and the p values calculated for each D based on a chi-squared distribution are also provided. The bottom row of Table 2 provides the number of parameters in each model and the model number that was used for the pairwise comparisons. As shown here, the models with non-stationary location and scale relying on both precipitation and temperature as covariates have the best (i.e., lowest) NLLH scores for both stations and are a statistically significant improvement over the other models listed in Table 2. Figure 2 plots stationary and non-stationary location and scale models with histograms of observed flow for both gauges. Qualitatively, the stationary model fits well with the center of the distribution but overestimates the tails. The non-stationary models overestimate the median values but are a closer fit to the extreme values.

The coefficients for Eqs. (2) and (3) for the selected models are provided in Table 3. Using the coefficients determined above, the location and scale parameters are calculated for every climate projection (i.e., 234) and flood season month (i.e., November to April 1950 to 2099) based on the downscaled precipitation and temperature values detailed in Sect. 2 (note that the shape parameter remains fixed). Thus, for every future month there is a separate GEV distribution curve for each of the 234 climate projections.

To address uncertainty of model parameters (namely the model coefficients, β in Eqs. 2 and 3), models of the same form (i.e., non-stationary location and scale with precipitation and temperature as covariates) were also fit to the historical simulation period (1950–1999) using downscaled precipitation and temperature from all 234 climate projections.

Table 2. Negative log likelihood (NLLH) and Akaike information criterion (AIC) scores for each model, as well as the deviance statistics (D) of pairwise comparisons of different model configurations (P is precipitation only, T is temperature only and P & T are both) and the p values of each D score based on a chi-squared distribution. The number of parameters in each model and the models used for comparison are listed at the bottom of the table. The selected model for each station is in bold.

Station	Metric	Stationary	Non-stationary location			Non-stationary scale			Non-stationary location and scale		
			P & *T*	*P*	*T*	*P* & *T*	*P*	*T*	*P* & *T*	*P*	*T*
		1	2	3	4	5	6	7	8	9	10
Farad	NLLH	508.9	422.9	467.1	499.7	487.3	500.9	506.5	**416.4**	462.2	496.9
	AIC	1023.7	855.9	942.3	1007.4	984.6	1009.8	1021.1	**846.8**	934.4	1003.8
	D		171.8	83.4	18.3	43.1	15.9	4.7	**13.0**	9.9	5.7
	p value of *D*		< 0.05	< 0.05	< 0.05	< 0.05	< 0.05	< 0.05	< 0.05	< 0.05	< 0.05
Reno	NLLH	505.4	418.4	462.5	496.0	484.4	497.6	503.1	**408.8**	457.4	493.2
	AIC	1016.8	846.8	932.9	1000.0	978.8	1003.2	1016.1	**831.7**	924.8	996.5
	D		174.0	85.9	18.8	42.0	15.6	4.7	**19.1**	10.1	5.5
	p value of *D*		< 0.05	< 0.05	< 0.05	< 0.05	< 0.05	< 0.05	< 0.05	< 0.05	< 0.05
# of model parameters		3	5	4	4	5	4	4	7	5	5
Model # compared to for *p* value			1	1	1	1	1	1	2	3	4

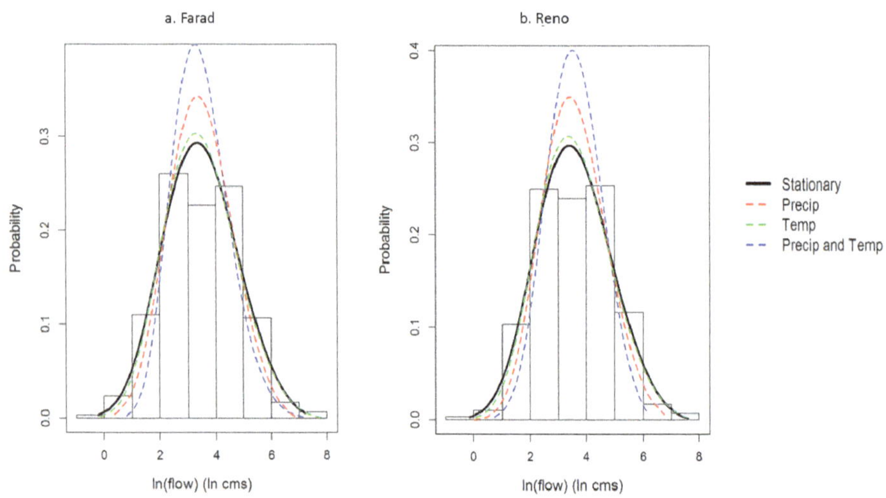

Figure 2. Probability density function of fitted stationary (solid black) and non-stationary (dashed) GEV models compared to historical VIC-simulated flow histogram.

Because each climate projection seeks to reproduce similar behavior over the historical 1950–1999 period, the variability between projections in this time frame is a measure of uncertainty in model coefficients given the representation of the same physical system. This differs from the variability between climate projections in future periods (i.e., after 1999), which is a measure of uncertainty in future forcing conditions. Table 3 shows the interquartile range of model coefficients calculated from the 234 historical GCM simulations.

Using these parameters, the return period of the design flood at Reno (37 600 cfs, 1065 cms) was calculated for every set of model parameters using observed historical precipitation and temperature. The observed model estimates a return period of 45 years while the interquartile range (IQR) using the simulated model parameters (i.e., the model parameters estimated from each of the 234 historical GCMs) with observed precipitation and temperature varies from 28 to 247 years. Note that the return period of 45 years estimated from observed meteorology is within the IQR of 28 to 247 years. Although the IQR is large, it should be kept in mind that some of the uncertainty in this range is a result of the performance of individual GCMs in simulating historical climate and in the BCSD downscaling methodology. The monthly BCSD algorithm used for downscaling GCM climate only constrains the monthly precipitation and temperature statistics (total precipitation and mean monthly tempera-

Table 3. Summary of derived model covariates for Eqs. (2) and (3) based on historical observations (historical observed) and using historical simulated data from the 234 CMIP-5 projections (historical simulated interquartile range, IQR).

	Farad			Reno		
	Historical observed	Historical simulated IQR		Historical observed	Historical simulated IQR	
$\beta_{0\mu}$	2.155	1.738	4.794	2.582	2.135	4.827
$\beta_{1\mu}$	0.175	0.053	0.148	0.180	0.066	0.152
$\beta_{2\mu}$	0.115	0.046	0.138	0.105	0.046	0.124
$\beta_{0\sigma}$	0.211	0.517	1.673	0.530	0.569	1.748
$\beta_{1\sigma}$	−0.013	−0.020	0.006	−0.018	−0.023	0.008
$\beta_{2\sigma}$	0.027	−0.012	0.022	0.017	−0.015	0.019
Shape (ξ)	−0.178	−0.389	−0.094	−0.275	−0.389	−0.070

ture) over the historical 1950–1999 period. Furthermore, uncertainty is introduced when monthly total precipitation and mean temperature are disaggregated to daily values. Thus, the estimated IQR implicitly captures climate simulation and downscaling uncertainties, in addition to explicitly representing model parameter uncertainty. The need to consider uncertainties at each and every step of the process starting with, for example, downscaling methods (statistical, dynamical or some combination of statistical and dynamical methods) is a topic of ongoing research.

4.2 Hydrologic and GEV model validation

Since we used modeled VIC flows for flood analysis, there are two considerations for model validation. First, we compare VIC-simulated 1-day flood events to the observed unregulated flow estimates (i.e., validating that our calibrated VIC model is accurately simulating flood flows). Second, we compare the GEV-modeled floods to the VIC-simulated 1-day flood events and the observed unregulated flow estimates (i.e., validating that the GEV models we fit to the simulated data match both the observed unregulated flows and the VIC-simulated flows).

Although unregulated flows are not available for the entire period of record, 1-day maximum unregulated flow estimates are available at Reno for six historical floods (USACE, 2013b). Figure 3 plots the observed flow (blue triangle) with the 1-day VIC flow that was simulated using historical observed forcings from Maurer et al. (2002) (red triangle) and a box plot of the non-stationary GEV distribution for the same month generated using the same monthly historical precipitation and temperature (i.e., Maurer et al., 2002). Comparing first the 1-day maximum VIC-simulated flow with the observed flow the maximum percent difference between the natural logarithm of simulated and observed flows is 12 %. There does appear to be a slight positive bias in the VIC simulations (i.e., VIC-simulated flows are greater than observed flood flows). Still, the simulated flood values (red circles) generally fall within the interquartile range of the GEV

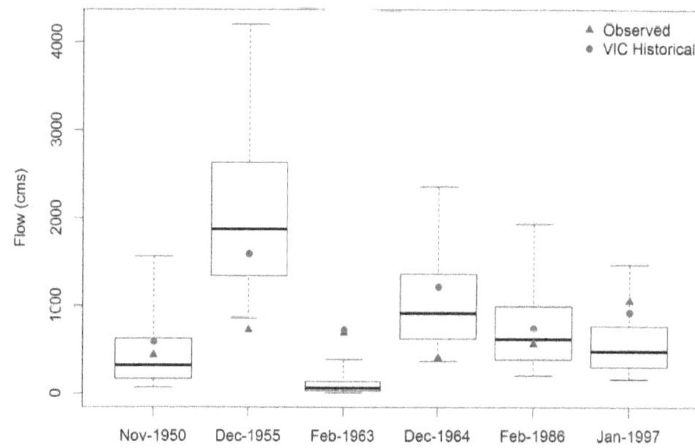

Figure 3. "Observed" unregulated flow estimated from gauge records (blue triangle) compared to VIC-simulated flow (red circles) and the simulated GEV distribution. Boxes span the 25th to 75th percentile of the GEV distribution for a given month and the whiskers extend to the 5th and 95th percentiles.

distribution except in the case of the 2 February 1963 flood and the 2 January 1997 flood.

In these instances the VIC simulation matches very closely (percent difference in the natural logarithm of flows are 0.5 and 1.2 %, respectively) with the observed flow; however, the GEV model underestimates the events. This discrepancy is caused by the flood timing. In both cases the flood occurs at the very beginning of the month. In the GEV framework, the precipitation and temperature are used as covariates for the flow of the same month. However, for these storms, flooding is linked to precipitation and temperature in the month of flooding and the preceding month. Therefore, the GEV model simulates the flood in the preceding month and/or underestimates the flood magnitude if the precipitation is split between 2 months. While this is a limitation for matching individual historical events (primarily timing), it should not be a major concern for future projections. This is because, for the purposes of risk calculations, it does not matter in what month the GEV model simulates the flood event as long as it realistically captures flood magnitude behavior.

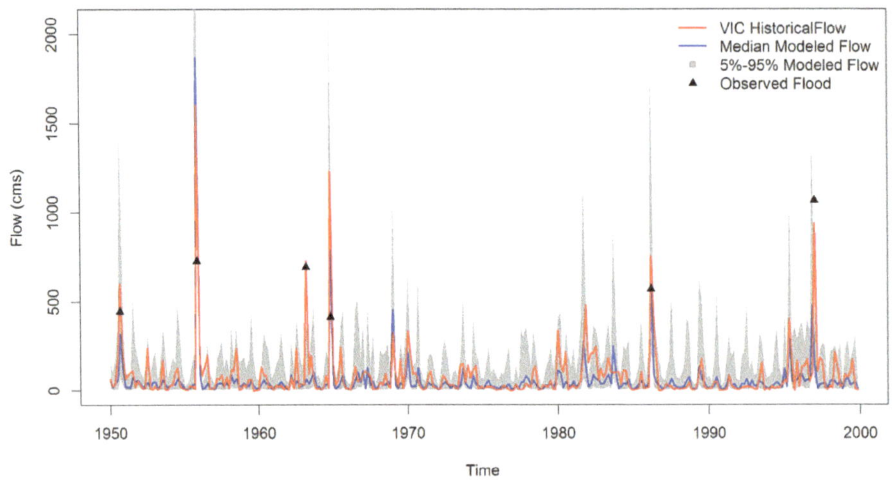

Figure 4. VIC-simulated 1-day flood maximums for November through April 1950 to 1999 (red lines) compared with the historical GEV distributions (blue line is median and grey shading is the 5th to 95th percentile range) and the six observed flow rates.

Comparing the GEV model distribution to the other observed floods (blue triangles), the distribution encompasses the observed flood magnitude (within the 5th and 95th percentile) for all except two of the floods (1955 and 1963). For 1963, the VIC-simulated and observed floods are in close agreement (the difference between the natural logarithm of simulated and observed flows is the smallest of any event at 0.5 %), and the discrepancy with the GEV model is consistent with the flood timing described above. The 1955 flood resulted from 38 cm of melted snow combined with 33 cm of rainfall over a 3-day period (O'Hara et al., 2007). In the historical forcings used to drive the VIC model, December 1955 has 75 cm of precipitation, which is the highest December precipitation value in the historical period. In this instance, the VIC-simulated flow falls within the interquartile range of the GEV model, but the high monthly precipitation results in an overestimation of the flood magnitude. Again, this is a limitation of using monthly forcings because the total December precipitation is used as a covariate and not a storm-specific value, though in many cases the storm-specific values constitute the bulk of the monthly precipitation totals.

Figure 4 is a time series plot of VIC historical simulated flow along with the median and 5th to 95th percentile flow of the GEV model. As would be expected from the model fit demonstrated in Figs. 2 and 3, Fig. 4 shows that the VIC-simulated flows are generally close to the median GEV-modeled flow and nearly always fall within the 5th to 95th percentile range. Although there are differences in the simulation of individual events discussed above, the median simulated flood magnitudes are only greater than the maximum observed flood in two instances of the 300 historical months simulated.

In general, Figs. 3 and 4 show that the VIC-simulated flows match closely with the observed floods (based on percent difference in the natural logarithm of flows) and that the interquartile range of the GEV distributions encompasses the observed and simulated flows in most instances. Figure 3 does illustrate some of the complications in matching individual events. However, based on analysis of the driving forces behind each individual event we are able to explain and document the sources of these discrepancies. Based on this analysis we conclude that the VIC model behavior has a reasonable match with the natural system.

4.3 Future flood risk

Future flood risk is calculated using Eq. (5b) from Table 1. For the first part of this analysis we define "flood" as 1-day flow exceeding 1065 cms (37 600 cfs). This is the maximum historical unregulated flow at Reno from the 2 January 1997 event and is considered to be the design flood for flood protection infrastructure design. For each simulation month (1950–2099 November–April) exceedance probabilities are calculated for every climate projection (234 in total) using the selected non-stationary GEV models from Table 3 (fit to the historical observations) and the projected monthly precipitation and temperature. As detailed in Sect. 3.2, when exceedance probabilities are time dependent, the flood risk (refer to Eq. 5b, Table 1) is a function of both the length of the design life and the period of operation. Figure 5 plots the risk of flood versus project life for three time periods: 1950 to 1999, 2000 to 2049 and 2050 to 2099. In other words, this is the risk of a flood exceeding 1065 cms in the next n years from 1950, 2000 or 2050. The median and interquartile ranges show the distribution of the 234 climate projections simulated. Here we use the interquartile range, as opposed to the 5th and 95th percentile, to focus on the central tendencies of each time period. Note that the ranges presented here express the variability between climate projections. Uncertainty of future VIC model simulations is not investigated

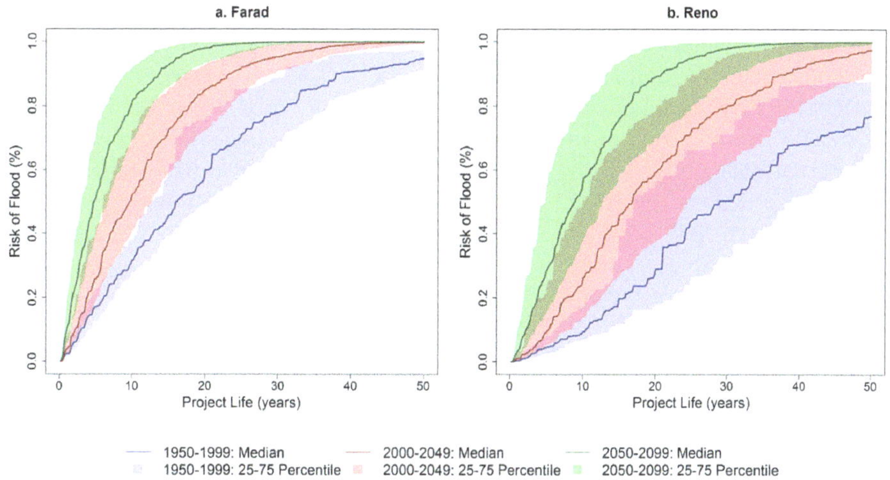

Figure 5. Probability of 1-day flood exceeding historical maximum of 1065 cms (risk) at Farad and Reno. Solid lines represent the median risk of the 234 climate projections and shading covers the interquartile range (i.e., 25th to 75th percentile).

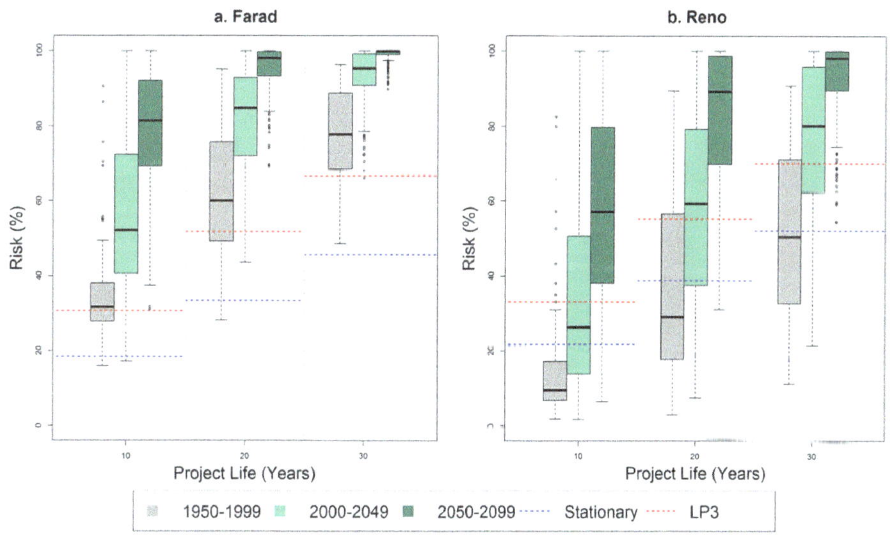

Figure 6. Box plots of the probability of a 1-day flood exceeding 1065 cms (risk) for three project life lengths (10, 20 and 30 years). Results are grouped by time period (1950–1999, 2000–2049 and 2050–2099). Blue dashed lines show the flood risk calculated from the stationary GEV model fit to the historical data.

here. For a detailed analysis of uncertainty in VIC simulations the reader is referred to Elsner et al. (2014).

For both Farad and Reno there is a clear positive shift in flood risk between the three time periods. In all cases the median risk for each subsequent time period falls outside the interquartile range of the preceding time period although the prediction spread for Reno is greater than that for Farad. It is important to note that the flood risk is actually higher at Farad than Reno in both the historical and future periods despite the fact that the observed flow distributions at the two stations are very similar (refer to Fig. 2). This shift between Farad and Reno is caused by the differences in the shape parameters (refer to Table 3). Farad has a relatively heavier tailed distribution (i.e., the GEV shape parameter for Farad is greater than the shape parameter for Reno) and therefore flood risks are increased. The sensitivity of the model parameters (and the associated flood risk) to small differences in the flow and covariate distributions is further demonstrated by Fig. 6.

Figure 6 presents the project life risk from Fig. 5 for three project life periods (10, 20 and 30 years). Box plots show the non-stationary model results for the 234 climate projections with the different time periods compared side by side. Also, the risk calculated using a stationary GEV model and a stationary LP3 model (i.e., the distribution prescribed by Bulletin 17B and fitted using L-moments; IACWD, 1982) fit to the historical flow data are plotted for reference (blue and red dashed lines, respectively). Comparing these three approaches (non-stationary GEV, stationary GEV and stationary LP3) provides information on the sensitivity of results to the modeling approach and non-stationary parameters. For instance, both stationary models are fit to the same historical simulated flows (one using MLE and the other using L-moments) so differences between the stationary lines reflect the impact of model choice and fitting approach on estimated risk. Conversely the stationary GEV model (blue line) and the historical non-stationary models (grey box plot) have the same model form and cover the same time period; the only difference is the addition of covariates to estimate model parameters. Thus differences between these two show the effect of model parameter changes from the non-stationary approach. Finally, variability between the box plots for a given design period demonstrates the evolution of risk over time (i.e., the impact of climate change on risk). The latter (i.e., changing risk over time) is the purpose of this analysis; however, before assessing change over time, we must first discuss the impact of model choice and parameters on risk estimates.

For both of the stationary methods, the risk increases with project life following Eq. (5a) from Table 1. The distinction between these lines and the non-stationary approaches is that, with the stationary approach, a single exceedance probability is calculated for the given flood magnitude, and this probability is assumed to remain constant throughout the design life. Also, for both stationary approaches the model is fit directly to the historical 1-day maximum flow distribution and no covariates are required (note that stationary models are not fit to the future time periods because this would require future simulated flows). Comparing the GEV (blue line) and the LP3 (red line) stationary models, there is a 10–20 % increase in risk between the two models. This difference is purely a function of model form and highlights the sensitivity of the risk calculations to model choice.

Contrasting the difference between the stationary (blue line) and the non-stationary GEV for the historical time period (grey box plot) illustrates the effect of adding non-stationary parameters to a given model form. Recall that in both cases the GEV model is fit to the historical simulated flows. However, for the stationary approach, model fitting results in a single set of parameters (location, scale and shape), whereas with the non-stationary approach we derive the shape parameter and a set of coefficients for linear models to determine the location and scale parameters based on precipitation and temperature values. Thus, for the non-stationary approach, different location and scale parameters are calculated for every historical cool season month and GCM model (234).

Overall, there is close agreement between the stationary (S) and average non-stationary (NS) location parameters (6.55 S vs. 6.64 NS at Farad and 6.63 S vs. 6.78 NS at Reno). However, for both gauges the scale parameter is lower with the non-stationary approach (1.30 S vs. 0.94 NS at Farad and 1.28 S vs. 0.96 NS at Reno). At Reno the shape parameter is similar (−0.24 S vs. −0.27 NS), but at Farad the difference is somewhat larger (−0.24 S vs. −0.18 NS). Differences in model parameters are reflected in the distance between the stationary GEV model (blue line) and the median historical non-stationary GEV box plots (center of the grey box plots) in Fig. 6. For Reno, the stationary line is closer to the historical box plots. However, at Farad the non-stationary box plots are consistently higher than the stationary line. The larger differences between the stationary and non-stationary models for Farad result from changes in the shape parameter between the stationary and non-stationary model fits. This change demonstrates the sensitivity of model results to changes in model parameters.

As with Fig. 5, Fig. 6 shows significant increases in risk moving into the future and subsequently larger differences between the stationary and non-stationary approach. By the second future period the differences between the stationary and non-stationary models can be as much as 50 % or more. For both gauges, difference in risk between the non-stationary and stationary approaches grows over time, indicating greater potential to underestimate the future risk if non-stationary parameters are not considered.

Results were also grouped by RCPs to analyze connections between greenhouse gas emission rates and changes in flood risk. As shown in Fig. 7, we observed no clear trend in flood risk based on the different RCPs. This indicates that, for this flood statistic in this basin, the variability between GCM model form and initial conditions likely overwhelms the influence of greenhouse gas emissions when comparing between scenarios. Although we caution that this is not a general finding, for this application we show that the variability between projections within any RCP scenario is larger than the difference between RCP scenarios. Harding et al. (2012) also noted similar behavior in their study of the Colorado River basin.

Given the sensitivity of projected risk to model parameters, an obvious question is whether increases in risk over time are similarly sensitive. For the 1065 cms flood plotted in Fig. 6, the increased risk with added project life (i.e., 20 years vs. 10 years) is greater with the non-stationary models than the stationary one at both stations. This is intuitive, given the increased flood risk with time demonstrated in Fig. 5 for the non-stationary models. Although Farad has higher overall risk, the relative increase in risk between time periods is similar between the two stations. For example, the median 10-year flood risk when comparing the first (1950–1999) and second (2000–2049) time periods increases by 21 % for Farad and 29 % for Reno.

Next, analysis is expanded to a range of flood magnitudes. Figure 8 plots the flood risk over a 10-year project life starting in 1950, 2000 and 2050 for flood values ranging from 283 to 1416 cms (10 000 to 50 000 cfs). As would be expected, the 10-year flood risk decreases with increasing flood rate. The shapes of the curves are slightly different between Farad and Reno; flood risk decreases more sharply with increased flow at Reno than Farad. Again, this behavior is a function of the shape parameter of the respective GEV distri-

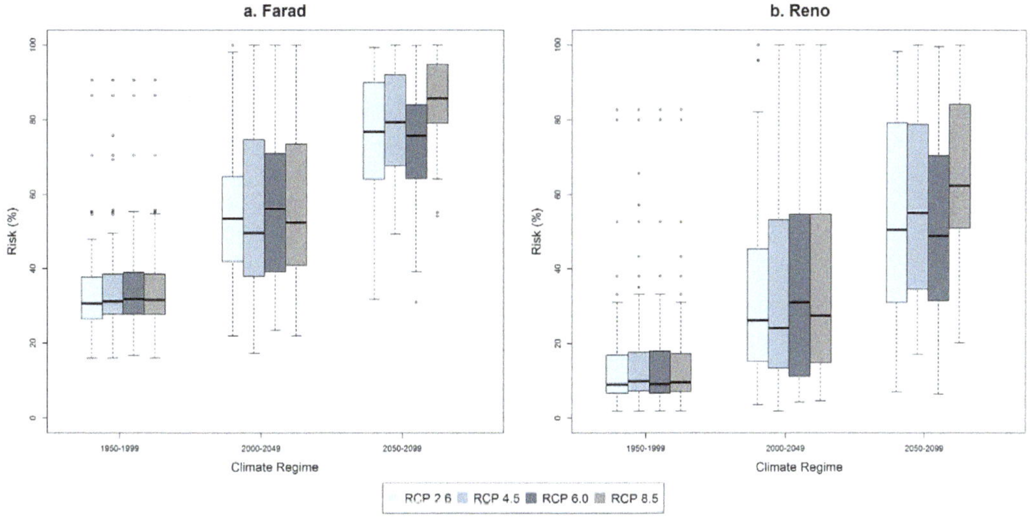

Figure 7. Box plots of the probability of a 1-day flood exceeding 1065 cms (risk) in 10 years for three 50-year periods. Results are grouped by the representative concentration pathways (RCPs) used to drive the GCM projection. RCP 8.5 has the largest increase in greenhouse gas concentrations and RCP 2.6 the smallest.

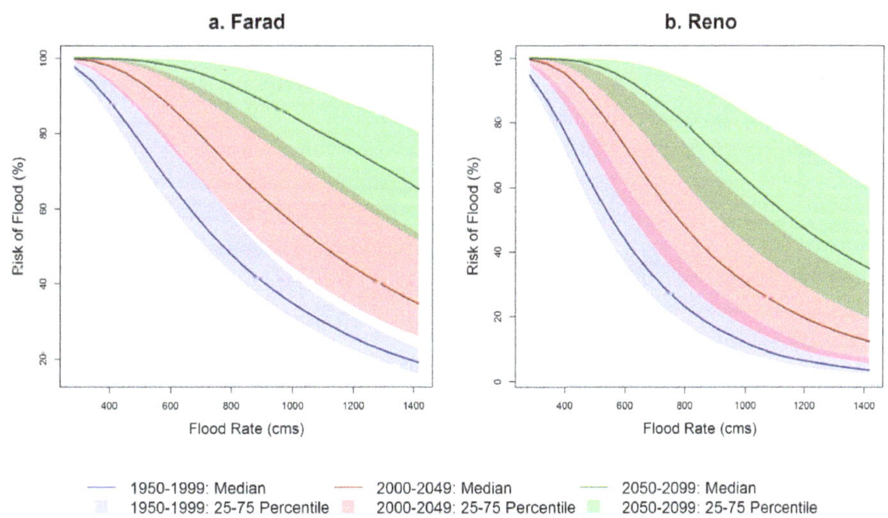

Figure 8. Probability of flood in a 10-year project life (risk) vs. median 1-day flood rate at **(a)** Farad and **(b)** Reno for three time periods 1950–1999 (blue), 2000–2049 (red) and 2050–2099 (green). Solid lines represent the median of the 234 climate projections and shading covers the interquartile range (i.e., 25th to 75th percentile).

butions. Despite these differences, both gauges display clear shifts between time periods similar to Fig. 5. Again, the median risk for each subsequent period consistently falls outside the interquartile range of the preceding period.

Changes in the median flood risk (i.e., differences between the solid lines in Fig. 8) between each future period and the historical period are plotted in Fig. 9 for both gauges. As would be expected based on the qualitative differences in Fig. 8, the shape of the Farad and Reno difference curves are slightly different. However, the salient point for this analysis is that the increased risk between periods and the two stations is generally within 10 %. Overall, the increased risk between the first future period (2000–2050) and the historical period (1950–1999) is between 10 and 20 % for flows from 600 to 1200 cms. Similarly, the increased risk from the historical period to the second future period (2050–2099) is between 30 and 50 %. Differences for the highest and lowest flows are difficult to assess because the median is skewed for high and low flow values by the fact that the risk values are bound between 0 and 100 %.

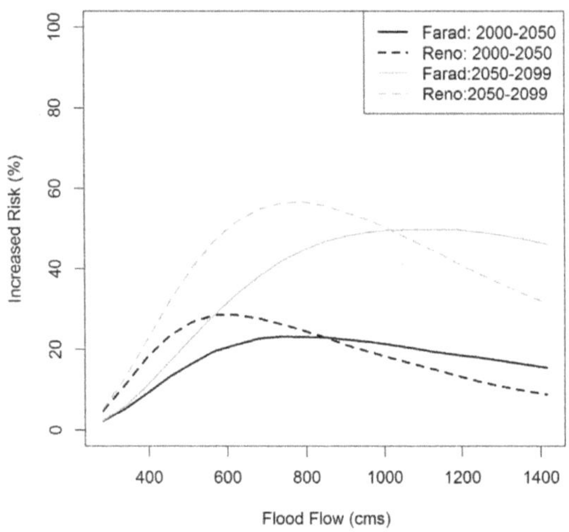

Figure 9. Increased probability of flood occurrence for a 10-year project life (risk) from the historical period (1950–1999) to each of the two future periods, 2000–2050 (black) and 2050–2099 (grey). Farad is plotted with a solid line and Reno is a dashed line.

5 Summary and conclusions

The analysis presented is unique in its incorporation of non-stationary GEV analysis using CMIP-5 projections and the design-life level risk assessment concepts. We present our findings as a relevant case study and an example application of recent developments in non-stationary flood assessment. Lacking sufficient unregulated flow data we simulate historical floods using the VIC model. Subsequently we use the simulated floods to fit non-stationary GEV models, with downscaled monthly precipitation and temperature as covariates. Although there are some discrepancies between individual simulated and observed flood events, we demonstrate that the VIC model adequately captures the range of flood magnitudes. Furthermore, we show that the GEV-modeled historical floods are in good agreement with both the VIC-simulated floods and the published flood events (USACE, 2013b).

Discrepancies between historical and simulated events often result from the monthly time step used for covariates. This can affect the ability to model floods that are generated by precipitation that occurs in 2 months. Also, because the climate variables are monthly aggregates, and not event based, large floods can be generated in months with high precipitation even if that precipitation does not occur in one concentrated event. Despite these differences, comparison with historical flood events demonstrates that the GEV model does reasonably well at simulating historical flood magnitudes, even if some individual historical events are not matched exactly.

Using the derived non-stationary GEV models, we generate flood distributions for 234 CMIP5 climate projections from 1950 to 2099. For the historical 1-day design flood magnitude of 1065 cms, results show significant increases in the frequency of high-flow events in the future. From a water management standpoint, this finding translates directly to increased flood risk. For example, we calculate a 21 % (29 %) increase risk of a 1065 cms flood over a 10-year design life for Farad (Reno) from the historical time period to the first future period and similar increases from the first future period to the second. Increased risk between time periods is also relatively consistent for longer design life periods and similar shifts in flood risk are noted across a range of flood magnitudes. For both stations, the increased risk from the historical to the first future period is between 10 and 20 % and from the historical to the second future period is between 30 and 50 % for floods ranging from 600 to 1200 cms.

The significant increases in flood risk through time indicate the importance of non-stationary flood frequency analysis for future infrastructure planning, and the potential to underestimate risk when stationarity is assumed. For both stations the difference between the stationary and non-stationary approach increases over time. By the second future period (2000–2049), differences in risk calculations between the stationary and non-stationary models can be 50 % or larger. This finding is in keeping with a number of recent studies (e.g., Griffis and Stedinger, 2007; Katz et al., 2002; Towler et al., 2010) that have highlighted potential applications for non-stationary analysis of flood frequency.

An important consideration for this approach is the sensitivity of results to model parameters. In all cases, the flood risk is higher at Farad than Reno due to the relatively heavier tailed distribution that was fit. Estimated model parameters differed by station despite the fact that the flow, precipitation and temperature distributions for both locations are very similar. While these changes affected the overall risk projections, the relative increase in risk over time remained consistent between stations. This indicates that the more robust metric from this analysis is the relative increase in flood risk and not the absolute values. This finding is further supported by the fact that absolute flood risk estimates could be impacted by model bias. By focusing on differences in risk, we specifically highlight the impact of non-stationarity on risk assessment as opposed to parameter sensitivity. Similarly, it is important to note that this analysis is based on natural flow estimates and does not include infrastructure development or operation. Therefore, results indicate the potential increase in the underlying natural flood risk in the UTRB over the 21st century.

Acknowledgements. The authors would like to acknowledge the Reclamation WaterSMART Basin Study program for funding the Truckee River basin study which supported this publication. In addition, we would like to thank the reviewers for their constructive comments that helped to improve the manuscript.

Edited by: R. Merz

References

Akaike, H.: New look at statistical-model identification, IEEE Trans. Autom. Control, 19, 716–723, 1974.

Allan, R. P.:Climate change: Human influence on rainfall, Nature, 470, 344–345, 2011.

Cayan, D. R., Redmond, K. T., and Riddle, L. G.: ENSO and Hydrologic Extremes in the Western United States, J. Climate, 12, 2881–2893, 1999.

Cayan, D. R., Kammerdiener, S. A., Dettinger, M. D., Caprio, J. M., and Peterson, D. H.: Changes in the Onset of Spring in the Western U.S., B. Am. Meteorol. Soc., 82, 399–415, 2001.

Christensen, N. S. and Lettenmaier, D. P.: A multimodel ensemble approach to assessment of climate change impacts on the hydrology and water resources of the Colorado River Basin, Hydrol. Earth Syst. Sci., 11, 1417–1434, doi:10.5194/hess-11-1417-2007, 2007.

Christensen, N. S., Wood, A. W., Lettenmaier, D. P., and Palmer, R. N.: Effects of climate change on the hydrology and water resources of the Colorado river basin, Climate Change, 62, 337–363, 2004.

Das, T., Pierce, D. W., Cayan, D. R., Vano, J. A., and Lettenmaier, D. P.: The importance of warm season warming to western U.S. streamflow changes, Geophys. Res. Lett., 38, L23403, doi:10.1029/2011GL049660, 2011.

Dettinger, M. D. and Cayan, D. R.: Large-scale Atmospheric Forcing of Recent Trends toward Early Snowmelt Runoff in California, J. Climate, 8, 606–623, 1995.

Douglas, E. M., Vogel, R. M., and Kroll, C. N.: Trends in floods and low flows in the United States: impact of spatial correlation, J. Hydrol., 240, 90–105, 2000.

Easterling, D. R., Meehl, G. A., Parmesan, C., Changnon, S. A., Karl, T. R., and Mearns, L. O.: Climate Extremes: Observations, Modeling and Impacts, Science, 289, 2068–2074, 2000.

Elsner, M. M., Gangopadhyay, S. G., Pruitt, T., Brekke, L. D., Mizukami, N., and Clark, M. P.: How does the choice of distributed meteorological data affect hydrologic model calibration and streamflow simulations?, J. Hydrometeorol., 15, 1384–1403, 2014.

Franks, S. W.: Identification of a change in climate state using regional flood data, Hydrol. Earth Syst. Sci., 6, 11–16, doi:10.5194/hess-6-11-2002, 2002.

Gangopadhyay, S., Pruitt, T., Brekke, L. D., and Raff, D. A.: Hydrologic Projections for the Western United States, Eos, 92, 441–452, 2011.

Gilleland, E. and Katz, R. W.: New software to analyze how extremes change over time, Eos, 92, 13–14, 2011.

Gilroy, K. L. and McCuen, R. H.: A nonstationary flood frequency analysis method to adjust for future climate change and urbanization, J. Hydrol., 414–415, 40–48, 2012.

Griffis, V. and Stedinger, J. R.: Incorporating climate change and variability into Bulletin 17B LP3 model, paper presented at ASCE World Env. & Water Resour. Congress, Tampa, Florida, USA, 2007.

Gutowski, W. J., Hegerl, G. C., Holland, G. J., Knutson, T. R., Mearns, L. O., Stouffer, R. J., Webster, P. J., Wehner, M. F., and Zwiers, F. W.: Causes of Observed Changes in Extremes and Projections of Future Changes in Weather and Climate Extremes in a Changing Climate, Regions of Focus: North America, Hawaii, Caribbean, and U.S. Pacific Islands Rep., Washington, D.C., 2008.

Hall, J., Arheimer, B., Borga, M., Brázdil, R., Claps, P., Kiss, A., Kjeldsen, T. R., Kriaučiūnienė, J., Kundzewicz, Z. W., Lang, M., Llasat, M. C., Macdonald, N., McIntyre, N., Mediero, L., Merz, B., Merz, R., Molnar, P., Montanari, A., Neuhold, C., Parajka, J., Perdigão, R. A. P., Plavcová, L., Rogger, M., Salinas, J. L., Sauquet, E., Schär, C., Szolgay, J., Viglione, A., and Blöschl, G.: Understanding flood regime changes in Europe: a state-of-the-art assessment, Hydrol. Earth Syst. Sci., 18, 2735–2772, doi:10.5194/hess-18-2735-2014, 2014.

Harding, B. L., Wood, A. W., and Prairie, J. R.: The implications of climate change scenario selection for future streamflow projection in the Upper Colorado River Basin, Hydrol. Earth Syst. Sci., 16, 3989–4007, doi:10.5194/hess-16-3989-2012 2012.

Hirsch, R. M. (2011), A perspective on nonstationarity and water management, *Journal of the American Water Resources Association*, 47(), 436-446.

IACWD – Interagency Advisory Committee on Water Data: Guidelines for determining flood flow frequency: Bulletin 17B of the Hydrology Subcommittee, Office of Water Data Coordination, US Geological Survey, Reston, VA, p. 183, 1982.

Jain, S. and Lall, U.: Floods in a changing climate: Does the past represent the future?, Water Resour. Res., 37, 3193–3205, 2001.

Katz, R. W.: Statistics of extremes in climate change, Climate Change, 100, 71–76, 2010.

Katz, R. W., Parlange, M. B., and Naveau, P.: Statistics of extremes in hydrology, Adv. Water Resour., 25, 1287–1304, 2002.

Kunkel, K. E.: North American Trends in Extreme Precipitation, Nat. Hazards, 29, 291–305, 2003.

Liang, X., Lettenmaier, D. P., Wood, E. F., and Burges, S. J.: A simple hydrologically based model of land surface water and energy fluxes for general circulation models, J. Geophys. Res., 99, 14415–14428, 1994.

Liang, X., Wood, E. F., and Lettenmaier, D. P.: Surface soil moisture parameterization of the VIC-2L model: Evaluation and modification, Global Planet. Change, 13, 195–206, 1996.

Madsen, T. and Figdor, E.: When it Rains it Pours – Global Warming and the Rising Frequency of Extreme Precipitaiton in the U.S., Environmental America Research and Policy Center, http://www.environmentamerica.org/reports/ (last access: January 2015), 2007.

Maidment, D. R.: Handbook of Hydrology, McGraw-Hill, 1993.

Mailhot, A. and Duchesne, S.: Design criteria of urban drainage infrastructures under climate change, J. Water Resour. Pl. Manage., 136, 201–208, 2010.

Maurer, E. P., Brekke, L. D., Pruitt, T., and Duffy, P. B.: Fine-resolution climate projections enhance regional climate change impact studies, Eos Tran. AGU, 88, 504, 2007.

Maurer, E. P., Wood, A. W., Adam, J. C., Lettenmaier, D. P., and Nijssen, B.: A Long-Term Hydrologically-Based Dataset of Land Surface Fluxes and States for the Conterminous United States, J. Climate, 15, 3237–3252, 2002.

Meehl, G. A., Karl, T., Easterling, D. R., Changnon, S., Pielke Jr., R., Changnon, D., Evans, J., Groisman, P. Y., Knutson, T. R., Kunkel, K. E., Mearns, L. O., Parmesan, C., Pulwarty, R., Root, T., Sylves, R. T., Whetton, P., and Zwiers, F.: An Introduction to Trends in Extreme Weather and Climate Events: Observations, Socioeconomic Impact, Terrestrial Ecological Impacts, and Model Projections, B. Am. Meterorol. Soc., 81, 413–416, 2000.

Merz, B., Vorogushyn, S., Uhlemann, S., Delgado, J., and Hundecha, Y.: HESS Opinions "More efforts and scientific rigour are needed to attribute trends in flood time series", Hydrol. Earth Syst. Sci., 16, 1379-1387, doi:10.5194/hess-16-1379-2012, 2012.

Milly, P. C. D., Betancourt, J., Falkenmark, M., Hirsch, R. M., Kundzewicz, Z. W., Lettenmaier, D. P., and Stouffer, R. J.: Stationarity Is Dead: Whither Water Management, Science, 319, 573–574, 2008.

Min, S.-K., Zhang, X., Zwiers, F. W., and Hegerl, G. C.: Human contribution to more-intense precipitation extremes, Nature, 470, 378–381, 2011.

Mote, P. W., Hamlet, A. F., Clark, M. P., and Lettenmaier, D. P.: Declining Mountain Snowpack In Western North America, B. Am. Meterorol. Soc., 86, 39–49, doi:10.1175/BAMS-86-1-39, 2005.

Mullet, C. J., O'Gorman, P. A., and Back, L. E.: Intensification of precipitation extremes with warming in a cloud resolving model, J. Climate, 24, 2784–2800, 2011.

Nijssen, B., Lettenmaier, D. P., Liang, X., Wetzel, S. W., and Wood, E. F.: Streamflow simulation for continental-scale river basins, Water Resour. Res., 33, 711–724, 1997.

Obeysekera, J. and Salas, J. D.: Quantifying the uncertainty of design floods under nonstationary conditions, J. Hydrol. Eng;, 19, 1438–1446, 2014.

O'Gorman, P. A. and Schneider, T.: The physical basis for increases in precipitaiton extremes in simulations of 21st century climate change, P. Natl. Acad. Sci., 106, 14773–14777, 2009.

O'Hara, B. F., Barbato, G. E., James, J. W., Angeloff, H. A., and Cylke, T.: Weather and climate of the Redno-Carson City-Lake Tahoe Region, Nevada Special Publication 34, Bureau of Mines and Geology, Reno, Nevada, USA, 2007.

Pall, P., Aina, T., Stone, D. A., Stott, P. A., Nozawa, T., Hilberts, A. G. J., Lohmann, D., and Allen, M. R.: Anthropogenic greenhouse gas contribution to flood risk in England and Wales in autumn 2000, Nature, 470, 382–385, 2011.

Payne, J. T., Wood, A. W., Hamlet, A. F., Palmer, R. N., and Lettenmaier, D. P.: Mitigating the effects of climate change on the water resources of the Columbia River basin, Climate Change, 62, 233–256, 2004.

Pierce, D. W., Barnett, T. P., Santer, B. D., and Gleckler, P. J.: Selecting global climate models for regional climate change studies, P. Natl. Acad. Sci., 106, 8441–8446, 2009.

Pierce, D. W., Das, T., Cayan, D. R., Maurer, E. P., Miller, N. L., Bao, Y., Kanamitsu, M., Yoshimura, K., Snyder, M. A., Sloan, L. C., Franco, G., and Tyree, M.: Probabilistic estimates of future changes in California temperature and precipitation using statistical and dynamical downscaling, Clim. Dynam., 40, 839–856, 2012.

Raff, D. A., Pruitt, T., and Brekke, L. D.: A framework for assessing flood frequency based on climate projection information, Hydrol. Earth Syst. Sci., 13, 2119–2136, doi:10.5194/hess-13-2119-2009, 2009.

Ralph, F. M. and Dettinger, M. D.: Historical and National Perspectives on Extreme West Coast Precipitation Associated with Atmospheric Rivers during December 2010, B. Am. Meterorol. Soc., 93, 783–790, 2012.

R Core Team: R: A language and environment for statistical computing. R Foundation for Statistical Computing, Vienna, Austria, http://www.R-project.org/ (last access: 26 November 2014), 2012.

Reclamation: Truckee River Basin Study, Fact Sheet, available at: http://www.usbr.gov/WaterSMART/bsp/docs/fy2010/Truckee_BasinFactsheetFinal.pdf (last access: 22 September 2014), 2010.

Reclamation: West-wide climate risk assessments: Bias-corrected and spatially downscaled surface water projections, Tech. Memo. 86-68210-2011-01, US Department of the Interior, Bureau of Reclamation, Technical Service Center, Denver, Colorado, 138 pp., 2011.

Reclamation: Downscaled CMIP3 and CMIP5 Climate Projections: Release of Downscaled CMIP5 Climate Projections, Comparison with Preceding Information, and Summary of User Needs, Downscaled CMIP3 and CMIP5 Climate and Hydrology Projections, US Department of the Interior, Bureau of Reclamation, Technical Service Center, Denver, Colorado, 47 pp., 2013.

Reclamation: Downscaled CMIP3 and CMIP5 Climate and Hydrology Projections: Release of Hydrology Projections, Comparison with preceding Information, and Summary of User Needs, prepared by the US Department of the Interior, Bureau of Reclamation, Technical Service Center, Denver, Colorado, 110 pp., 2014.

Regonda, S. K., Rajagopalan, B., Clark, M., and Pitlick, J.: Seasonal Cycle Shifts in Hydroclimatology Over the Western U.S., J. Climate, 18, 372–384, 2005.

Rootzén, H. and Katz, R. W.: Design Life Level: Quantifying risk in a changing climate, Water Resour. Res., 49, 5964–5972, 2013.

Salas, J. and Obeysekera, J.: Revisiting the Concepts of Return Period and Risk for Nonstationary Hydrologic Extreme Events, J. Hyrol. Eng., 19, 554–568, 2014.

Slack, J. R., Lumb, A. M., and Landwehr, J. M.: Hydroclimatic data network (HCDN): A U.S. Geological Survey streamflow data set for the United Sates for the study of climate variation 1874–1988, USGS Water Resour. Invest. Rep. 93-4076, USGS, Reston, Virginia, USA, 1993.

Small, D., Islam, S., and Vogel, R. M.: Trends in precipitation and streamflow in the eastern U.S.: Paradox or perception?, Geophys. Res. Lett., 33, L03403, doi:10.1029/2005GL024995, 2006.

Stedinger, J. R. and Griffis, V. W.: Getting from here to where? Flood frequency analysis and climate, J. Am. Water Resour. Ass., 47, 506–513, 2011.

Stokes, J.: Draft Farad Diversion Dam Replacement Project Environmental Impact Report Rep., State Water Resources Control Board, Sacramento, CA, 2002.

Sun, Y., Solomon, S., Dai, A., and Portmann, R. W.: How Often Will it Rain?, J. Climate, 20, 4801–4818, 2007.

Taylor, K. E., Stouffer, R. J., and Meehl, G. A.: A Summary of the CMIP5 Experiment Design, B. Am. Meteorol. Soc., 93, 485–498, 2012.

Towler, E., Rajagopalan, B., Gilleland, E., Summers, R. S., Yates, D., and Katz, R. W.: Modeling hydrologic and water quality extremes in a changing climate: A statistical approach based on extreme value theory, Water Resour. Res., 46, W11504, doi:10.1029/2009WR008876, 2010.

USACE: Final environmental impact statement for the Truckee Meadows Flood Control Project: General Reevaluation Report Volume 1, US Army Corps of Engineers, Sacramento, 2013a.

USACE: Truckee Meadows Flood Control Project, Nevada: Draft General Reevaluation Report Rep., US Army Corps of Engineers, Sacramento, 2013b.

Van Rheenen, N. T., Wood, A. W., Palmer, R. N., and Lettenmaier, D. P.: Potential implications of PCM climate change scenarios for Sacramento-San Joaquin River Bain hydrology and water resources, Climate Change, 62, 257–281, 2004.

Villarini, G., Serinaldi, F., Smith, J. A., and Krajewski, W. F.: On the stationarity of annual flood peaks in the continental United States during the 20th century, Water Resour. Res., 45, W08417, doi:10.1029/2008WR007645, 2009.

Vogel, R. M., Yaindl, C., and Walter, M.: Nonstationarity: Flood magnification and recurrence reduction factors in the United States, J. Am. Water Resour. Assoc., 47, 464–474, 2011.

Walter, M. and Vogel, R. M.: Increasing trends in peak flows in the northeastern United States and their impacts on design, paper presented at 2nd Joint Federal Interagency Conference, Las Vegas, NV, 2010.

Wood, A. W., Leung, L. R., Sridhar, V., and Lettenmaier, D. P.: Hydrologic implications of dynamical and statistical approaches to downscaling climate model outputs, Climatic Change, 62, 189–216, 2004.

Wu, H., Adler, R. F., Tian, Y., Juffman, G. J., Li, H., and Wang, J. J.: Real-time global flood estimation using satellite-based precipitation and a coupled land surface routing model, Water Resour. Res., 50, 2693–2717, 2014.

6

Satellite radar altimetry for monitoring small rivers and lakes in Indonesia

Y. B. Sulistioadi[1,5], K.-H. Tseng[2], C. K. Shum[3,9], H. Hidayat[4,7], M. Sumaryono[5], A. Suhardiman[5,6], F. Setiawan[7], and S. Sunarso[8]

[1]Hydrological Sciences Laboratory, NASA Goddard Space Flight Center, Greenbelt, MD, USA
[2]Center of Space & Remote Sensing Research, National Central University, Taoyuan, Taiwan
[3]Division of Geodetic Science, School of Earth Sciences, the Ohio State University, Columbus, OH, USA
[4]Hydrology and Quantitative Water Management Group, Wageningen University, Wageningen, the Netherlands
[5]Department of Forest Science, University of Mulawarman, Samarinda, Indonesia
[6]Department of Global Agricultural Sciences, the University of Tokyo, Tokyo, Japan
[7]Research Center for Limnology, Indonesian Institute of Sciences, Cibinong, Indonesia
[8]PT Vale Indonesia, Tbk, Sorowako, Indonesia
[9]Institute of Geodesy and Geophysics, Chinese Academy of Sciences, Wuhan, China

Correspondence to: Y. B. Sulistioadi (y.b.sulistioadi@nasa.gov)

Abstract. Remote sensing and satellite geodetic observations are capable of hydrologic monitoring of freshwater resources. Although satellite radar altimetry has been used in monitoring water level or discharge, its use is often limited to monitoring large rivers (> 1 km) with longer interval periods (> 1 week) because of its low temporal and spatial resolutions (i.e., satellite revisit period). Several studies have reported successful retrieval of water levels for small rivers as narrow as 40 m. However, processing current satellite altimetry signals for such small water bodies to retrieve water levels accurately remains challenging. Physically, the radar signal returned by water bodies smaller than the satellite footprint is most likely contaminated by non-water surfaces, which may degrade the measurement quality. In order to address this scientific challenge, we carefully selected the waveform shapes corresponding to the range measurement resulting from standard retrackers for the European Space Agency's (ESA's) Envisat (Environmental Satellite) radar altimetry. We applied this approach to small (40–200 m in width) and medium-sized (200–800 m in width) rivers and small lakes (extent < 1000 km^2) in the humid tropics of Southeast Asia, specifically in Indonesia. This is the first study that explored the ability of satellite altimetry to monitor small water bodies in Indonesia.

The major challenges in this study include the size of the water bodies that are much smaller than the nominal extent of the Envisat satellite footprint (e.g., ~ 250 m compared to ~ 1.7 km, respectively) and slightly smaller than the along-track distance (i.e., ~ 370 m). We addressed this challenge by optimally using geospatial information and optical remote sensing data to define the water bodies accurately, thus minimizing the probability of non-water contamination in the altimetry measurement. Considering that satellite altimetry processing may vary with different geographical regions, meteorological conditions, or hydrologic dynamic, we further evaluated the performance of all four Envisat standard retracking procedures.

We found that satellite altimetry provided a good alternative or the only means in some regions of measuring the water level of medium-sized rivers and small lakes with high accuracy (root mean square error (RMSE) of 0.21–0.69 m and a correlation coefficient of 0.94–0.97). In contrast to previous studies, we found that the commonly used Ice-1 retracking algorithm was not necessarily the best retracker among the four standard waveform retracking algorithms for Envisat radar altimetry observing inland water bodies. As a recommendation, we propose to include the identification and selection of standard waveform shapes to complete the use of standard waveform retracking algorithms for Envisat radar altimetry data over small and medium-sized rivers and small lakes.

1 Introduction

A number of small to medium-sized rivers are poorly gauged (Alsdorf and Lettenmaier, 2003). Small rivers are defined as those with 40–200 m in width and 10–100 $m^3 s^{-1}$ average discharge, while medium rivers are defined as those with 200–800 m in width and 100–1000 $m^3 s^{-1}$ average discharge (Meybeck et al., 1996). The installation and operation of in situ measurement such as permanent gauging is costly and not a priority for developing countries such as Indonesia. However, there is an increasing interest in continuous satellite-based monitoring of hydrologic bodies, including narrow or small rivers. Therefore, with the absence of continuously operating in situ measurements, it is a scientific and social challenge to develop a complementary water resource monitoring system with water level and discharge as the essential variables.

Space geodesy and satellite remote sensing are viable sources of observation to complement or replace in situ measured data that are lacking or unavailable. A number of researchers have demonstrated the ability of remote sensing to measure hydrological variables (Tang et al., 2009). Initiatives to develop a global river and lake water level database exist to date, but none of them accounts for small to medium-sized rivers and lakes in the humid tropics.

Satellite altimetry missions were initially aimed to support oceanographic studies (Brown and Cheney, 1983). However, scientists were able to use altimetry data to retrieve the water surface elevation of large rivers and lakes. These studies include those utilizing early satellite altimetry missions (Wingham and Rapley, 1987; Koblinsky et al., 1993; Morris and Gill, 1994), as well as recent ones (e.g., Birkett, 1998; Benveniste and Defrenne, 2003; Kouraev et al., 2004; Calmant and Seyler, 2006; Frappart et al., 2006; Cretaux et al., 2011).

The application of satellite altimetry to monitor inland waters has several limitations. The long satellite repeat cycle makes the satellite potentially miss important hydrological events (e.g., flash flood) between the repeats. For instance, the repeat period for TOPEX/Poseidon and Jason-1/2 is 10 days, 35 days for ERS-1/2, Envisat and SARAL/Altika, and 91 days for ICESat. The low spatial resolution of the radar altimeter, as represented by the radar altimeter footprint (about 1.7 to 3 km for calm waters), limits the measurement only to wide rivers, due to interference of the returned radar signal by non-water features. Earlier studies showed that satellite radar altimetry was useful for monitoring large rivers with widths greater than 1 km (Birkett, 1998; Birkett et al., 2002). However, recent studies demonstrated successful retrieval of water levels of small rivers (< 100 m width) (Kuo and Kao, 2011; Michailovsky et al., 2012). Nonetheless, the processing of satellite altimetry measurement for small water bodies remains challenging because of its spatial and temporal limitations.

Early studies of satellite altimetry to retrieve water levels of a river used waveform shapes to match the specular characteristics that exclusively belong to the signals returned by the river (Koblinsky et al., 1993). "Specular" refers to a reflection characteristic where a signal reflects in one direction, thus matching the reflection by a mirror (e.g., Torrance and Sparrow, 1967). In the context of radar signal processing, this occurs when the radar signal hits a calm or smooth water surface, which is represented as a peak in the return signal power (e.g., as represented by a power spectrum). Along with this principle, scientists developed non-ocean retrackers in the last decade. These include the offset center of gravity (OCOG) or Ice-1 (Wingham et al., 1986), the volume scattering retracker (Davis, 1997), the Sea Ice retracker (Laxon, 1994), the NASA ß retracker (Zwally, 1996), the surface/threshold retracker (Davis, 1997) and Ice-2 (Legresy and Remy, 1997). The offset center of gravity (OCOG) or Ice-1 (Wingham et al., 1986) is a simple but robust retracker that only requires the statistics of the waveform samples and does not require any model (model-free retracker) (Bamber, 1994). The Ice-2 algorithm modifies the Ocean retracker (Brown, 1977) by adding a scattering distribution coefficient that describes the vertical profile of the reflecting surfaces. This coefficient accounts for the interference of the default scattering pattern as generated by snow, ice sheets, sand or vegetation (Legresy and Remy, 1997). Laxon (1994) introduced the Sea Ice algorithm specifically to study sea ice elevation by (1) characterizing the power and shapes of the radar return, (2) classifying the sea ice and determining the waveform parameters, and (3) correcting the retracked range. Ice-1, Ice-2 and Sea Ice along with the Ocean retracker (that is exclusively developed for ocean studies) were the standard retrackers for the European Space Agency (ESA)'s Envisat (Environmental Satellite) until the satellite was decommissioned in June 2012. Recent developments of inland water retracking methods include the improvements to the threshold retracker (Davis, 1997) by Lee (2008) and Bao et al. (2009), sub-waveform analysis (e.g., Hwang et al., 2006; Fenoglio-Marc et al., 2009) and sub-waveform filtering and track offset correction (Tseng, 2012).

For inland water studies of rivers and lakes, Frappart et al. (2006) found Ice-1 to be the best retracker for large rivers (e.g., the Amazon River) over the other standard retrackers for Envisat (i.e., Ocean, Ice-2 and Sea Ice). None of these retrackers is specifically developed for inland waters. Satellite altimetry processing also varies depending on geographical regions, meteorological conditions, and the hydrological dynamics of the water bodies. Up to this point, no "one size fits all" method for satellite altimetry waveform retracking is readily available to measure the water level of small (40–

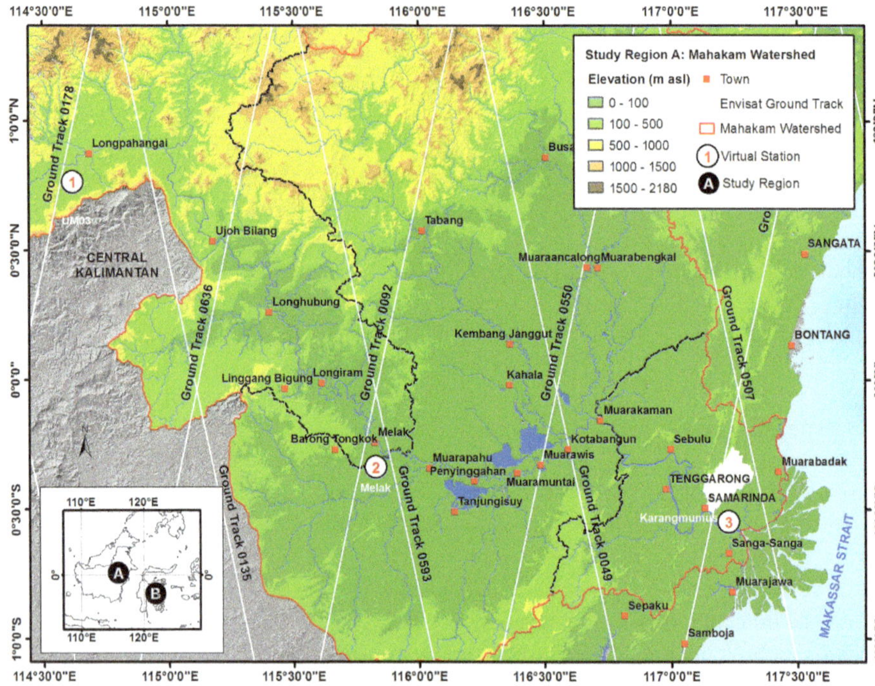

Figure 1. Study sites at the Mahakam watershed, East Kalimantan, Indonesia.

200 m in width) and medium-sized (200–800 m in width) rivers and lakes. Hence, there is a need to develop a specific algorithm or additional procedures for satellite altimetry applications to study inland waters. Furthermore, there is also a need to evaluate the commonly used Ice-1-based retracker in different regions of interest.

Since the size of the water bodies is smaller than the satellite footprint, the surrounding non-water surface often contaminates the satellite altimetry's returned radar signal. In this study, we solved this issue by integrating geospatial information and optical remote sensing with satellite altimetry measurement to monitor small water bodies. Our study indicates that careful demarcation of water bodies reduces the contamination of a return radar signal caused by the presence of a non-water surface, thus improving the quality of the measurement.

In this study, we processed the results of Envisat standard waveform retracking procedures (Ocean, Ice-1, Ice-2 and Sea Ice) to monitor the water level of a small river, a medium river and two lakes in the tropics. In addition to the standard waveform retracking procedures, we performed careful spatial and waveform shape selection and outlier detection to screen out low-quality data. We then evaluated the results against in situ measured water levels to assess their accuracy.

2 Study area

This study was conducted in the following water bodies in Indonesia (Figs. 1 and 2): the Mahakam and Karangmumus rivers in East Kalimantan Province (Borneo Island), and Lake Matano and Lake Towuti in South Sulawesi Province (Sulawesi Island). The Karangmumus River is a tributary downstream of the Mahakam River, while Lake Matano and Lake Towuti are part of the Malili lake complex. These water bodies represent different geomorphology, climate and anthropogenic situations, as described below.

2.1 Mahakam and Karangmumus rivers

The Mahakam watershed is located at 113°40′ to 117°30′ E longitude and 1°00′ S to 1°45′ N latitude. Mahakam is the second largest river in the country, which stretches to ~920 km and drains an area of 77 095 km^2. The Mahakam River rises in the mountainous forest ranges with dramatic elevation drops in the first hundreds of kilometers of the main stem, where the formation of rolling hills and steep slopes forms the upstream part of this watershed. The middle Mahakam Lake and wetlands form up starting from about 500 km downstream from the headwater and transform into the Mahakam Delta estuary in the last hundred kilometers of the Mahakam River (MacKinnon et al., 1996). The upstream part of the Mahakam River has a narrow channel with 40–100 m in width, 5 to 10 m average depth, and a river bed slope greater than 2 %. Forest and small patches of subsidence farmlands dominate the land use of this upstream por-

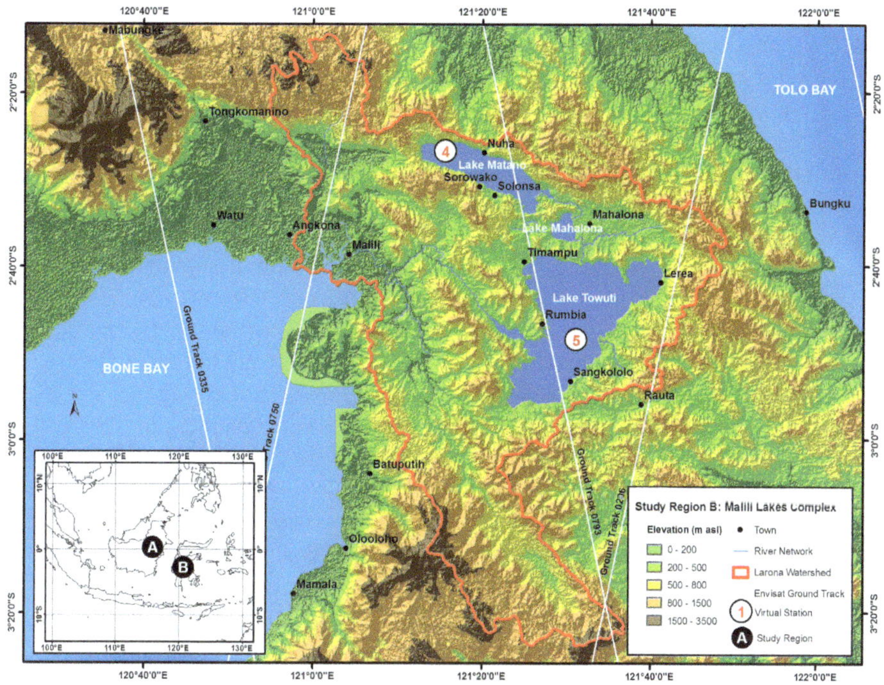

Figure 2. Study sites at the Malili lake complex, South Sulawesi, Indonesia.

tion. The middle part has a medium-sized channel with 100–300 m in width, 10–24 m in depth and a 0.5–2 % slope. Extensive lowland and agricultural areas spread about everywhere along with country-style residential areas, lakes and swampy shrubs. The lower part and the Mahakam Delta have a wide channel of 500–850 m in width, 10–24 m in depth and a 0–0.5 % slope. The lower sub-watershed is typically a developed area with residential areas, scarce forest patches and heavily inhabited land (Estiaty et al., 2007).

The Karangmumus River is a narrow channel (3 to 45 m in width) that is an important waterway for the residents of Samarinda in East Kalimantan Province. The Karangmumus sub-watershed often experiences gradual increases and steady high water levels during simultaneous heavy rainfall and backwater intrusion from ocean tides through the Mahakam Delta.

2.2 Lake Matano and Lake Towuti

Lake Matano is located at 121°12′ to 121°29′ E longitude and 2°23′ to 2°34′ N latitude. This lake counts as the seventh deepest lake of the world (Herdendorf, 1982), despite its small extent (164 km^2). With the maximum depth of 595 m and mean water surface elevation measured at 392 m, Lake Matano represents a cryptodepression (i.e., the lake bed is below the mean sea level) (Hehanussa and Haryani, 1999). Originating by tectonic processes since 2–3 million years ago, this lake is one of the oldest lakes of the world. The lake hosts endemic faunas that provide remarkable examples of ecological diversification and speciation (Cristescu et al., 2010). The basins in the surrounding areas of Lake Matano are formed by the hardness of the rocks and the softness of a uplift tectonic fault that forms a limited number of alluvial plains. Lake Matano also has two flat depressions separated by a saddle. It drains through the Petea River into Lake Mahalona, which is located in the same Malili lake complex (Vaillant et al., 2011).

Lake Towuti is recognized as the largest tectonic lake in Indonesia (Russel and Bijaksana, 2012). Located at the downstream end of the Malili lake complex, this lake covers an extent of 562 km^2 with 206 m depth. Similar to Lake Matano, Lake Towuti carries locally endemic fauna, since this lake is also one of the ancient lakes.

3 Materials and methods

3.1 Envisat radar altimetry

In this study we used satellite radar altimeter measurements from the European Space Agency (ESA)'s Envisat Radar Altimeter (RA-2) during the period of July 2002 to October 2010, corresponding to cycles 6 to 93 (ESA, 2007). The RA-2 determines the two-way delay of radar echo from the Earth's surface at a very high precision of less than a nanosecond. In addition, it measures the power and shape of the reflected radar pulses, which are represented by the waveforms. The RA-2 on-board signal processor calculates the average of approximately 100 measurements of individual echo burst at ~1800 Hz. These data, along with the waveforms, are averaged into the 18 measurements per second (18 Hz). The 18 Hz data correspond to an along-track sampling interval of ~ 350 m (ESA, 2011). The averaged 18 Hz waveforms are arranged into 128 gates with 3.125 nanosecond temporal resolution, and presents the default tracking gate at #46 (ESA, 2007). We also utilized the Envisat RA-2/Microwave Radiometer (MWR) Sensor Geophysical Data Record (SGDR) (hereafter, RA-2/MWR SGDR) Level-2 product. The RA-2/MWR SGDR contains parameters for time tagging, geo-location, output from re-trackers (i.e., range, wind speed, significant wave height) at 1 Hz, and other 18 Hz parameters such as range and orbital altitude. The RA-2/MWR SGDR also contains the 18 Hz waveforms that we used in the waveform shape selection procedure. We used the 18 Hz re-tracked range to infer the water surface elevation. Before comparing the altimetry with in situ measurements, we first corrected the instrumental (i.e., Doppler shift and oscillator drift), the geophysical (i.e., inverse barometer, polar and solid Earth tides) and the media (i.e., ionosphere and dry/wet troposphere) range in order to match the standard retracker range (Ocean, Ice-1, Ice-2 and Sea Ice) produced from the Level-2 radar altimeter product.

Satellite radar altimetry measures water surface elevation with respect to the reference ellipsoid. Due to the uncertainty in the relationship between the elevations of the field gauge benchmark relative to the local vertical datum, we used the water level anomaly in our analysis. The anomaly was calculated by subtracting the water level mean over the study period (July 2002–October 2010) from the observed level. Hence, it represents the fluctuation of a water level relative to its mean level. In order to test the current assumption of Ice-1 as the best retracking algorithm for inland waters (Frappart et al., 2006), we compared the water level anomaly obtained from water surface elevation measured by the Ocean, Ice-1, Ice-2 and Sea Ice retrackers with those obtained from the in situ gauge measurement.

3.2 Optical remote sensing and geospatial data set

We applied standard optical remote sensing data processing techniques in order to obtain imageries with a precise position and better contrast ratio between land and water. The processing included geometric correction, development and contrast adjustment of the pseudo-natural color composite imagery from a red–green–blue combination (bands 5, 4 and 3 of Landsat 5 and Landsat 7, or bands 6, 5 and 4 for the recently launched Landsat 8). We then measured river and lake width through visual interpretation of the remote sensing imagery (i.e., through the dark-blue color reflected by the water bodies in the pseudo-natural color composite of Landsat imagery) and marked the boundaries. When the object was too small to detect using visual inspection of remote sensing images, we used medium-scale (1 : 50 000) topographic maps released by the Indonesian Geospatial Agency to identify and mark the boundary.

Previous work (Sarmiento and Khan, 2010) showed that satellite altimetry measurements were less accurate when the center of the satellite altimetry footprint was closer to the lakeshore. In order to test this hypothesis, we created masks with varying distances to the lakeshore (i.e., 0–500, 500–1000 and > 1000 m). The footprint diameter of the Envisat RA-2 over a smooth surface is about 1.7 km (Rees, 1990; ESA, 2007). We assumed that the Envisat altimeter measurements within the last mask (i.e., > 1000 m from the lakeshore) were not influenced by the surrounding non-water surface. We then analyzed the performance of altimeter measurements based on these masks. As for the river, we created a mask with a 5 m buffer distance to the riverbank, in order to reduce the land surface–waveform contamination and to tolerate any geo-referencing and projection errors of the satellite imagery and topographic maps.

3.3 In situ water level data

Indonesia's Ministry of Public Works provided the data sets used for validation of water levels of the Mahakam River at the Melak site (2002–2004) and the Karangmumus River (2008–2010), while PT Vale Indonesia provided validation data for Lake Matano and Lake Towuti (2002–2012). Similar to the satellite altimetry data, we transformed the water level time series into a water level anomaly by removing the mean water surface elevation over the period of observation.

3.4 Waveform shape analysis

The presence of variable land cover (e.g., vegetation on the riverbank, lakeshore or coastline, as well as islands or sandbanks within the river or lake) affects the returned radar signal in altimetry measurement (e.g., Deng and Featherstone, 2006; Berry et al., 2005). Therefore, we analyzed the waveform shapes considering that the radar pulse reflected by the small water bodies might be influenced by other surfaces within the projected radar footprint. For the lakes, a 1 km distance to the lakeshore was sufficient, since the radius of the Envisat footprint (half of its diameter) is about 850 m. However, this issue becomes more challenging for small and medium-sized rivers (40–800 m in width), rendering the waveform produced by the processed radar pulse return unpredictable.

Due to the fact that an inland water surface is smoother than the ocean (Birkett, 1998), we assumed that a (quasi-)specular shape is the standard waveform shape for radar pulse returns reflected by inland water bodies, in contrast to the ocean-reflected diffuse shape (Koblinsky et al., 1993). Additional shapes of Envisat RA-2 returned radar pulses over inland water include (Berry et al., 2005) (i) a quasi-Brown shape representing a transition from land to water, (ii) a flat-patch shape denoting an intermediate surface, and

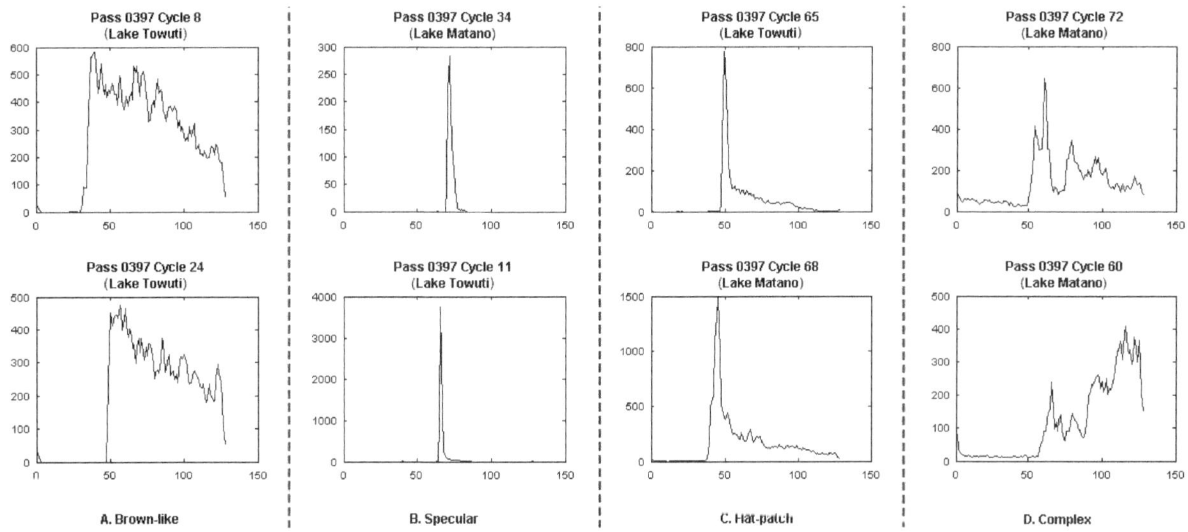

Figure 3. General categories of waveform shapes.

(iii) a complex shape indicating a mixture of water and vegetation (Dabo-Niang et al., 2007). In this study, we considered (quasi-)specular, quasi-Brown and flat-patch shapes as qualified waveforms to perform reliable range measurement and discard complex and non-classified shapes from further processing. We assumed that the mixture of water, vegetation and/or shoreline provides less accurate elevation measurements as compared to the radar signal returned by water-dominated surfaces. Some examples of actual waveforms that are classified as "Brown-like", specular, flat-patch, as well as complex and non-classified shapes, are presented in Fig. 3a–d, respectively. In practice, we displayed the standard waveform shapes (Brown-like, specular, flat-patch), with another window showing waveform shapes from each measurement along with their IDs. Then, we noted down the IDs of measurements that matched waveform shapes for further processing. It is interesting that, in order to select the most appropriate waveforms that are less contaminated by land surfaces, another study offered the highest weight for waveforms originating by water surfaces, and assigned a lower weight for waveforms reflected by other land surfaces (Michailovsky et al., 2012). Operationally, the implementation of straightforward waveform shape qualification as presented in this study offers slightly more efficient waveform processing, especially when the algorithm for waveform geometry processing can be developed.

3.5 Outlier removal, validation and performance evaluation

Although the altimetry measurements that carry non-qualified waveform shapes were excluded, some measurements remained far beyond the mean and median values. In order to obtain a data set with minimum influences from outliers, we excluded mild outliers – defined as any values outside of 1.5 times the inter-quartile range (IQR) (Kenney and Keeping, 1947; Panik, 2012). IQR is defined as the range between the 25 % quartile value ($Q_{0.25}$) and the 75 % quartile value ($Q_{0.75}$). If we denoted $\mathrm{WSE}_{\min}$ and $\mathrm{WSE}_{\max}$ as the minimum and maximum water surface elevations from the Envisat radar altimetry, respectively, then

$$\mathrm{IQR} = Q_{0.75} - Q_{0.25} \text{ therefore } \begin{array}{l} \mathrm{WSE}_{\min} = Q_{0.25} - 1.5 \times \mathrm{IQR} \\ \mathrm{WSE}_{\max} = Q_{0.75} + 1.5 \times \mathrm{IQR}. \end{array} \quad (1)$$

Consequently, we discarded any measurements below the $\mathrm{WSE}_{\min}$ and above the $\mathrm{WSE}_{\max}$ threshold in further processing.

We used root mean square error (RMSE) and the coefficient of correlation (r) as measures of performance (or validation) between satellite altimetry water level measurements and the virtual stations where in situ measurements were available. The RMSE is a measure of how far the estimated measures are from the "truth" values. It is defined as (e.g., Nagler, 2004; Li, 2010)

$$\mathrm{RMSE} = \sqrt{\sum_{i=1}^{n} \frac{(x_i - y_i)^2}{n}}, \quad (2)$$

where x_i is the Envisat water level anomaly and y_i is the in situ measured water level anomaly.

The Pearson correlation coefficient is the standard measure of association for continuous types of data (de Sa, 2007). Therefore, we used it to measure the association between satellite altimetry and in situ water level measurements as described in the following equation:

$$r = \frac{S_{xy}}{S_x S_y} \text{ with } S_{xy} = \sum_{i=1}^{n} \frac{(x_i - \overline{x})(y_i - \overline{y})}{(n-1)}, \quad (3)$$

where S_x and S_y are variances for each sample and n is the number of observations. The correlation coefficient (r) value

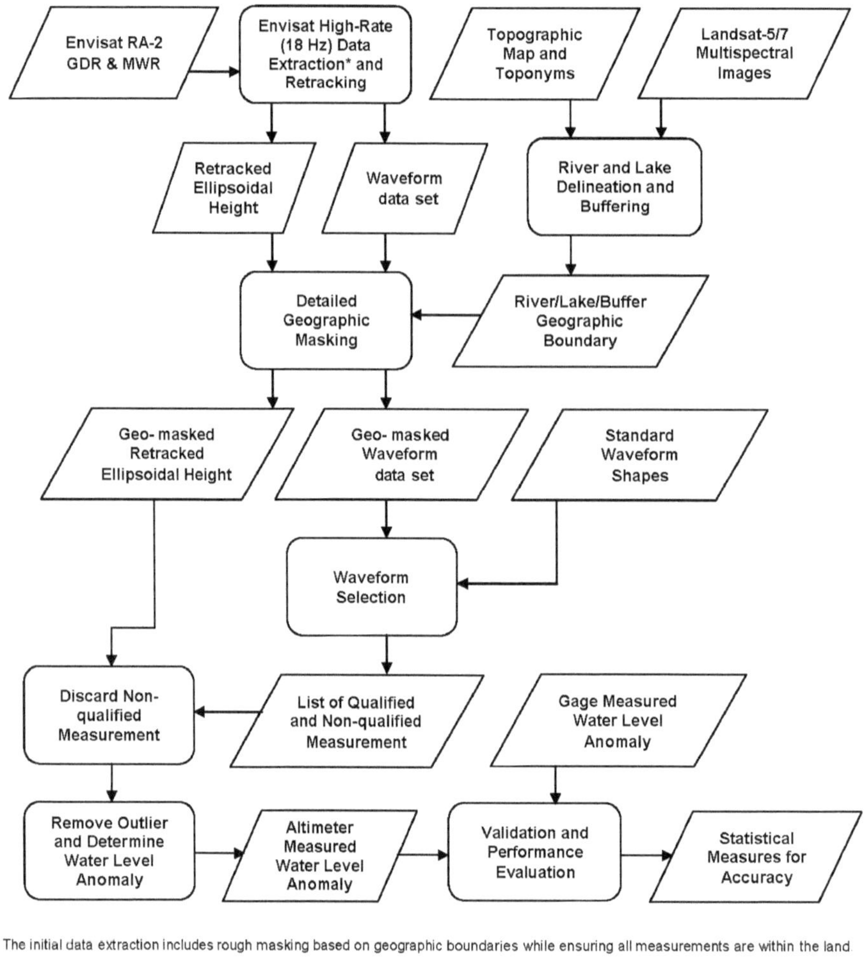

Figure 4. Data processing workflow.

falls within the interval $[-1, 1]$, where a coefficient of 0 indicates no correlation between two measurements, $+1$ indicates the total correlation in the same direction (proportional relationship) and -1 indicates the total correlation in the opposite direction (inverse relationship).

In order to provide a comprehensive understanding of the data processing sequences in this study, Fig. 4 shows each data processing step and their relationships.

4 Results and discussion

4.1 Mahakam and Karangmumus rivers

Table 1 shows that most of the radar pulse returns from both small-sized rivers (40–200 m in width) and medium-sized rivers (200–800 m in width) produced qualified waveforms to infer water level fluctuation. The percentages of qualified waveforms relative to all measurements within the water bodies were high (90–97 %) even for a small river at virtual station UM03 (river width 54 m). Interestingly, there were more missing cycles – regular satellite repeat schedule without available measurements within the water bodies – in the smaller river (UM03 site) than in the wider rivers (Melak01 and Melak02 sites).

For the water level measurements at Melak, we combined two virtual stations (i.e., Melak01 and Melak02), since they were just separated by a 14–40 km distance and there was no drastic change in the terrain and configuration of the channel (e.g., no reservoir or steep gradient) based on the topographic map and digital elevation model. Having two different satellite tracks nearby in fact increased the spatial and temporal sampling intensity for this location. Figure 5 shows the location of the Ministry of Public Works' gauge station, which was right in between these two virtual stations. Figure 5 also indicates dynamic channel morphology in this area. The channel was heavily meandering just before and along the Melak01 virtual station, which then changed into a 13 km straight channel along the heavily populated town of Melak before it was back to a lightly meandering channel. Figure 6 shows the combined water level anomaly from the two virtual stations, along with the water level anomaly observed by the gauge station for the period of 2002–2004.

Table 1. The number of qualified and non-qualified altimeter measurements and outliers for study sites at Mahakam River.

Site name	Cycles	No. of missing cycles	Measurements in water body	Qualified measurement		Non-qualified measurement		No. of outliers	River width
				(No.)	(%)	(No.)	(%)		(m)
UM03	9–93	34	51	46	90.2	5	9.8	N/A	54 m
Melak01	7–93	8	225	220	97.8	5	2.2	8	247 m
Melak02	7–93	11	148	134	90.5	14	9.5	0	294 m

Table 2. Performance evaluation of Envisat RA-2 radar altimetry measurements over Melak virtual stations at Mahakam River (width 247 m).

Site name	Cycles covered	Validated measurements	Number of passes	Retracker	RMSE (m)	Correlation coefficient
Melak	7–33	46	2	Ocean	0.885	0.955
				Ice 1	0.720	0.962
				Ice-2	0.724	0.966
				Sea Ice	0.685	0.970

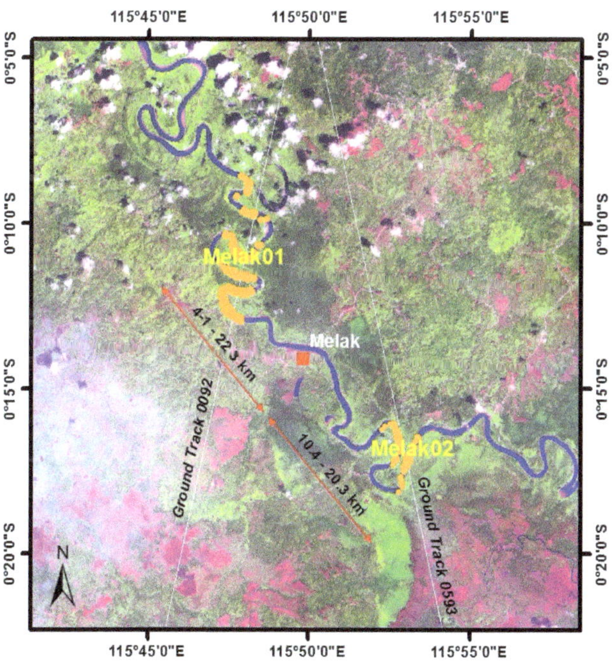

Figure 5. Location of Envisat virtual stations and in situ water level gauge stations at Melak.

To facilitate visual investigation, we presented scatter plots between the water level anomaly obtained from gauge measurements and those derived from the four different retracking algorithms. We found that Ice-1 was not the best retracking algorithm for inland water body elevation measurement (Table 2). The sea ice retracking algorithm outperformed the other three standard retrackers (Table 2). With a correlation coefficient of up to 0.97, satellite radar altimetry was a more suitable alternative for monitoring of the medium-sized river (200–800 m in width), even for a poorly gauged basin such as the Mahakam Watershed. Compared to other studies, the magnitude of the root mean square error (RMSE) from our study (i.e., 0.69) was just about the average RMSE obtained from other studies that deal with medium-sized rivers (200–800 m in width) (Table 3).

It is important to note that we did not adjust the magnitude of the satellite altimetry range measurements in any way. Apart from the spatial selection of the range measurements with the projected nadir footprint center within the water body and the removal of outliers, the only manipulation we performed was selecting the range measurements based on its waveform shape to strictly follow the standard waveform shape for an inland water body as described in the previous studies (Koblinsky et al., 1993; Birkett, 1988; Berry et al., 2005; Dabo-Niang et al., 2007). Therefore, there are several possibilities for improvement to increase the accuracy of the satellite altimetry measurement of river water levels, especially for this study area. Examples are the use of other altimetry missions (e.g., Jason-1, ICESat), more detailed evaluation of retracked water elevation within a cycle, and inclusion of the actual river slope in the processing.

In this study, we found that Envisat altimetry showed a potential to observe small-sized rivers. Satellite altimetry crossing at the UM03 virtual station returned a high percentage of qualified measurement even with fewer measurements within the water body (i.e., 46 over 51) compared to that of other virtual stations. Figure 8 indicates the water level fluctuation at this virtual station, while Fig. 9 shows variable gaps that existed between the measurements, with an average of 84 days and a maximum gap that lasted for 300 days (~ 10 months). This temporal gap was a serious problem for hydrological applications, especially those requiring the measurement of hydrological variables at short intervals. Furthermore, there

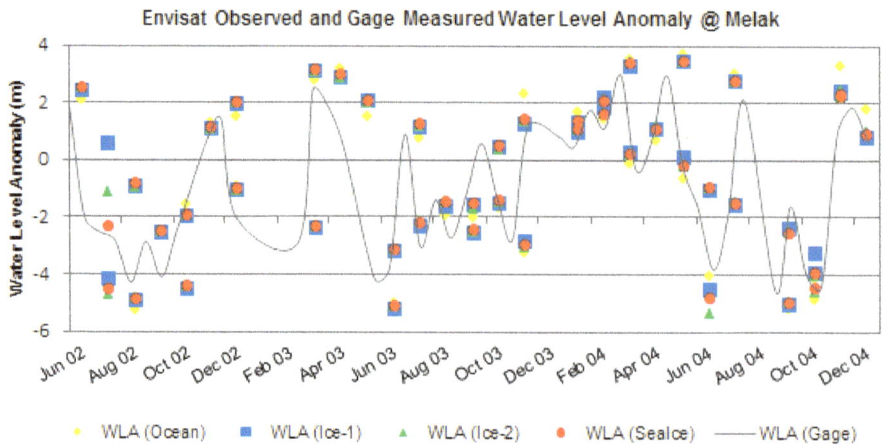

Figure 6. Water level anomaly at Melak as observed by two Envisat passes and retracked by four retrackers, compared with the in situ water level anomaly.

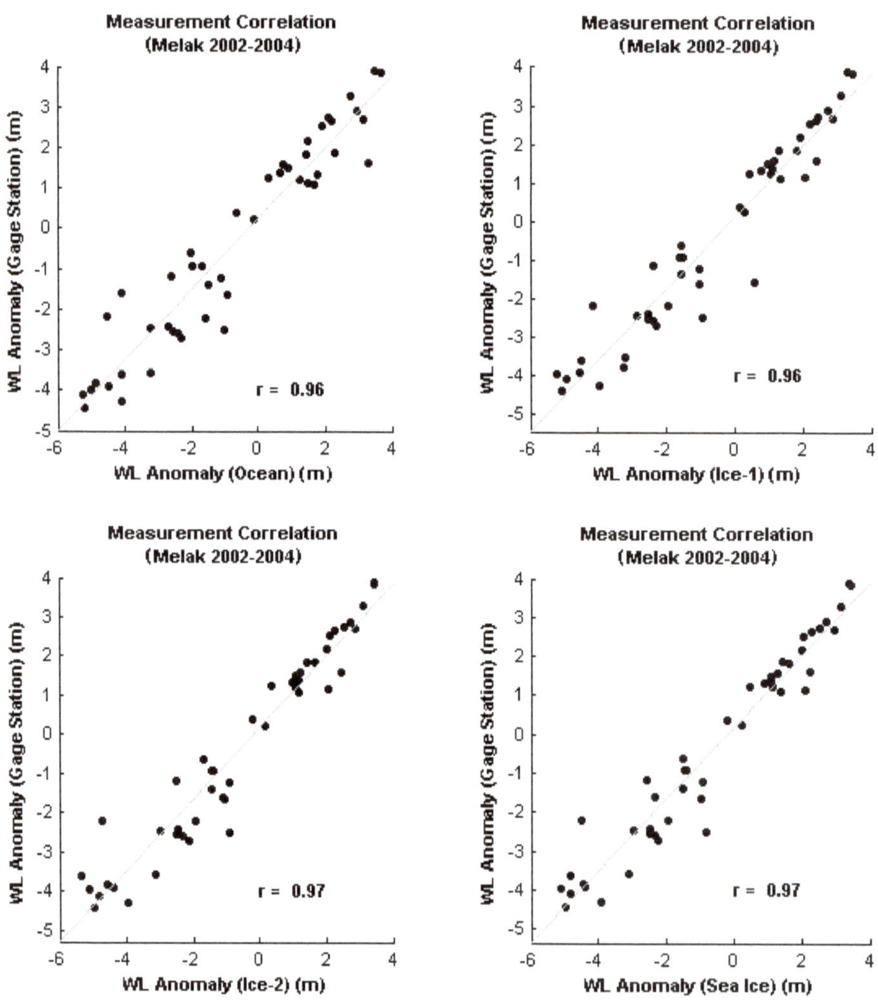

Figure 7. Correlation between the water level anomaly measured by the Envisat altimeter and processed with the Ocean (top left panel), Ice-1 (top right panel), Ice-2 (bottom left panel) and Sea Ice (bottom right panel) retrackers and the in situ water level measurement over Melak.

Table 3. Summary of studies on satellite radar altimetry for water level over river.

Reference	Location	River width	Satellite/ sensor	Reported error (m)
Koblinsky et al. (1993)	Amazon Basin	N/A	Geosat	STDE*: 0.31–1.68 m
Birkett (1998)	Amazon Basin	3–9 km	T/P	RMSE*: 0.11–0.60 m
Birkett et al. (2002)	Amazon Basin	2–6 km	T/P	RMSE*: 0.40–0.60 m
Kouraev et al. (2004)	Ob' River	3 km	T/P	%*: 8 % (discharge)
Frappart et al. (2006)	Mekong River	450 m	Envisat, T/P	RMSE*: 0.23 m RMSE*: 0.15 m
Birkinshaw et al. (2010)	Mekong River	400 m– 1.7 km	ERS-2, Envisat	RMSE*: 0.44–1.24 m
Kuo and Kao (2011)	Bajhang River	100 m	Jason-2	STDE*: 0.31 m
Michailovsky et al. (2012)	Zambezi River	40–380 m	Envisat	RMSE*: 0.27–1.07 m
This study (2013)	Mahakam River	240–279 m	Envisat	RMSE*: 0.69 m

* STDE = standard deviation of error, % = % difference, RMSE = root mean square error.

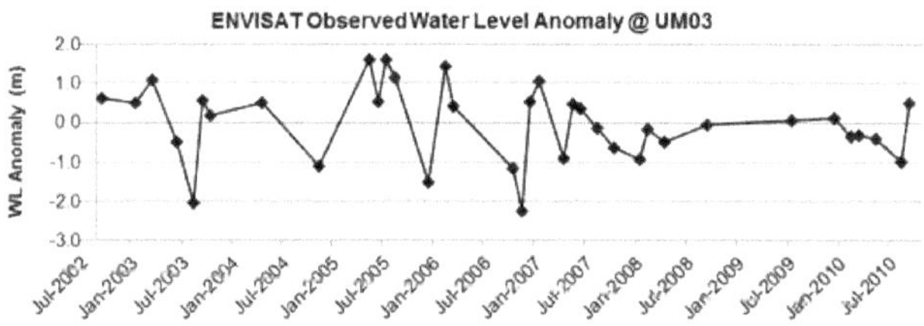

Figure 8. ENVISAT observed water level anomaly at site UM03 (river width 54 m) as measured by Envisat RA-2 and processed by the Ice-1 retracker.

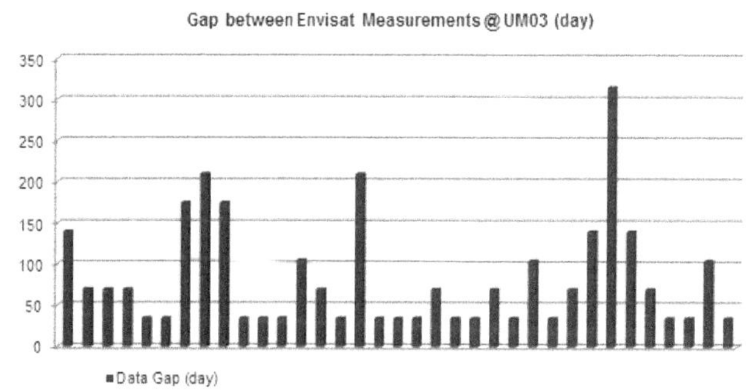

Figure 9. Gaps between Envisat observations of water levels over site UM03.

was no in situ gauge station in the vicinity that provided validation data for this particular virtual station (UM03). Although we could not validate the water level retrievals at this location, this experiment showed the potential of satellite altimetry for monitoring small rivers (40–200 m in width).

We conducted another experiment of satellite altimetry measurement over the narrow Karangmumus River (width 8–45 m). The northeast–southwest orientation of this river made it difficult to find the crossing with Envisat ground tracks. However, a high-resolution IKONOS image (1 m ground resolution) allowed detailed selection of the altimeter ground tracks that fall within its narrow channel. Still, the ultra-narrow channel width seriously hampered successful satellite radar altimetry measurement of this study site. After careful spatial filtering and a waveform shape selection procedure, we extracted only 11 water surface elevations from the Karangmumus River. Figure 10 depicts the location of this experiment, while Table 4 summarizes the qualified measurements.

Figure 11 shows the time series of the Karangmumus River water level anomaly from 2004 to 2006, and it is obvious that the number of retrieved water level anomalies was very limited. In addition, the in situ measurement record from the nearest available gauge stations (i.e., Pampang, Muang, Gununglingai and the outlet of the Karangmumus River) were available only during years 2008–2010. Therefore, we could not evaluate the performance of satellite altimetry measurements over this very small river. However, this result serves as a preliminary indication of the range of water level magnitude in this river.

Presently, only a few other studies indicated successful exploitation of the rivers with 100 m in width or less. Michailovsky et al. (2012) extracted 13 useful water level measurements from a river with 40 m in width and Kuo and Kao (2011) revealed the water level of the Bajhang River in Taiwan with less than 100 m in width with a standard deviation of error of 0.31 m.

We therefore urge further exploration of satellite altimetry observations to monitor small rivers supported by complete validation data.

To conclude this section, we demonstrated that medium-sized rivers as narrow as 240 m can still be monitored and validated, provided that the water surface boundary was identified accurately. This result expands the capability of the satellite altimetry, since previous studies showed that 1 km seems to be the ideal width from which to expect typical altimetry radar returns from the water surface (Birkett,

Table 4. Qualified Envisat RA-2 altimetry measurements for the Karangmumus River.

Cycle	Date	ID	Longitude	Latitude	Water level anomaly*	Remarks
8	23 Jul 2002	KM08	117.181540	−0.404124	−0.07 m	
9	27 Aug 2002	KM10	117.194581	−0.408362	−4.52 m	Benanga Reservoir
13	13 Jan 2003	KM11	117.195384	−0.407573	2.94 m	Benanga Reservoir
23	30 Dec 2003	KM01	117.157190	−0.507934	−1.92 m	
23	30 Dec 2003	KM02	117.157910	−0.504634	−2.32 m	
28	22 Jun 2004	KM09	117.188367	−0.405981	3.63 m	47 m to field gauge
37	3 May 2005	KM06	117.169721	−0.448573	−0.11 m	
37	3 May 2005	KM07	117.170441	−0.445263	−0.12 m	
39	12 Jul 2005	KM03	117.158610	−0.503317	−2.28 m	
42	25 Oct 2005	KM05	117.171486	−0.452076	4.12 m	
49	27 Jun 2006	KM04	117.159139	−0.501533	−0.93 m	

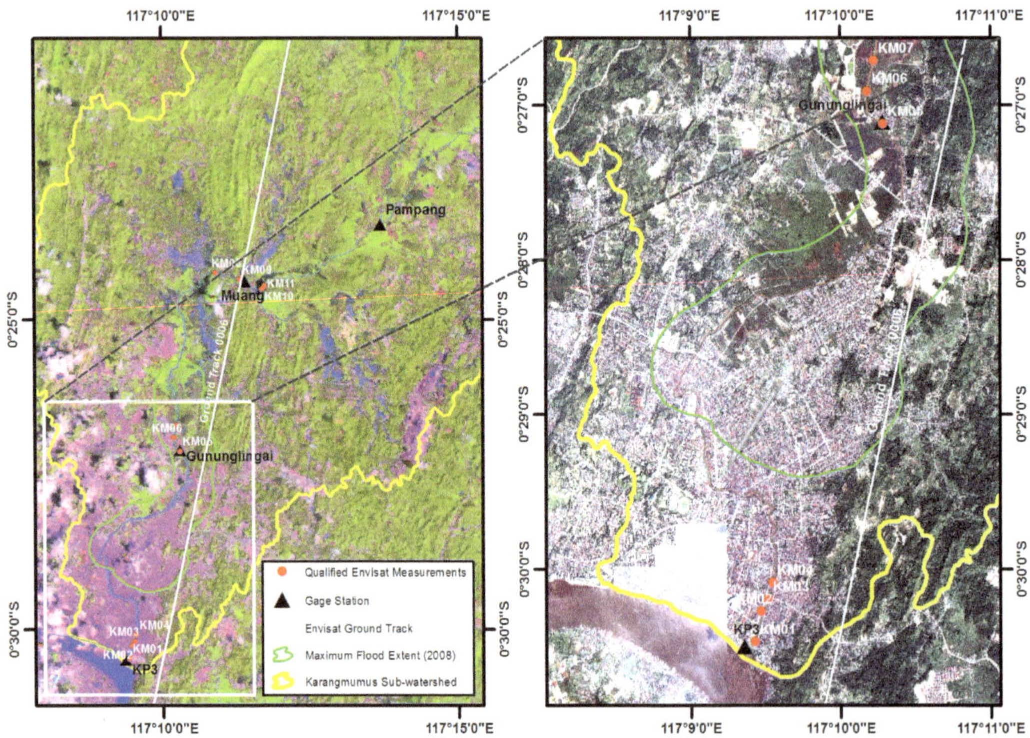

Figure 10. The Karangmumus sub-watershed. Left panel: overview of the sub-watershed with Landsat-7 image of January 2007 as the background. Right panel: detailed view of the urbanized downstream part with IKONOS image of February 2002 as the background. Envisat ground tracks appear as white lines in both panels.

1998; Birkett et al., 2002). We also emphasize that successful retrieval of qualified satellite radar altimetry measurement in this research was very much supported by detailed geographic masking, which carefully excluded all altimetry measurements with a projected nadir position outside of the water bodies.

4.2 Lake Matano and Lake Towuti

Inland water has been known to produce different, sometimes irregular, waveform shapes and patterns as compared to that of the ocean. In particular is the difference with respect to their responses to radar pulse signals transmitted by satellite-based active sensors. Some examples of distinguished waveform shapes from Lake Matano and Lake Towuti at different buffer distances from the lakeshore are presented in Fig. 12.

Table 5. The number of qualified and non-qualified altimeter measurements and outliers over Lake Matano and Lake Towuti.

Location	Width	Cycle	Distance to shore	Measurement within water body	Qualified		Non-qualified		No. of outliers
					No.	%	No.	%	
Lake Matano	8159	8–79	< 500 m	453	416	91.8	37	8.2	42
			500 m–1 km	253	215	85.0	38	15.0	26
			> 1 km	989	805	81.4	184	18.6	115
Lake Towuti	28 818	8–79	< 500 m	1314	786	59.8	528	40.2	79
			500 m–1 km	1328	764	57.5	564	42.5	64
			> 1 km	2450	1353	54.3	1137	45.7	156

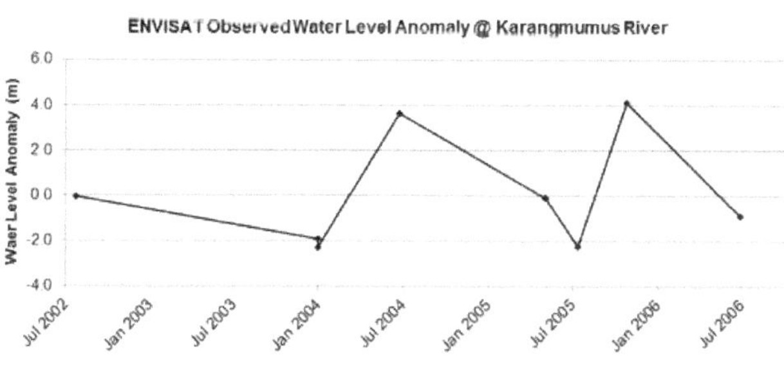

Figure 11. Water level anomaly of the Karangmumus River from Envisat RA-2, processed by Ice-1 retracker.

Our findings indicated that the waveform shapes resulting from satellite altimetry measurement over the lakes had more variability compared to those over the small to medium-sized rivers. We suspect this was due to the fact that lakes possess a larger extent of water surface and are much more influenced by wind that may develop waves with some height. Figure 12 shows the typical ocean-like, multi- and low-peak, gradually rising and many other kinds of irregular patterns that were not present in the data set from small and medium-sized rivers. Up to now, a systematic and verified classification of waveform shapes, especially for inland waters, has not existed, except for the early development such as presented by Dabo-Niang et al. (2007): hence the need to study this subject further.

Table 5 summarizes the results of satellite altimetry waveform selection over Lake Matano and Lake Towuti. Similar to the result of satellite altimetry measurements for small to medium-sized rivers in the previous section, most of the radar pulse returns produced qualified waveforms that were subsequently used to compute the water level anomalies at these two lakes. Our findings suggested that the separation distance from the lakeshore did not significantly affect the number of qualified waveforms. For instance, the percentage of qualified waveforms for the lake surface with a distance from the lakeshore of more than 1 km in Lake Matano and Lake Towuti was lower than those closer to the lakeshore (Table 5). This complex result calls for further investigation in the field of satellite altimetry application for small and medium lakes in the tropics, given the fact that the land cover does not always influence the shapes of the returned altimeter waveform.

Upon the completion of waveform sorting, we processed the range measurements performed by the Ocean, Ice-1, Ice-2 and Sea Ice retrackers and evaluated them against observed water levels from the in situ gauge station. Figures 13 and 14 show the satellite altimetry and in situ measured water level anomaly at Lake Matano and Lake Towuti. These plots visually indicate that the satellite altimetry-observed water level anomalies closely matched the in situ gauged water level anomaly. From Figs. 13 and 14, we estimated the range of the water level anomaly at Lake Matano to be in the magnitude of 1.2 m, while that of Lake Towuti only ranged in the magnitude of 1.4 m. Figures 15 and 16 show the correlation between the Envisat radar altimeter measurements as processed by the Ocean, Ice-1, Ice-2 and Sea Ice retrackers with the gauge-measured water level anomaly for Lake Matano and Lake Towuti, respectively.

In terms of performance, Envisat radar altimetry measurements over Lake Matano and Lake Towuti performed equally well, as reflected by the lowest RMSE obtained by the best retracker for each lake (0.21, see Table 6). Based on the performance evaluation (Table 6), our results could not verify the hypothesis that the shorter distance to the lakeshore was associated with a lower accuracy of satellite altimetry measurement. The satellite altimetry measurements of the water level anomaly over Lake Matano indicated better accuracy (lower RMSE and higher correlation coefficient) as the distance between the altimeter footprint and the lakeshore increased, whereas measurements over Lake Towuti showed the opposite (see Figs. 17 and 18). These inconclusive results further suggest the use of a sample classification based on the distance to the lakeshore for future investigation.

Intercomparison between the available retrackers (i.e., Ocean, Ice-1, Ice-2 and Sea Ice) also cannot convincingly suggest any single retracker to infer the water level of the small lakes, since the Ocean retracker surprisingly performed best for Lake Matano, while the Ice-1 retracker

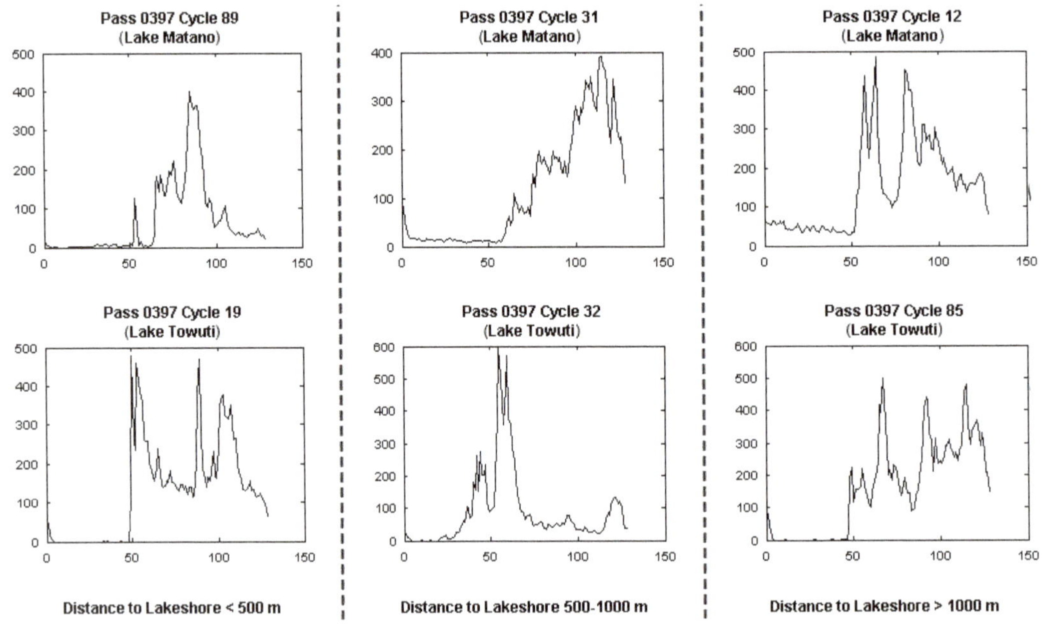

Figure 12. Distinguished waveform shapes as reflected by Lake Matano and Lake Towuti at different buffer distances to the lakeshore.

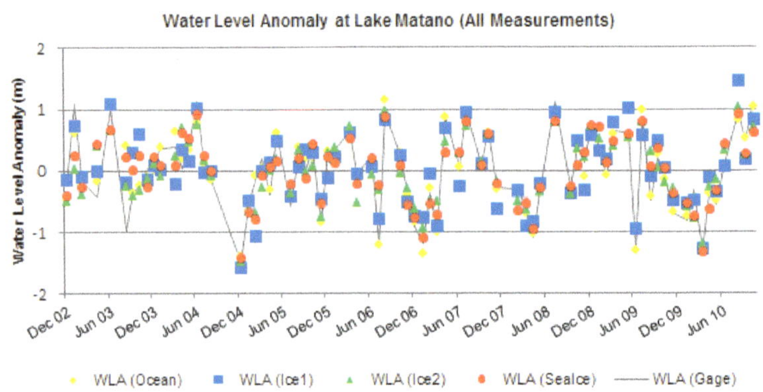

Figure 13. Water level anomaly at Lake Matano as measured by Envisat RA-2 and processed by all retrackers, compared with in situ measurement.

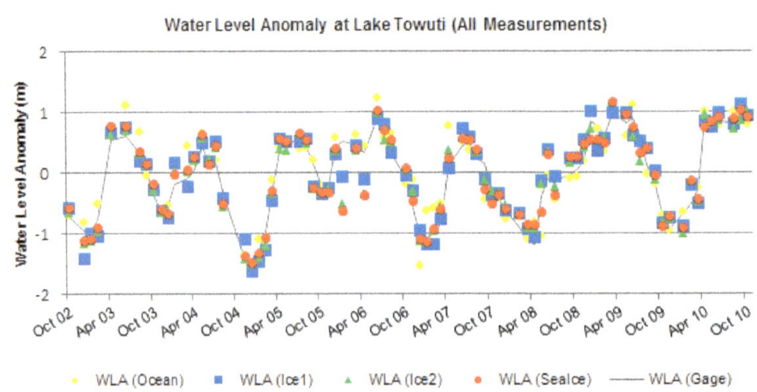

Figure 14. Water level anomaly at Lake Towuti as measured by Envisat RA-2 and processed by all retrackers, compared with in situ measurement.

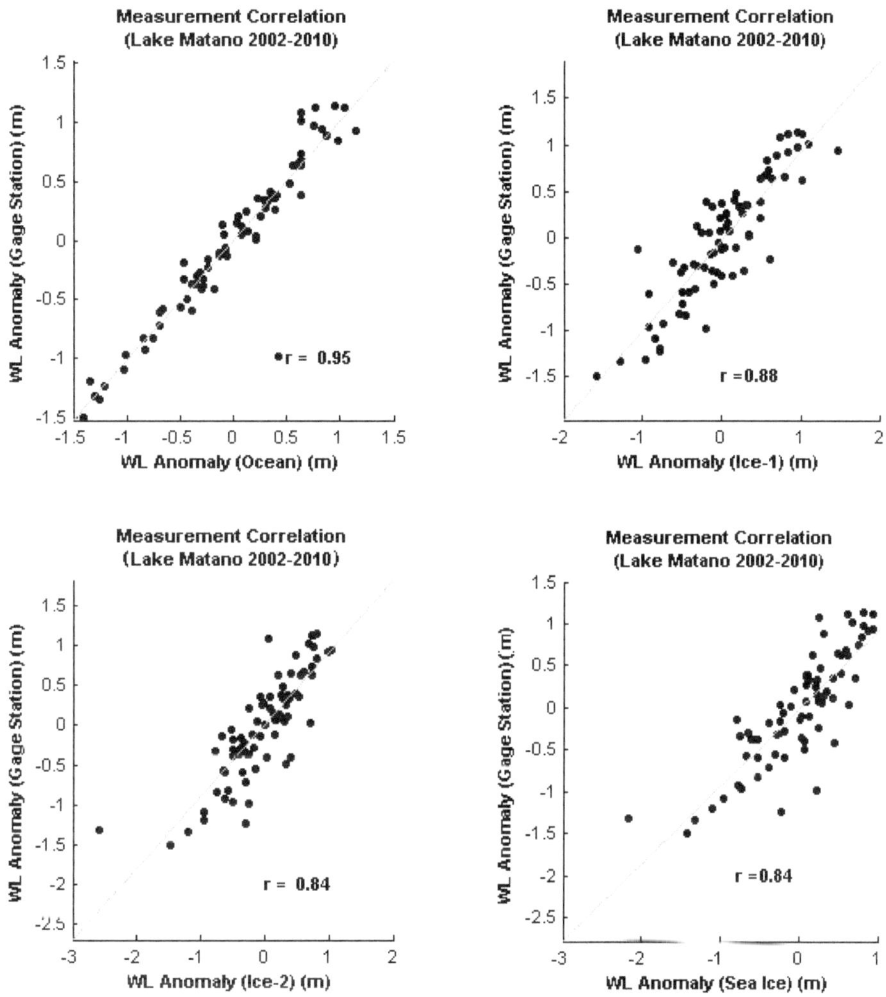

Figure 15. Correlation between the water level anomaly at Lake Matano as measured by the Envisat RA-2 altimeter and processed with the Ocean (top left panel), Ice-1 (top right panel), Ice-2 (bottom left panel) and Sea Ice (bottom right panel) retrackers.

performed best for Lake Towuti. An important conclusion from this study is that Ice-1 is not necessarily the best retracker for measuring water level anomaly over small to medium lakes.

The best RMSE obtained from measurements of water level anomaly in this study (0.21 m at both Lake Matano and Lake Towuti) was quite close to the lowest RMSE in other similar studies (e.g., Coe and Birkett, 2004; Munyaneza et al., 2009; Cai and Ji, 2009). Table 7 states that satellite altimetry measurements over small lakes produced RMSE magnitudes in the range of 30 to 50 cm, as compared to large lakes that produced an RMSE as low as 3 cm. Lake Matano is in fact the smallest among all the lakes listed in Table 7.

5 Conclusions

In this study we demonstrated the ability of satellite altimetry to monitor the water level of medium-sized (200–800 m in width) rivers in Southeast Asia's humid tropics with high accuracy (correlation coefficient of 0.97 and an RMSE of 0.685 m). Despite its performance variability, the water level anomaly inferred by Envisat radar altimetry through the standard waveform retracking method was validated in this study. These results thus confirmed its ability to monitor water level fluctuations in medium-sized rivers. In addition to the medium-sized rivers, we found that small rivers (40–200 m in width) are *potentially* observable through satellite altimetry, as indicated by the high percentage of qualified range measurements that we filtered based on the waveform shapes. It is important to note however that there could possibly be a variation in the measurement capability and accuracy across different regions; therefore, a specific approach should be developed for each region, as part of the development of a permanent monitoring effort in those regions.

In contrast to what was previously found (Frappart et al., 2006), Ice-1 is not necessarily the best retracker for monitoring small water bodies, especially for the Southeast Asia hu-

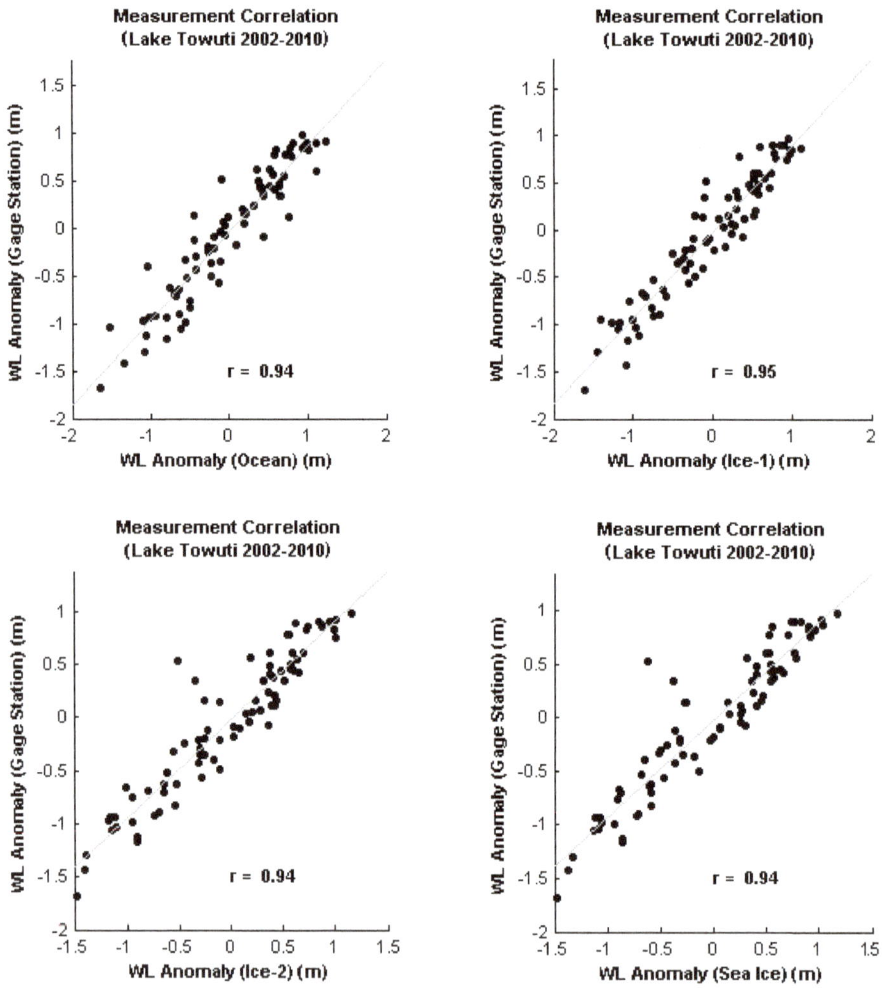

Figure 16. Correlation between the water level anomaly at Lake Towuti as measured by the Envisat RA-2 altimeter and processed with the Ocean (top left panel), Ice-1 (top right panel), Ice-2 (bottom left panel) and Sea Ice (bottom right panel) retrackers.

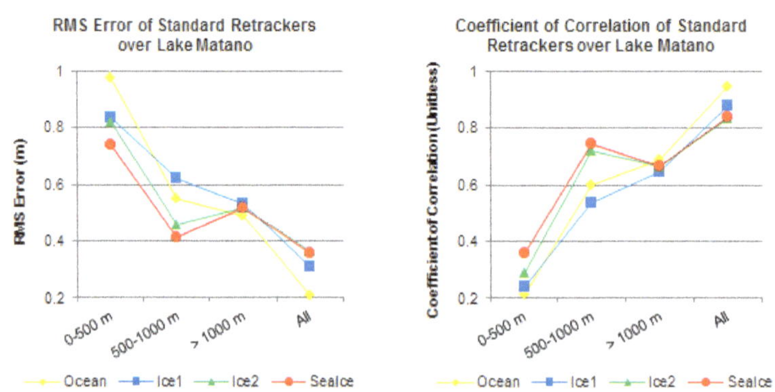

Figure 17. The performance of Envisat RA-2 radar altimetry measurements over Lake Matano, classified by the distance to the lakeshore.

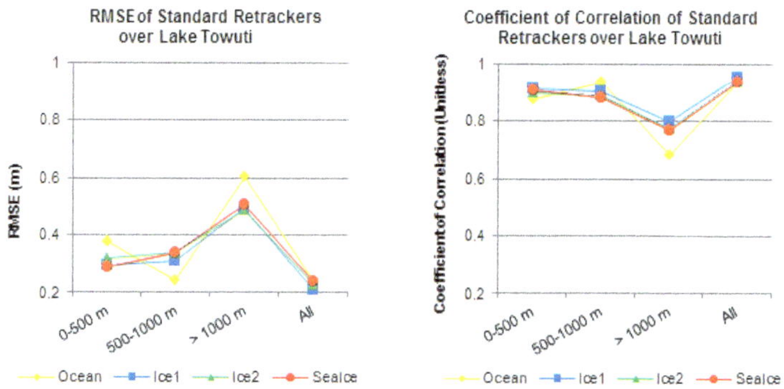

Figure 18. The performance of Envisat RA-2 radar altimetry measurements over Lake Towuti, classified by the distance to the lakeshore.

Table 6. Performance evaluation of Envisat RA-2 radar altimetry measurements over Lake Matano and Lake Towuti.

Site	Lake width (m)	Cycles	Validated measurement	Re-tracker	Correlation coefficient	RMSE (m)	No./% of outliers
Lake Matano	8159	8–79					
0–500 m			75	Ocean	0.214	0.981	42/387
				Ice-1	0.242	0.835	10.85 %
				Ice-2	0.290	0.819	
				Sea Ice	0.358	0.743	
500–1000 m			71	Ocean	0.605	0.555	26/214
				Ice-1	0.538	0.624	12.15 %
				Ice-2	0.723	0.458	
				Sea Ice	0.745	0.417	
> 1000 m			73	Ocean	0.692	0.493	115/805
				Ice-1	0.647	0.535	14.29 %
				Ice-2	0.667	0.518	
				Sea Ice	0.666	0.518	
All			75	Ocean	0.948	0.209	183/1406
				Ice-1	0.881	0.311	13.02 %
				Ice-2	0.837	0.364	
				Sea Ice	0.839	0.359	
Lake Towuti	28 818	8–79					
0–500 m			77	Ocean	0.880	0.380	79/786
				Ice-1	0.917	0.296	10.05 %
				Ice-2	0.898	0.321	
				Sea Ice	0.911	0.291	
500–1000 m			79	Ocean	0.942	0.244	64/764
				Ice-1	0.903	0.312	8.38 %
				Ice-2	0.890	0.339	
				Sea Ice	0.887	0.341	
> 1000 m			79	Ocean	0.689	0.608	156/1353
				Ice-1	0.802	0.494	11.53 %
				Ice-2	0.777	0.490	
				Sea Ice	0.774	0.507	
All			80	Ocean	0.940	0.241	299/2903
				Ice-1	0.953	0.212	10.30 %
				Ice-2	0.941	0.231	
				Sea Ice	0.938	0.239	

Table 7. Summary of studies on satellite radar altimetry for water levels over lakes.

Reference	Location	Lake extent	Satellite/ sensor	Reported error
Morris and Gill (1994a)	Superior, Ontario Michigan, Huron Erie Lake St. Clair	Large Large	Geosat Geosat Geosat Geosat	RMSE*: 0.09 m RMSE*: 0.11 m RMSE*: 0.13 m RMSE*: 0.17 m
Morris and Gill (1994b)	Great Lakes		Topex/ Poseidon	RMSE*: 0.03 m
Korotaev et al. (2001)	Black Sea	436 402 km^2	T/P, ERS-1	RMSE*: 0.03 m
Mercier et al. (2002)	Victoria, Tanganyika Malawi and Turkana	131–390 × 10^3	TOPEX/ Poseidon	RMSE*: 0.10 m
	Rukwa and Kyoga	75–80 × 10^3	TOPEX/ Poseidon	RMSE*: 0.50 m
Coe and Birkett (2004)	Lake Chad	2.5 × 10^6 km^2	TOPEX/ Poseidon	RMSE*: 0.21 m
Zhang et al. (2006)	Dongting Lake	2623 km^2	TOPEX/ Poseidon	RMSE*: 0.08 m
Medina et al. (2008)	Lake Izabal	717 km^2	Envisat	RMSE*: 0.09 m
Munyaneza et al. (2009)	Lake Kivu	2400 km^2	Envisat	RMSE*: 0.30 m
Cai and Ji (2009)	Poyang Lake	20 290 km^2	Envisat	Mean Error: 0.31 m
Guo et al. (2009)	Hulun Lake	2339 km^2	TOPEX/ Poseidon	RMSE*: 0.13 m
Troitskaya et al. (2012)	Gorki Reservoir	1358 km^2	T/P, Jason-1	RMSE*: 0.15 m
Tseng et al. (2013)	Qinghai Lake	4186 km^2	Envisat	RMSE*: 0.06 m
This study	Lake Matano Lake Towuti	164 km^2 562 km^2	Envisat Envisat	RMSE*: 0.21 m RMSE*: 0.21 m

* RMSE = root mean square error.

mid tropics area. We also found that the Ocean retracker surprisingly performed best for retracking small lakes (i.e., Lake Matano), as well as Sea Ice for the Mahakam River and Ice-1 for Lake Towuti.

The RMSE of satellite altimetry measurement of Lake Matano and Lake Towuti (0.21 m for both locations) falls within the range of RMSE of small lakes observed by satellite altimetry throughout the world (e.g., between 0.03 and 0.50 m). It is worth noting that Lake Matano is the smallest water body analyzed from satellite altimetry.

Considering the results of this study, we recommend the following: (1) in addition to the use of standard retrackers, we propose the selection of altimetry measurements based on the waveform shape to filter out returned radar signals contaminated by non-water surfaces. We recommend the selection to strictly follow the standard waveform shape for inland water bodies (Koblinsky et al., 1993; Birkett, 1998; Berry et al., 2005; Dabo-Niang et al., 2007), especially for studies involving small (40–200 m in width) to medium rivers (200–800 m in width), as well as small lakes (e.g., those with an extent of less than 1000 km^2), and, (2) over lakes, we do not recommend analyzing the performance of the satellite altimetry retrackers based on the distance from the satellite altimetry measurements to the lakeshore.

Lastly, we found that the geographic orientation of the river affected the application of satellite altimetry for monitoring small rivers. For instance, small (40–200 m in width) and medium-sized (200–800 m in width) rivers with a north–south orientation suffered from the satellite altimetry orbit deviation, which ranges from ±1 km relative to its theoretical orbit.

Acknowledgements. This research was primarily supported by the Fulbright PhD Presidential Scholarship administered by the American Indonesian Exchange Foundation (AMINEF) and the Institute for International Education (IIE). In addition, this study was partially funded by grants from NASA's Application Science Program under the SERVIR project (NNX12AM85G), and by the Chinese Academy of Sciences/SAFEA International Partnership Program for Creative Research Teams (grant no. KZZD-EW-TZ-05). The authors greatly appreciate the Ministry of Public Works of Republic of Indonesia and PT Vale Indonesia, Tbk for providing the in situ water level data used in this research. The first author is supported by an appointment to the NASA Postdoctoral Program at the NASA Goddard Space Flight Center, administered by Oak Ridge Associated Universities through a contract with NASA. We thank the editor, Paola Passalacqua, anonymous referees and Radina Soebiyanto for their help in improving this manuscript.

Edited by: P. Passalacqua

References

Alsdorf, D. E. and Lettenmaier, D. P.: Tracking fresh water from space, Science, 301, 1491–1494, 2003.

Bamber, J. L.: Ice sheet altimeter processing scheme, Int. J. Remote Sens., 15, 925–938, 1994.

Bao, L., Lu, Y., and Wang, Y.: Improved retracking algorithm for oceanic altimeter waveforms, Prog. Nat. Sci., 19, 195–203, 2009.

Benveniste, J. and Defrenne, D.: Radar Altimetry Processing for Inland Waters: Introduction and Background Review, Slides presented at Workshop on Hydrology from Space, 29 September–1 October 2003, Touluse, 2003.

Berry, P. A. M., Garlick, J. D., Freeman, J. A., and Mathers, E. L.: Global inland water monitoring from multi-mission altimetry, Geophys. Res. Lett., 32, L16401, doi:10.1029/2005GL022814, 2005.

Birkett, C. M.: Contribution of the TOPEX NASA radar altimeter to the global monitoring of large rivers and wetlands, Water Resour. Res., 34, 1223–1239, 1998.

Birkett, C. M., Mertes, L. A. K., Dunne, T., Costa, M. H., and Jasinski, M. J.: Surface water dynamics in the Amazon Basin: application of satellite radar altimetry, J. Geophys. Res., 107, 8059, doi:10.1029/2001JD000609, 2002.

Birkinshaw, S. J., O'Donnell, G. M., Moore, P., Kilsby, C. G., Fowler, H. J., and Berry, P. A. M.: Using satellite altimetry data to augment flow estimation techniques on the Mekong River, Hydrol. Process., 24, 3811–3825, doi:10.1002/hyp.7811, 2010.

Brown, G. S.: The average impulse response of a rough surface and its applications, IEEE T. Antennas Propag., 25, 67–74, doi:10.1109/TAP.1977.1141536, 1977.

Brown, O. B. and Cheney, R. E.: Advances in satellite oceanography, Rev. Geophys., 21, 1216–1230, 1983.

Cai, X. and Ji, W.: Wetland hydrologic application of satellite altimetry – A ase study in the Poyang Lake watershed, Prog. Nat. Sci., 19, 1781–1787, doi:10.1016/j.pnsc.2009.07.004, 2009.

Calmant, S. and Seyler, F.: Continental surface waters from satellite altimetry, internal geophysics (space physics), C. R. Geoscience, 338, 1113–1122, 2006.

Coe, M. T. and Birkett, C. M.: Calculation of river discharge and prediction of lake height from satellite radar altimetry: Example for the Lake Chad basin, Water Resour. Res., 40, W10205, doi:10.1029/2003WR002543, 2004.

Cretaux, J.-F., Jelinski, W., Calmant, S., Kouraev, A., Vuglinski, V., Berge-Nguyen, M., Gennero, M.-C., Nino, F., Abarca Del Rio, R., Cazenave, A., and Maisongrande, P.: SOLS: A lake database to monitor in the near real time water level and storage variations from remote sensing data, Adv. Space Res., 47, 1497-1507, 2011.

Cristescu, M. E., Adamowicz, S. J., Vaillant, J. J., and Haffner, D. G.: Ancient lakes revisited: from the ecology to the genetics of speciation, Mol. Ecol., 19, 4837–4851, doi:10.1111/j.1365-294X.2010.04832.x, 2010.

Dabo-Niang, S., Ferraty, F., and Vieu, P.: On the using of modal curves for radar waveforms classification, Comput. Stat. Data Anal., 51, 4878–4890, 2007.

Davis, C. H.: A robust threshold retracking algorithm for measuring ice-sheet surface elevation change from satellite radar altimeters, IEEE T. Geosci. Remote, 35, 974–979, 1997.

Deng, X. and Featherstone, W. E.: A coastal retracking system for satellite radar altimeter waveforms: Application to ERS-2 around Australia, J. Geophys. Res., 111, C06012, doi:10.1029/2005JC003039, 2006.

De Sa, J. P. M.: Applied Statistics using SPSS, Statistica, MATLAB and R, Springer-Verlag, Berlin, Heidelberg, 2007.

ESA – European Space Agency: Envisat RA2/MWR Product Handbook, European Space Agency, 27 February 2007.

ESA – European Space Agency: Envisat RA-2/MWR Level 2 User Manual, Envisat Altimetry Quality Working Group, Ver. 1.4, 8 September 2011.

Estiaty, L. M., Susilowati, Y., Harsono, E., and Tjiptasamara, T.: Pemodelan Spasial Fluks Polutan pada Sistem Daerah Aliran Sungai dan Angkutan Polutan pada Sistem Sungai, Studi Kasus: DAS Mahakam, Pusat Penelitian Geoteknologi, Lembaga Ilmu Pengetahuan Indonesia, Bandung, Indonesia, 2007.

Fenoglio-Marc, L., Fehlau, M., Ferri, L., Becker, M., Gao, Y., and Vignudelli, S.: Coastal sea surface heights from improved altimeter data in the Mediterranean Sea, Proceedings GGEO2008, IAG Symposia, Springer Verlag, Boston, MA, 2009.

Frappart, F., Do Minh, K., L'Hermitte, J., Cazenave, A., Ramillien, G., Le Toan, T., and Mognard-Campbell, N.: Water volume change in the lower Mekong from satellite altimetry and imagery data, Geophys. J. Int., 167, 570–584, 2006.

Guo, J., Chang, X., Gao, Y., Sun, J., and Hwang, C.: Lake Level Variations Monitored With Satellite Altimetry Waveform Retracking, IEEE J. Select. Top. Appl. Earth Obs. Rem. S., 2, 80–86, doi:10.1109/JSTARS.2009.2021673, 2009.

Hehanussa, P. E. and Haryani, G. S.: Morfologi cekungan dan profil suhu Danau Matano dan Danau Poso, Limnotek, 6, 13–23, 1999.

Herdendorf, C. E.: Large Lakes of the World, J. Great Lakes Res., 8, 379–412, 1982.

Hwang, C., Guo, J. Y., Deng, X. L., Hsu, H. Y., and Liu, Y. T.: Coastal gravity anomalies from retracked Geosat/GM altimetry: improvement, limitation and the role of airborne gravity data, J. Geodyn., 80, 204–216, 2006.

Kenney, J. F. and Keeping, E. S.: Mathematics of statistics, in: Mathematics of Statistics, Van Nostrand, http://books.google.com/books?id=UdlLAAAAMAAJ (last access: January 2015), 1947,

Koblinsky, C. J., Clarke, R. T., Brenner, C. A., and Frey, H.: Measurement of river water levels with satellite altimetry, Water Resour. Res., 29, 1839–1848, 1993.

Kouraev, A. V., Zakharova, E. A., Samain, O., Mognard, N. M., and Cazenave, A.: Ob' river discharge from TOPEX/Poseidon satellite altimetry (1992–2002), Remote Sens. Environ., 93, 238–245, 2004.

Kuo, C.-Y. and Kao, H.-C.: Retracked Jason-2 altimetry over small water bodies: case study of Bajhang River, Taiwan, Mar. Geodyn., 34, 382–392, 2011.

Laxon, S.: Sea ice altimeter processing scheme at the EODC, Int. J. Remote Sens., 15, 915–924, doi:10.1080/01431169408954124, 1994.

Lee, H.: Radar altimetry methods for solid earth geodynamics studies, PhD thesis, School of Earth Sciences, The Ohio State University, Columbus, Ohio, 2008.

Legresy, B. and Remy, F.: Altimetric observations of surface characteristics of the Antarctic ice sheet, J. Glaciol., 43, 265–275, 1997.

Li, Y.: Root Mean Square Error, in: Encyclopedia of Research Design, edited by: Salkind, N. J., SAGE Publications Inc., Thousand Oaks, CA, 1288–1289, 2010.

McKinnon, K., Hatta, G., Halim, H., and Mangalik, A.: The Ecology of Kalimantan: Indonesian Borneo, The Ecology of Indonesia Series, Vol. 3, Periplus, Singapore, 1996.

Medina, C. E., Gomez-Enri, J., Alonso, J. J., and Villares, P.: Water level fluctuations derived from ENVISAT Radar Altimeter (RA-2) and in-situ measurements in a subtropical waterbody: Lake Izabal (Guatemala), Remote Sens. Environ., 112, 3604–3617, doi:10.1016/j.rse.2008.05.001, 2008.

Meybeck, M., Friedrich, G., Thomas, R., and Chapman, D. (Eds.): Rivers, in: Water Quality Assessments – a Guide to Use of Biota, Sediments and Water in Environmental Monitoring, 2nd Edn., UNESCO/WHO/UNEP. 1992, Taylor and Francis, London, New York, 1996.

Michailovsky, C. I., McEnnis, S., Berry, P. A. M., Smith, R., and Bauer-Gottwein, P.: River monitoring from satellite radar altimetry in the Zambezi River basin, Hydrol. Earth Syst. Sci., 16, 2181–2192, doi:10.5194/hess-16-2181-2012, 2012.

Morris, C. S. and Gill, S. K.: Evaluation of the TOPEX/POSEIDON altimeter system over the Great Lakes, J. Geophys. Res., 99, 24527–24539, 1994.

Munyaneza, O., Wali, U. G., Uhlenbrook, S., and Maskey, S., and McArd, J. M.: Water level monitoring using radar remote sensing data: Application to Lake Kivu, central Africa, Phys. Chem. Earth P. A/B/C, 34, 722–728, doi:10.1016/j.pce.2009.06.008, 2009.

Nagler, J.: Root Mean Square, in: The SAGE Encyclopedia of Social Science Research Methods, edited by: Lewis-Beck, M. S., Bryman, A., and Liao, T. F., SAGE Publications, Inc., Thousand Oaks, CA, 978–979, doi:10.4135/9781412950589.n871, 2004.

Panik, M. J.: Statistical Inference: a Short Course, John Wiley & Sons, Hoboken, 2012.

Rees, G.: Physical Principles of Remote Sensing, Cambridge University Press, Cambridge, England, 1990.

Russel, J. and Bijaksana, S.: The Towuti Drilling Project: paleoenvironments, biological evolution, and geomicrobiology of a tropical lake, Scient. Drilling, 14, 68–71, doi:10.2204/iodp.sd.14.11.2012, 2012.

Sarmiento, S. E. and Khan, S. D.: Spatial-Temporal Variability of Great Slave Lake Levels From Satellite Altimetry, IEEE Geosci. Remote Sens. Lett., 7, 426–429, doi:10.1109/LGRS.2009.2038178, 2010.

Tang, Q., Gao, H., Lu, H., and Lettenmaier, D.: Remote sensing: hydrology, Prog. Phys. Geogr., 33, 490–509, 2009.

Torrance, K. E. and Sparrow, E. M.: Theory for Off-Specular Reflection From Roughened Surfaces J. Opt. Soc. Am., 57, 1105–1112, doi:10.1364/JOSA.57.001105, 1967.

Troitskaya, Y. I., Rybushkina, G. V., Soustova, I. A., Balandina, G. N., Lebedev, S. A., Kostyanoi, A. G., Panyutin, A. A., and Filina, L. V.: Satellite altimetry of inland water bodies, Water Resour., 39, 184–199, doi:10.1134/S009780781202008X, 2012.

Tseng, K.-H.: Satellite Altimetry and Radiometry for Inland Hydrology, Coastal Sea-Level and Environmental Studies, School of Earth Sciences, The Ohio State University, Columbus, Ohio, 2012.

Tseng, K.-H., Shum, C. K., Yi, Y., Fok, H. S., Kuo, C.-Y., Lee, H., Cheng, X., and Wang, X.: Envisat Altimetry Radar Waveform Retracking of Quasi-Specular Echoes over the Ice-Covered Qinghai Lake, Terr. Atmos. Ocean. Sci., 24, 615–627, doi:10.3319/TAO.2012.12.03.01(TibXS), 2013.

Vaillant, J. J., Haffner, G. D., and Cristescu, M. E.: The Ancient Lakes of Indonesia: towards Integrated Research on Speciation, Integr. Comp. Biol., 51, 634–643, 2011.

Wingham, D. J. and Rapley, C. G.: Saturation effects in the Seasat altimeter receiver, Int. J. Remote Sens., 8, 1163–1173, 1987.

Wingham, D. J., Rapley, C. G., and Griffiths, H. G.: New techniques in satellite altimeter tracking systems, III IGARRS 1986 Symposium, Zurich, Proceedings, Noordwijk, ESTEC, Scientific and Technical Publications Branch, ESA SP-254, 1339–1344, 1986.

Zhang, J., Xu, K., Yang, Y., Qi, L., Hayashi, S. and Watanabe, M.: Measuring Water Storage Fluctuations in Lake Dongting, China, by Topex/Poseidon Satellite Altimetry, Environ. Monitor. Assess., 115, 23–37, doi:10.1007/s10661-006-5233-9, 2006.

Zwally, H. J.: The GSFC Retracking Algorithms, NASA Goddard Space Flight Center, Cryospheric Sciences Laboratory, http://icesat4.gsfc.nasa.gov/radar_data/data_processing/gsfcretrackdoc.960725.php (last access: January 2015), 1996.

Modelling hyporheic processes for regulated rivers under transient hydrological and hydrogeological conditions

D. Siergieiev[1], L. Ehlert[1,2], T. Reimann[2], A. Lundberg[1], and R. Liedl[2]

[1]Division of Geosciences and Environmental Engineering, Luleå University of Technology, Luleå, Sweden
[2]Institute for Groundwater Management, Faculty of Environmental Sciences, Technische Universtität Dresden, Dresden, Germany

Correspondence to: D. Siergieiev (dmysie@ltu.se)

Abstract. Understanding the effects of major hydrogeological controls on hyporheic exchange and bank storage is essential for river water management, groundwater abstraction, restoration and ecosystem sustainability. Analytical models cannot adequately represent complex settings with, for example, transient boundary conditions, varying geometry of surface water–groundwater interface, unsaturated and overland flow, etc. To understand the influence of parameters such as (1) sloping river banks, (2) varying hydraulic conductivity of the riverbed and (3) different river discharge wave scenarios on hyporheic exchange characteristics such as (a) bank storage, (b) return flows and (c) residence time, a 2-D hydrogeological conceptual model and, subsequently, an adequate numerical model were developed. The numerical model was calibrated against observations in the aquifer adjacent to the hydropower-regulated Lule River, northern Sweden, which has predominantly diurnal discharge fluctuations during summer and long-lasting discharge peaks during autumn and winter. Modelling results revealed that bank storage increased with river wave amplitude, wave duration and smaller slope of the river bank, while maximum exchange flux decreased with wave duration. When a homogeneous clogging layer covered the entire river–aquifer interface, hydraulic conductivity positively affected bank storage. The presence of a clogging layer with hydraulic conductivity $< 0.001\ \mathrm{m\,d^{-1}}$ significantly reduced the exchange flows and virtually eliminated bank storage. The bank storage return/fill time ratio was positively related to wave amplitude and the hydraulic conductivity of the interface and negatively to wave duration and bank slope. Discharge oscillations with short duration and small amplitude decreased bank storage and, therefore, the hyporheic exchange, which has implications for solute fluxes, redox conditions and the potential of riverbeds as fish-spawning locations. Based on these results, river regulation strategies can be improved by considering the effect of certain wave event configurations on hyporheic exchange to ensure harmonious hydrogeochemical functioning of the river–aquifer interfaces and related ecosystems.

1 Introduction

Surface water–groundwater interfaces have recently received growing research interest (Sophocleous, 2002) and have become the focus of multiple water resources management policies (Klöve et al., 2011). The hyporheic zone that harbours river–aquifer interactions plays a key role in riverine and riparian ecosystem functioning (e.g. Krause et al., 2011). The reactive nature of this zone maintains exchange and transformation of solutes along the pathways between surface water and groundwater. The hyporheic zone is an appreciated habitat for hyporheos, microorganisms and bacteria occupying the space below and along the river channel (Boulton et al., 1998). Besides having negative impacts on the ecosystem of the zone itself, with changes in hyporheic water composition (Calles et al., 2007; Siergieiev et al., 2014c), alteration of the hyporheic functionality due to surface water–aquifer disconnection can also severely modify neighbouring ecosystems. As a result, riparian zones and adjacent wetlands may experience changes in groundwater table, water and nutrient fluxes. Furthermore, restricted hyporheic exchange limits mobilisation of solutes from the riparian zone and their fluxes into the river during key hydrological events as a result of river–aquifer disconnection (Burt and Pinay, 2005), which can in turn affect surface water quality (Valett et al., 1996).

A large number of rivers worldwide are obstructed by dams (Nilsson et al., 2005) and are therefore subject to artificial discharge fluctuations. These fluctuations stress hyporheic exchange flows, which often results in degradation of the river–aquifer continuum. A major impact is the alteration of river sediment transport and, subsequently, increased colmation of the riverbed (Brunke and Gonser, 1997; Blaschke et al., 2003), which deteriorates river–aquifer hydraulic connectivity (Burt and Pinay, 2005) and controls functional changes in the hyporheic zone (Siergieiev et al., 2014c). In addition, inundation of the river banks by construction of reservoirs changes the shape of the river–aquifer interface, successively affecting the hyporheic exchange (Doble et al., 2012a).

Therefore, it is necessary to understand the hydrogeological functioning of this interface under artificial conditions, such as hydropower-regulated rivers, in order to incorporate this knowledge into water resource management and thereby improve the functional behaviour of the hyporheic zone by optimising river discharge strategies (e.g. Hanrahan, 2008).

The hyporheic zone size, bank storage volume and bank fluxes vary with river stage fluctuations (amplitude, duration) and river bank conditions (slope, hydraulic conductivity). Todd (1955) provided a first theoretical analysis of flood-induced bank storage. This work served as the foundation for later analytical solutions and numerical simulations of floodplain hydrology. The dynamics of bank storage were later estimated using analytical models (e.g. Cooper and Rorabaugh, 1963) based on the following simplifications: single flood wave, homogeneous aquifer and a fully penetrating vertical river bank. During base flow conditions, hydrological river–aquifer interactions can be described by precipitation–runoff models (Butturini et al., 2002). However, these often neglect the distributed effects of e.g. unsaturated zone processes or topography, resulting in residual unexplained variability in bank storage. Chen and Chen (2003) used a numerical approach to simulate the bank storage response to changes in river stage and riverbed hydraulic conductivity for a partially penetrating river with vertical banks in a fully saturated aquifer. They pointed out the importance of subsurface anisotropy, which governs the directions of hyporheic zone development. Simultaneous consideration of seepage and a variably saturated aquifer showed that unsaturated zone processes have a pronounced effect on bank storage (Li et al., 2008; Doble et al., 2012a). Inclusion of the unsaturated zone in bank storage simulations decreased the modelled storage and improved the return flows (Doble et al., 2012b). Furthermore, models that consider vertical river banks for sloping banks under-estimate bank storage (Doble et al., 2012a). Understanding of bank storage processes can improve hyporheic ecotone and river–aquifer continuum concepts, with positive implications for e.g. base flow separation techniques (McCallum et al., 2010) and ecosystem sustainability (Schneider et al., 2011).

For several seasons, hyporheic exchange was studied in the hydropower regulated Lule River, northern Sweden (Siergieiev et al., 2014a, c). Low hydraulic conductivity of the riverbed and daily varying river discharge have resulted in depleted hyporheic exchange flows across the river–aquifer interface (Siergieiev et al., 2014a). Deteriorated river water quality as a result of regulation (Smedberg et al., 2009; Siergieiev et al., 2014b) may partly depend on suppression of hyporheic processes due to regulation (Valett et al., 1996). Improved understanding of the major hydrogeological controls of hyporheic exchange has legacy effects on understanding geochemical fluxes between surface water and groundwater (Fritz and Arntzen, 2007) and can provide a platform for implementation of environmental flows and improved management and ecological status of regulated rivers.

The aim of this study was therefore to provide a set of scenarios with variable river discharge schemes (wave duration and amplitude), river bank slope and riverbed hydraulic conductivity, in order to investigate the effects of these parameters on fluxes across the river–aquifer interface, bank storage volume, fill/return time ratio and residence time. A realistic case study was used to set up a conceptual and a numerical model and justify the use of the method for further scenario simulations. The major difference of this work to earlier studies is an attempt to demonstrate possible limits of the surface water–groundwater models and the data necessity. Furthermore, application of numerical surface water–groundwater exchange modelling in a highly dynamic environment, common for many other regulated rivers in the world, was evaluated. This kind of modelling can be used in different applications such as river restoration, implementation of environmental flows, integration with regional models etc. The topic is also of interest from the perspective of improvement of ecological status of impacted watercourses as requested by Water Framework Directive. Understanding and coupling of river–aquifer functioning to the transfer of nutrients across their interface and the impact on sediment transport and biological migration are essential for improved ecological status of regulated rivers, which are today severely impacted by disruption of hydrological connectivity and related biogeochemical consequences.

2 Site description

The measurement profile orthogonal to the river included an observation station in the river and two groundwater wells (Fig. 1) with hourly registration of water level during 2010–2011 and every 15 min during 2012. The riverbed at the site slopes gently towards the middle of the channel and is composed of silty-clayey material with layers of highly conductive stratum and laterally spread sparse patches of sand and

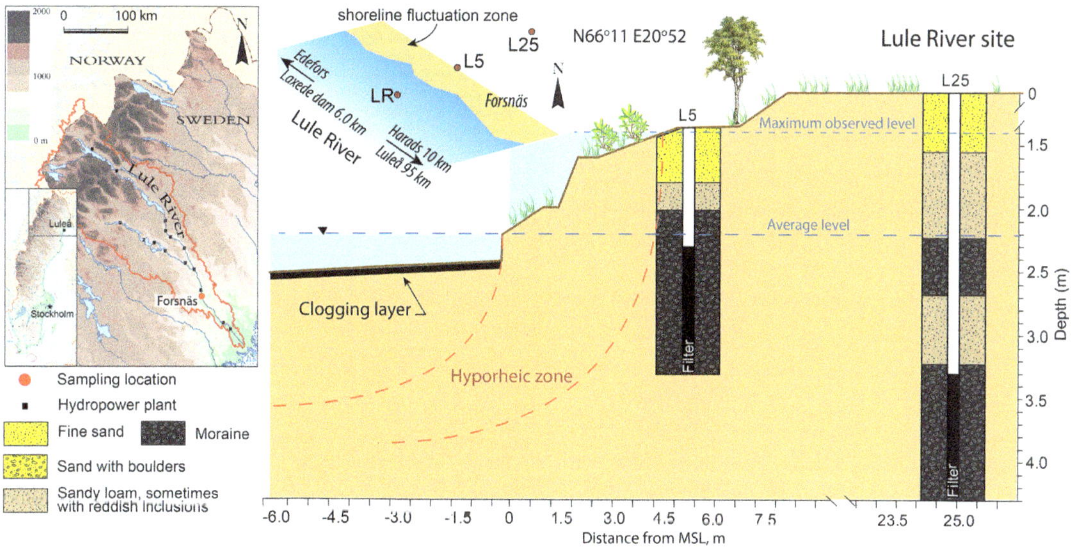

Figure 1. Location of the observation site and cross section of the aquifer with groundwater wells, soil depth profiles and extent of the hyporheic zone. Numbers in well names indicate distance to the mean shoreline in metres.

gravel. The former floodplain is limited by an earth embankment from the river side and stretches for over 150 m where it meets the foot of the hillslope. The depth to the bedrock varies from none to 60 m with an average depth of around 10 m and thinnest overburden towards the hillslope. The flat topography of the floodplain and the hillslope orientation which is normal to the river suggest orthogonal groundwater flow towards the channel. Orientation of the groundwater well profile was chosen accordingly.

3 Materials and methods

3.1 Data collection

The soil was visually inspected during installation of groundwater wells. Samples were collected at 0.3 m interval and sieving analysis was performed on three samples from each of the two locations. Unsaturated flow parameters were estimated on these six selected samples using pressure pot experiments to obtain water-holding characteristics (Ehlert, 2014). To assess saturated hydraulic conductivity, repeatable slug tests (three in each well) using both falling and rising hydraulic head were carried out in the wells, while a direct push piezometer (two repeatable tests at two locations; 1 and 2.5 m from the shoreline) using a falling head was performed at the riverbed at 0.1 m depth interval down to 0.7 m depth (Siergieiev et al., 2014a). A harmonic mean over the top 0.3 m least permeable layers was used as hydraulic conductivity of the clogging layer in the model.

3.2 Conceptual model

The data collected at the site did not allow development of a highly distributed model. Parameter values obtained in the field and in the laboratory were therefore averaged for the saturated and unsaturated zone and the clogging layer. The vertical 2-D conceptual model considered a homogeneous aquifer, partially penetrating river, sloping banks, unsaturated zone and clogging layer that covered the entire river–aquifer interface (Fig. 2). The Dirichlet boundary condition at the riverside was varied according to the measured water level time series. The top of the model was represented by a constant flux boundary to consider recharge, which was assumed to be 50 % of annual precipitation (Lemmelä, 1990) of 470 mm yr^{-1}, resulting in 6.5×10^{-4} m d^{-1} recharge. No flow boundaries were assigned to the remaining borders. The distance to the right boundary was set to ensure that influences of river stage fluctuations did not reach this boundary for the longest fluctuation period. A first approximation of the distance of influence based on an analytical solution (Sawyer et al., 2009) assumed negligible riverbed resistance and is therefore questionable in the present case (e.g. Singh, 2004). This estimate of the maximum extent (180 m for a 1-month fluctuation period) was further tested using the numerical model.

The following assumptions were used in the model:

- Two-dimensional model space
- Simplified geometry, neglecting microtopography
- Constant recharge, representing both groundwater recharge and regional gradient
- Isotropic and homogeneous aquifer and clogging layer

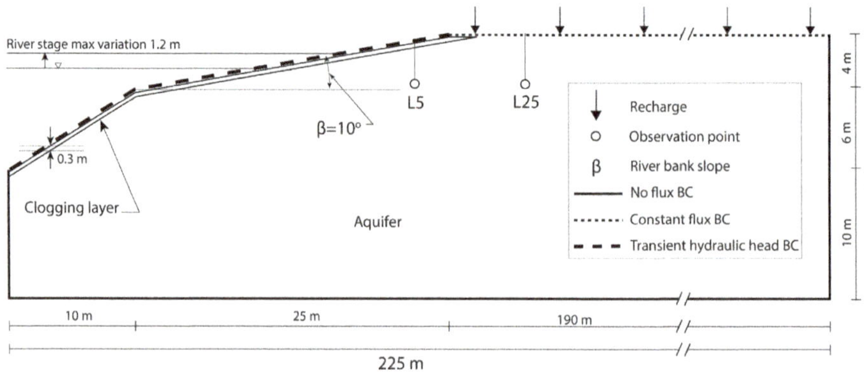

Figure 2. Conceptual model of the site showing boundary conditions (BC), clogging layer and observation points.

- The same unsaturated parameters for the entire model domain
- Surface flow resistance and hydraulic effects of the river processes due to variable discharge were neglected
- Viscosity effects (temperature and solute concentration differences between the river and the aquifer) were neglected.

3.3 Numerical model

The numerical modelling code FEFLOW 6.2 (Diersch, 2014) was used to simulate variably saturated flow during river–aquifer interaction by solving Richards' equation using the PARDISO solver (Schenk and Gärtner, 2004). The model domain was discretised using the triangle mesh generator. The time step was set to 30 min intervals to keep the computational time within reasonable limits and was increased to 1 hour during the validation run. To enable inverse parameter estimation for the van Genuchten model, a plug-in for coupling FePEST (graphical user interface for PEST by Doherty et al., 2011) with FEFLOW was developed (Ehlert, 2014).

3.4 Model calibration

The model was sequentially calibrated against measurements in L5 and L25 collected during June–October 2012. First, only hydraulic conductivity, specific storage and porosity were calibrated, followed by unsaturated van Genuchten parameters and maximum and residual saturation. Finally, all parameters were calibrated together. To track improvement of the fit between modelled and observed hydraulic head, the regression coefficient (R^2), Nash–Sutcliffe index (NS) and root mean square error (RMSE) were calculated for each calibration run.

The conceptual understanding of hydrogeological processes is often erroneous in terms of boundary and initial conditions (Bredehoeft, 2005). While the initial state of models is often calibrated, the importance of other conceptualisation aspects seems to be rarely verified, even though their effects have been debated by different authors (e.g. Refsgaard et al., 2006). To test our assumptions on hydrogeological conditions in the area, a sensitivity analysis for model boundary conditions was carried out. The distance from the river to the aquifer boundary was varied from the reference distance (200 m) to 50, 500, 1000 and 5000 m, keeping the recharge at 50 % of annual precipitation. Afterwards, the recharge rate was changed from the assumed 50 to 30 and 70 %, keeping the distance from the river to the aquifer boundary at 200 m.

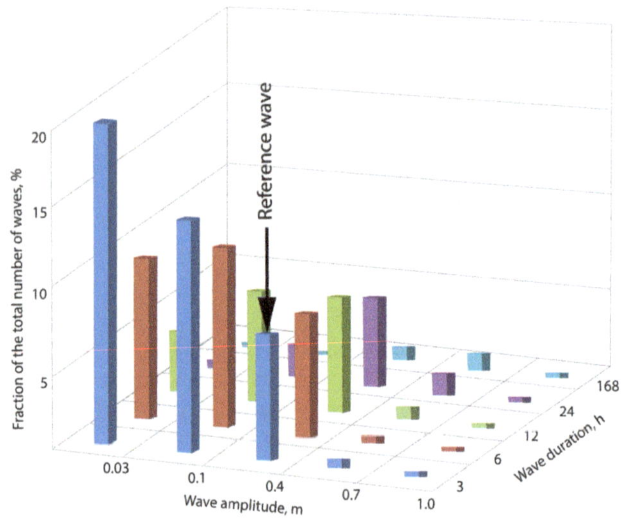

Figure 3. Summary matrix of wave amplitude (m) and duration (h) as a fraction of all observed wave events.

3.5 Modelling scenarios

Based on the calibrated model domain, the effect of multiple hydrogeological parameters on hyporheic exchange was evaluated, varying one parameter at a time. Artificial river stage variations were applied according to the distribution of commonly observed amplitudes and durations during 2012 (Fig. 3). The head boundary on the river side was varied as a cosine-shaped wave between $t = 0$ and $t = t'$, with ampli-

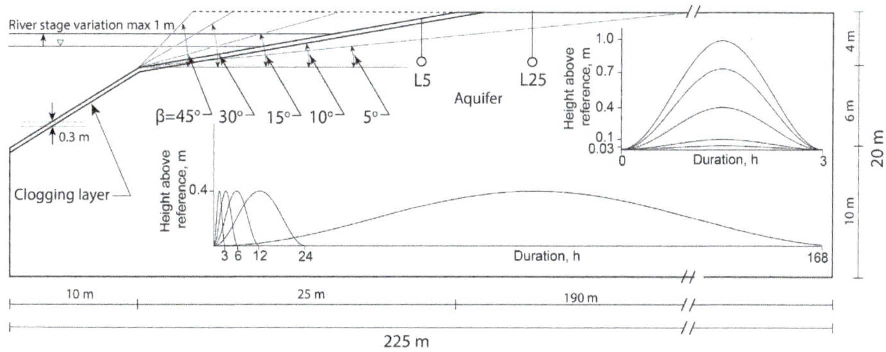

Figure 4. Graphical summary of modelling scenarios for varying river bank slope, wave duration and amplitude. For hydraulic conductivity scenarios, see text.

tude $h_{\max} - h_0$ (McCallum et al., 2010):

$$h(t) = h_0 + \frac{(h_{\max} - h_0)}{2}\left(1 - \cos\left(2\pi \frac{t}{t'}\right)\right), \tag{1}$$

where h is the hydraulic head (m), t is time (h), t' is the duration of the stage oscillation (h), h_0 is the head at $t = 0$, and $h_{\max}$ is the maximum head (at $t = t'/2$). All scenarios used a single wave event and were terminated after steady-state conditions were reached.

The sensitivity of the model to various scenarios was evaluated using the flux across the river–aquifer interface, bank storage and the ratio between the time to fill and to empty the bank storage. A reference simulation was based on the calibrated model (hydraulic conductivity of the aquifer $2.14\,\mathrm{m\,d^{-1}}$ and of the clogging layer $0.01\,\mathrm{m\,d^{-1}}$), the actual conditions at the site (bank slope 10°) and an input wave with 3 h duration and 0.4 m amplitude. Other scenarios included varying hydraulic conductivity of the clogging layer (0.001, 0.01, 0.1, $1\,\mathrm{m\,d^{-1}}$), river bank slope (5, 10, 15, 30, 45°), wave amplitude (0.03, 0.1, 0.4, 0.7, 1.0 m) and duration (3, 6, 12, 24, 168 h) (Fig. 4). The initial conditions were generated by running a transient simulation with constant hydraulic head at the riverside, no flow at the aquifer side and constant distributed diffuse recharge rate at the top for the time sufficient to recreate steady-state conditions. The flux into the aquifer was considered to be positive. The exchange flux per metre riverbed width ($\mathrm{m^2\,d^{-1}}$) was identified as the difference between inflow and outflow rates across the river boundary. The bank storage ($\mathrm{m^2}$) was the cumulative exchange flux multiplied by the time step assuming the flux being either positive or negative for in- and outflow from the model, respectively, while bank storage always positive. The fill time was the time required for the bank storage to reach its maximum, whereas the return time was the time between the maximum and zero bank storage on the falling limb. Residence time was calculated as the sum of the fill and return times.

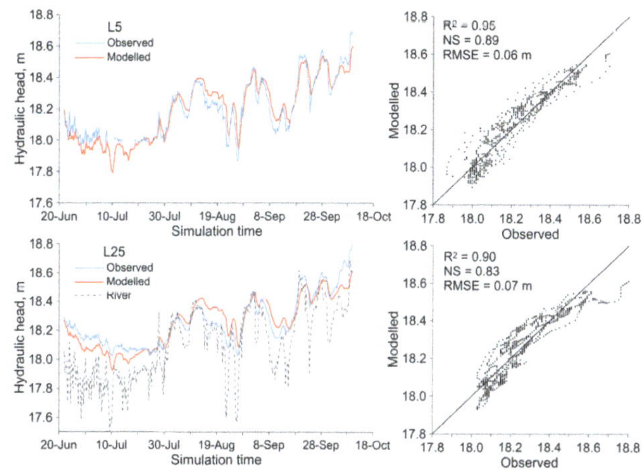

Figure 5. Hydraulic head at observation wells L5 (above) and L25 and the river (below) compared with the simulated results using calibrated parameters (see Table 1).

4 Results

4.1 Model calibration

Sequential calibration using FePEST yielded minor underestimation of hydraulic head in the beginning of the simulation and minor over-estimation in later parts (Fig. 5). Generally, the fit was better for the well closer to the river, i.e. L5. The resulting R^2, NS, RMSE were 0.95, 0.89, 0.06 m, respectively, for observation well L5 and 0.89, 0.77, 0.08 m, respectively, for well L25. The calibrated model (Table 1) was validated using observation data from 2010 with R^2, NS, RMSE of 0.85, 0.97, 0.09 m, respectively, for well L5 and 0.74, −17.85, 0.43 m, respectively, for well L25 (data not shown).

Numerical simulation results verified that an aquifer boundary located 200 m away from the river was out of reach of influences induced by the river stage fluctuations used here. Sensitivity analysis of the boundary conditions resulted in substantial over-estimation of measured hydraulic head for

Table 1. Model parameters measured and calibrated using FePEST.

Parameter	Symbol	Units	Measured*	Calibrated
Saturated K aquifer	K_{aq}	$m\,d^{-1}$	0.13 ± 0.08	2.14
Saturated K clogging layer	K_{cl}	$m\,d^{-1}$	0.04 ± 0.01	0.01
Specific storage	S	m^{-1}	–	0.001
Effective porosity	n	–	–	0.56
Maximum saturation	θ_s	–	0.92 ± 0.04	0.95
Residual saturation	θ_r	–	0.14 ± 0.08	0.14
Anisotropy ratio	K_v/K_h	–	–	1
van Genuchten parameters	α	m^{-1}	0.003 ± 0.001	0.015
	n	–	2.1 ± 0.4	2.1

* Mean of all measurements ± standard deviation.

a model domain size of 5000 m. The fit improved slightly for a domain size of 500 and 1000 m compared with the 200 m long domain. However, the larger model domains were discarded to keep the computation time reasonable. A model domain of 50 m tended to under-estimate the measured hydraulic head for both L5 and L25. The highest and lowest recharge rates over- and under-estimated the observed data, respectively. Therefore, the recharge rate taken as 50 % of precipitation was the most suitable solution for the present case. Overall, observation well L25 was more affected by changes in the recharge than well L5, indicating a strong influence of groundwater gradient on L25 and of the river boundary on L5. However, it was recognised here that the final influence of recharge and distance to the boundary on calibration results is a combined effect rather than a single effect of one of these.

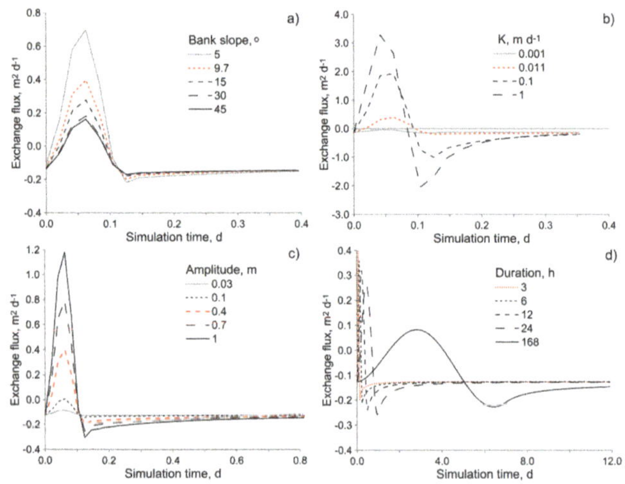

Figure 6. Exchange flux for different bank slope **(a)**, hydraulic conductivity **(b)**, wave amplitude **(c)** and duration **(d)** scenarios, with the reference case (red) and insets of the input wave pulse. Note different scale.

4.2 Scenarios

The simulated scenarios of varying river bank slope, clogging layer hydraulic conductivity and input wave amplitude and duration were compared based on their effect on the resulting exchange fluxes, bank storage and residence time.

4.2.1 Exchange fluxes

There were variations in exchange fluxes across the river–aquifer interface as a result of varying forcing parameters (Fig. 6). The maximum exchange flux decreased with river bank slope (Fig. 6a). A change in the bank slope from 5 to 10° caused a similar decrease in the exchange flux as a change from 10 to 45°. The exchange flux increased with hydraulic conductivity of the interface (Fig. 6b). However, for the scenario with $K = 0.001\,m\,d^{-1}$, the fluxes were always directed towards the river, indicating no bank storage and thus no hyporheic exchange. The increase in exchange flux was logarithmically proportional to the change in hydraulic conductivity, e.g. a 100-fold increase in hydraulic conductivity generated only a 7-fold rise in exchange flux. The wave amplitude was related positively to the exchange flux across the river–aquifer interface (Fig. 6c). There were no positive (towards the aquifer) fluxes for amplitude 0.1 m and lower. Every further 0.3 m increase in amplitude resulted in a $0.4\,m^2\,d^{-1}$ increase in the maximum simulated fluxes. The maximum exchange flux was higher for shorter wave duration times, e.g. $0.40\,m^2\,d^{-1}$ for a 3 h wave and $0.09\,m^2\,d^{-1}$ for a 168 h wave (Fig. 6d).

4.2.2 Bank storage

The effects of bank slope, hydraulic conductivity of the clogging layer and input wave amplitude and duration on bank storage volume were plotted as a function of time (Fig. 7). Bank storage increased with lower bank slope (Fig. 7a). For example, an almost 5-fold increase in bank slope from approx. 10 to 45° reduced bank storage by less than 50 %. Meanwhile, a decrease in bank slope from 10 to 5° doubled bank storage (Fig. 7b), indicating the importance of small

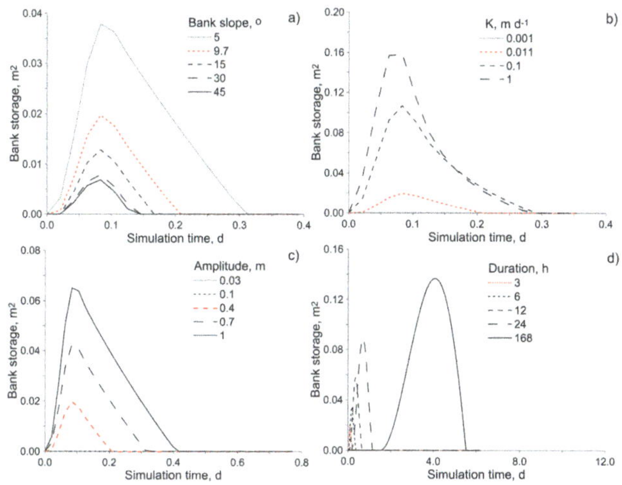

Figure 7. Bank storage for different bank slope (a), hydraulic conductivity (b), wave amplitude (c) and duration (d) scenarios, with the reference case (red) and insets of the input wave pulse. Note different scale.

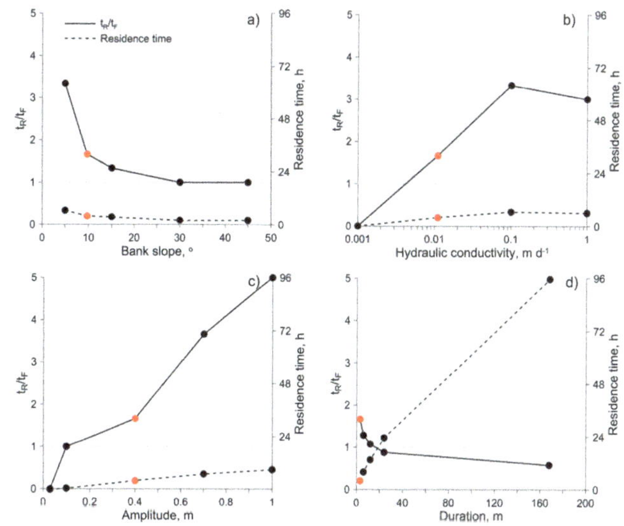

Figure 8. Return/fill time ratio (t_R/t_F) and residence time for different bank slope (a), hydraulic conductivity (b), wave amplitude (c) and duration (d) scenarios, with the reference case (red) and insets of the input wave pulse.

slope for river–aquifer exchange. Overall, the bank storage for 5° was 5-fold higher than that for 45°. A 10-fold increase in hydraulic conductivity of the riverbed from the reference scenario (bank storage $= 0.02\,m^2$) improved bank storage by 500 % ($0.10\,m^2$), which further increased to $0.16\,m^2$ with another 10-fold increase in hydraulic conductivity. As was the case for exchange flux, virtually no bank storage occurred for the scenarios with 0.03 and 0.1 m wave amplitude (Fig. 7c), which were the most common amplitudes at the observation site (Fig. 3). An approximately 60 % rise in wave amplitude (from 0.4 to 0.7 m) resulted in a 30 % increase in maximum bank storage and a 50 % increase in maximum exchange flux. Duration of the river stage oscillation also positively affected bank storage (Fig. 7d). A 56-fold rise in duration (from 3 to 168 h) resulted in a 7-fold increase in maximum bank storage (from 0.02 to $0.14\,m^2$).

4.2.3 Residence time

The timing of bank storage (residence time and return/fill ratio) was examined under different modelling scenarios (Fig. 8). The residence time and return/fill time ratio decreased with increasing bank slope (Fig. 8a). The return time always exceeded the fill time except for the slopes above 30°, which indicated $t_R/t_F = 1$. Increased hydraulic conductivity of the river–aquifer interface increased the return time, which positively affected the overall residence time of river water in the subsurface (Fig. 8b). Nonetheless, the residence time was highest (6.5 h) for the scenario with hydraulic conductivity of the interface of $0.1\,m\,d^{-1}$, marginally exceeding the scenario with the highest hydraulic conductivity (6 h). Return time increased with rising wave amplitude, as did the residence time, ranging from 0 h for the smallest wave to 9.6 h for the largest (Fig. 8c). The ratio between the change in residence time and the change in amplitude varied between 0.3 and 0.5 and was higher at lower amplitudes. The return time of bank storage was longer than the fill time for the waves with duration below 24 h and decreased with wave duration (Fig. 8d). The return/fill time ratio decreased by almost two-thirds from the shortest wave duration (1.7) to the longest (0.6), whereas the residence time increased by more than one order of magnitude from the shortest wave to the longest (from 4.8 to 96 h).

5 Discussion

5.1 Conceptual and numerical models and calibration

The model calibration resulted in hydraulic conductivity one order of magnitude higher than estimated via field tests. This difference is attributable to the limitations of slug tests, which provide point data for saturated hydraulic conductivity around the well filter. According to the observed soil profile (Fig. 1), grain size decreased with depth and can be a reason for lower measured and higher calibrated hydraulic conductivity of the aquifer. It was beyond the scope of this work to analyse whether the calibrated parameter set converged around a local or global optimum. However, there was a tendency for over-estimation of both saturated hydraulic conductivity and effective porosity (Table 1). Hydraulic conductivity and porosity are related through the hydraulic diffusivity term that controls the connectivity of a high-permeability flow path (Knudby and Carrera, 2006). Hydraulic diffusivity remains virtually the same with proportional change in both hydraulic conductivity and porosity. Therefore, even if over-estimation of both parameters by model calibration took place, this would have had a limited effect on the results.

Although the validation results showed a good fit for hydraulic head at observation well L5, major under-prediction of hydraulic head at L25 was observed. This could be related to the substantially different precipitation patterns during the 2 years (2012 was used for calibration and 2010 for validation) and the fact that L5 is more influenced by the river and L25 by the aquifer. Recharge on top of the model, used for both calibration and validation runs, was assumed to be constant and equal to 50 % of mean annual precipitation. In the following sections, the implications of the modelling results for hyporheic exchange and the limitations due to the assumptions used are discussed.

5.2 Implications for hyporheic exchange

5.2.1 General reflections

The implications of sloping banks for numerical modelling have been discussed previously, e.g. by Doble et al. (2012a). In terms of functioning of the river–aquifer interface, low bank slope angles increase the contact area between river and aquifer and therefore result in enlarged volume of bank storage and size of the hyporheic zone. This also positively affects the residence time, which is primarily governed by the penetration distance of river water and the return time necessary for it to discharge back into the river. Enhanced hyporheic exchange due to river bank slope would be true for many regulated rivers, as construction of reservoirs is associated with river floodplain inundation, and therefore with the formation of gently sloping banks along the channel. Consequently, lower bank slope angles have the potential to improve e.g. spawning conditions and species richness (Hanrahan, 2008). However, this appears not to be the case in several regulated temperate and boreal rivers due to the effect of colmation (Brunke and Gonser, 1997; Blaschke et al., 2003; Calles et al., 2007; Siergieiev et al., 2014a). In the present study, bank storage decreased with decreasing hydraulic conductivity of the river–aquifer interface (Fig. 7b). Therefore, in rivers with clear interstices, flat river banks contribute greatly to an increase in bank storage and hyporheic exchange, as demonstrated by the simulations, whereas this effect is hampered in rivers with a clogged riverbed.

River wave duration and amplitude were positively related to bank storage (Fig. 7c, d), but only amplitude positively affected exchange flux (Fig. 6c). As opposed to bank storage, exchange flux is more dependent on soil properties than on input wave configuration. This is supported by the fact that the peak exchange flux decreased with the prolonged wave duration (Fig. 6d), due to smaller hydraulic gradients at the river–aquifer interface, whereas the maximum bank storage increased (Fig. 7d).

A linear relationship between maximum bank storage and the product of wave amplitude and period has been reported previously by Todd (1955). However, this was only valid for a fully saturated homogeneous aquifer adjacent to a fully penetrating river. Using the results of the modelling scenarios in the present study, it was possible to show that there is a relationship between the ratio of wave duration/amplitude and bank storage or residence time for waves with amplitude exceeding 0.1 m (Fig. 9). This indicates that there is an optimal wave configuration (duration and amplitude) for every specific set of hydrogeological conditions that accounts for the highest bank storage and can potentially improve hyporheic exchange and minimise energy losses in hydropower-regulated rivers.

The hysteresis patterns observed for the t_R/t_F ratio for different modelling scenarios illustrate that the process of filling the pores of an aquifer is different from that of draining them and depends on hydraulic gradient and river–aquifer contact area (Fig. 8). The former is a function of the river wave configuration, while the latter depends on the river bank slope. The contribution of bank storage to river runoff is complex and of high importance in catchment hydrology (Harr, 1977; Turton et al., 1992; McGlynn et al., 2004). The results presented here suggest that with decreasing bank slope, the contribution of bank storage to the river extends in time, prolonging the falling limb of the river hydrograph. The same effect occurs with rising amplitude, which generates steeper hydraulic gradients across the river–aquifer interface. However, it requires less time to return the bank storage to the river with prolonged wave duration. A wave duration exceeding 24 h indicates faster return than the time required to fill the soil moisture deficit. These modelling results were obtained for a one-time wave event and no repeated wetting process was simulated. However, it is known that the hysteresis pattern can change direction over time (McGuire and McDonnell, 2010), which implies that the patterns observed here may differ for initially wet soil. Generally, a simple estimate of water residence time included in our scenarios is a useful proxy for hyporheic geochemical processing.

Investigation of theoretical scenarios creates a platform for development of more environmental friendly river regulation strategies as requested by Water Framework Directive.

5.2.2 Site-specific implications

Hyporheic exchange at the observation site was mainly characterised by hindered water flow across the river–aquifer interface and had residence time sufficient to establish suboxic conditions in the subsurface due to a number of reasons. The observed hydrograph for the Lule River was mainly dominated by short-term regulation with daily discharge peaks during July–early August (first 40 days of the simulation) and by long-term regulation with extended discharge waves during late August–October. A geochemical investigation of the hyporheic zone at the site revealed a basically suboxic environment, with elevated dissolved concentrations of Fe, Mn, NH_4 and organic carbon (Siergieiev et al., 2014c). These

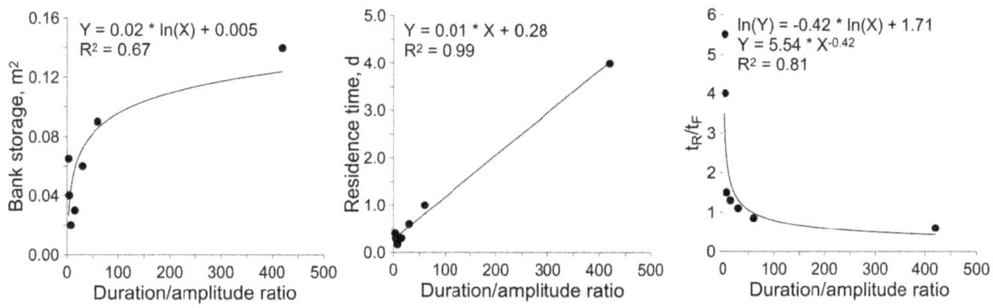

Figure 9. Duration/amplitude ratio in relation to bank storage, residence time and return/fill time ratio (t_R/t_F) for all waves exceeding 0.1 m amplitude.

conditions suggest that the area along the banks and below the bed at the site experiences deficiencies in river water intrusion, which are primarily caused by the river discharge and the clogging layer. Using nitrogen as an example, the hyporheic zone is a nitrate source at low residence time and a nitrate sink at high residence time (Zarnetske et al., 2011). Consequently, biogeochemical activity in the hyporheic zone is controlled by exchange fluxes and bank storage (Gu et al., 2012). The combination of a rapidly rising discharge limb with long duration time favours intensive intrusion of river water into the subsurface, transfer of oxygen and dissolved organic carbon, and therefore promotes nitrification. Note that wave amplitude had a higher influence on the maximum flux across the river–aquifer interface, whereas wave duration affected total bank storage, i.e. the subsurface volume available for hyporheic exchange. This is explained by a steep hydraulic gradient across the river–aquifer interface and thus increased exchange flows due to a rapid rise in river discharge. To provide stable conditions for these ecologically important flows, an extended discharge wave with a sharp rising limb is required. At the Lule River study site, bank hyporheic exchange was triggered by 40 % of the wave events during 2012. These conditions had a potential to reset pore water geochemistry but only a part of it might satisfy sufficient residence time for reactions to occur and thus guarantee an effective biogeochemical exchange. The validity of this relationship requires further testing by e.g. a sediment transport survey, among other techniques, which can form the basis for implementation of environmental flows in restoration programmes (Schneider et al., 2011).

It is not only the hyporheic zone intimately connected to the river that can be affected by fluctuating river water stages, but also the distant groundwater. The simulation results indicated that groundwater head was affected by pressure propagation beyond observation well L25 (25 m distance to the river). This can have an impact on oxidation–reduction conditions in the aquifer due to changes in the redox potential during wetting and drying cycles of the soil (Reddy and Patrick, 1975; Cavanaugh et al., 2006). The relationship between the depth to the groundwater and groundwater composition in observation well L25, sampled during the period May–October 2011, was investigated. Based on nine water quality samples and principal component analysis, the first two significant components explained 77 % of the data variance, indicating a positive correlation between depth to groundwater and NO_3 concentration and a negative correlation between depth to groundwater and Fe and Al concentration. A positive correlation between Mn and alkalinity and depth to groundwater explained only 13 % of the data variance and P showed no relationship. For the significant correlations, a rising groundwater level promoted a more reduced environment and was associated with higher Fe and lower NO_3 concentrations. This suggests that transient changes in river water stages in response to hydropower management can force time-dependent alterations in groundwater quality, with further potential impacts on riparian soils.

5.3 Limitations

The assumptions made in this modelling study resulted in the following limitations:

- Because of the vertical 2-D conceptualisation perpendicular to the river, longitudinal fluxes parallel to the river were neglected. A 3-D model is required for proper consideration of these processes. Bates et al. (2000) argued that the contribution of the longitudinal component is most important at the beginning and end of an event, implying confidence about timing but not about the absolute value of computed peak bank storage and fluxes using a 2-D approach.

- In aquifers with a clogging layer, hydraulic pressure propagation will always be ahead of water flow that follows oscillations at the river–aquifer interface (Welch et al., 2014). Assuming homogeneous subsurface media, solute travel time may exceed that of the pressure, resulting in over-estimated bank storage. In addition, it was assumed that all return flow came from bank storage, even though it contains a mixture of old water from the unsaturated zone and groundwater (Burt and Pinay, 2005). This is crucial for chemical fluxes through the hyporheic zone (McDonnell, 1990) and for chemical hydrograph separation (McCallum et al., 2010). Because the response of solute fluxes to bank storage is dependent on heterogeneity, verification of the fluxes obtained by pressure propagation using measurements of solute concentrations, e.g. electrical conductivity (Welch et al., 2014), or measurements of temperature (Anibas et al., 2012) may be required.

- Lateral variability in riverbed hydraulic conductivity at the site was simplified by implementing a continuous low hydraulic conductivity layer. In field settings, however, a riverbed with variable sediment composition is much more likely (Hancock and Boulton, 2005; Siergieiev et al., 2014a), which suggests that hyporheic exchange seeks more conductive patches. This assumption is likely to result in under-estimated hyporheic exchange (Kalbus et al., 2009) and partially compensate for using homogeneous media (see above).
- Hydraulic conductivity of the riverbed affected the initial distribution of hydraulic gradients. However, the difference in the initial conditions was negligible (4 % between the extreme scenarios) compared with the differences in bank storage caused by the presence of a clogging layer with variable hydraulic conductivity. Therefore, possible effects on bank storage can be ignored.
- In order to avoid over-parameterisation of the model and to limit the effect of sparse information availability regarding regional groundwater gradients, precipitation and evapotranspiration were approximated using a constant recharge flux term.
- The hydraulic effects of the river flow were not included in the model.
- Viscosity effects on hydraulic conductivity (Ma and Zheng, 2010) were excluded due to the small temperature difference between river and groundwater (max. 10 °C) at the site. Ehlert (2014) has shown that a 24 % increase in hydraulic conductivity is possible due to this temperature difference. However, field measurements indicated solely conductive heat transport (Siergieiev et al., 2014a), due to attenuation of the advective–dispersive heat transfer by the clogging layer.

6 Conclusions

Bank hyporheic exchange was simulated using a field case scenario in an alluvial aquifer adjacent to the hydropower-regulated Lule River. The modelling showed that ecosystem requirements in terms of river–aquifer exchange flux are satisfied during 40 % of all wave events during the studied year. Discharge waves with longer duration and increased amplitude are essential for this site to improve hydrological exchange across the river–aquifer interface and bank hyporheic water quality. The combination of realistic and theoretical models improved current process understanding of hyporheic exchange in free-flowing and regulated rivers. Hypothetical scenarios included variable river discharge wave (duration and amplitude), river bank slope and hydraulic conductivity of the river–aquifer interface. Bank storage increased with lower bank slope, indicating the necessity of correct data on geometry of the river–aquifer interface when modelling surface water–groundwater interactions. Hydraulic conductivity of the riverbed positively affected bank storage. However, the influence on the residence time was not always consistent. Higher amplitude and longer wave duration increased bank storage, although larger maximum fluxes were observed for shorter waves at a given amplitude. There will be always a unique relationship between bank storage or residence time and the duration/amplitude wave ratio, which depends on the hydrogeological conditions. Hence, hyporheic exchange suppressed by colmation processes or flow manipulation can be improved by periodically releasing river discharge waves that are optimised for the specific river reach. This type of modelling offers a platform for understanding the transfer of nutrients across river–aquifer interfaces in natural and hydropower-impacted catchments. It further demonstrates the importance of conceptualisation (e.g. process consideration, boundary conditions, data necessity) and the limitations of numerical models.

Author contributions. D. Siergieiev and L. Ehlert collected the data, developed the model and performed the simulations. T. Reimann contributed to model development. D. Siergieiev prepared the paper with contributions from all co-authors.

Acknowledgements. This work was carried out as part of the GENESIS project on groundwater systems (http:/www.thegenesisproject.eu), funded by European Commission 7FP contract 226536. The Kempe Foundation and Formas (project no. 242-210-1187) are acknowledged for additional financial support. DHI-WASY is acknowledged for providing at no cost an academic licence for FEFLOW. Data acquisition for wave amplitude and duration scenarios by Lino Nilsson is appreciated. The authors thank Rebecca Doble and an anonymous reviewer for valuable comments that improved the manuscript.

Edited by: H. Kupfersberger

References

Anibas, C., Verbeiren, B., Buis, K., Chormański, J., De Doncker, L., Okruszko, T., Meire, P., and Batelaan, O.: A hierarchical approach on groundwater-surface water interaction in wetlands along the upper Biebrza River, Poland, Hydrol. Earth Syst. Sci., 16, 2329–2346, doi:10.5194/hess-16-2329-2012, 2012.

Bates, P. D., Stewart, M. D., Desitter, A., Anderson, M. G., Renaud, J.-P., and Smith, J. A.: Numerical simulation of floodplain hydrology, Water Resour. Res., 36, 2517–2529, doi:10.1029/2000WR900102, 2000.

Blaschke, A. P., Steiner, K. H., Schmalfuss, R., Gutknecht, D., and Sengschmitt, D.: Clogging processes in hyporheic interstices of an impounded river, the Danube at Vienna, Austria, Int. Rev. Hydrobiol., 88, 397–413, 2003.

Boulton, A. J., Findlay, S., Marmonier, P., Stanley, E. H., and Valett, H. M.: The functional significance of the hyporheic zone in streams and rivers, Annual Rev. Ecol. Syst., 29, 59–81, 1998.

Bredehoeft, J.: The conceptualization model problem – surprise, Hydrogeol. J., 13, 37–46, doi:10.1007/s10040-004-0430-5, 2005.

Brunke, M. and Gonser, T.: The ecological significance of exchange processes between rivers and groundwater, Freshwater Biol., 37, 1–33, 1997.

Burt, T. P. and Pinay, G.: Linking hydrology and biogeochemistry in complex landscapes, Progress Phys. Geogr., 29, 297–316, 2005.

Butturini, A., Bernal, S., Sabater, S., and Sabater, F.: The influence of riparian-hyporheic zone on the hydrological responses in an intermittent stream, Hydrol. Earth Syst. Sci., 6, 515–526, doi:10.5194/hess-6-515-2002, 2002.

Calles, O., Nyberg, L., and Greenberg, L.: Temporal and spatial variation in quality of hyporheic zone in one unregulated and two regulated boreal rivers, Riv. Res. Appl., 23, 829–842, 2007.

Cavanaugh, J. C., Richardson, W. B., Strauss, E. A., and Bartsch, L. A.: Nitrogen dynamics in sediment during water level manipulation on the Upper Mississippi River, River Res. Applic., 22, 651–666, doi:10.1002/rra.926, 2006.

Chen, X. and Chen, X.: Stream water infiltration, bank storage, and storage zone changes due to stream-stage fluctuations, J. Hydrol., 280, 246–264, 2003.

Cooper, H. H. and Rorabaugh, M. I.: Ground-water movements and bank storage due to flood stages in surface streams US Government Printing Office, 1963.

Diersch, H. J.: Finite element modeling of flow, mass and heat transport in porous and fractured media, Springer, Berlin, p. 996, 2014.

Doble, R., Brunner, P., McCallum, J., and Cook, P. G.: An analysis of river bank slope and unsaturated flow effects on bank storage, Ground Water, 50, 77–86, 2012a.

Doble, R., Crosbie, R. S., Smerdon, B. D., Peeters, L., and Cook, F. J.: Groundwater recharge from overbank floods, Water Resour. Res., 48, W09522, doi:10.1029/2011WR011441, 2012b.

Doherty, J., Brebber, L., and Whyte, P.: PEST: Model-independent parameter estimation. Watermark Computing, Corinda, Australia 122, 2011.

Ehlert, L.: Sensitivity of hyporheic exchange processes to hydrogeological conditions at the river-aquifer interface, Master thesis, Faculty of Environmental Sciences, Institute for Groundwater Management, Technische Universität Dresden, Dresden, Germany, 2014.

Fritz, B. G. and Arntzen, E. V.: Effect of rapidly changing river stage on uranium flux through the hyporheic zone, Ground Water, 45, 753–760, 2007.

Gu, C., Anderson, W., and Maggi, F.: Riparian biogeochemical hot moments induced by stream fluctuations, Water Resour. Res., 48, W09546, doi:10.1029/2011WR011720, 2012.

Hancock, P. J. and Boulton, A. J.: The effects of an environmental flow release on water quality in the hyporheic zone of the Hunter River, Australia, Hydrobiol., 552, 75–85, 2005.

Hanrahan, T. P.: Effects of river discharge on hyporheic exchange flows in salmon spawning areas of a large gravel-bed river, Hydrol. Process., 22, 127–141, doi:10.1002/hyp.6605, 2008.

Harr, R. D: Water flux in soils and subsoils on a steep forested slope, J. Hydrol., 33, 37–58, 1977.

Kalbus, E., Schmidt, C., Molson, J. W., Reinstorf, F., and Schirmer, M.: Influence of aquifer and streambed heterogeneity on the distribution of groundwater discharge, Hydrol. Earth Syst. Sci., 13, 69–77, doi:10.5194/hess-13-69-2009, 2009.

Kløve, B., Ala-aho, P., Bertrand, G., Boukalova, Z., Ertürk, A., Goldscheider, N., Illmonen, J., Karakaya, N., Kupfersberger, H., Kvarner, J., Lundberg, A., Mileusnic, M., Moszczynska, A., Muotka, T., Preda, E., Rossi, P., Siergieiev, D., Simek, J., Vadineanu, A., Wachniew, P., and Widerlund, A.: Groundwater dependent ecosystems. Part I: Hydroecological status and trends, Environ. Sci. Policy, 14, 770–781, 2011.

Knudby, C. and Carrera, J.: On the use of apparent hydraulic diffusivity as an indicator of connectivity, J. Hydrol., 329, 377–389, 2006.

Krause, S., Hannah, D. M., Fleckenstein, J. H., Heppell, C. M., Kaeser, D., Pickup, R., Pinay, G., Robertson, A. L., and Wood, P. J.: Inter-disciplinary perspectives on processes in the hyporheic zone, Ecohydrol., 4, 481–499, doi:10.1002/eco.176, 2011.

Lemmelä, R.: Water balance of sandy aquifer at Hyrylä in southern Finland, Dissertation thesis, 340 pp., University of Turku, Turku, 1990.

Li, H., Boufadel, M. C., and Weaver, J. W.: Quantifying bank storage of variably saturated aquifers, Groundwater, 46, 841–850, 2008.

Ma, R. and Zheng, C.: Effects of density and viscosity in modeling heat as a groundwater tracer, Ground Water, 48, 380–389, 2010.

McCallum, J. L., Cook, P. G., Brunner, P., and Berhane, D.: Solute dynamics during bank storage flows and implications for chemical base flow separation, Water Resour. Res., 46, W07541, doi:10.1029/2009WR008539, 2010.

McDonnell, J. J.: A rationale for old water discharge through macropores in a steep, humid catchment, Water Resour. Res., 26, 2821–2832, 1990.

McGlynn, B. L., McDonnell, J. J., Seibert, J., and Kendall, C.: Scale effects on headwater catchment runoff timing, flow sources, and groundwater-streamflow relations, Water Resour. Res., 40, W07504, doi:10.1029/2003WR002494, 2004.

McGuire, K. J. and McDonnell, J. J.: Hydrological connectivity of hillslopes and streams: Characteristic time scales and nonlinearities, Water Resour. Res., 46, W10543, doi:10.1029/2010WR009341, 2010.

Nilsson, C., Reidy, C. A., Dynesius, M., and Revenga, C.: Fragmentation and flow regulation of the world's large river systems, Sciene, 308, 405–408, 2005.

Reddy, K. R. and Patrick, W. H.: Effect of alternate aerobic and anaerobic conditions on redox potential, organic matter decomposition and nitrogen loss in a flooded soil, Soil Biol. Biochem., 7, 87–94, 1975.

Refsgaard, J. C., van der Sluijs, J. P., Brown, J., and van der Keur, P.: A framework for dealing with uncertainty due to model structure error, Adv. Water Res., 29, 1586–1597, 2006.

Sawyer, H. A., Cardenas, M. B., Bomar, A., and Mackey, M.: Impact of dam operations on hyporheic exchange in the riparian zone of a regulated river, Hydrol. Process., 23, 2129–2137, 2009.

Schenk, O. and Gärtner, K.: Solving unsymmetric sparse systems of linear equations with PARDISO, Future Generat. Comp. Syst. 20, 475–487, 2004.

Schneider, P., Vogt, T., Schirmer, M., Doetsch, J., Linde, N., Pasquale, N., Perona, P., and Cirpka, O. A.: Towards improved instrumentation for assessing river-groundwater interactions in a restored river corridor, Hydrol. Earth Syst. Sci., 15, 2531–2549, doi:10.5194/hess-15-2531-2011, 2011.

Siergieiev, D., Lundberg, A., and Widerlund, A.: Hyporheic water exchange in a large hydropower regulated boreal river – directions and rates, Hydrol. Res., 45, 334–348, doi:10.2166/nh.2013.011, 2014a.

Siergieiev, D., Widerlund, A., Lundberg, A., Almqvist, L., Collomp, M., Ingri, J., and Öhlander, B.: Impact of hydropower regulation on river water composition in Northern Sweden, Aq. Geochem., 20, 59–80, doi:10.1007/s10498-013-9215-6, 2014b.

Siergieiev, D., Widerlund, A., Ingri, J., Lundberg, A., and Öhlander, B.: Flow regulation effects on the hydrogeochemistry of the hyporheic zone in boreal rivers, Sci. Total Environ., 499, 424–436, 2014c.

Singh, S. K.: Aquifer response to sinusoidal or arbitrary stage of semipervious stream, J. Hydraul. Eng., 130, 1108–1118, 2004.

Smedberg, E., Humborg, C., Jakobsson, M., and Mörth, C.-M.: Landscape elements and river chemistry as affected by river regulation – a 3-D perspective, Hydrol. Earth Syst. Sci., 13, 1597–1606, doi:10.5194/hess-13-1597-2009, 2009.

Sophocleous, M.: Interactions between groundwater and surface water: The state of the science, Hydrogeol. J., 10, 52–67, 2002.

Todd, D. K.: Ground-water flow in relation to a flooding stream, Am. Soc. Civ. Eng., 81, 628.1–628.20, 1955.

Turton, D. J., Haan, C. T., and Miller, E. L.: Subsurface flow responses of a small forested catchment in the Ouachita Mountains, Hydrol. Process., 6, 111–125, doi:10.1002/hyp.3360060110, 1992.

Valett, H. M., Morrice, J. A., Dahm, C. N., and Campana, M. E.: Parent lithology, surface-groundwater exchange, and nitrate retention in headwater streams, Limnol. Oceanogr., 41, 333–345, 1996.

Welch, C., Harrington, G. A., Leblanc, M., Batlle-Aguilar, J., and Cook, P. G.: Relative rates of solute and pressure propagation into heterogeneous alluvial aquifers following river flow events, J. Hydrol., 511, 891–903, 2014.

Zarnetske, J. P., Haggerty, R., Wondzell, S. M., and Baker, M. A.: Dynamics of nitrate production and removal as a function of residence time in the hyporheic zone, J. Geophys. Res.-Biogeosci., 116, G01025, doi:10.1029/2010JG001356, 2011.

8

The skill of seasonal ensemble low-flow forecasts in the Moselle River for three different hydrological models

M. C. Demirel[1,*], M. J. Booij[1], and A. Y. Hoekstra[1]

[1]Water Engineering and Management, Faculty of Engineering Technology, University of Twente, P.O. Box 217, 7500 AE Enschede, the Netherlands
[*]current address: Portland State University, Department of Civil & Environmental Engineering, 1930 S. W. 4th Avenue, Suite 200, Portland, OR 97201, USA

Correspondence to: M. C. Demirel (demirel@pdx.edu)

Abstract. This paper investigates the skill of 90-day low-flow forecasts using two conceptual hydrological models and one data-driven model based on Artificial Neural Networks (ANNs) for the Moselle River. The three models, i.e. HBV, GR4J and ANN-Ensemble (ANN-E), all use forecasted meteorological inputs (precipitation P and potential evapotranspiration PET), whereby we employ ensemble seasonal meteorological forecasts. We compared low-flow forecasts for five different cases of seasonal meteorological forcing: (1) ensemble P and PET forecasts; (2) ensemble P forecasts and observed climate mean PET; (3) observed climate mean P and ensemble PET forecasts; (4) observed climate mean P and PET and (5) zero P and ensemble PET forecasts as input for the models. The ensemble P and PET forecasts, each consisting of 40 members, reveal the forecast ranges due to the model inputs. The five cases are compared for a lead time of 90 days based on model output ranges, whereas the models are compared based on their skill of low-flow forecasts for varying lead times up to 90 days. Before forecasting, the hydrological models are calibrated and validated for a period of 30 and 20 years respectively. The smallest difference between calibration and validation performance is found for HBV, whereas the largest difference is found for ANN-E. From the results, it appears that all models are prone to over-predict runoff during low-flow periods using ensemble seasonal meteorological forcing. The largest range for 90-day low-flow forecasts is found for the GR4J model when using ensemble seasonal meteorological forecasts as input. GR4J, HBV and ANN-E under-predicted 90-day-ahead low flows in the very dry year 2003 without precipitation data. The results of the comparison of forecast skills with varying lead times show that GR4J is less skilful than ANN-E and HBV. Overall, the uncertainty from ensemble P forecasts has a larger effect on seasonal low-flow forecasts than the uncertainty from ensemble PET forecasts and initial model conditions.

1 Introduction

Rivers in western Europe usually experience low flows in late summer and high flows in winter. These two extreme discharge phenomena can lead to serious problems. For example, high-flow events are sudden and can put human life at risk, whereas streamflow droughts (i.e. low flows) develop slowly and can affect a large area. Consequently, the economic loss during low-flow periods can be much bigger than during floods (Pushpalatha et al., 2011; Shukla et al., 2012). In the River Rhine, severe problems for freshwater supply, water quality, power production and river navigation were experienced during the dry summers of 1976, 1985 and 2003. For that reason forecasting seasonal low flows (Towler et al., 2013; Coley and Waylen, 2006; Li et al., 2008) and understanding low-flow indicators (Vidal et al., 2010; Fundel et al., 2013; Demirel et al., 2013a; Wang et al., 2011; Saadat et al., 2013; Nicolle et al., 2014) have both societal and scientific value. The seasonal forecast of water flows is therefore listed as one of the priority topics in the EU's Horizon 2020 research programme (EU, 2013). Further, there is an increasing interest in incorporating seasonal flow forecasts in decision support systems for river navigation and power plant operation during low-flow periods. We are interested in forecasting low flows with a lead time of 90 days, and in presenting the effect of ensemble meteorological forecasts for three hydrological models.

Generally, two approaches are used in seasonal hydrological forecasting. The first one is a statistical approach, making use of data-driven models based on relationships between river discharge and hydroclimatological indicators (Wang et al., 2011; van Ogtrop et al., 2011; Förster et al., 2014). The second one is a dynamic approach running a hydrological model with forecasted climate input.

The first approach is often preferred in regions where significant correlations between river discharge and climatic indicators exist, such as sea surface temperature anomalies (Chowdhury and Sharma, 2009), AMO – Atlantic Multidecadal Oscillation (Ganguli and Reddy, 2014; Giuntoli et al., 2013), PDO – Pacific Decadal Oscillation (Soukup et al., 2009) and warm and cold phases of the ENSO – El Niño Southern Oscillation – index (Chiew et al., 2003; Kalra et al., 2013; Tootle and Piechota, 2004). Kahya and Dracup (1993) identified the lagged response of regional streamflow to the warm phase of ENSO in the southeastern United States. In the Rhine Basin, no teleconnections have been found between climatic indices, e.g. NAO and ENSO, and river discharges (Rutten et al., 2008; Bierkens and van Beek, 2009). However, Demirel et al. (2013a) found significant correlations between hydrological low-flow indicators and observed low flows. They also identified appropriate lags and temporal resolutions of low-flow indicators (e.g. precipitation, potential evapotranspiration, groundwater storage, lake levels and snow storage) to build data-driven models.

The second approach is the dynamic seasonal forecasting approach which has long been explored (Wang et al., 2011; Van Dijk et al., 2013; Gobena and Gan, 2010; Fundel et al., 2013; Shukla et al., 2013; Pokhrel et al., 2013) and has led to the development of the current ensemble streamflow prediction system (ESP) used by different national climate services like the National Weather Service in the United States. The seasonal hydrologic prediction systems are most popular in regions with a high risk of extreme discharge situations like hydrological droughts (Robertson et al., 2013; Madadgar and Moradkhani, 2013). Well-known examples are the NOAA Climate Prediction Centre's seasonal drought forecasting system (available from: http://www.cpc.ncep.noaa.gov), the University of Washington's Surface Water Monitoring system (Wood and Lettenmaier, 2006), Princeton University's drought forecast system (available from: http://hydrology.princeton.edu/forecast) and the University of Utrecht's global monthly hydrological forecast system (Yossef et al., 2012). These models provide indications about the hydrologic conditions and their evolution across the modelled domain using available weather ensemble inputs (Gobena and Gan, 2010; Yossef et al., 2012). Moreover, Dutra et al. (2014) showed that global seasonal forecasts of meteorological drought onset are feasible and skilful using the standardized precipitation index (SPI) and two data sets as initial conditions.

Many studies have investigated the seasonal predictability of low flows in different rivers such as the Thames and different other rivers in the UK (Bell et al., 2013; Wedgbrow et al., 2002, 2005), the Shihmen and Tsengwen rivers in Taiwan (Kuo et al., 2010), the River Jhelum in Pakistan (Archer and Fowler, 2008), more than 200 rivers in France (Sauquet et al., 2008; Giuntoli et al., 2013), five semi-arid areas in South Western Queensland, Australia (van Ogtrop et al., 2011), five rivers including Limpopo basin and the Blue Nile in Africa (Dutra et al., 2013; Winsemius et al., 2014), the Bogotá River in Colombia (Felipe and Nelson, 2009), the Ohio in the eastern USA (Wood et al., 2002; Luo et al., 2007; Li et al., 2009), the North Platte in Colorado, USA (Soukup et al., 2009), large rivers in the USA (Schubert et al., 2007; Shukla and Lettenmaier, 2011) and the Thur River in the northeastern part of Switzerland (Fundel et al., 2013). The common result of the above-mentioned studies is that the skill of the seasonal forecasts made with global and regional hydrological models is reasonable for lead times of 1–3 months (Shukla and Lettenmaier, 2011; Wood et al., 2002) and these forecasting systems are all prone to large uncertainties as their forecast skills mainly depend on the knowledge of initial hydrologic conditions and weather information during the forecast period (Shukla et al., 2012; Yossef et al., 2013; Li et al., 2009; Doblas-Reyes et al., 2009). In a recent study, Yossef et al. (2013) used a global monthly hydrological model to analyse the relative contributions of initial conditions and meteorological forcing to the skill of seasonal streamflow forecasts. They included 78 stations in large basins in the world including the River Rhine for forecasts with lead times up to 6 months. They found that improvements in seasonal hydrological forecasts in the Rhine depend on better meteorological forecasts, which underlines the importance of meteorological forcing quality particularly for forecasts beyond lead times of 1–2 months.

Most of the previous River Rhine studies use only one hydrological model, e.g. PREVAH (Fundel et al., 2013) or PCR-GLOBWB (Yossef et al., 2013), to assess the value of ensemble meteorological forcing, whereas in this study, we compare three hydrological models with different structures varying from data-driven to conceptual models. The two objectives of this study are to contrast data-driven and conceptual modelling approaches and to assess the effect of ensemble seasonal forecasted precipitation and potential evapotranspiration on low-flow forecast quality and skill scores. By comparing three models with different model structures we address the issue of model structure uncertainty, whereas the latter objective reflects the benefit of ensemble seasonal forecasts. Moreover, the effect of initial model conditions is partly addressed using climate mean data in one of the cases.

The analysis complements recent efforts to analyse the effects of ensemble weather forecasts on low-flow forecasts

Table 1. Overview of observed data used.

Variable	Name	Number of stations/ sub-basins	Period	Annual range (mm)	Time step (days)	Spatial resolution	Source
Q	Discharge	1	1951–2006	163–550	1	Point	GRDC
P	Precipitation	26	1951–2006	570–1174	1	Basin average	BfG
PET	Potential evapotranspiration	26	1951–2006	512–685	1	Basin average	BfG
h	Mean altitude	26	–	–	–	Basin average	BfG

Table 2. Overview of ensemble seasonal meteorological forecast data.

Data	Spatial resolution	Ensemble size	Period	Time step (days)	Lead time (days)
Forecasted P	$0.25 \times 0.25°$	39 + 1 control	2002–2005	1	1–90
Forecasted PET	$0.25 \times 0.25°$	39 + 1 control	2002–2005	1	1–90

with a lead time of 10 days using two conceptual models (Demirel et al., 2013b), by studying the effects of seasonal ensemble weather forecasts on 90-day low-flow forecasts using not only conceptual models but also data-driven models.

The outline of the paper is as follows. The study area and data are presented in Sect. 2. Section 3 describes the model structures, their calibration and validation set-ups and the methods employed to estimate the different attributes of the forecast quality. The results are presented in Sect. 4 and discussed in Sect. 5, and the conclusions are summarized in Sect. 6.

2 Study area and data

2.1 Study area

The study area is the Moselle River basin, the largest sub-basin of the Rhine River basin. The Moselle River has a length of 545 km. The river basin has a surface area of approximately 27 262 km^2. The altitude in the basin varies from 59 to 1326 m, with a mean altitude of 340 m (Demirel et al., 2013a). There are 26 sub-basins with surface areas varying from 102 to 3353 km^2. Approximately 410 mm ($\sim$130 m^3 s^{-1}) discharge is annually generated in the Moselle Basin (Demirel et al., 2013b). The outlet discharge at Cochem varies from 14 m^3 s^{-1} in dry summers to a maximum of 4000 m^3 s^{-1} during winter floods.

The Moselle River has been heavily regulated by dams, power plants, weirs and locks. There are around 12 hydropower plants between Koblenz and Trier producing energy since the 1960s (Bormann, 2010). Moreover, there are 12 locks only on the German part of the river (Bormann et al., 2011).

2.2 Data

2.2.1 Observed data

Observed daily data on precipitation (P), potential evapotranspiration (PET) and the mean altitudes (h) of the 26 sub-basins have been provided by the German Federal Institute of Hydrology (BfG) in Koblenz, Germany (Table 1). PET is estimated using the Penman–Wendling equation (ATV-DVWK, 2002) and both variables have been spatially averaged by BfG over 26 Moselle sub-basins using areal weights. Observed data from 12 meteorological stations in the Moselle Basin (as part of 49 stations over the Rhine Basin), mainly provided by the CHR, the DWD and Météo France, are used to estimate the basin-averaged input data (Görgen et al., 2010). Observed daily discharge (Q) data at Cochem (station #6336050) are provided by the Global Runoff Data Centre (GRDC), Koblenz. The daily observed data (P, PET and Q) are available for the period 1951–2006.

2.2.2 Ensemble seasonal meteorological forecast data

The ensemble seasonal meteorological forecast data, comprising 40 members, are obtained from the European Centre for Medium-Range Weather Forecasts (ECMWF) seasonal forecasting archive and retrieval system, i.e. MARS system 3 (ECMWF, 2012). This data set contains regular $0.25 \times 0.25°$ latitude–longitude grids and each ensemble member is computed for a lead time of 184 days using perturbed initial conditions and model physics (Table 2). We estimated the PET forecasts using the Penman–Wendling equation requiring forecasted surface solar radiation and temperature at 2 m above the surface, and the altitude of the sub-basin (ATV-DVWK, 2002). The PET estimation is consistent with the observed PET estimation carried out by BfG (ATV-DVWK,

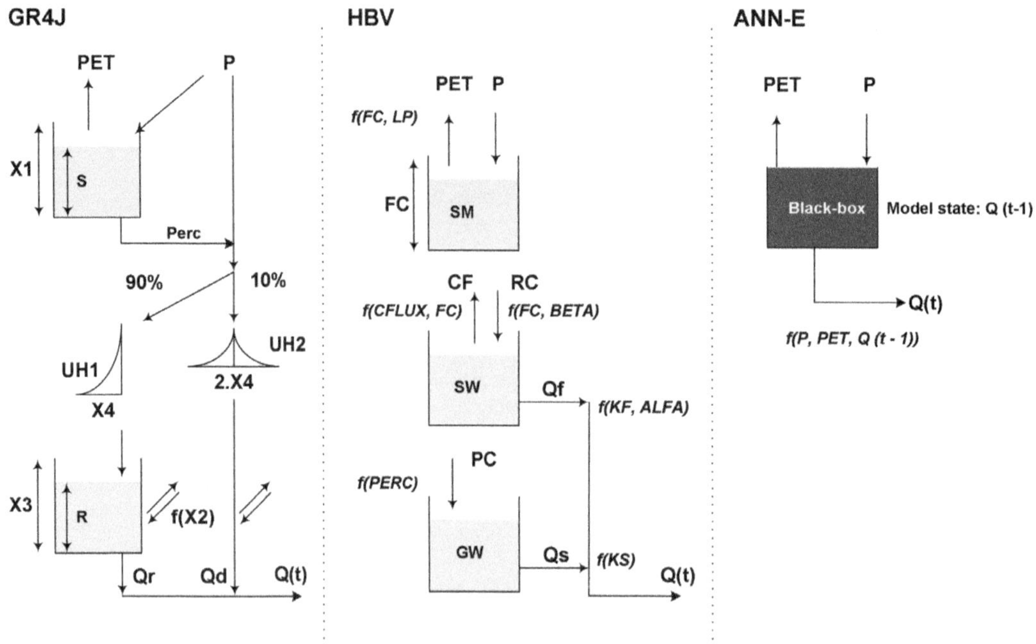

Figure 1. Schematization of the three models. PET is potential evapotranspiration, *P* is precipitation, *Q* is discharge and *t* is the time (day).

2002). The grid-based *P* and PET ensemble forecast data are firstly interpolated over 26 Moselle sub-basins using areal weights. These sub-basin averaged data are then aggregated to the Moselle basin level.

3 Methodology

3.1 Overview of model structures and forecast scheme

The three hydrological models (GR4J, HBV and ANN-E) are briefly described in Sects. 3.1.1–3.1.3. Figure 1 shows the simplified model structures. The calibration and validation of the models is described in Sect. 3.1.4. Five cases with different combinations of ensemble meteorological forecast input and climate mean input are introduced in Sect. 3.1.5. We provide a detailed description for each parameter of the three models in Sect. 4.1.

3.1.1 GR4J

The GR4J model (Génie Rural à 4 paramètres Journalier) is used as it has a parsimonious structure with only four parameters. The model has been tested over hundreds of basins worldwide, with a broad range of climatic conditions from tropical to temperate and semi-arid basins (Perrin et al., 2003). GR4J is a conceptual model and the required model inputs are daily time series of *P* and PET (Table 3). All four parameters (Fig. 1a) are used to calibrate the model. The upper and lower limits of the parameters are selected based on previous works (Perrin et al., 2003; Pushpalatha et al., 2011; Tian et al., 2014).

3.1.2 HBV

The HBV conceptual model (Hydrologiska Byråns Vattenbalansavdelning) was developed by the Swedish Meteorological and Hydrological Institute (SMHI) in the early 1970s (Lindström et al., 1997). The HBV model consists of four subroutines: a precipitation and snow accumulation and melt routine, a soil moisture accounting routine and two runoff generation routines. The required input data are daily *P* and PET. The snow routine and daily temperature data are not used in this study as the Moselle basin is a rain-fed basin. Eight parameters (see Fig. 1b) in the HBV model are calibrated (Engeland et al., 2010; Van den Tillaart et al., 2013; Tian et al., 2014). The eight parameters are selected for calibration and the parameter ranges are selected based on previous works (Booij, 2005; Eberle, 2005; Tian et al., 2014).

3.1.3 ANN-E

An Artificial Neural Network (ANN) is a data-driven model inspired by functional units (neurons) of the human brain (Elshorbagy et al., 2010). A neural network is a universal approximator capable of learning the patterns and relation between outputs and inputs from historical data and applying it for extrapolation (Govindaraju and Rao, 2000). A three-layer feed-forward neural network (FNNs) is the most widely preferred model architecture for prediction and forecasting of hydrological variables (Adamowski et al., 2012; Shamseldin, 1997; Kalra et al., 2013). Each of these three layers has an important role in processing the information. The first layer receives the inputs and multiplies them with a weight (adds a

Table 3. Model descriptions. PET is potential evapotranspiration, P is precipitation and Q is discharge.

Model type		Input	Temporal	Lag between	Model	Model
Conceptual	Data-driven		resolution of input	forecast issue day and final day of temporal averaging (days)	time step	lead time (days)
GR4J		P: ensemble PET: ensemble Q: state update	Daily P Daily PET	P: 0 PET: 0 Q: 1	Daily	1 to 90
HBV		P: ensemble PET: ensemble Q: state update	Daily P Daily PET	P: 0 PET: 0 Q: 1	Daily	1 to 90
	ANN-E	P: ensemble PET: ensemble Q: state update	Daily P Daily PET Daily Q	P: 0 PET: 0 Q: 1	Daily	1 to 90

bias if necessary) before delivering them to each of the hidden neurons in the next layer (Gaume and Gosset, 2003). The weights determine the strength of the connections. The number of nodes in this layer corresponds to the number of inputs. The second layer, the hidden layer, consists of an activation function (also known as transfer function) which nonlinearly maps the input data to output target values. In other words, this layer is the learning element of the network which simulates the relationship between inputs and outputs of the model. The third layer, the output layer, gathers the processed data from the hidden layer and delivers the final output of the network.

A hidden neuron is the processing element with n inputs $(x_1, x_2, x_3, \ldots, x_n)$, and one output y using

$$y = f(x_1, x_2, x_3, \ldots, x_n) = \text{logsig}\left[\left(\sum_{i=1}^{n} x_i w_i\right) + b\right], \quad (1)$$

where w_i are the weights, b is the bias and logsig is the logarithmic sigmoid activation function. We tested the tansig and logsig activation functions and the latter was selected for this study as it gave better results for low flows. ANN model structures are determined based on the forecast objective. In this study, we used a conceptual type ANN model structure, ANN-Ensemble (ANN-E), which requires daily P, PET and historical Q as input. Observed discharge on the forecast issue day is used to update the model states (Table 3). In other words, the ANN-E model receives $Q_{\text{obs}}(t)$ as input on the time step t when the forecast is issued, and then receives the streamflow forecast of the previous time step as input for lead times larger than 1 day. Further, forecasted Q for time step $t + j$ is used as input to forecast Q at $t + j + 1$.

This is a 1 day memory which also exists in the conceptual models, i.e. GR4J and HBV (Fig. 1). The ANN-E is assumed to be comparable with the conceptual models with similar model structures. The determination of the optimal number of hidden neurons in the second layer is an important issue in the development of ANN models. Three common approaches are ad hoc (also known as trial and error), global and stepwise (Kasiviswanathan et al., 2013). We used a global approach (i.e. Genetic Algorithm) to avoid local minima (De Vos and Rientjes, 2008) and tested the performance of the networks with one, two and three hidden neurons corresponding to a number of parameters (i.e. number of weights and biases) of 6, 11 and 16, respectively. Based on the parsimonious principle, testing ANNs only up to three hidden neurons is assumed to be enough as the number of parameters increases exponentially for every additional hidden neuron.

3.1.4 Calibration and validation of models

A global optimization method, i.e. Genetic Algorithm (GA) (De Vos and Rientjes, 2008), and historical Moselle low flows for the period from 1971–2001 are used to calibrate the models used in this study. The 30-year calibration period is carefully selected as the first low-flow forecast is issued on 1 January 2002. The first 3 years are used as warm-up period for the hydrological model. For all GA simulations, we use 100 as population size, 5 as reproduction elite count size, 0.7 as crossover fraction, 2000 as maximum number of iterations and 5000 as the maximum number of function evaluations based on the studies by De Vos and Rientjes (2008) and Kasiviswanathan et al. (2013). The evolution starts from the population of 100 randomly generated individuals. The population in each iteration is called a generation and the fitness of every individual in the population is evaluated using the objective function. The best 70 % of the population (indicated as crossover fraction) survives in the process of 2000 iterations.

The validation period spans 1951–1970. The definition of low flows, i.e. discharges below the Q_{75} threshold of $\sim 113\,\mathrm{m^3\,s^{-1}}$, is based on previous work by Demirel et al. (2013a). Prior parameter ranges and deterministic equations used for dynamic model state updates of the conceptual models based on observed discharges on the forecast issue day are based on the study by Demirel et al. (2013b). In this study, we use a hybrid Mean Absolute Error (MAE) based on only low flows ($\mathrm{MAE_{low}}$) and inverse discharge values ($\mathrm{MAE_{inverse}}$) as objective function (see Eq. 4):

$$\text{Mean absolute error}_{\text{low}}: \frac{1}{m}\sum_{j=1}^{m}\left|Q_{\text{sim}}(j)-Q_{\text{obs}}(j)\right|, \tag{2}$$

where Q_{obs} and Q_{sim} are the observed and simulated values for the jth observed low-flow day (i.e. $Q_{\text{obs}} < Q_{75}$) and m is the total number of low-flow days:

$$\text{Mean absolute error}_{\text{inverse}}: \frac{1}{n}\sum_{i=1}^{n}\left|\frac{1}{Q_{\text{sim}}(i)+\epsilon}-\frac{1}{Q_{\text{obs}}(i)+\epsilon}\right|, \tag{3}$$

where n is the total number of days (i.e. $m < n$), and ϵ is 1 % of the mean observed discharge to avoid infinity during zero discharge days (see Pushpalatha et al., 2012). The hybrid Mean Absolute Error is defined as

$$\mathrm{MAE_{hybrid}} = \mathrm{MAE_{low}} + \mathrm{MAE_{inverse}}. \tag{4}$$

The $\mathrm{MAE_{low}}$ and $\mathrm{MAE_{inverse}}$ were not normalized to calculate $\mathrm{MAE_{hybrid}}$ metric. It should be noted that we did not fully neglect the high and intermediate flows using $\mathrm{MAE_{inverse}}$, whereas only low-flow periods are considered in $\mathrm{MAE_{low}}$. This is one of the advantages of using the $\mathrm{MAE_{hybrid}}$ metric and also avoids redundancy.

3.1.5 Model storage update procedure for HBV and GR4J models

The storages in the two conceptual models are updated based on the observed discharge on the forecast issue day. In our previous study (Demirel et al., 2013b), we derived empirical relations between the simulated discharge and the fast runoff for each model to divide the observed discharge between the fast and slow runoff components:

$$\text{k\_GR4J} = \frac{Qd}{Qr+Qd} \tag{5}$$

$$\text{k\_HBV} = \frac{Qf}{Qf+Qs}. \tag{6}$$

The Qf and Qs in the HBV model, and Qr and Qd in the GR4J model are estimated using the fractions above and the observed discharge value on the forecast issue day. The routing storage (R) in the GR4J model is updated for a given value of the $X3$ parameter using Eq. (7). Moreover, the surface water (SW) and groundwater (GW) storages in the HBV model are updated for given values of KF, ALFA and KS parameters using Eqs. (8) and (9):

$$Qr = R\left\{1-\left[1+\left(\frac{R}{X3}\right)^{4}\right]^{-1/4}\right\} \tag{7}$$

$$\mathrm{SW} = \left(\frac{Qf}{\mathrm{KF}}\right)^{\left(\frac{1}{(1+\mathrm{ALFA})}\right)} \tag{8}$$

$$\mathrm{GW} = \frac{Qs}{\mathrm{KS}}. \tag{9}$$

The remaining two storages S (in GR4J) and SM (in HBV) are updated using the calibrated model run until the forecast issue day (i.e. top-down approach).

3.1.6 Case description

In this study, three hydrological models are used for the seasonal forecasts. Five ensemble meteorological forecast input cases for ANN-E, GR4J and HBV models are compared: (1) ensemble P and PET forecasts, (2) ensemble P forecasts and observed climate mean PET, (3) observed climate mean P and ensemble PET forecasts, (4) observed climate mean P and PET, and (5) zero P and ensemble PET forecasts (Table 4). P and PET forecasts are joint forecasts in our modelling practice. For example, if the first ensemble member is called from P then the first member from PET is also called to force the hydrological model.

Cases 1–4 are the different possible combinations of ensemble and climate mean meteorological forcing. Case 5 is analysed to determine to which extent the precipitation forecast in a very dry year (2003) is important for seasonal low-flow forecasts. It should be noted that all available historical data (1951–2006) were used to estimate the climate mean. For example the climate mean for January 1st is estimated by the average of 55 January 1st values in the available period (1951–2006).

3.2 Forecast skill scores

Three probabilistic forecast skill scores (Brier Skill Score, reliability diagram, hit and false alarm rates) and one deterministic forecast skill score (Mean Forecast Score) are used to analyse the results of low-flow forecasts with lead times of 1–90 days. Forecasts for each day in the test period (2002–2005) are used to estimate these scores. The Mean Forecast Score focusing on low flows is introduced in this study, whereas the other three scores have been often used in meteorology (WMO, 2012) and flood hydrology (Velázquez et al., 2010; Renner et al., 2009; Thirel et al., 2008). For the three models, i.e. GR4J, HBV and ANN-E, the forecast probability for each forecast day is estimated as the ratio of the number of ensemble members non-exceeding the preselected thresh-

Table 4. Details of the five input cases.

Case	Precipitation (P)	The number of ensemble members (P)	Potential evapotranspiration (PET)	The number of ensemble members (PET)
1	Ensemble forecast	40	Ensemble forecast	40
2	Ensemble forecast	40	Climate mean	1
3	Climate mean	1	Ensemble forecast	40
4	Climate mean	1	Climate mean	1
5	Zero	0	Ensemble forecast	40

Table 5. Contingency table for the assessment of low-flow events based on the Q_{75}.

	Observed	Not observed
Forecasted	*hit*: the event forecasted to occur and did occur	*false alarm*: event forecasted to occur, but did not occur
Not forecasted	*miss*: the event forecasted not to occur, but did occur	*correct negative*: event forecasted not to occur and did not occur

olds (here Q_{75}) and the total number of ensemble members (i.e. 40 members) for that forecast day.

3.2.1 Brier skill score (BSS)

The Brier Skill Score (BSS) (Wilks, 1995) is often used in hydrology to evaluate the quality of probabilistic forecasts (Devineni et al., 2008; Hartmann et al., 2002; Jaun and Ahrens, 2009; Roulin, 2007; Towler et al., 2013):

$$\text{Brier skill score}: 1 - \frac{\text{BS}_{\text{forecast}}}{\text{BS}_{\text{climatology}}}, \tag{10}$$

where the $\text{BS}_{\text{forecast}}$ is the Brier Score (BS) for the forecast, defined as

$$\text{Brier score}: \frac{1}{n}\sum_{t=1}^{n}(F_t - O_t)^2, \tag{11}$$

where F_t refers to the forecast probability, O_t refers to the observed probability ($O_t = 1$ if the observed flow is below the low-flow threshold, 0 otherwise), and n is the sample size. $\text{BS}_{\text{climatology}}$ is the BS for the climatology, which is also calculated from Eq. (11) for every year using climatological probabilities. BSS values range from minus infinity to 1 (perfect forecast). Negative values indicate that the forecast is less accurate than the climatology and positive values indicate more skill compared to the climatology.

3.2.2 Reliability diagram

The reliability diagram is used to evaluate the performance of probabilistic forecasts of selected events, i.e. low flows. A reliability diagram represents the observed relative frequency as a function of forecasted probability and the 1 : 1 diagonal shows the perfect reliability line (Velázquez et al., 2010; Olsson and Lindström, 2008). This comparison is important as reliability is one of the three properties of a hydrological forecast (WMO, 2012). A reliability diagram shows the portion of observed data inside preselected forecast intervals.

In this study, exceedance probabilities of 50, 75, 85, 95 and 99 % are chosen as thresholds to categorize the discharges from mean flows to extreme low flows. The forecasted probabilities are then divided into bins of probability categories; here, five bins (categories) are chosen: 0–20, 20–40, 40–60, 60–80 and 80–100 %. The observed frequency for each day is chosen to be 1 if the observed discharge is below the low-flow threshold, or 0, if not.

3.2.3 Hit and false alarm rates

We used hit and false alarm rates to assess the effect of ensembles on low-flow forecasts for varying lead times. The hit and false alarm rates indicate respectively the proportion of events for which a correct warning was issued, and the proportion of non-events for which a false warning was issued by the forecast model. These two simple rates can be easily calculated from contingency tables (Table 5) using Eqs. (12) and (13). These scores are often used for evaluating flood forecasts (Martina et al., 2006); however, they can also be used to estimate the utility of low-flow forecasts as they indicate the model's ability to correctly forecast the occurrence or non-occurrence of preselected events (i.e. Q_{75} low flows). There are four cases in a contingency table as shown in Table 5:

Table 6. Parameter ranges and calibrated values of the pre-selected three models.

Parameter	Unit	Range	Calibrated value	Description
				GR4J model
$X1$	[mm]	10–2000	461.4	Capacity of the production store
$X2$	[mm]	−8 to +6	−0.3	Groundwater exchange coefficient
$X3$	[mm]	10–500	80.8	One day ahead capacity of the routing store
$X4$	[d]	0–4	2.2	Time base of the unit hydrograph
				HBV model
FC	[mm]	200–800	285.1	Maximum soil moisture capacity
LP	[−]	0.1–1	0.7	Soil moisture threshold for reduction of evapotranspiration
BETA	[−]	1–6	2.2	Shape coefficient
CFLUX	[mm d^{-1}]	0.1–1	1.0	Maximum capillary flow from upper response box to soil moisture zone
ALFA	[−]	0.1–3	0.4	Measure for non-linearity of low flow in quick runoff reservoir
KF	[d^{-1}]	0.005–0.5	0.01	Recession coefficient for quick flow reservoir
KS	[d^{-1}]	0.0005–0.5	0.01	Recession coefficient for base flow reservoir
PERC	[mm d^{-1}]	0.3–7	0.6	Maximum flow from upper to lower response box
				ANN-E model
$W1$	[−]	−10 to +10	−2.3	Weight of connection between 1st input node (P) and hidden neuron
$W2$	[−]	−10 to +10	0.03	Weight of connection between 2nd input node (PET) and hidden neuron
$W3$	[−]	−10 to +10	−0.02	Weight of connection between 3rd input node ($Q(t-1)$) and hidden neuron
$W4$	[−]	−10 to +10	3.7	Weight of connection between hidden neuron and output node
$B1$	[−]	−10 to +10	0.02	Bias value in hidden layer
$B2$	[−]	−10 to +10	1.1	Bias value in output layer

$$\text{hit rate} = \frac{\text{hits}}{(\text{hits} + \text{misses})} \tag{12}$$

$$\text{false alarm rate} = \frac{\text{false alarms}}{(\text{correct negatives} + \text{false alarms})}. \tag{13}$$

3.2.4 Mean forecast score (MFS)

The mean forecast score (MFS) is a new skill score which can be derived from either probabilistic or deterministic forecasts. The probabilities are calculated for the days when low flow occurred. In this study we used a deterministic approach for calculating the observed frequency for all three models. For all three models, ensembles are used for estimating forecast probabilities. The score is calculated as below only for deterministic observed low flows:

$$\text{Mean forecast score}: \ \frac{1}{m}\sum_{j=1}^{m} F_j \tag{14}$$

where F_j is the forecast probability for the jth observed low-flow day (i.e. $O_j \leq Q_{75}$) and m is the total number of low-flow days. The probability of a deterministic forecast can be 0 or 1, whereas it varies from 0 to 1 for ensemble members. For instance, if 23 of the 40 ensemble forecast members indicate low flows for the jth low-flow day then $F_j = 23/40$. It should be noted that this score is not limited to low flows as it has a flexible forecast probability definition which can be adapted to any type of discharge. MFS values range from zero to 1 (perfect forecast).

4 Results

4.1 Calibration and validation

Table 6 shows the parameter ranges and the best-performing parameter sets of the three models. The GR4J and HBV models have both well-defined model structures; therefore, their calibration was more straightforward than the calibration of the ANN models. Calibration of the ANN-E model was done in two steps. First, the number of hidden neurons was determined by testing the performance of the ANN-E model with one, two and three hidden neurons.

Second, daily P, PET and Q are used as three inputs for the tested ANN-E model with one, two and three hidden neurons due to the fact that these inputs are comparable with the inputs of the GR4J and HBV models. Figure 2a shows that the performance of the ANN-E model does not improve with additional hidden neurons. Based on the performance in the validation period, one hidden neuron is selected. GR4J and HBV are also calibrated. The results of the three models used in this study are presented in Fig. 2b.

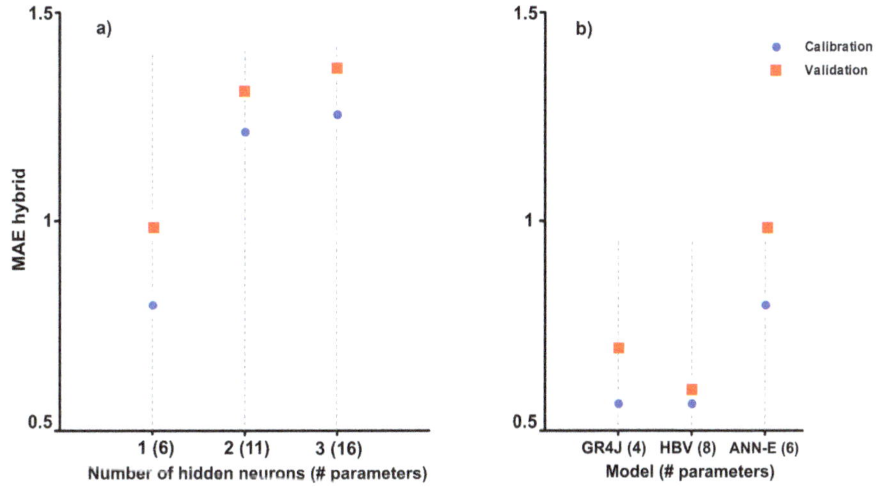

Figure 2. Calibration and validation results of **(a)** the ANN-E model with one, two and three hidden neurons and **(b)** the three models used in this study. The same calibration (1971–2001) and validation (1951–1970) periods are used for both plots.

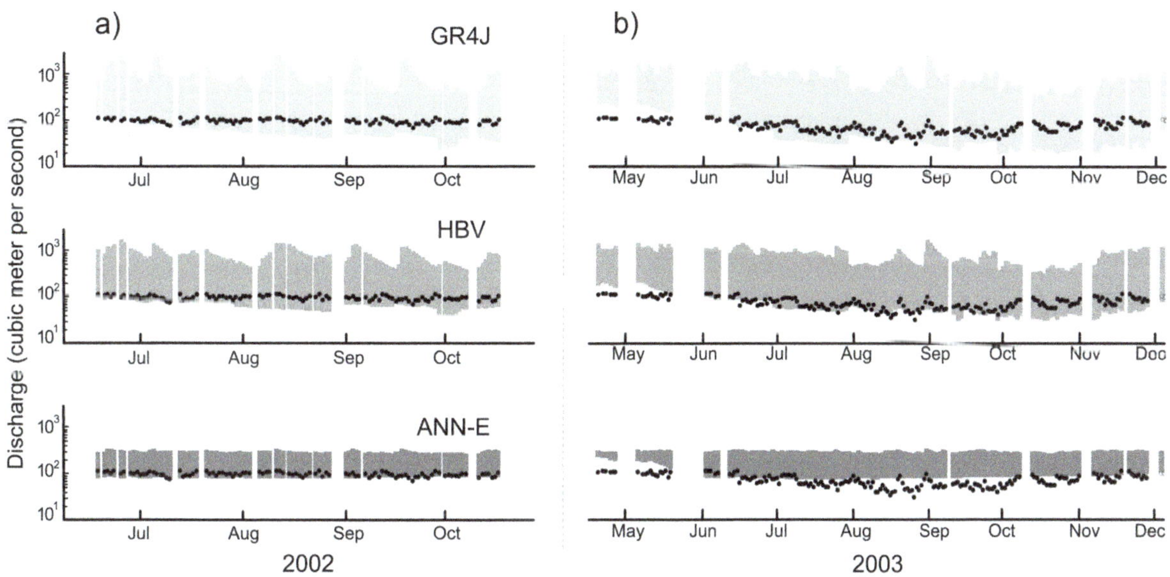

Figure 3. Range (shown as grey shade) of low-flow forecasts in **(a)** 2002 (the wettest year of the test period with 101 low-flow days) and **(b)** 2003 (the driest year of the test period with 192 low-flow days) for a lead time of 90 days using ensemble *P* and PET as input for GR4J, HBV and ANN-E models (case 1 – 2002 and 2003). The gaps in the figures indicate non-low-flow days (i.e. censored).

The performances of GR4J and HBV are similar in the calibration period, whereas HBV performs better in the validation period (Fig. 2b). This is not surprising, since HBV has a more sophisticated model structure than GR4J.

It should be noted that the effect of anthropogenic activities (e.g. flood preventive regulations and urbanization) on the alteration of flow magnitude and dynamics is not obvious, as we found weak positive trends in all *P*, PET and *Q* series ($p < 0.025$ for the three variables using the Mann–Kendall method) which might be caused by climatic changes. Other studies reported that the trends in flood stages in Moselle River were not significant (Bormann et al., 2011).

4.2 Effect of ensembles on low-flow forecasts for 90-day lead time

The effect of ensemble *P* and PET on GR4J, HBV and ANN-E is presented as a range bounded by the lowest and highest forecast values in Fig. 3a and b. The 2 years, i.e. 2002 and 2003, are carefully selected as they represent a relatively wet year and a very dry year respectively. Figure 3a shows that there are significant differences between the three model results. The 90-day-ahead low flows in 2002 are mostly over-predicted by the ANN-E model, whereas GR4J and HBV over-predict low flows observed after August. The over-

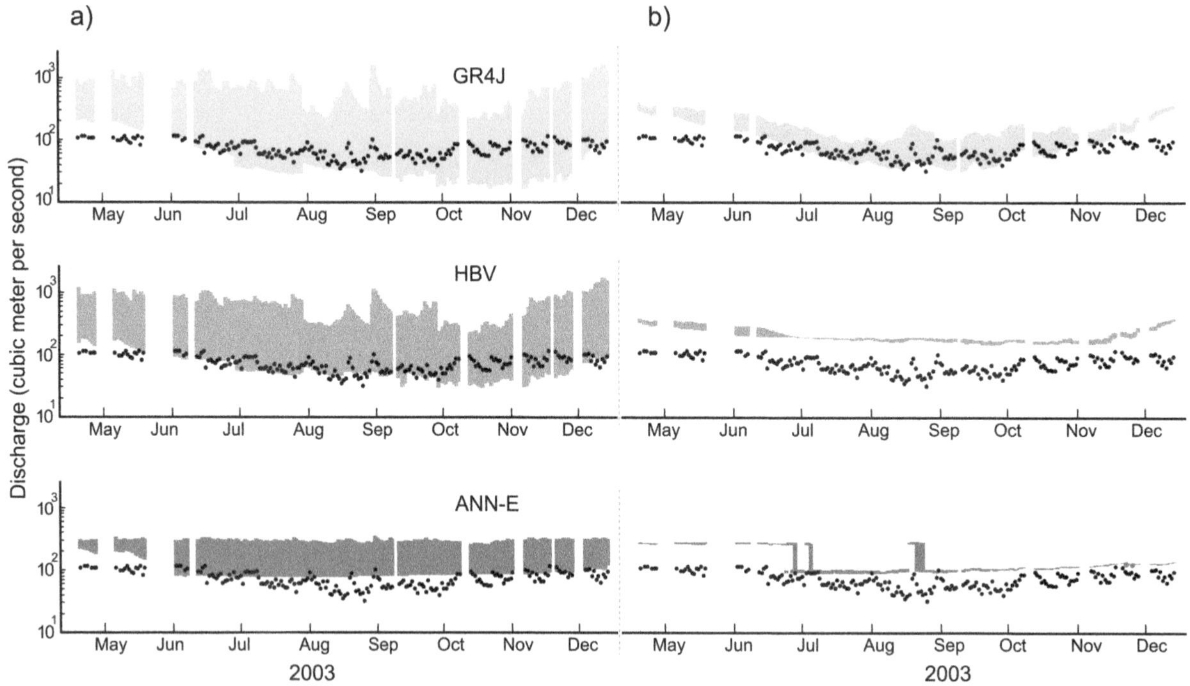

Figure 4. Range (shown as grey shade) of low-flow forecasts in 2003 for a lead time of 90 days using **(a)** ensemble *P* and climate mean PET (case 2) and **(b)** climate mean *P* and ensemble PET as input for GR4J, HBV and ANN-E models (case 3). The gaps in the figures indicate non-low-flow days (i.e. censored).

prediction of low flows is more pronounced for GR4J than for the other three models. The over-prediction of low flows by ANN-E is mostly at the same level. This less sensitive behaviour of ANN-E to the forecasted ensemble inputs shows the effect of the logarithmic sigmoid transfer function on the results. Due to the nature of this algorithm, input is rescaled to a small interval [0, 1] and the gradient of the sigmoid function at large values approximates zero (Wang et al., 2006). Further, ANN-E is also not sensitive to the initial model conditions updated on every forecast issue day. The less pronounced over-prediction of low flows by HBV compared to GR4J may indicate that the slow responding groundwater storage in HBV is less sensitive to different forecasted ensemble *P* and PET inputs (Demirel et al., 2013b).

The results for 2003 are slightly different than those for 2002. As can be seen from Fig. 3b the number of low-flow days has increased in the dry year, i.e. 2003, and the low flows between August and November are not captured by any of the 40-ensemble forecasts using ANN-E. The most striking result in Fig. 3b is that the low flows observed in the period between April and May are not captured by any of the three models, i.e. GR4J, HBV and ANN-E. The poor performance of the models during the spring period can be explained by the high precipitation amount in this period. The poor simulation of high flows in the preceding winter months can have an effect on the forecasts too. The 90-day low flows between October and November are better forecasted by GR4J and HBV than the ANN-E model. The two hydrological models used in this study have well-defined surface- and groundwater components. Therefore, they react to the weather inputs in a physically meaningful way. However, in black box models, the step functions (transfer functions or activation functions) may affect the model behaviour. The ANN model will then react to a certain range of inputs based on the objective function. This feature of ANN is the main reason for the erratic behaviour in Fig. 4b and the small (and uniform) uncertainty range in the figures (e.g. Fig. 3).

For the purpose of determining to which extent ensemble *P* and PET inputs and different initial conditions affect 90-day low-flow forecasts, we ran the models with different input combinations such as ensemble *P* or PET and climate mean *P* or PET and zero precipitation. Figure 4a shows the forecasts using ensemble *P* and climate mean PET as input for three models. The picture is very similar to Fig. 3b as most of the observed low flows fall within the constructed forecast range by GR4J and HBV. The forecasts issued by GR4J are better than those issued by the other two models. However, the range of forecasts using GR4J is larger than for the other models showing the sensitivity of the model for different precipitation inputs. It is obvious that most of the range in all forecasts is caused by uncertainties originating from ensemble precipitation input.

Figure 4b shows the forecasts using climate mean *P* and ensemble PET as input for three models, i.e. GR4J, HBV and ANN-E. Interestingly, only GR4J could capture the 90-day low flows between July and November using climate mean *P*

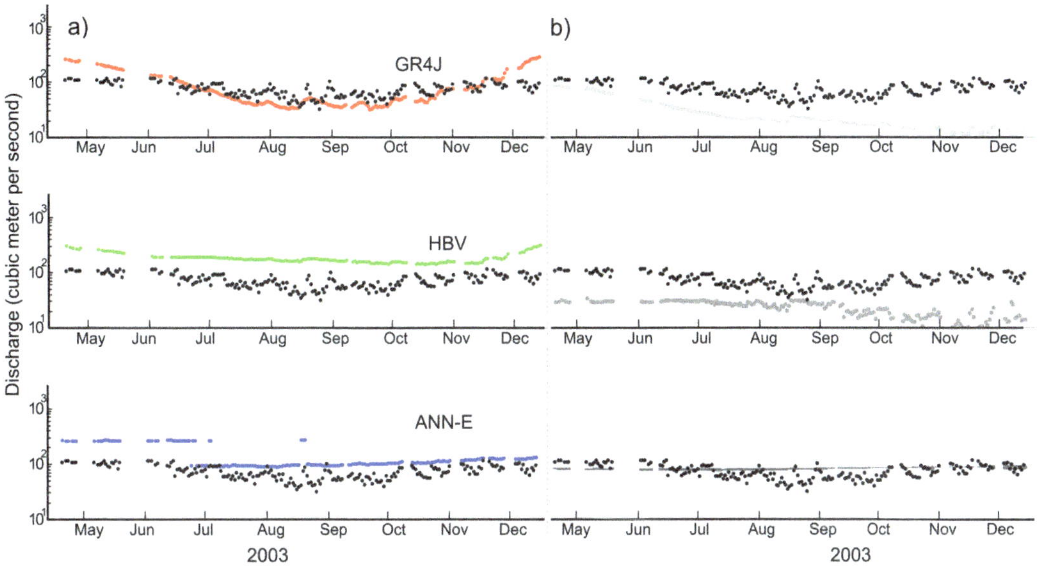

Figure 5. Low-flow forecasts in 2003 for a lead time of 90 days using **(a)** both climate mean *P* and PET (case 4) and **(b)** zero *P* and ensemble PET (case 5) as input for GR4J, HBV and ANN-E models. The gaps in the figures indicate non-low-flow days (i.e. censored).

and ensemble PET showing the ability of the model to handle the excessive rainfall. None of the low flows were captured by HBV, whereas very few low-flow events were captured by ANN-E (Fig. 4b). The precipitation information is crucial for the conceptual models to forecast low flows for a lead time of 90 days. The narrow uncertainty band indicates that the effect of the PET ensemble on the forecasts is less pronounced as compared to the effect of the *P* ensemble.

Figure 5a shows the forecasts using climate mean *P* and PET as input for three models. The results are presented by point values without a range since only one deterministic forecast is issued. There are significant differences in the results of the three models. For instance, all 90-day-ahead low flows in 2003 are over-predicted by HBV, whereas the over-prediction of low flows is less pronounced for ANN-E. It is remarkable that GR4J can forecast a very dry year accurately using the climate mean. The low values of the calibrated maximum soil moisture capacity and percolation parameters of HBV (FC and PERC) can be the main reason for over-prediction of all low flows as the interactions of parameters with climate mean *P* input can result in higher model outputs.

We also assessed the seasonal forecasts using zero *P* and ensemble PET as inputs for three models (Fig. 5b). Not surprisingly, both GR4J and HBV under-predicted most of the low flows when they are run without precipitation input. The results of case 5 confirm that the *P* input is crucial for improving low-flow forecasts although obviously less precipitation is usually observed in a low-flow period compared to other periods.

Figure 6 shows the performance of the three models in the test period using perfect *P* and PET forecasts as input. This is an idealistic case showing that GR4J model performs better than the other two models. It is interesting to note that the ANN-E model does not produce constant predictions as in the previous figures, showing the ability of this black box model to perform comparable to the conceptual models when configured and trained properly.

We also show the minimum and maximum prediction errors for each case in Table 7. There are large differences in cases 1 and 2 as compared to the other cases. It is also obvious that the uncertainty range is larger in case 1 than in case 2 for the conceptual models. This is also what we see in Figs. 3 and 4 above.

4.3 Effect of ensembles on low-flow forecast skill scores

Figure 7 compares the three models and the effect of ensemble *P* and PET on the skill of probabilistic low-flow forecasts with varying lead times. In this figure, four different skill scores are used to present the results of probabilistic low-flow forecasts issued by GR4J, HBV and ANN-E. From an operational point of view, the main purpose of investigating the effect of ensembles and model initial conditions on ensemble low-flow forecasts with varying lead times is to improve the forecast skills (e.g. hit rate, reliability, BSS and MFS) and to reduce false alarms and misses. From Fig. 7 we can clearly see that the results of GR4J show the lowest BSS, MFS and hit rate. The false alarm rate of forecasts using GR4J is also the lowest compared to those using other models. The decrease in false alarm rates after a lead time of 20 days shows the importance of initial condition uncertainty for short lead time forecasts. The limit is around 20 days for ANN-E and shorter for the other two models. When the forecast is issued on day (t), the model states are updated us-

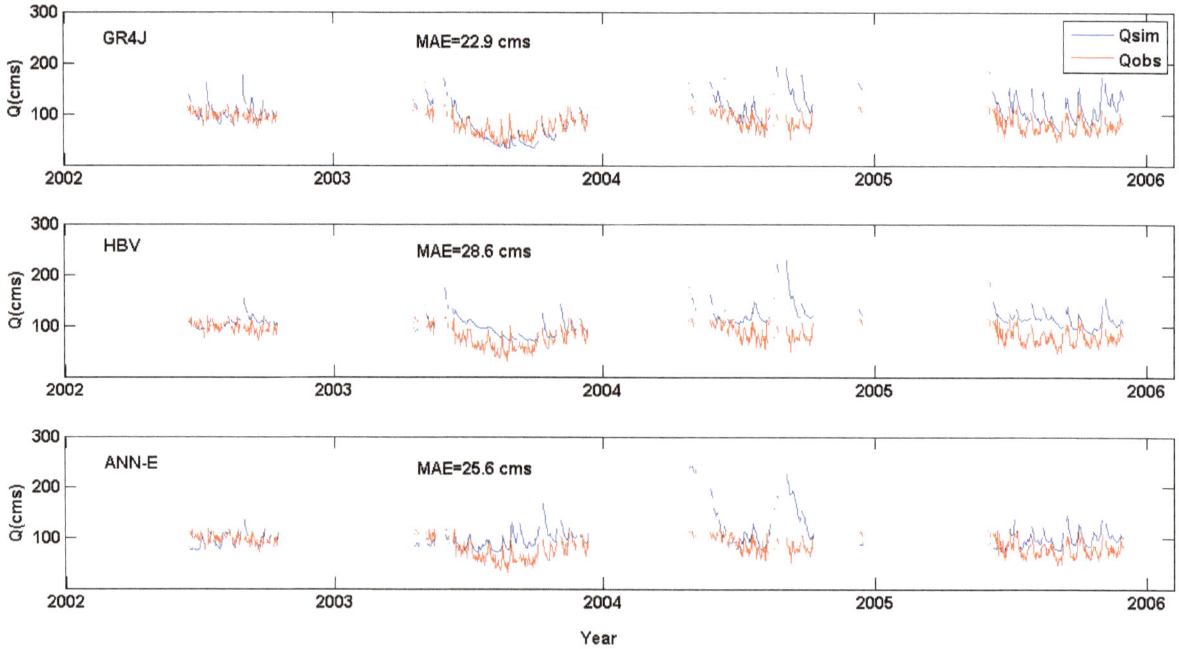

Figure 6. Benchmark reference forecasts using the three models (GR4J, HBV and ANN-E) using observed P and PET (i.e. perfect forecasts).

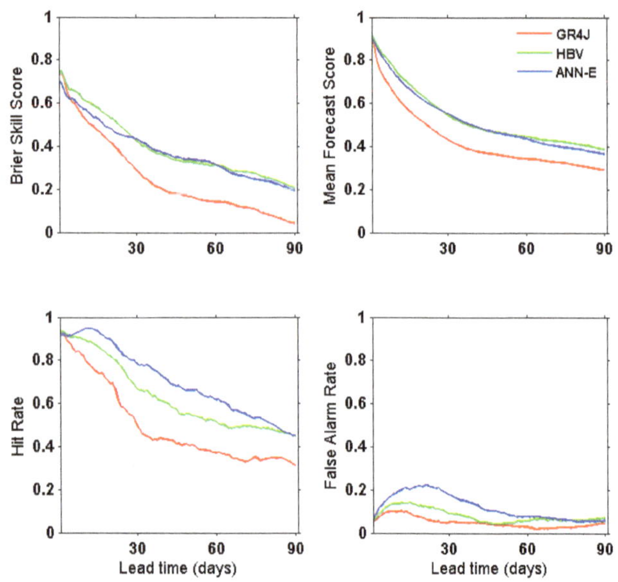

Figure 7. Skill scores for forecasting low flows at different lead times for three different hydrological models for the test period 2002–2005. Note that all forecasts (including high- and low-flow time steps) are used to estimate these skill scores.

ing the observed discharge on that day (t). For GR4J and HBV we used the deterministic state update procedure described in Sect. 3.1.5. However, the models probably spin-up after some days and the results for false alarm rate are improved. For longer lead times the error is better handled by the models. We further analysed the forecasted meteorological forcing data (P and PET) to see if there is any difference between the short lead time ($\sim$ 20 days) and long lead time (e.g. 90 days). This is done for three different lead times for each model when the false alarm rate was highest (i.e. 12, 15 and 21 days based on the false alarm rates of GR4J, HBV and ANN-E, respectively). We compared the boxplots from these problematic lead times with the 90-day lead time (not shown here but available in the review reports). It is interesting to note that the ranges for P and PET are larger at 90-day lead time as compared to shorter lead times. However, the observed P and PET values (i.e. perfect forecasts) are covered by the large ranges resulting in higher hit rates (i.e. lower false alarm rates). In other words, for short lead times, 12, 15 and 21 days in particular, the ranges for P and PET are smaller than those for the 90-day lead time but the observed P and PET values are usually missed, causing higher false alarm rates in the results.

It appears from the results that ANN-E and HBV show a comparable skill in forecasting low flows up to a lead time of 90 days.

Figure 8 compares the reliability of probabilistic 90-day low-flow forecasts below different thresholds (i.e. Q_{75}, Q_{90} and Q_{95}) using ensemble P and PET as input for three models. The figure shows that the Q_{75} and Q_{90} low-flow forecasts issued by the HBV model are more reliable compared to the other models. Moreover, all three models under-predict most of the forecast intervals. It appears from Fig. 8c that very critical low flows (i.e. Q_{99}) are under-predicted by the GR4J model.

Table 7. Minimum and maximum prediction errors for low-flow forecasts for a lead time of 90 days during the test period 2002–2005.

Model	Minimum, median and maximum MAE ($m^3 s^{-1}$)				
	Case 1	Case 2	Case 3	Case 4	Case 5
HBV	[23 101 785]	[23 72 600]	[108 119 135]	[105 105 105]	[57 57 57]
GR4J	[33 122 906]	[36 75 646]	[46 61 111]	[44 44 44]	[55 58 59]
ANN-E	[17 94 227]	[18 72 221]	[65 73 80]	[65 65 65]	[16 16 17]

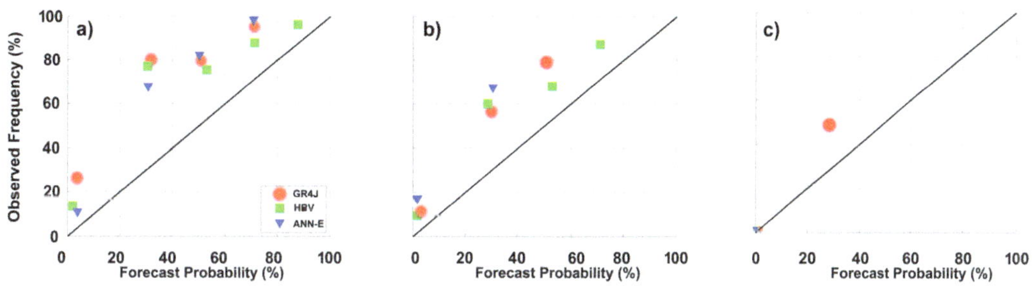

Figure 8. Reliability diagram for different low-flow forecasts. **(a)** Low flows below Q_{75} threshold (584 observed events in the test period 2002–2005). **(b)** Low flows below Q_{90} threshold (250 observed events). **(c)** Low flows below Q_{99} threshold (20 observed events). The forecasts are issued for a lead time of 90 days for the test period 2002–2005 using ensemble P and PET as input for GR4J, HBV and ANN-E models.

5 Discussion

To compare data-driven and conceptual modelling approaches and to evaluate the effects of seasonal meteorological forecasts on low-flow forecasts, 40-member ensembles of ECMWF seasonal meteorological forecasts were used as input for three low-flow forecast models.

These models were calibrated using a hybrid low-flow objective function. Although combining two metrics offered a selective evaluation of low flows, we have noted an important caveat using the second component of the hybrid metric as it is less sensitive as compared to the first part of the hybrid metric resulting in higher (optimistic) values for most cases. The different units had no effect on our calibration results as the ultimate calibration target value is zero (i.e. unit independent). Other studies also combined different metrics with different units (Nash–Sutcliffe, RMSE, R^2 and NumSC, i.e. the number of sign changes in the errors) into one objective function (Hamlet et al., 2013). However, the modellers should carefully use the hybrid function introduced in this study, in particular when comparing different model results. Plotting the two parts of this hybrid function as a Pareto front can lead to a clearer picture than simply summing the two metrics.

In this study, different input combinations were compared to distinguish between the effects of ensemble P and PET and model initial conditions on 90-day low-flow forecasts. The models could reasonably forecast low flows when ensemble P was introduced into the models. This result is in line with that of Shukla and Lettenmaier (2011) who found that seasonal meteorological forecasts have a greater influence than initial model conditions on the seasonal hydrological forecast skills. Moreover, our analyses show that the better forecast performance for longer lead times is an obvious artefact since the higher hit rates are the result of more uncertain (larger range) forecasts. The probabilistic skill scores focuses on the forecasts; the uncertainty in the meteorological forcing data should be carefully scrutinized using different quantitative screening methods, e.g. boxplots.

Two other related studies also showed that the effect of a large spread in ensemble seasonal meteorological forecasts is larger than the effect of initial conditions on hydrological forecasts with lead times longer than 1–2 months (Li et al., 2009; Yossef et al., 2013). The encouraging results of low-flow forecasts using ensemble seasonal precipitation forecasts for the hydrological models confirm the utility of seasonal meteorological forcing for low-flow forecasts. Shukla et al. (2012) also found useful forecast skills for both runoff and soil moisture forecasting at seasonal lead times using the medium-range weather forecasts.

In this study, we also assessed the effects of ensemble P and PET on the skill scores of low-flow forecasts with varying lead times up to 90 days. In general, the four skill scores show similar results. Not surprisingly, all models under-predicted low flows without precipitation information (zero P). The most evident two patterns in these scores are that first, the forecast skill drops sharply until a lead time of 30 days and second, the skill of probabilistic low-flow forecasts issued by GR4J is the lowest, whereas the skill of forecasts issued by ANN-E is the highest compared to the other two models. Further, our study showed that data-driven models can be good alternatives to conceptual models for issuing seasonal low-flow forecasts (e.g. Fig. 6).

The two hydrological models used in this study have well-defined surface- and groundwater components. Therefore, they react to the weather inputs in a physically meaningful way. However, in black box models, the step functions (transfer functions or activation functions) may limit model sensitivity after the training. The ANN model will then react to a certain range of inputs based on the objective function. This feature of an ANN is the main reason for the small (and uniform) uncertainty range in the figures. The over-prediction of the models is closely related to the over-prediction of the P by the ensembles. Low flows are usually over-predicted by the models for the entire period. However, there are under-predictions of low flows for some days in November–December as well. Before June, none of the low flows are captured by the ensemble members. The best-performing period is the fall and the worst-performing period is the spring period for the models. The poor performance of the models during the spring period can be explained by the high precipitation amount in this period. Since the first part of the objective function used in this study solely focuses on low flows, the high-flow period is less important in the calibration. The low flows occurring in the spring period are, therefore, missed in the forecasts. The simulation of snow cover during winter and snow melt during the spring can both have effects on the forecasts too.

6 Conclusions

Three hydrological models have been compared regarding their performance in the calibration, validation and forecast periods, and the effect of seasonal meteorological forecasts on the skill of low-flow forecasts has been assessed for varying lead times. The comparison of three different models help us to contrast data-driven and conceptual models in low-flow forecasts, whereas running the models with different input combinations, e.g. climate mean precipitation and ensemble potential evapotranspiration, help us to identify which input source led to the largest range in the forecasts. A new hybrid low-flow objective function, comprising the mean absolute error of low flows and the mean absolute error of inverse discharges, is used for comparing low-flow simulations, whereas the skill of the probabilistic seasonal low-flow forecasts has been evaluated based on the ensemble forecast range, Brier Skill Score, reliability, hit/false alarm rates and Mean Forecast Score. The latter skill score (MFS) focusing on low flows is firstly introduced in this study. In general our results showed that:

- Based on the results of the calibration and validation, one hidden neuron in ANN was found to be enough for seasonal forecasts as additional hidden neurons did not increase the simulation performance. The difference between calibration and validation performances was smallest for the HBV model, i.e. the most sophisticated model used in this study.
- Based on the results of the comparison of different model inputs for 2 years (i.e. 2002 and 2003), the largest range for 90-day low-flow forecasts is found for the GR4J model when using ensemble seasonal meteorological forecasts as input. Moreover, the uncertainty arising from ensemble precipitation has a larger effect on seasonal low-flow forecasts than the effects of ensemble potential evapotranspiration. All models are prone to over-predict low flows using ensemble seasonal meteorological forecasts. However, the precipitation forecasts in the forecast period are crucial for improving the low-flow forecasts. As expected, all three models, i.e. GR4J, HBV and ANN-E, under-predicted 90-day-ahead low flows in 2003 without rainfall data.
- Based on the results of the comparison of forecast skills with varying lead times, the false alarm rate of GR4J is the lowest indicating the ability of the model of forecasting non-occurrence of low-flow days. The low-flow forecasts issued by HBV are more reliable compared to the other models. The hit rate of ANN-E is higher than that of the two conceptual models used in this study. Overall, the ANN-E and HBV models are the best-performing two of the three models using ensemble P and PET.

Further work should examine the effect of model parameters and initial conditions on the seasonal low-flow forecasts as the values of the maximum soil moisture and percolation related parameters of conceptual models can result in over- or under-prediction of low flows. The uncertainty increases in seasonal meteorological forecasts can lead to better skill scores as an artefact of large ranges in input. Therefore, the quality of the model inputs should be assessed in addition to the model outputs. It is noteworthy to mention that the data-driven model developed in this study, i.e. ANN-E, can be applied to other large river basins elsewhere in the world. Surprisingly, ANN-E and HBV showed a similar skill for seasonal forecasts, where a priori we expected that the two conceptual models, GR4J and HBV, would show similar results up to a lead time of 90 days.

Acknowledgements. We acknowledge the financial support of the Dr. Ir. Cornelis Lely Stichting (CLS), Project No. 20957310. The research is part of the programme of the Department of Water Engineering and Management at the University of Twente and it supports the work of the UNESCO-IHP VII FRIEND-Water programme. Discharge data for the River Rhine were provided by the Global Runoff Data Centre (GRDC) in Koblenz (Germany). Areal precipitation and evapotranspiration data were supplied by the Federal Institute of Hydrology (BfG), Koblenz (Germany). REGNIE grid data were extracted from the archive of the Deutscher Wetterdienst (DWD: German Weather Service), Offenbach (Germany).

ECMWF ENS data used in this study have been obtained from the ECMWF seasonal forecasting system, i.e. Mars System 3. We thank Dominique Lucas from ECMWF who kindly guided us through the data retrieval process. The GIS base maps with delineated 134 basins of the Rhine basin were provided by Eric Sprokkereef, the secretary general of the Rhine Commission (CHR). The GR4J and HBV model codes were provided by Ye Tian. We are grateful to the members of the Referat M2 – Mitarbeiter/innen group at BfG, Koblenz, in particular Peter Krahe, Dennis Meißner, Bastian Klein, Robert Pinzinger, Silke Rademacher and Imke Lingemann, for discussions on the value of seasonal low-flow forecasts.

The constructive review comments of Kerstin Stahl (Associate Editor), Renata Romanowicz, Stefanie Jörg-Hess and one anonymous reviewer significantly improved this paper.

Edited by: K. Stahl

References

Adamowski, J., Chan, H. F., Prasher, S. O., Ozga-Zielinski, B., and Sliusarieva, A.: Comparison of multiple linear and nonlinear regression, autoregressive integrated moving average, artificial neural network, and wavelet artificial neural network methods for urban water demand forecasting in Montreal, Canada, Water Resour. Res., 48, W01528, doi:10.1029/2010wr009945, 2012.

Archer, D. R. and Fowler, H. J.: Using meteorological data to forecast seasonal runoff on the River Jhelum, Pakistan, J. Hydrol., 361, 10–23, doi:10.1016/j.jhydrol.2008.07.017, 2008.

ATV-DVWK: Verdunstung in Bezug zu Landnutzung, Bewuchs und Boden, Merkblatt ATV-DVWK-M 504, Hennef, 2002.

Bell, V. A., Davies, H. N., Kay, A. L., Marsh, T. J., Brookshaw, A., and Jenkins, A.: Developing a large-scale water-balance approach to seasonal forecasting: application to the 2012 drought in Britain, Hydrol. Process., 27, 3003–3012, doi:10.1002/hyp.9863, 2013.

Bierkens, M. F. P. and van Beek, L. P. H.: Seasonal Predictability of European Discharge: NAO and Hydrological Response Time, J. Hydrometeorol., 10, 953–968, doi:10.1175/2009jhm1034.1, 2009.

Booij, M. J.: Impact of climate change on river flooding assessed with different spatial model resolutions, J. Hydrol., 303, 176–198, doi:10.1016/j.jhydrol.2004.07.013, 2005.

Bormann, H.: Runoff regime changes in German rivers due to climate change, Erdkunde, 64, 257–279, doi:10.3112/erdkunde.2010.03.04, 2010.

Bormann, H., Pinter, N., and Elfert, S.: Hydrological signatures of flood trends on German rivers: Flood frequencies, flood heights and specific stages, J. Hydrol., 404, 50–66, doi:10.1016/j.jhydrol.2011.04.019, 2011.

Chiew, F. H. S., Zhou, S. L., and McMahon, T. A.: Use of seasonal streamflow forecasts in water resources management, J. Hydrol., 270, 135–144, 2003.

Chowdhury, S. and Sharma, A.: Multisite seasonal forecast of arid river flows using a dynamic model combination approach, Water Resour. Res., 45, W10428, doi:10.1029/2008wr007510, 2009.

Coley, D. M. and Waylen, P. R.: Forecasting dry season streamflow on the Peace River at Arcadia, Florida, USA, J. Am. Water Resour. Assoc., 42, 851–862, 2006.

Demirel, M. C., Booij, M. J., and Hoekstra, A. Y.: Identification of appropriate lags and temporal resolutions for low flow indicators in the River Rhine to forecast low flows with different lead times, Hydrol. Process., 27, 2742–2758, doi:10.1002/hyp.9402, 2013a.

Demirel, M. C., Booij, M. J., and Hoekstra, A. Y.: Effect of different uncertainty sources on the skill of 10 day ensemble low flow forecasts for two hydrological models, Water Resour. Res., 49, 4035–4053, doi:10.1002/wrcr.20294, 2013b.

Devineni, N., Sankarasubramanian, A., and Ghosh, S.: Multimodel ensembles of streamflow forecasts: Role of predictor state in developing optimal combinations, Water Resour. Res., 44, W09404, doi:10.1029/2006wr005855, 2008.

De Vos, N. J. and Rientjes, T. H. M.: Multiobjective training of artificial neural networks for rainfall-runoff modeling, Water Resour. Res., 44, W08434, doi:10.1029/2007wr006734, 2008.

Doblas-Reyes, F. J., Weisheimer, A., Déqué, M., Keenlyside, N., McVean, M., Murphy, J. M., Rogel, P., Smith, D., and Palmer, T. N.: Addressing model uncertainty in seasonal and annual dynamical ensemble forecasts, Q. J. Roy. Meteorol. Soc., 135, 1538–1559, doi:10.1002/qj.464, 2009.

Dutra, E., Di Giuseppe, F., Wetterhall, F., and Pappenberger, F.: Seasonal forecasts of droughts in African basins using the Standardized Precipitation Index, Hydrol. Earth Syst. Sci., 17, 2359–2373, doi:10.5194/hess-17-2359-2013, 2013.

Dutra, E., Pozzi, W., Wetterhall, F., Di Giuseppe, F., Magnusson, L., Naumann, G., Barbosa, P., Vogt, J., and Pappenberger, F.: Global meteorological drought – Part 2: Seasonal forecasts, Hydrol. Earth Syst. Sci., 18, 2669–2678, doi:10.5194/hess-18-2669-2014, 2014.

Eberle, M.: Hydrological Modelling in the River Rhine Basin Part III – Daily HBV Model for the Rhine Basin BfG-1451, Institute for Inland Water Management and Waste Water Treatment (RIZA) and Federal Institute of Hydrology (BfG) Koblenz, Germany, 2005.

ECMWF: Describing ECMWF's forecasts and forecasting system, ECMWF newsletter 133, available from: http://old.ecmwf.int/publications/manuals/mars/ (last access: 26 July 2014), 2012.

Elshorbagy, A., Corzo, G., Srinivasulu, S., and Solomatine, D. P.: Experimental investigation of the predictive capabilities of data driven modeling techniques in hydrology – Part 1: Concepts and methodology, Hydrol. Earth Syst. Sci., 14, 1931–1941, doi:10.5194/hess-14-1931-2010, 2010.

Engeland, K., Renard, B., Steinsland, I., and Kolberg, S.: Evaluation of statistical models for forecast errors from the HBV model, J. Hydrol., 384, 142–155, 2010.

EU: Horizon 2020 – Work Programme 2014–2015: Water 7_2015: Increasing confidence in seasonal-to-decadal predictions of the water cycle, http://www.aber.ac.uk/en/media/departmental/researchoffice/funding/UKRO-Horizon-2020_climatechangedraftwp.pdf, last access: 4 September 2013.

Felipe, P.-S. and Nelson, O.-N.: Forecasting of Monthly Streamflows Based on Artificial Neural Networks, J. Hydrol. Eng., 14, 1390–1395, 2009.

Förster, K., Meon, G., Marke, T., and Strasser, U.: Effect of meteorological forcing and snow model complexity on hydrological simulations in the Sieber catchment (Harz Mountains, Germany), Hydrol. Earth Syst. Sci., 18, 4703–4720, doi:10.5194/hess-18-4703-2014, 2014.

Fundel, F., Jörg-Hess, S., and Zappa, M.: Monthly hydrometeorological ensemble prediction of streamflow droughts and corresponding drought indices, Hydrol. Earth Syst. Sci., 17, 395–407, doi:10.5194/hess-17-395-2013, 2013.

Ganguli, P. and Reddy, M. J.: Ensemble prediction of regional droughts using climate inputs and SVM-copula approach, Hydrol. Process., 28, 4989–5009, doi:10.1002/hyp.9966, 2014.

Gaume, E. and Gosset, R.: Over-parameterisation, a major obstacle to the use of artificial neural networks in hydrology?, Hydrol. Earth Syst. Sci., 7, 693–706, doi:10.5194/hess-7-693-2003, 2003.

Giuntoli, I., Renard, B., Vidal, J. P., and Bard, A.: Low flows in France and their relationship to large-scale climate indices, J. Hydrol., 482, 105–118, doi:10.1016/j.jhydrol.2012.12.038, 2013.

Gobena, A. K. and Gan, T. Y.: Incorporation of seasonal climate forecasts in the ensemble streamflow prediction system, J. Hydrol., 385, 336–352, doi:10.1016/j.jhydrol.2010.03.002, 2010.

Görgen, K., Beersma, J., Brahmer, G., Buiteveld, H., Carambia, M., de Keizer, O., Krahe, P., Nilson, E., Lammersen, R., Perrin, C., and Volken, D.: Assessment of Climate Change Impacts on Discharge in the Rhine River Basin: Results of the RheinBlick 2050 Project, Lelystad, CHR, p. 211, available from: http://www.news.admin.ch/NSBSubscriber/message/attachments/20770.pdf (last access: 30 October 2014), 2010.

Govindaraju, R. S. and Rao, A. R.: Artificial Neural Networks in Hydrology, Kluwer Academic Publishers Norwell, MA, USA, 329 pp., 2000.

Hamlet, A. F., Elsner, M. M., Mauger, G. S., Lee, S.-Y., Tohver, I., and Norheim, R. A.: An Overview of the Columbia Basin Climate Change Scenarios Project: Approach, Methods, and Summary of Key Results, Atmos.-Ocean, 51, 392–415, doi:10.1080/07055900.2013.819555, 2013.

Hartmann, H. C., Pagano, T. C., Sorooshian, S., and Bales, R.: Confidence builders: Evaluating seasonal climate forecasts from user perspectives, B. Am. Meteorol. Soc., 83, 683–698, 2002.

Jaun, S. and Ahrens, B.: Evaluation of a probabilistic hydrometeorological forecast system, Hydrol. Earth Syst. Sci., 13, 1031–1043, doi:10.5194/hess-13-1031-2009, 2009.

Kahya, E. and Dracup, J. A.: U.S. streamflow patterns in relation to the El Niño/Southern Oscillation, Water Resour. Res., 29, 2491–2503, doi:10.1029/93wr00744, 1993.

Kalra, A., Ahmad, S., and Nayak, A.: Increasing streamflow forecast lead time for snowmelt-driven catchment based on large-scale climate patterns, Adv. Water Resour., 53, 150–162, doi:10.1016/j.advwatres.2012.11.003, 2013.

Kasiviswanathan, K. S., Raj, C., Sudheer, K. P., and Chaubey, I.: Constructing prediction interval for artificial neural network rainfall runoff models based on ensemble simulations, J. Hydrol., 499, 275–288, doi:10.1016/j.jhydrol.2013.06.043, 2013.

Kuo, C.-C., Gan, T. Y., and Yu, P.-S.: Seasonal streamflow prediction by a combined climate-hydrologic system for river basins of Taiwan, J. Hydrol., 387, 292–303, 2010.

Li, H., Luo, L., and Wood, E. F.: Seasonal hydrologic predictions of low-flow conditions over eastern USA during the 2007 drought, Atmos. Sci. Lett., 9, 61–66, 2008.

Li, H., Luo, L., Wood, E. F., and Schaake, J.: The role of initial conditions and forcing uncertainties in seasonal hydrologic forecasting, J. Geophys. Res., 114, D04114, doi:10.1029/2008jd010969, 2009.

Lindström, G., Johansson, B., Persson, M., Gardelin, M., and Bergstrom, S.: Development and test of the distributed HBV-96 hydrological model, J. Hydrol., 201, 272–288, 1997.

Luo, L., Wood, E. F., and Pan, M.: Bayesian merging of multiple climate model forecasts for seasonal hydrological predictions, J. Geophys. Res., 112, D10102, doi:10.1029/2006jd007655, 2007.

Madadgar, S. and Moradkhani, H.: A Bayesian Framework for Probabilistic Seasonal Drought Forecasting, J. Hydrometeorol., 14, 1685–1705, doi:10.1175/JHM-D-13-010.1, 2013.

Martina, M. L. V., Todini, E., and Libralon, A.: A Bayesian decision approach to rainfall thresholds based flood warning, Hydrol. Earth Syst. Sci., 10, 413–426, doi:10.5194/hess-10-413-2006, 2006.

Nicolle, P., Pushpalatha, R., Perrin, C., François, D., Thiéry, D., Mathevet, T., Le Lay, M., Besson, F., Soubeyroux, J.-M., Viel, C., Regimbeau, F., Andréassian, V., Maugis, P., Augeard, B., and Morice, E.: Benchmarking hydrological models for low-flow simulation and forecasting on French catchments, Hydrol. Earth Syst. Sci., 18, 2829–2857, doi:10.5194/hess-18-2829-2014, 2014.

Olsson, J. and Lindström, G.: Evaluation and calibration of operational hydrological ensemble forecasts in Sweden, J. Hydrol., 350, 14–24, 2008.

Perrin, C., Michel, C., and Andréassian, V.: Improvement of a parsimonious model for streamflow simulation, J. Hydrol., 279, 275–289, 2003.

Pokhrel, P., Wang, Q. J., and Robertson, D. E.: The value of model averaging and dynamical climate model predictions for improving statistical seasonal streamflow forecasts over Australia, Water Resour. Res., 49, 6671–6687, doi:10.1002/wrcr.20449, 2013.

Pushpalatha, R., Perrin, C., Moine, N. L., Mathevet, T., and Andréassian, V.: A downward structural sensitivity analysis of hydrological models to improve low-flow simulation, J. Hydrol., 411, 66–76, 2011.

Pushpalatha, R., Perrin, C., Moine, N. L., and Andréassian, V.: A review of efficiency criteria suitable for evaluating low-flow simulations, J. Hydrol., 420–421, 171–182, doi:10.1016/j.jhydrol.2011.11.055, 2012.

Renner, M., Werner, M. G. F., Rademacher, S., and Sprokkereef, E.: Verification of ensemble flow forecasts for the River Rhine, J. Hydrol., 376, 463–475, 2009.

Robertson, D. E., Pokhrel, P., and Wang, Q. J.: Improving statistical forecasts of seasonal streamflows using hydrological model output, Hydrol. Earth Syst. Sci., 17, 579–593, doi:10.5194/hess-17-579-2013, 2013.

Roulin, E.: Skill and relative economic value of medium-range hydrological ensemble predictions, Hydrol. Earth Syst. Sci., 11, 725–737, doi:10.5194/hess-11-725-2007, 2007.

Rutten, M., van de Giesen, N., Baptist, M., Icke, J., and Uijttewaal, W.: Seasonal forecast of cooling water problems in the River Rhine, Hydrol. Process., 22, 1037–1045, 2008.

Saadat, S., Khalili, D., Kamgar-Haghighi, A., and Zand-Parsa, S.: Investigation of spatio-temporal patterns of seasonal streamflow droughts in a semi-arid region, Nat. Hazards, 69, 1697–1720, doi:10.1007/s11069-013-0783-y, 2013.

Sauquet, E., Lerat, J., and Prudhomme, C.: La prévision hydrométéorologique à 3-6 mois, Etat des connaissances et applications, La Houille Blanche, 6, 77–84, doi:10.1051/lhb:2008075, 2008.

Schubert, S., Koster, R., Hoerling, M., Seager, R., Lettenmaier, D., Kumar, A., and Gutzler, D.: Predicting Drought on Seasonal-to-Decadal Time Scales, B. Am. Meteorol. Soc., 88, 1625–1630, doi:10.1175/bams-88-10-1625, 2007.

Shamseldin, A. Y.: Application of a neural network technique to rainfall-runoff modelling, J. Hydrol., 199, 272–294, doi:10.1016/s0022-1694(96)03330-6, 1997.

Shukla, S. and Lettenmaier, D. P.: Seasonal hydrologic prediction in the United States: understanding the role of initial hydrologic conditions and seasonal climate forecast skill, Hydrol. Earth Syst. Sci., 15, 3529–3538, doi:10.5194/hess-15-3529-2011, 2011.

Shukla, S., Voisin, N., and Lettenmaier, D. P.: Value of medium range weather forecasts in the improvement of seasonal hydrologic prediction skill, Hydrol. Earth Syst. Sci., 16, 2825–2838, doi:10.5194/hess-16-2825-2012, 2012.

Shukla, S., Sheffield, J., Wood, E. F., and Lettenmaier, D. P.: On the sources of global land surface hydrologic predictability, Hydrol. Earth Syst. Sci., 17, 2781–2796, doi:10.5194/hess-17-2781-2013, 2013.

Soukup, T. L., Aziz, O. A., Tootle, G. A., Piechota, T. C., and Wulff, S. S.: Long lead-time streamflow forecasting of the North Platte River incorporating oceanic-atmospheric climate variability, J. Hydrol., 368, 131–142, 2009.

Thirel, G., Rousset-Regimbeau, F., Martin, E., and Habets, F.: On the Impact of Short-Range Meteorological Forecasts for Ensemble Streamflow Predictions, J. Hydrometeorol., 9, 1301–1317, doi:10.1175/2008jhm959.1, 2008.

Tian, Y., Booij, M. J., and Xu, Y.-P.: Uncertainty in high and low flows due to model structure and parameter errors, Stoch. Environ. Res. Risk A., 28, 319–332, doi:10.1007/s00477-013-0751-9, 2014.

Tootle, G. A. and Piechota, T. C.: Suwannee River Long Range Streamflow Forecasts Based On Seasonal Climate Predictors, J. Am. Water Resour. Assoc., 40, 523–532, 2004.

Towler, E., Roberts, M., Rajagopalan, B., and Sojda, R. S.: Incorporating probabilistic seasonal climate forecasts into river management using a risk-based framework, Water Resour. Res., 49, 4997–5008, doi:10.1002/wrcr.20378, 2013.

Van den Tillaart, S. P. M., Booij, M. J., and Krol, M. S.: Impact of uncertainties in discharge determination on the parameter estimation and performance of a hydrological model, Hydrol. Res., 44, 454–466 2013.

Van Dijk, A. I. J. M., Peña-Arancibia, J. L., Wood, E. F., Sheffield, J., and Beck, H. E.: Global analysis of seasonal streamflow predictability using an ensemble prediction system and observations from 6192 small catchments worldwide, Water Resour. Res., 49, 2729–2746, doi:10.1002/wrcr.20251, 2013.

van Ogtrop, F. F., Vervoort, R. W., Heller, G. Z., Stasinopoulos, D. M., and Rigby, R. A.: Long-range forecasting of intermittent streamflow, Hydrol. Earth Syst. Sci., 15, 3343–3354, doi:10.5194/hess-15-3343-2011, 2011.

Velázquez, J. A., Anctil, F., and Perrin, C.: Performance and reliability of multimodel hydrological ensemble simulations based on seventeen lumped models and a thousand catchments, Hydrol. Earth Syst. Sci., 14, 2303–2317, doi:10.5194/hess-14-2303-2010, 2010.

Vidal, J.-P., Martin, E., Franchistéguy, L., Habets, F., Soubeyroux, J.-M., Blanchard, M., and Baillon, M.: Multilevel and multiscale drought reanalysis over France with the Safran-Isba-Modcou hydrometeorological suite, Hydrol. Earth Syst. Sci., 14, 459–478, doi:10.5194/hess-14-459-2010, 2010.

Wang, E., Zhang, Y., Luo, J., Chiew, F. H. S., and Wang, Q. J.: Monthly and seasonal streamflow forecasts using rainfall-runoff modeling and historical weather data, Water Resour. Res., 47, W05516, doi:10.1029/2010wr009922, 2011.

Wang, W., Gelder, P. H. A. J. M. V., Vrijling, J. K., and Ma, J.: Forecasting daily streamflow using hybrid ANN models, J. Hydrol., 324, 383–399, doi:10.1016/j.jhydrol.2005.09.032, 2006.

Wedgbrow, C. S., Wilby, R. L., Fox, H. R., and O'Hare, G.: Prospects for seasonal forecasting of summer drought and low river flow anomalies in England and Wales, Int. J. Climatol., 22, 219–236, doi:10.1002/joc.735, 2002.

Wedgbrow, C. S., Wilby, R. L., and Fox, H. R.: Experimental seasonal forecasts of low summer flows in the River Thames, UK, using Expert Systems, Clim. Res., 28, 133–141, 2005.

Wilks, D. S.: Statistical Methods in the Atmospheric Sciences, Elsevier, New York, 1995.

Winsemius, H. C., Dutra, E., Engelbrecht, F. A., Archer Van Garderen, E., Wetterhall, F., Pappenberger, F., and Werner, M. G. F.: The potential value of seasonal forecasts in a changing climate in southern Africa, Hydrol. Earth Syst. Sci., 18, 1525–1538, doi:10.5194/hess-18-1525-2014, 2014.

WMO: Forecastverification – issues, methods and faq. WWRP/WGNE, Joint Working Group on Verification, available at: www.cawcr.gov.au/projects/verification (last access: 24 September 2013), 2012.

Wood, A. W. and Lettenmaier, D. P.: A Test Bed for New Seasonal Hydrologic Forecasting Approaches in the Western United States, B. Am. Meteorol. Soc., 87, 1699-1712, doi:10.1175/bams-87-12-1699, 2006.

Wood, A. W., Maurer, E. P., Kumar, A., and Lettenmaier, D. P.: Long-range experimental hydrologic forecasting for the eastern United States, J. Geophys. Res, 107, 4429, doi:10.1029/2001JD000659, 2002.

Yossef, N. C., van Beek, L. P. H., Kwadijk, J. C. J., and Bierkens, M. F. P.: Assessment of the potential forecasting skill of a global hydrological model in reproducing the occurrence of monthly flow extremes, Hydrol. Earth Syst. Sci., 16, 4233–4246, doi:10.5194/hess-16-4233-2012, 2012.

Yossef, N. C., Winsemius, H., Weerts, A., van Beek, R., and Bierkens, M. F. P.: Skill of a global seasonal streamflow forecasting system, relative roles of initial conditions and meteorological forcing, Water Resour. Res., 49, 4687–4699, doi:10.1002/wrcr.20350, 2013.

9

Drivers of spatial and temporal variability of streamflow in the Incomati River basin

A. M. L. Saraiva Okello[1,3], I. Masih[1], S. Uhlenbrook[1,2], G. P. W. Jewitt[3], P. van der Zaag[1,2], and E. Riddell[3]

[1]UNESCO-IHE, Institute for Water Education, P.O. Box 3015, 2601 DA Delft, the Netherlands
[2]Delft University of Technology, Department of Water Resources, P.O. Box 5048, 2600 GA Delft, the Netherlands
[3]Centre for Water Resources Research, School of Agriculture, Earth and Environmental Science, University of KwaZulu-Natal, Private Bag X01, Scottsville, 3209, South Africa

Correspondence to: A. M. L. Saraiva Okello (a.saraiva@unesco-ihe.org)

Abstract. The Incomati is a semi-arid trans-boundary river basin in southern Africa, with a high variability of streamflow and competing water demands from irrigated agriculture, energy, forestry and industries. These sectors compete with environmental flows and basic human water needs, resulting in a "stressed" water resource system. The impacts of these demands, relative to the natural flow regime, appear significant. However, despite being a relatively well-gauged basin in South Africa, the natural flow regime and its spatial and temporal variability are poorly understood and remain poorly described, resulting in a limited knowledge base for water resource planning and management decisions. Thus, there is an opportunity to improve water management, if it can be underpinned by a better scientific understanding of the drivers of streamflow availability and variability in the catchment.

In this study, long-term rainfall and streamflow records were analysed. Statistical analysis, using annual anomalies, was conducted on 20 rainfall stations, for the period 1950–2011. The Spearman test was used to identify trends in the records on annual and monthly timescales. The variability of rainfall across the basin was confirmed to be high, both intra- and inter-annually. The statistical analysis of rainfall data revealed no significant trend of increase or decrease. Observed flow data from 33 gauges were screened and analysed, using the Indicators of Hydrologic Alteration (IHA) approach. Temporal variability was high, with the coefficient of variation of annual flows in the range of 1 to 3.6. Significant declining trends in October flows, and low flow indicators, were also identified at most gauging stations of the Komati and Crocodile sub-catchments; however, no trends were evident in the other parameters, including high flows. The trends were mapped using GIS and were compared with historical and current land use. These results suggest that land use and flow regulation are larger drivers of temporal changes in streamflow than climatic forces. Indeed, over the past 40 years, the areas under commercial forestry and irrigated agriculture have increased over 4 times.

1 Introduction

Global changes, such as climate change, population growth, urbanisation, industrial development and the expansion of agriculture, put huge pressure on natural resources, particularly water (Jewitt, 2006a; Milly et al., 2008; Vörösmarty et al., 2010; Miao et al., 2012; Montanari et al., 2013). In order to manage water in a sustainable manner, it is important to have a sound understanding of the processes that control its existence, the variability in time and space, and our ability to quantify that variability (Jewitt et al., 2004; Hu et al., 2011; Montanari et al., 2013; Hughes et al., 2014).

Water is critically important to the economies and social well-being of the predominantly rural populations of southern Africa, where environmental sustainability issues are increasingly coming into conflict with human development objectives, and where data are also scarce. The local economies and livelihoods of many southern African communities are strongly dependent on agriculture and fisheries, and water availability remains one of the main constraints to development in Africa (Jewitt, 2006a; Pollard and du Toit, 2009). Hydropower is also locally important, while a substantial amount of foreign income is derived from wildlife tourism in some countries of the region (Hughes et al., 2014).

Climate change intensifies the global hydrological cycle, leading to more frequent and variable extremes. For southern Africa, recent studies forecast an increase in the occurrence of drought due to decreased rainfall events (Shongwe et al., 2009; Rouault et al., 2010; Lennard et al., 2013). Furthermore, it is expected that temperatures will rise, and thus the hydrological processes driven by them will intensify (Kruger and Shongwe, 2004; Schulze, 2011). Compounding the effect of climate change are the increased pressures on land and water use, owing to increased population and the consequent requirements for food, fuel and fibre (Rockström et al., 2009; Warburton et al., 2010, 2012). Areas of irrigated agriculture and forestry have been expanding steadily over the past decades. Urbanisation also brings with it an increase in impervious areas and the increased abstraction of water for domestic, municipal and industrial purposes (Schulze, 2011).

In southern Africa, these pressures have led to changes in natural streamflow patterns. However, not many studies are available concerning the magnitude of such changes and what the main drivers are (Hughes et al., 2014). Projections on the impact of climate change on the water resources of South Africa were investigated by Schulze (2012) and streamflow trends of some southern Africa rivers have been analysed (Fanta et al., 2001; Love et al., 2010), but no such studies are available for the Incomati basin.

The Incomati is a semi-arid trans-boundary river basin in southern Africa, which is water-stressed because of high competing demands from, amongst others, irrigated agriculture, forestry, energy, environmental flow and basic human needs provision (DWAF, 2009b; TPTC, 2010). The impact of these demands, relative to the natural flow regime, is significant. Hence, there is an opportunity to improve water management, if a better scientific understanding of water resource availability and variability can be provided (Jewitt, 2006a).

The goal of this paper is to determine whether or not there have been significant changes in rainfall and streamflow during the time of record, and what the potential reasons for and implications of such changes are. The main research questions are the following.

- Does the analysis of precipitation and streamflow records reveal any persistent trends?
- What are the drivers of these trends?
- What are the implications of these trends for water management?

The variability and changes in rainfall and streamflow records were analysed and the possible drivers of changes were identified from the literature. The spatial variation of

Figure 1. Map of the location of the study area, illustrating the main sub-catchments, the hydrometric and rainfall stations analysed, and the basin topography and dams.

trends in streamflow and their possible linkages with the main drivers are analysed. Based on the findings, approaches and alternatives for improved water resource management and planning are proposed.

2 Methodology

2.1 Study area

The Incomati River basin is located in the south-eastern part of Africa and it is shared by the Kingdom of Swaziland, the Republic of Mozambique and the Republic of South Africa (Fig. 1). The total basin area is approximately 46 750 km^2, of which 2560 km^2 (5.5 %) are in Swaziland, 15 510 km^2 (33.2 %) in Mozambique and 28 681 km^2 (61.3 %) in South Africa. The Incomati watercourse includes the Komati, Crocodile, Sabie, Massintoto, Uanetse and Mazimechopes rivers and the estuary. The Komati, Crocodile and Sabie are the main sub-catchments, contributing about 94 % of the natural discharge, with an area of 61 % of the basin. The Incomati River rises in the Highveld and escarpment (2000 m above sea level) in the west of the basin and drops to the coastal plain in Mozambique. The general climate in the Incomati River basin varies from a warm to hot humid climate in Mozambique to a cooler dry climate in South Africa in the west. The mean annual precipitation of about 740 mm a^{-1} falls entirely during the summer months (October to March). The Incomati (Fig. 1) can be topographically and climatically divided into three areas (TPTC, 2010):

- high escarpment, with a high rainfall (800 to 1600 mma^{-1}), low temperatures (mean annual average of 10 to 16 °C) and lower potential evaporation (1600 to 2000 mm a^{-1});

Table 1. Summary of estimated natural streamflow, water demands in the Incomati basin in 10^6 m^3 per year (TPTC, 2010) and major dams (> 10^6 m^3) (Van der Zaag and Vaz, 2003).

	Natural MAR	First priority supplies	Irrigation supplies	Afforestration	Total water use
Komati	1332	141.5	621	117	879.5
Crocodile	1124	74.7	482	158	714.7
Sabie	668	30	98	90	218
Massintoto	41	0.3	0	0	0.3
Uanetse	33	0.3	0	0	0.3
Mazimechopes	20	0	0	0	0
Lower Incomati	258	1.5	412.8	0	414.3
Mozambique	325		412.8		
South Africa	2663		961		
Swaziland	488		240		
Total	3476	248	1614	365	2227

Tributary	Country	Major dam	Year commissioned	Storage capacity (10^6 m^3)
Komati	South Africa	Nooitgedacht	1962	81
Komati	South Africa	Vygeboom	1971	84
Komati	Swaziland	Maguga	2002	332
Komati	Swaziland	Sand River	1966	49
Lomati	South Africa	Driekoppies	1998	251
Crocodile	South Africa	Kwena	1984	155
Crocodile	South Africa	Witklip	1979	12
Crocodile	South Africa	Klipkopje	1979	12
Sabie	South Africa	Da Gama	1979	14
Sabie	South Africa	Injaka	2001	120
Sabie	Mozambique	Corumana	1988	879
Total				1989

* First priority supplies include domestic and industrial uses.

- Highveld and middle Lowveld, which lie between the Drakensberg and the Lebombo Mountains, warmer than the escarpment (mean annual average of 14 to 22 °C), with rainfall that decreases towards the east (400 to 800 mm a^{-1}) and high potential evaporation (2000 to 2200 mm a^{-1});

- coastal plain, located mostly in Mozambique, with higher temperatures (mean annual average of 20 to 26 °C) and lower rainfall (400 to 800 mm a^{-1}) in the west, increasing eastward towards the coast, where there is also high potential evaporation (2200 to 2400 mm a^{-1}).

The geology is complex, characterised by sedimentary, volcanic, granitic and dolomitic rocks, as well as Quaternary and recent deposits (Van der Zaag and Vaz, 2003). The soils in the basin are highly variable, ranging from moderately deep clayey loam in the west, to moderately deep sandy loam in the central areas and moderately deep clayey soils in the east. The dominant land uses in the catchment are commercial forest plantations of exotic trees (pine, eucalyptus) in the escarpment region, dryland crops (maize) and grazing in the Highveld region, and irrigated agriculture (sugarcane, vegetables and citrus) in the Lowveld (DWAF, 2009b; Riddell et al., 2013). In the Mozambican coastal plains, sugarcane and subsistence farming dominate. A substantial part of the basin has been declared a conservation area, which includes the recently established Greater Limpopo Transfrontier Park (the Kruger National Park in South Africa and the Limpopo National Park in Mozambique are part of it) (TPTC, 2010).

The level of water abstraction in the Incomati River is very high, and the actual water demand is projected to increase in the future, as a result of further economic development and population growth (Nkomo and van der Zaag, 2004; LeMarie et al., 2006; Pollard et al., 2011). The consumptive use of surface water amounts to more than 1880 million cubic metres per annum (10^6 m^3 a^{-1}), which represents 51 % of the average amount of surface water generated in the basin (Van der Zaag and Vaz, 2003). The major water consumers (see Tables 1 and 2), accounting for 91 % of all consumptive water use, are the irrigation and forestry sectors, followed by inter-basin water transfers to the Umbeluzi basin and the Olifants catchment in the Limpopo basin (Van der Zaag and Vaz, 2003; DWAF, 2009b; TPTC, 2010). Since the 1950s, the area of irrigated agriculture and forestry has increased steadily, particularly in the Komati and Crocodile systems (Table 2).

Table 2. Land-use and water-use change from the 1950s to 2004 in the Komati, Crocodile and Sabie sub-catchments. Source: adapted from TPTC (2010).

		1950s	1970s	1996	2004
Komati	Irrigation area (km^2)	17.6	144.1	385.1	512.4
	Afforested area (km^2)	247	377	661	801
	Domestic water use (10^6 m^3 a^{-1})	0.5	7.7	15.5	19.7
	Industrial and mining water use (10^6 m^3 a^{-1})	0	0	0.5	0.5
	Water transfers out (10^6 m^3 a^{-1}):				
	To power stations in South Africa	3.4	103	98.1	104.7
	To irrigation in Swaziland outside Komati	0	111.8	122.2	121.8
Crocodile	Irrigation area (km^2)	92.8	365.8	427	510.7
	Afforested area (km^2)	375	1550	1811	1941
	Domestic water use (10^6 m^3 a^{-1})	3	12.2	33.6	52.4
	Industrial and mining water use (10^6 m^3 a^{-1})	0.1	7.5	19.8	22.3
Sabie	Irrigation area (km^2)	27.7	68.4	113.4	127.6
	Afforested area (km^2)	428	729	708	853
	Domestic water use (10^6 m^3 a^{-1})	2.4	5.3	13	26.7
	Industrial and mining water use (10^6 m^3 a^{-1})	0	0	0	0

2.2 Data and analysis

2.2.1 Rainfall

Annual, monthly and daily rainfall data for southern Africa for the period of 1905 to 2000 were extracted from the Lynch (2003) database. The database consists of daily precipitation records for over 12 000 stations in southern Africa, and data quality was checked and some data were patched. The main custodians of rainfall data are SAWS (South African Weather Service), SASRI (South Africa Sugarcane Research Institute) and ISCW (Institute for Soil Climate and Water). About 20 stations out of 374 available for Incomati were selected for detailed analysis. The selection criteria were the quality of data, evaluated by the percentage of reliable data in the database, and the representative spatial coverage of the basin. Eight of the 20 stations' time series were extended up to 2012, using new data collected from the SAWS.

The spatial and temporal heterogeneity in rainfall across the study area was characterised using statistical analysis and annual anomalies. The time series of annual and monthly rainfall from each station were subjected to the Spearman test in order to identify trends for the periods of 1950–2000 and 1950–2011. Two intersecting periods were chosen to evaluate the consistency of the trends. Due to natural climatic variability, there are sequences of wetter and drier periods, so some trends appearing in a specific period might be absent when a longer or shorter period is considered. The Pettitt test (Pettitt, 1979) is used to detect abrupt changes in the time series. Potential change points divide the time series into two sub-series. Then, the significance of the change in the mean and variance of the two sub-series is evaluated by F and t tests. Potential change points were evaluated with a 0.8 probability threshold and significance of change was assessed with F and t tests at the 95 % confidence level (Zhang et al., 2008; Love et al., 2010). The annual and monthly time series were also analysed for the presence of serial correlation. Tests were carried out using SPELL-stat v.1.5.1.0B (Guzman and Chu, 2003).

2.2.2 Streamflow

Streamflow data of 104 gauging stations in South Africa (obtained from the Department of Water Affairs – DWA) and two gauging stations in Mozambique (obtained from ARA-Sul) were used in this study. A long time series of flow data was not available in Swaziland. Based on the quality of the data, time series length, influence of infrastructure (dams, canals) and spatial distribution, 33 stations were selected for detailed analysis (Table 3 and Fig. 2). As this catchment is highly modified, very few stations could be considered not impacted by human interventions. Data from pristine catchments can reveal the dynamics of natural variability of streamflow and isolate the impacts of climate change on streamflow. An analysis of the indicators of hydrologic alteration was conducted to identify patterns and trends of the streamflow record (a single period analysis for the entire time series and for the period of 1970–2011), as well as to assess the impact of infrastructure on the streamflow (two-period analysis, before and after the major infrastructure development).

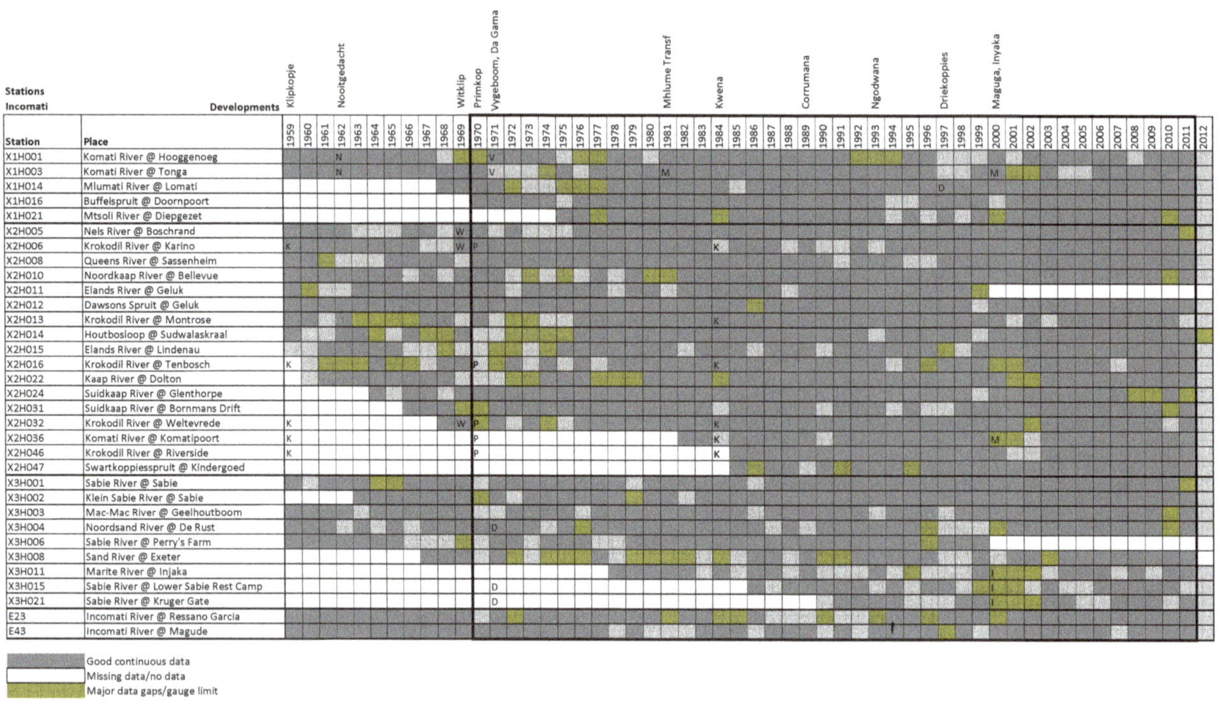

Figure 2. Streamflow data used on this study, with indication of time series length, data quality, and missing data. Major developments in the basin, such as dams, are on the top horizontal line in the year they were commissioned; indication is made of the gauges affected by the developments by the initial letter of the dam.

2.2.3 Indicators of hydrologic alteration

The US Nature Conservancy developed a statistical software program known as the Indicators of Hydrologic Alteration (IHA) for assessing the degree to which human activities have changed flow regimes. The IHA method (Richter et al., 1996, 2003; Richter and Thomas, 2007) is based upon the concept that hydrologic regimes can be characterised by five ecologically relevant attributes, listed in Table 4: (1) magnitude of monthly flow conditions; (2) magnitude and duration of extreme flow events (e.g. high and low flows); (3) the timing of extreme flow events; (4) frequency and duration of high low flow pulses; and (5) the rate and frequency of changes in flows. It consists of 67 parameters, which are subdivided into two groups – 33 IHA parameters and 34 environmental flow component parameters. These hydrologic parameters were developed based on their ecological relevance and their ability to reflect human-induced changes in flow regimes across a broad range of influences including dam operations, water diversions, groundwater pumping, and landscape modification (Mathews and Richter, 2007). 33 selected gauges from the Incomati basin were analysed with this method using daily flow data. Many studies successfully applied the methodology of Indicators of Hydrologic Alteration, in order to assess impacts on streamflow caused by anthropogenic drivers (Maingi and Marsh, 2002; Taylor et al., 2003; Mathews and Richter, 2007; De Winnaar and Jewitt, 2010; Masih et al., 2011). In the case of the present study, the indicators of the magnitude of monthly flow, the magnitude and duration of extreme flow, as well as timing, were analysed for the period 1970–2011, to assess whether consistent trends of increase or decrease of the hydrological indicators were present.

The IHA software was used to identify trends of the streamflow time series, based on the regression of least squares. This trend is evaluated with the P value, and only trends with $P \leq 0.05$ were considered significant trends, the value of the slope of the trend line indicating an increasing or decreasing trend. This information was compiled and mapped for the various hydrological indicators using ArcGIS 9.3.

2.2.4 Land-use analysis

Land use was analysed based on secondary data because most of the current forestry plantations were already established on the 1970s, when remote sensing maps became available. Additionally, a map of current land use (2011) (Jarmain et al., 2013) and the land use of 2000 (Van den Berg et al., 2008) were compared with the maps of trends of indicators of hydrologic alteration. Where the occurrence of trends in a flow regime was consistent with the changes in land use, this the temporal evolution of the land-use changes was further investigated.

Table 3. Hydrometric stations analysed, location, catchment area, data length and missing data.

	Station	Latitude	Longitude	River and location	Catchment area (km^2)	Data available	Period analysed for trends	Missing data
Komati	X1H001	−26.04	31.00	Komati River, Hooggenoeg	5499	1909–2012	1970–2011 (42 years)	8.0 %
	X1H003	−25.68	31.78	Komati River, Tonga	8614	1939–2012	1970–2011 (42 years)	6.8 %
	X1H014	−25.67	31.58	Mlumati River, Lomati	1119	1968–2012	1978–2011 (34 years)	0.5 %
	X1H016	−25.95	30.57	Buffelspruit, Doornpoort	581	1970–2012	1970–2011 (42 years)	3.4 %
	X1H021	−26.01	31.08	Mtsoli River, Diepgezet	295	1975–2012	1976–2011 (36 years)	2.7 %
Crocodile	X2H005	−25.43	30.97	Nels River, Boschrand	642	1929–2012	1970–2011 (42 years)	0.8 %
	X2H006	−25.47	31.09	Krokodil River, Karino	5097	1929–2012	1970–2011 (42 years)	0.1 %
	X2H008	−25.79	30.92	Queens River, Sassenheim	180	1948–2012	1970–2011 (42 years)	0.5 %
	X2H010	−25.61	30.87	Noordkaap River, Bellevue	126	1948–2012	1970–2011 (42 years)	5.7 %
	X2H011	−25.65	30.28	Elands River, Geluk	402	1956–1999	1957–1999 (43 years)	0.9 %
	X2H012	−25.66	30.26	Dawsons Spruit, Geluk	91	1956–2012	1970–2011 (42 years)	0.3 %
	X2H013	−25.45	30.71	Krokodil River, Montrose	1518	1959–2012	1970–2011 (42 years)	1.6 %
	X2H014	−25.38	30.70	Houtbosloop, Sudwalaskraal	250	1958–2012	1970–2011 (42 years)	5.1 %
	X2H015	−25.49	30.70	Elands River, Lindenau	1554	1959–2012	1970–2011 (42 years)	3.1 %
	X2H016	−25.36	31.96	Krokodil River, Tenbosch	10 365	1960–2012	1970–2011 (42 years)	5.6 %
	X2H022	−25.54	31.32	Kaap River, Dolton	1639	1960 2012	1970–2011 (42 years)	5.7 %
	X2H024	−25.71	30.84	Suidkaap River, Glenthorpe	80	1964–2012	1970–2011 (42 years)	1.7 %
	X2H031	−25.73	30.98	Suidkaap River, Bornmans Drift	262	1966–2012	1966–2011 (46 years)	5.0 %
	X2H032	−25.51	31.22	Krokodil River, Weltevrede	5397	1968–2012	1970–2011 (42 years)	2.4 %
	X2H036	−25.44	31.98	Komati River, Komatipoort	21 481	1982–2012	1983–2011 (28 years)	4.1 %
	X2H046	−25.40	31.61	Krokodil River, Riverside	8473	1985–2012	1986–2011 (26 years)	2.0 %
	X2H047	−25.61	30.40	Swartkoppiesspruit, Kindergoed	110	1985–2012	1986–2011 (26 years)	2.2 %
Sabie	X3H001	−25.09	30.78	Sabie River, Sabie	174	1948–2012	1970–2011 (42 years)	0.8 %
	X3H002	−25.09	30.78	Klein Sabie River, Sabie	55	1963–2012	1970–2011 (42 years)	0.4 %
	X3H003	−24.99	30.81	Mac-Mac River, Geelhoutboom	52	1948–2012	1970–2011 (42 years)	0.5 %
	X3H004	−25.08	31.13	Noordsand River, De Rust	200	1948–2012	1970–2011 (42 years)	3.9 %
	X3H006	−25.03	31.13	Sabie River, Perry's Farm	766	1958–2000	1970–1999 (30 years)	2.6 %
	X3H008	−24.77	31.39	Sand River, Exeter	1064	1967–2011	1968–2011 (43 years)	15.5 %
	X3H011	−24.89	31.09	Marite River, Injaka	212	1978–2012	1979–2011 (32 years)	7.6 %
	X3H015	−25.15	31.94	Sabie River, Lower Sabie Rest Camp	5714	1986–2012	1988–2011 (24 years)	8.2 %
	X3H021	−24.97	31.52	Sabie River, Kruger Gate	2407	1990–2012	1991–2011 (21 years)	10.8 %
Lower Incomati	E23	−25.44	31.99	Incomati River, Ressano Garcia	21 200	1948–2011	1970–2011 (42 years)	9.0 %
	E43	−25.03	32.65	Incomati River, Magude	37 500	1952–2011	1970–2011 (42 years)	3.5 %

Table 4. Hydrologic parameters used in the range of variability approach (Richter et al., 1996).

Indicators of hydrologic alteration group	Regime characteristics	Hydrological parameters
Group 1: magnitude of monthly water conditions	Magnitude timing	Mean value for each calendar month
Group 2: magnitude and duration of annual extreme water conditions	Magnitude duration	Annual minima and maxima based on 1, 3, 7, 30 and 90 day means
Group 3: timing of annual extreme water conditions	Timing	Julian date of each annual 1 day maximum and minimum
Group 4: frequency and duration of high/low pulses	Frequency and duration	No. of high and low pulses each year Mean duration of high and low pulses within each year (days)
Group 5: rate/frequency of water condition changes	Rates of change of frequency	Means of all positive and negative differences between consecutive daily values No. of rises and falls

Table 5. Description of rainfall stations analysed for trends, the long-term mean annual precipitation (MAP) in mm a^{-1}, the standard variation, and the detection of the trend (confidence level of 95 % using the Spearman test) and occurrence change point (using the Pettitt test followed by the T test of stability of the mean and the F test of the stability of variance).

							Analysis for the period 1950 to 2011			
Name	Station ID	Latitude	Longitude	Altitude (m a.s.l.)	MAP (mm)	P reliable (%)	Mean (mm a^{-1})	SD (mm a^{-1})	Trend Spearman	Pettitt
Machadodorp	0517430 W	−25.67	30.25	1563	781	79.6	773	134		
Badplaas (Pol)	0518088 W	−-25.97	30.57	1165	829	90.6	817	153		
Kaapsehoop	0518455 W	−25.58	30.77	1564	1443	78.5	1461	286		Decr (1975)
Mac-Mac (Bos)	0594539 W	−24.98	30.82	1295	1463	75.1	1501	287		
Spitskop (Bos)	0555579 W	−25.15	30.83	1395	1161	68.5	1197	266	Decr	Decr (1978)*
Alkmaar	0555567 W	−25.45	30.83	715	830	95.2	874	172		
Oorschot	0518859 W	−25.80	30.95	796	787	92.2	775	185		
Bosbokrand (Pol)	0595110 W	−24.83	31.07	778	982	82.4	919	297		Decr (1978)*
Pretoriuskop	0556460 W	−25.17	31.18	625	707	60.0	734	188		
Riverbank	0519310 W	−25.67	31.23	583	683	70.5	782	163	Incr	Incr (1977)**
Piggs Pig	0519448 A	−25.97	31.25	1029	1024	40.1	1075	315	Decr	Decr (1978)*
Skukuza	0596179 W	−25.00	31.58	300	560	63.1	566	140		
Riverside	0557115 W	−25.42	31.60	315	547	66.5	520	187		
Satara	0639504 W	−24.40	31.78	257	568	42.1	602	151	Incr	Incr (1971)
Fig Tree	0520589 W	−25.82	31.83	256	591	63.4	594	145	Decr	Decr (1978)*
Tsokwane	0596647 W	−24.78	31.87	262	540	66.1	544	134		Incr (1971)*
Krokodilbrug	0557712 W	−25.37	31.90	192	584	62.9	590	147		
Moamba	P821 M	−25.60	32.23	108	632	63.9	633	185		
Xinavane	P10 M	−25.07	32.87	18	853	76.2	773	241		
Manhica	P63 M	−25.40	32.80	33	883	86.2	903	275		Incr (1970)**

* Significant change with 2.5 % significance level with T test of stability of mean. ** Significant change with 2.5 % significance level with T test of stability of mean and F Test of stability of variance. Explanatory note: MAP is the mean annual precipitation, and P reliable is the percentage of reliable data for the rainfall station, as assessed by Lynch (2003) for the period 1905 to 1999. The mean refers to the average of the total annual precipitation for the period of 1950 to 2011. In the trend Spearman column, only stations that had trends significant at the 95 % confidence level are indicated with Decr or Incr, corresponding to a decreasing or increasing trend, respectively. In the Pettitt column, the direction of change and year are indicated, as well as the significance of the change point.

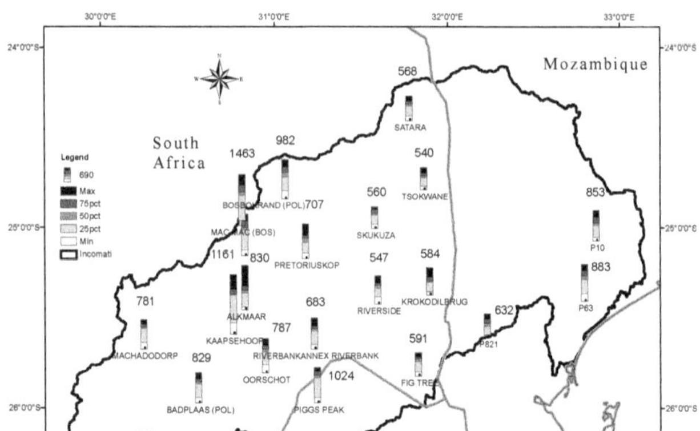

Figure 3. Map illustrating the spatial variation of annual rainfall across the Incomati basin (minimum, 25 %, median, 75 %, and maximum annual precipitation are shown in the bar; the number on top of the bar is the Mean Annual Precipitation for the period 1950–2000).

3 Results

3.1 Rainfall

Data series of 20 rainfall stations were statistically analysed for the period 1950–2011 (Table 5). The variability of rainfall across the basin was confirmed to be high, both intra- and inter-annually, with a wide range between years. This variability is highest for the stations located in mountainous areas. The variability across the basin is also significant, as illustrated by the boxplot of Fig. 3.

The Spearman trend test revealed that only 5 of the 20 investigated stations showed significant trends of increase (2 stations) and decrease (3 stations). However, the stations that presented significant trends are also stations with a lower percentage of reliability; thus, it is possible that the trend identified could be affected by data infilling procedures. There was no serial correlation of annual and monthly time series. Some change points were identified using the Pettitt test, mostly in the years 1971 and 1978 (Table 5). Only 2 stations out of the 20 studied showed significant change towards a wetter regime (Riverbank and Manhica).

Monthly rainfall also does not exhibit any clear trend at most stations. This is consistent with the larger-scale analyses conducted by Schulze (2012) for South Africa and Shongwe et al. (2009) for southern Africa.

Table 6. Hydrological indicators of main sub-catchments.

Streamflow indicators		Komati X1H003 – Tonga		Crocodile X2H016 – Tenbosh		Incomati X2H036 – Komatipoort		Sabie X3H015 – Lower Sabie		Incomati E43 – Magude	
Period of analysis:	Units	1970–2011 (42 years)		1970–2011 (42 years)		1983–2011 (28 years)		1988–2011 (24 years)		1970–2011 (42 years)	
Drainage area	km^2	8614		10 365		21 481		5714		37 500	
		Median	CD**	Median	CD**	Median	CD**	Median	CD**	Median	CD**
Annual*	$m^3\,s^{-1}$	16.94	2.14	21.35	1.97	34.28	2.11	17.35	2.31	47.44	2.01
October	$m^3\,s^{-1}$	3.95	1.47	2.54	1.88	2.24	1.87	3.08	0.92	8.72	1.21
November	$m^3\,s^{-1}$	5.72	1.94	5.75	2.35	7.09	3.88	4.81	1.09	16.14	1.49
December	$m^3\,s^{-1}$	11.46	2.09	15.07	1.48	18.79	2.63	10.83	1.49	22.91	2.90
January	$m^3\,s^{-1}$	17.26	1.82	20.68	1.47	34.47	1.52	18.52	1.35	37.96	1.35
February	$m^3\,s^{-1}$	25.09	1.95	31.37	2.01	29.77	2.80	16.33	1.84	45.09	3.21
March	$m^3\,s^{-1}$	18.33	1.74	27.15	1.63	42.15	1.90	19.51	2.30	51.75	2.32
April	$m^3\,s^{-1}$	11.64	1.74	19.82	1.37	24.10	2.13	13.69	1.13	34.90	2.03
May	$m^3\,s^{-1}$	8.03	1.41	9.11	1.68	9.98	2.16	7.04	1.64	17.85	1.86
June	$m^3\,s^{-1}$	4.96	1.90	5.66	1.62	7.10	2.45	5.64	1.25	14.04	1.44
July	$m^3\,s^{-1}$	3.77	1.98	4.56	1.48	4.72	2.28	3.79	1.18	10.41	1.47
August	$m^3\,s^{-1}$	2.67	1.63	2.63	1.71	2.51	1.35	3.40	1.08	8.46	1.41
September	$m^3\,s^{-1}$	2.43	1.47	2.08	1.81	2.24	1.51	2.69	1.15	7.06	1.11
1 day minimum	$m^3\,s^{-1}$	0.31	4.04	0.24	2.64	0.14	5.29	1.45	1.13	2.49	1.48
3 day minimum	$m^3\,s^{-1}$	0.38	3.38	0.32	2.16	0.25	3.76	1.53	1.08	2.71	1.76
7 day minimum	$m^3\,s^{-1}$	0.59	2.55	0.40	2.88	0.33	4.35	1.60	1.16	3.01	1.61
30 day minimum	$m^3\,s^{-1}$	1.46	2.13	1.52	1.79	1.29	2.08	2.01	1.12	4.84	1.37
90 day minimum	$m^3\,s^{-1}$	3.69	1.47	3.45	1.34	3.17	2.09	3.02	1.23	8.14	1.38
1 day maximum	$m^3\,s^{-1}$	134.4	1.26	142.2	1.38	274.3	1.00	113	2.51	381.5	1.80
3 day maximum	$m^3\,s^{-1}$	102.9	1.50	126.9	1.33	232.9	1.15	87.62	2.60	344.1	1.74
7 day maximum	$m^3\,s^{-1}$	81.79	1.59	107.4	1.20	201.4	1.13	62.55	2.27	273.7	1.56
30 day maximum	$m^3\,s^{-1}$	54.39	1.45	76.98	1.28	109.6	1.33	37.66	1.93	156.7	1.45
90 day maximum	$m^3\,s^{-1}$	39.19	1.33	45.08	1.16	68.69	1.71	28.06	1.47	102	1.32
Date of minimum	Julian Date	275	0.10	274	0.12	281.5	0.15	278.5	0.06	290.5	0.21
Date of maximum	Julian Date	38.5	0.16	33	0.11	35.5	0.19	20.5	0.17	39.5	0.14
Low pulse count	No.	6	1.63	4	1.63	5	1.55	4	1.00	3	1.33
Low pulse duration	Days	5.5	1.41	5	1.60	3.5	0.71	6.5	1.69	6.75	2.09
High pulse count	No.	6	0.75	4	1.25	5	0.95	4	0.69	4	0.75
High pulse duration	Days	4	1.31	4	2.13	4.5	1.28	5	2.10	8.5	1.03
Rise rate	$m^3\,s^{-1}$	0.7095	1.39	0.64	0.98	1.161	1.38	0.404	1.12	1.058	1.43
Fall rate	$m^3\,s^{-1}$	−0.7295	−0.98	−0.61	−0.78	−1.38	−1.28	−0.2398	−1.10	−0.6278	−2.31
Number of reversals	No	111.5	0.26	113	0.42	121	0.18	95	0.29	86	0.49

* In the annual statistics, the mean and coefficient of variation were used. **CD is the coefficient of dispersion. More details about CD are available in the text.

Mussá et al. (2013) studied the trends of annual and dry extreme rainfall using the Standardized Precipitation Index (SPI), and also found no significant trends in annual rainfall extremes across the Crocodile sub-catchment.

3.2 Variability of streamflow

The metrics of the different hydrologic indicators were compiled as an output of the IHA analysis, which is illustrated for the gauging stations located at the outlet (or the most downstream) of each main sub-catchment in Table 6. The variability is described, using non-parametric statistics (median and coefficient of dispersion), because the hydrological time series are not normally distributed, but are positively skewed. The coefficient of dispersion (CD) is defined as CD = (75th percentile − 25th percentile)/50th percentile. The larger the CD, the larger the variation of the parameter will be.

The flow patterns are consistent with the summer rainfall regime, with the highest flow and rainfall events associated with tropical cyclone activity in January–March.

A comparison of the flow normalised by area (Fig. 4) for the main sub-catchments reveals that the Sabie yields a higher runoff than the Komati and the Crocodile. This is the case because the observed streamflows include the impact of water abstractions and streamflow reduction activities

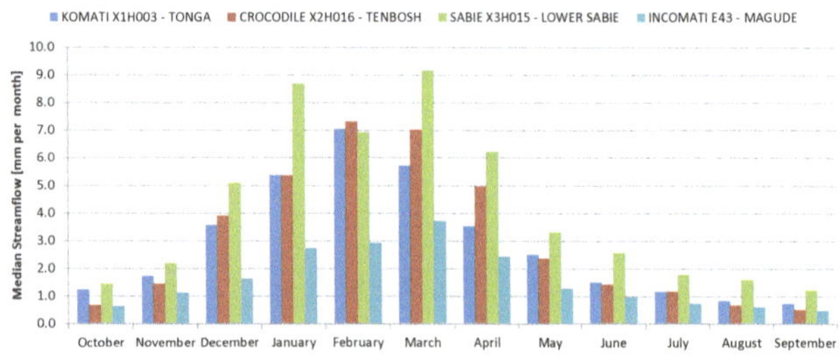

Figure 4. Median of observed daily streamflow for the gauges located at the outlets of the major sub-catchments: Komati, Crocodile, Lower Sabie and Incomati (based on daily time series from 1970 to 2011).

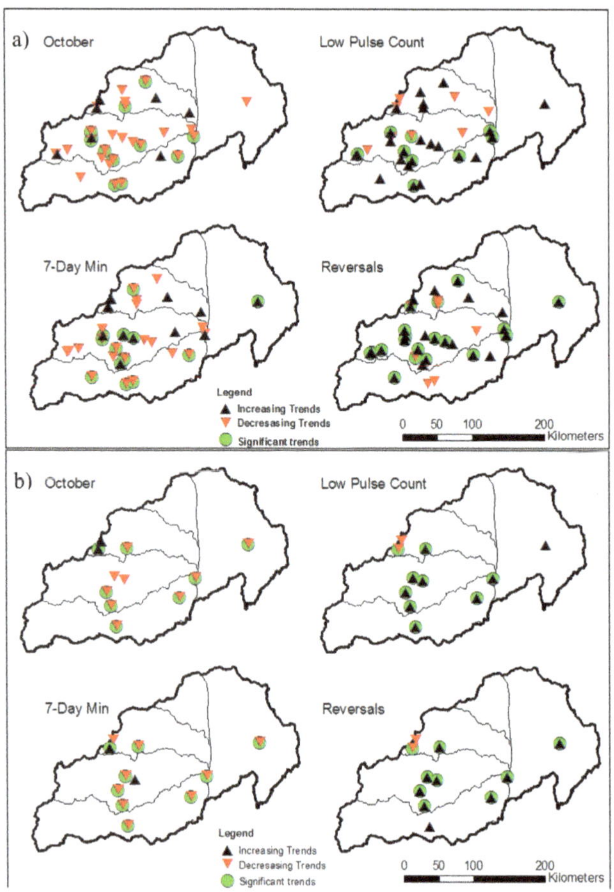

Figure 5. Trends of different indicators of streamflow: **(a)** for period 1970–2011; **(b)** for period 1950–2011.

(Hughes and Mallory, 2008; DWAF, 2009a), which are more intense in the Komati and Crocodile sub-catchments.

Another aspect to note is that the flows of February are likely to be higher than observed records, but high streamflow extremes are not fully captured by the current monitoring network, due to gauge limits.

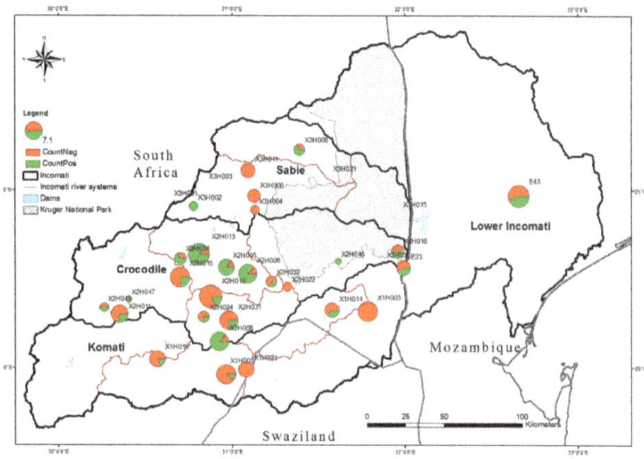

Figure 6. Count of significant trends. Declining trends are in red and increasing trends in green. The size of the pie is proportional to the total number of significant trends.

3.3 Trends in streamflow

Figure 5 presents a spatial plot of trends for selected hydrological indicators for the periods 1970–2011 (Fig. 5a) and 1950–2011 (Fig. 5b). The significant trends are highlighted with a circle. Table 7 presents the slope of the trend lines and P values for the gauges located at the outlet, or the most downstream point of each main sub-catchment. There is a significant trend of decreasing mean flow in October at almost all stations, especially the ones located on the main stem of the Crocodile and Komati rivers. October is the month of the start of the rainy season, when the dam levels are lowest and irrigation water requirements highest (DWAF, 2009a; ICMA, 2010).

This trend is consistent with the decreasing trends of minimum flows, as exemplified by the 7 day minimum. In contrast, it can be seen that the count of low pulses increased significantly in many gauges, which indicates the more frequent occurrence of low flows. Another striking trend is the significant increase in the number of reversals at almost all stations. Reversals are calculated by dividing the hydrologic record into "rising" and "falling" periods, which correspond to periods in which daily changes in flows are either positive or negative, respectively. The number of reversals is the number of times that the flow switches from one type of period to another. The observed increased number of reversals is likely due to the effect of flow regulation and water abstractions.

The significant trends (95 % confidence level) of the various indicators were counted per station and plotted on a map (Fig. 6). Most significant decreasing trends occur in the Komati and Crocodile systems, which are also the most stressed sub-catchments. An interesting aspect is that some of the trends cross-compensate each other. Some of the positive trends occurring in the tributaries of the Crocodile, for example, the October median flow and the baseflow, are cancelled out when moving down the main stem of the river.

Table 7. Trends of the hydrological indicators for the period 1970–2011. In bold are significant trends at the 95 % confidence level.

Streamflow indicators	Komati X1H003 – Tonga		Crocodile X2H016 – Tenbosh		Sabie X3H015 – Lower Sabie		Incomati X2H036 – Komatipoort		Incomati E43 – Magude	
Period of analysis	1970–2011 (42 years)		1970–2011 (42 years)		1988–2011 (24 years)		1983–2011 (28 years)		1970–2011 (42 years)	
Drainage area (km^2)	8614		10 365		5714		21 481		37 500	
	Slope	*P* value	Slope	*P* value	Slope	*P* value	Slope	*P* value	Slope	*P* value
October	−0.285	0.05	−0.052	0.5	0.017	0.5	−0.017	0.5	−0.313	0.25
November	−0.254	0.1	−0.006	0.5	0.263	0.5	0.020	0.5	−0.165	0.5
December	−0.194	0.5	−0.090	0.5	0.199	0.5	0.783	0.5	−0.087	0.5
January	−0.437	0.5	−0.023	0.5	1.493	0.25	1.979	0.25	−0.960	0.5
February	−1.027	0.1	−0.927	0.25	0.544	0.5	−0.486	0.5	**−2.847**	0.05
March	−0.360	0.5	−0.397	0.5	0.390	0.5	−0.112	0.5	−1.346	0.5
April	−0.082	0.5	−0.007	0.5	0.899	0.25	1.532	0.25	−0.195	0.5
May	−0.225	0.1	−0.045	0.5	0.416	0.5	0.788	0.5	−0.365	0.5
June	**−0.215**	0.025	0.059	0.5	0.270	0.5	0.470	0.5	−0.045	0.5
July	**−0.179**	0.005	0.060	0.5	0.219	0.5	0.171	0.5	−0.039	0.5
August	−0.074	0.1	0.105	0.5	0.134	0.25	0.312	0.5	0.090	0.5
September	−0.029	0.5	0.134	0.5	0.081	0.5	0.218	0.5	0.166	0.25
1 day minimum	**−0.027**	0.025	−0.015	0.25	0.061	0.1	0.003	0.5	**0.139**	0.001
3 day minimum	**−0.029**	0.025	−0.015	0.25	0.061	0.1	0.004	0.5	**0.127**	0.005
7 day minimum	**−0.038**	0.05	−0.015	0.5	0.064	0.1	0.004	0.5	0.094	0.05
30 day minimum	**−0.069**	0.025	−0.025	0.25	0.058	0.25	0.033	0.5	0.054	0.5
90 day minimum	**−0.115**	0.01	−0.059	0.25	0.131	0.1	0.038	0.5	−0.054	0.5
1 day maximum	−5.143	0.25	−5.425	0.25	−2.743	0.5	−12.070	0.25	**−10.580**	0.025
3 day maximum	−3.749	0.25	−3.670	0.25	−1.379	0.5	−8.171	0.5	**−9.254**	0.025
7 day maximum	−2.361	0.25	−2.427	0.25	0.014	0.5	−3.742	0.5	**−6.722**	0.05
30 day maximum	−1.022	0.25	−1.023	0.25	0.662	0.5	0.092	0.5	−3.400	0.1
90 day maximum	−0.671	0.25	−0.576	0.5	0.789	0.5	0.934	0.5	−2.147	0.25
Number of zero days	0.690	0.25	−0.005	0.5	0	0.5	0.032	0.5	−0.080	0.5
Base flow index	−0.001	0.25	0.000	0.5	0.004	0.5	0.001	0.25	**0.007**	0.001
Date of minimum	−0.686	0.5	0.354	0.5	0.548	0.5	−0.420	0.5	1.374	0.5
Date of maximum	0.817	0.5	0.347	0.5	−3.222	0.5	0.288	0.5	0.617	0.5
Low pulse count	0.132	0.1	**0.238**	0.001	−0.045	0.5	0.185	0.5	0.043	0.25
Low pulse duration	0.068	0.5	−0.140	0.5	−0.669	0.1	−0.297	0.25	−0.602	0.5
High pulse count	**−0.127**	0.005	0.007	0.5	−0.023	0.5	−0.096	0.25	−0.068	0.05
High pulse duration	0.029	0.5	**−1.263**	0.01	1.081	0.25	0.144	0.5	−0.103	0.5
Rise rate	−0.007	0.5	−0.008	0.5	0.005	0.5	0.017	0.5	**−0.034**	0.05
Fall rate	0.003	0.5	**−0.013**	0.05	−0.007	0.5	−0.012	0.5	−0.007	0.5
Number of reversals	0.574	0.1	**1.083**	0.01	0.723	0.5	0.560	0.5	**0.764**	0.005

The cross-compensation can also be observed on a basin scale in the Sabie, where the trends of decreasing flows are not so frequent or significant. It is likely that this occurs because the majority of the Sabie falls under the conservation area of the Kruger National Park (KNP), and therefore fewer abstractions occur compared to other sub-catchments, as illustrated in Table 1. The KNP has been playing an important role in the catchment management fora set up by the Inkomati Catchment Management Agency (ICMA), which concern the provision of environmental minimum flows, in order to maintain ecosystem services and biodiversity in the park (Pollard et al., 2012; Riddell et al., 2013).

Table 7 illustrates that many of the trends observed in the Sabie sub-catchment contrast with those observed in the Komati and Crocodile sub-catchments. Thus, the trends observed in downstream Magude (station E43) in Mozambique are the result of a combination of the positive effect of the conservation approach of KNP on the Sabie, and the negative effect of flow reductions on the Crocodile and the Komati.

The Komati sub-catchment (at Tonga gauge, X1H003) is where most negative trends occur, particularly significant during the months of October, June and July (Table 7 and Fig. 6). At the downstream end of the Crocodile (at Tenbosch gauge, X2H016), similar trends are not visible, because of cross-compensations: the Kaap and Elands tribu-

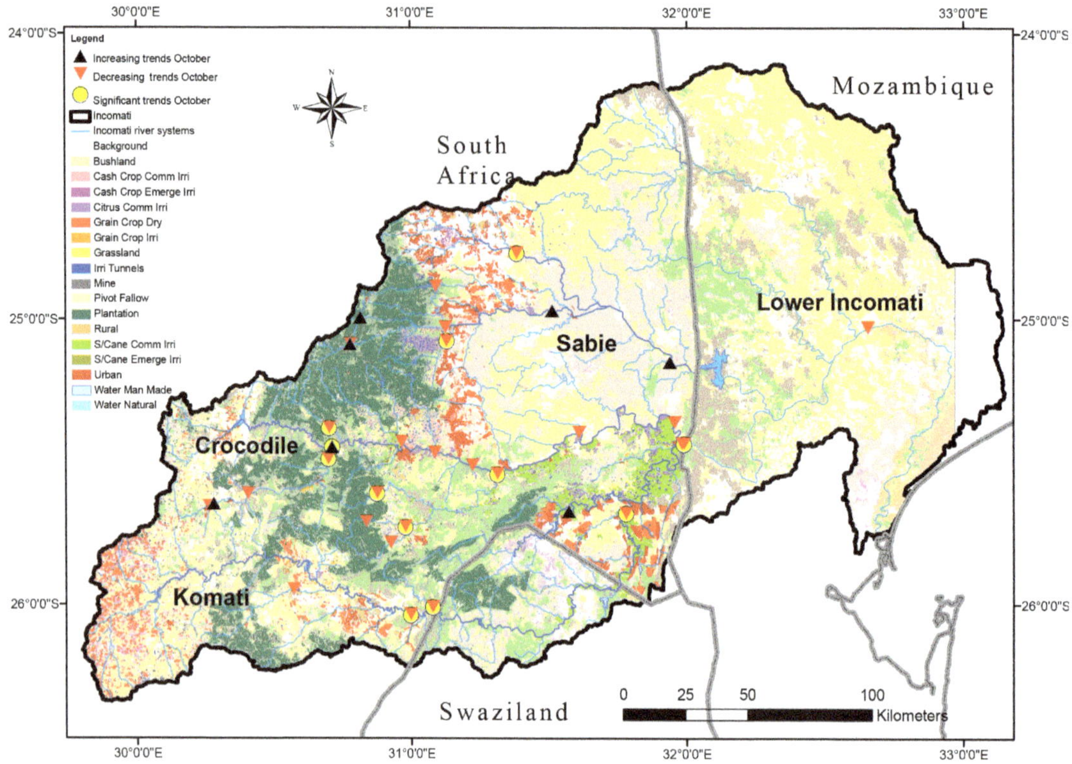

Figure 7. Land-use land-cover map of Incomati (ICMA, 2010; TPTC, 2010) and streamflow trends in the month of October.

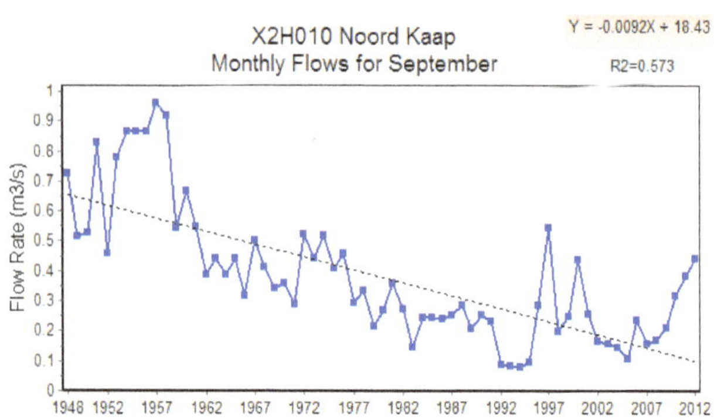

Figure 8. Plot of median monthly flows for September for the entire time series (1949–2011) on the Noord Kaap gauge, located in the Crocodile sub-catchment.

taries both have significant decreasing trends of their mean monthly flows, as well as the low flows; the Kwena Dam, located on the main stem of the Crocodile, on the other hand, is managed in a way so as to augment the flows during the dry season.

It is important to note that these trends are even more pronounced when longer time series are considered. Two examples from the Crocodile basin are presented below.

3.3.1 Example of decreasing trends: Noord Kaap X2H010

The Noord Kaap gauge (X2H010), located on a tributary of the Crocodile sub-catchment, displays the most intriguing trends. Out of the 33 IHA indicators, this gauge had 12 significant trends, 10 of them negative, indicating a major shift in flow regime. The decreasing trends occur in all months, but are more pronounced during low flow months, particularly September (Fig. 8) and October. There is a significant decrease in high flows and small floods and an increase in extreme low flows. However, there is no record of the presence of a dam or major infrastructure (DWAF, 2009d). The areal rainfall for the drainage area of this gauge does not show a significant decreasing trend, which suggests that the reduction observed in streamflow should be a result of land-use change, namely, conversion to forestry and irrigated land. Fig. 9 illustrates the comparison of median monthly flows for the two periods. From the analysis of land-use changes over time (Table 2), it can be seen that the sharp decrease in mean monthly flows during the 1960s coincides with an increase in the area under irrigated agriculture. During the 1970s, there was also a great increase in area under forestry, namely, eucalyptus (DWAF, 2009a). Commercial forestry consumes more water through evaporation than the native vegetation it replaces; therefore, under the South African National Water Act, it must be licensed as a water user, which is termed a

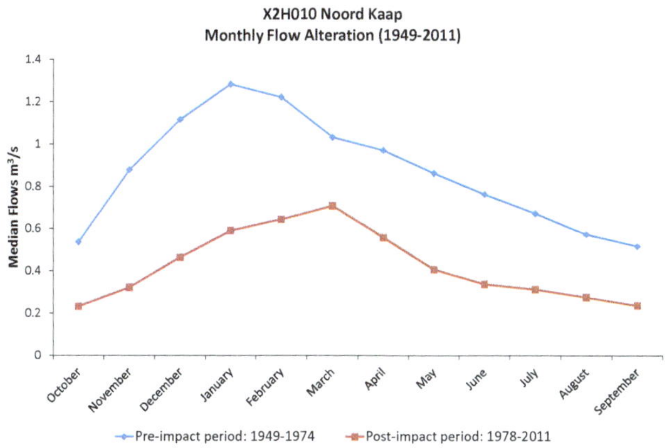

Figure 9. Plot of median monthly flows for two periods (1949–1974 and 1978–2011) on the Noord Kaap gauge, located in the Crocodile sub-catchment.

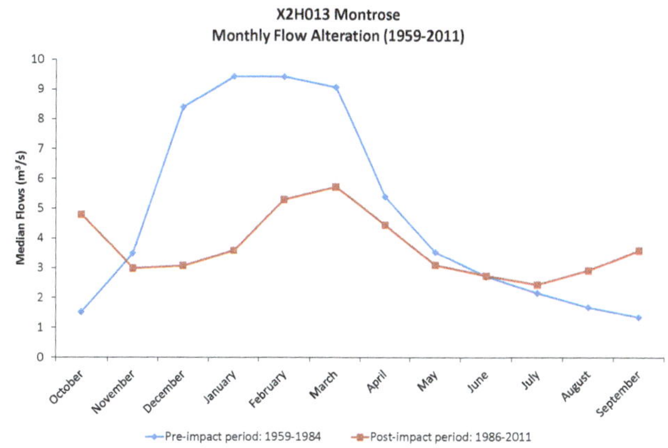

Figure 10. Impact of the Kwena Dam (commissioned in 1984) on streamflows of the Crocodile River, Montrose gauge X2H013.

streamflow reduction activity (SFRA) (Jewitt, 2002, 2006b). A recent study by van Eekelen et al. (2015) finds that streamflow reduction due to forest plantations may be 2 or even 3 times more than that allowed by the Interim IncoMaputo Agreement.

3.3.2 Impact of the Kwena Dam on streamflows of the Crocodile River

The Kwena Dam is the main reservoir in the Crocodile system, located upstream in the catchment and commissioned in 1984. The dam is used to improve the assurance of supply of water for irrigation purposes in the catchment. The Montrose gauge (X2H013) is located a few kilometres downstream of this dam. The two-period (1959–1984 and 1986–2011) analysis illustrates the main impacts of the Kwena Dam on the river flow regime, namely the dampening of peak flows and an increase in low flows (Fig. 10). These results are consistent with the analysis conducted by Riddell et al. (2013), which found significant alterations of the natural flow regime in the Crocodile basin over the past 40 years. Similar impacts were found in studies in different parts of the world (Richter et al., 1998; Bunn and Arthington, 2002; Maingi and Marsh, 2002; Birkel et al., 2014).

It can be seen that this reservoir is managed to augment the low flows and attenuate floods. This change in the flow regime influences the streamflow along the main stem of the Crocodile River, but as tributaries join and water is abstracted, the effect is reduced. At the outlet at Tenbosch gauge X2H016 (Fig. 6 and Table 7), the effects of flow regulation and water abstractions have counter-balanced the contrasting trends observed upstream.

3.3.3 Impact of anthropogenic actions

As can be seen from water-use information, the impacts of land-use change and water abstractions are the main drivers of changes in the flow regime on the Incomati. However, the situation is variable along the catchment. In the Sabie system, in spite of large areas of commercial forestry in the headwaters, the indicators of mean, annual and low flows do not show significant trends (Table 7). This can be explained by the fact that most of the forestry area was already established during the period of analysis (1970–2011) (DWAF, 2009c). The fact that a large proportion of the Sabie sub-catchment is under conservation land use (KNP and other game reserves) also plays an important role in maintaining the natural flow regime.

On the Crocodile, however, irrigated agriculture, forestry and urbanisation were the most important anthropogenic drivers. They affect the streamflow regime, the water quantity and possibly the water quality as well (beyond the scope of this analysis). This has important implications when environmental flow requirements and minimum cross-border flows need to be adhered to. Pollard and du Toit (2011a) and Riddell et al. (2013) have demonstrated that the Crocodile River is not complying with environmental flow requirements during most of the dry season at the outlet.

On the Komati, the strategic water uses, which have first priority (such as the water transfers to ESKOM plants in the Olifants catchment and to irrigation schemes in the Umbeluzi) (Nkomo and van der Zaag, 2004; DWAF, 2009b), have a high impact on streamflows. Because of other water allocations, for irrigation, forestry and other industries, steady trends of decreasing flows could be identified. This is another system where the environmental flows and cross-border requirements are often not met during the dry season (Pollard and du Toit, 2011a; Mukororira, 2012; Riddell et al., 2013).

4 Discussion

4.1 Limitations of this study

The available data series have some gaps, especially during high flow periods. Because of this, the analysis of high flow extremes is highly uncertain. For the trend analysis, the period of common data followed the construction of several impoundments and other developments.

Another challenge is the disparity of data availability across the different riparian countries. In Mozambique, only two gauges had reliable flow data for this analysis, representing the entire lower Incomati system. The rivers Massintoto, Uanetse and Mazimechopes in Mozambique do not have active flow gauges. There is definitely a need to strengthen the hydrometric monitoring network in the Mozambican part of the basin, as well as in the tributaries originating in the Kruger National Park.

4.2 What are the most striking trends and where do they occur?

The analysis resulted in the identification of major trends, including

- decreasing trends in the magnitude of monthly flow (significant for low flow months, e.g. October), minimum flow (1, 3, 7, 30 and 90 day minima) and the occurrence of high flow pulses,
- significant increasing trends of the magnitude of monthly flow (August and September) in some locations in the Crocodile and Sabie, and in the occurrence of flow reversals basin-wide, and
- Some gauges showed no significant change or no clear pattern of change in the parameters analysed. These are mainly gauges located on the Sabie, which by 1970 had already established the current land use.

In the Komati system, the flow regulation and water abstractions have strong impacts on streamflow. Most gauges are severely impacted and it is quite difficult to characterise natural flow conditions. Flow regulation has the largest impact on low flow and minimum flows. In the Komati, irrigated agriculture is significant, particularly sugarcane. The upstream dams of Nooitgedacht and Vygeboom are mainly used to supply cooling water to ESKOM power stations outside the basin; thus, this water is exported and not used within the basin.

In the Crocodile system, flow regulation by the Kwena Dam has attenuated extreme flow events. The high flows are reduced and the low flows generally increase, leading to reverse seasonality downstream. Reverse seasonality is the change in timing of hydrograph characteristics, for example, the occurrence of low flows in the wet season or high flows in the dry season. The Kwena Dam is used to improve the assurance of the supply of water for irrigation purposes in the catchment. However, Noord Kaap gauge X2H010, a headwater tributary of the Crocodile system, experiences a significant and dramatic reduction in flows, shown in the monthly flow, and the low flow parameters. This change was compared with the increase in the area under forestry in the sub-catchment, as well as with the increase in irrigation. The comparison revealed that the land-use change was the main driver of the flow alteration.

In the Sabie system, most gauges did not show significant trends. This is most likely due to fewer disturbances compared to the other main catchments: lower water demands, fewer water abstractions and larger areas under conservation. During the period 1970–2011 there were no clear impacts of climatic change (in terms of rainfall) on streamflow.

4.3 Implications of these findings for water resource management

The results of this study illustrate some hotspots where more attention should be put in order to ensure provision of water to society and the environment. When the analysis of trends is combined with the land use of the basin (Fig. 7), it is clear that the majority of gauges with decreasing trends are located in areas where forestry or irrigated agriculture dominates the land use and where conservation approaches are less prevalent. The presence of water management infrastructure (dams) greatly influences the flow regime.

For the management of water resources in the basin, it is important to note some clear patterns, illustrated by the Sabie, Crocodile and Komati. The Sabie flows generated in the upper parts of the catchment persist until the outlet, whilst in other rivers, flows are highly modified. This suggests that the use of the conservation approach through the strategic adaptive management of the Kruger National Park (KNP) and the Inkomati Catchment Management Agency (ICMA), which are stronger on the Sabie, can be very beneficial for keeping environmental flows in the system. It is important to consider not only the magnitude of flows, but their duration and timing as well.

Dams provide storage, generate hydropower and attenuate floods in the basin, but have impacts downstream, such as the change in mean monthly flows, the reversal of seasonality and the trapping of sediments, which can all hamper the health of downstream ecosystems. The recently concluded Mbombela Reconciliation Strategy (Beumer and Mallory, 2014) strongly recommends the construction of new dams in South Africa, including one at Mountain View in the Kaap sub-catchment. The plans for these developments happen when Swaziland is not yet fully utilising its allocation under the Piggs Peak Agreement and the Interim IncoMaputo Agreement (TPTC, 2010). Experiences of other countries around the world show that dam construction has many, often wide-ranging and long-term, social and ecological impacts that often are negative and that frequently are irreversible, including the social upheaval caused by the resettlement of communities, loss of ecosystems and biodiversity, increased sediment trapping, irreversible alteration of flow regimes and the prohibitive cost of decommissioning (see, for an overview, Tullos et al., 2009, and Moore et al., 2010). It is therefore important to explore alternative options fully before deciding on the construction of more large dams. So, alternative possibilities of restoring natural stream flows and/or increasing water storage capacity should be investigated further and adopted. These alternatives could include aquifer storage, artificial recharge, rainfall harvesting, decentralised storage, and reducing the water use of existing uses and users, including irrigation, industry and forest plantations. The operation rules of existing and future dams should also include objectives to better mimic crucial aspects of the system's natural variability.

Given the likely expansion of water demands due to urbanisation and industrial development, it is also important that water demand management and water conservation measures are better implemented in the basin. For example, there could be systems to reward users that use technology to improve their water-use efficiency and municipalities that encourage their users to lower water use.

This study also shows the complexity of water resource availability and variability. The complexity is even more relevant, considering that this is a trans-boundary basin and that there are international agreements regarding minimum cross-border flows and maximum development levels that have to be adhered to (Nkomo and van der Zaag, 2004; Pollard and Toit, 2011a, b; Riddell et al., 2013).

There is a great discrepancy in data availability between different riparian countries. It is important that Mozambique, in particular, improves its monitoring network, in order better to assess the impact of various management activities occurring upstream on the state of water resources. The monitoring of hydrological extremes should receive more attention, with the focus on increasing the accuracy of recording the flood events. The improvement of the monitoring network can be achieved by various means, such as

- water management institutions collaborating more intensely with academic and consultant institutions,
- developing realistic plans to improve monitoring and data management,
- learning from other countries/institutions that have adequate monitoring in place,
- using modern ICT and other technologies, which may become cheaper and more accessible, and
- involving more stakeholders and citizens in data collection.

5 Conclusions

The research conducted reveals the dynamics of streamflow and their drivers in a river basin.

The statistical analysis of rainfall data revealed no consistent significant trend of increase or decrease for the studied period. The analysis of streamflow, on the other hand, revealed significant decreasing trends of streamflow indicators, particularly the median monthly flows of September and October, and low flow indicators. This study concludes that land use and flow regulation are the largest drivers of streamflow trends in the basin. Indeed, over the past 40 years, the areas under commercial forestry and irrigated agriculture have increased over 4 times, increasing the consumptive water use basin-wide.

The study therefore recommends that the strategic adaptive management adopted by the Kruger National Park and the Inkomati Catchment Management Agency should be employed further in the basin. Water demand management and water conservation should be alternative options to the development of dams, and should be investigated further and established in the basin. Land-use practices, particularly forestry and agriculture, have a significant impact on the water quantity of the basin; therefore, stakeholders from these sectors should work closely with the water management institutions when planning for future developments and water allocation plans.

Considering the high spatial variability in the observed changes, no unified approach will work, but specific tailor-made interventions are needed for the most affected sub-catchments and main catchments. Future investigations should conduct a careful basin-wide assessment of benefits derived from water use, and assess the first priority water uses, including commercial forest plantations, which by default are priority users.

Acknowledgements. The authors would like to thank the UNESCO-IHE Partnership Research Fund (UPaRF), through the RISKOMAN project, for funding of this research. The Water Research Commission of South Africa, through project K5/1935, also contributed additional funding, which is appreciated. All partners of the RISKOMAN project (University of KwaZulu-Natal UKZN, ICMA, Komati Basin Water Authority KOBWA, Eduardo Mondlane University), including the reference group, are thanked for their valuable inputs. Streamflow and rainfall data were kindly provided by DWA, UKZN and SAWS.

Edited by: D. Mazvimavi

References

Beumer, J. and Mallory, S.: Water Requirements and Availability Reconciliation Strategy for the Mbombela Municipal Area. Final Reconcilaliation Strategy, Department of Water Affairs, South Africa, 2014.

Birkel, C., Soulsby, C., Ali, G., and Tetzlaff, D.: Assessing the cumulative impacts of hydropower regulation on the flow characteristics of a large atlantic Salmon river system, River Res. Applic., 30, 456–475, doi:10.1002/rra.2656, 2014.

Bunn, S. E. and Arthington, A. H.: Basic Principles and Ecological Consequences of Altered Flow Regimes for Aquatic Biodiversity, Environ. Manage., 30, 492–507, doi:10.1007/s00267-002-2737-0, 2002.

De Winnaar, G. and Jewitt, G.: Ecohydrological implications of runoff harvesting in the headwaters of the Thukela River basin, South Africa, Phys. Chem. Earth, Parts A/B/C, 35, 634–642, doi:10.1016/j.pce.2010.07.009, 2010.

DWAF – Department of Water Affairs and Forestry: Inkomati Water Availability Assessment Study, Water Requirements Volume 1, prepared by: Water for Africa Environmental, Engineering and Management Consultants, SRK Consulting and CPH20, PWMA 05/X22/00/0908, Pretoria, 2009a.

DWAF – Department of Water Affairs and Forestry: Inkomati Water Availability Assessment Study, Main Report, prepared by: Water for Africa Environmental, Engineering and Management Consultants, SRK Consulting and CPH20, PWMA 05/X22/00/0808, Pretoria, 2009b.

DWAF – Department of Water Affairs and Forestry: Inkomati Water Availability Assessment Study, Hydrology of Sabie River Volume 1, prepared by: Water for Africa Environmental, Engineering and Management Consultants, SRK Consulting and CPH20, PWMA 05/X22/00/1608, Pretoria, 2009c.

DWAF – Department of Water Affairs and Forestry: Inkomati Water Availability Assessment Study, Hydrology of Crocodile River Volume 1, prepared by: Water for Africa Environmental, Engineering and Management Consultants, SRK Consulting and CPH20, PWMA 05/X22/00/1508, Pretoria, 2009d.

Fanta, B., Zaake, B. T., and Kachroo, R. K.: A study of variability of annual river flow of the southern African region, Hydrol. Sci. J., 46, 513–524, doi:10.1080/02626660109492847, 2001.

Guzman, J. A. and Chu, M. L.: SPELL-stat v 15110 B. Grupo en Prediccion y Modelamiento Hidroclimatico Universidad Industrial de Santander, Colombia, Hydrological Modelling and Prediction Group, Industrial University of Santander, Colombia, available at: http://jguzman.info/legacy/index.html, 2003.

Hu, Y., Maskey, S., Uhlenbrook, S., and Zhao, H.: Streamflow trends and climate linkages in the source region of the Yellow River, China, Hydrol. Process., 25, 3399–3411, doi:10.1002/hyp.8069, 2011.

Hughes, D. A. and Mallory, S. J. L.: Including environmental flow requirements as part of real-time water resource management, River Res. Applic., 24, 852–861, doi:10.1002/rra.1101, 2008.

Hughes, D. A., Tshimanga, R. M., Tirivarombo, S., and Tanner, J.: Simulating wetland impacts on stream flow in southern Africa using a monthly hydrological model, Hydrol. Process., 28, 1775–1786, doi:10.1002/hyp.9725, 2014.

ICMA: The Inkomati Catchment Management Strategy: A First Generations Catchment Management Strategy for the Inkomati Water Management Area, Inkomati Catchment Management Agency, Nelspruit, 2010.

Jarmain, C., Dost, R. J. J., De Bruijn, E., Ferreira, F., Schaap, O., Bastiaanssen, W. G. M., Bastiaanssen, F., van Haren, I., van Haren, I. J., Wayers, T., Ribeiro, D., Pelgrum, H., Obando, E., and Van Eekelen, M. W.: Spatial Hydro-meteorological data for transparent and equitable water resources management in the Incomati catchment, Pretoria, South Africa, 2013.

Jewitt, G. P. W.: The 8 %–4 % debate: Commercial afforestation and water use in South Africa, S. Afr. Forest. J., 194, 1–5, 2002.

Jewitt, G.: Integrating blue and green water flows for water resources management and planning, Phys. Chem. Earth, Parts A/B/C, 31, 753–762, http://dx.doi.org/10.1016/j.pce.2006.08.033, 2006a.

Jewitt, G.: Water and Forests, in: Encyclopedia of Hydrological Sciences, John Wiley & Sons, Ltd, 2006b.

Jewitt, G. P. W., Garratt, J. A., Calder, I. R., and Fuller, L.: Water resources planning and modelling tools for the assessment of land use change in the Luvuvhu Catchment, South Africa, Phys. Chem. Earth, Parts A/B/C, 29, 1233–1241, 2004.

Kruger, A. C. and Shongwe, S.: Temperature trends in South Africa: 1960–2003, Int. J. Climatol., 24, 1929–1945, doi:10.1002/joc.1096, 2004.

LeMarie, M., van der Zaag, P., Menting, G., Baquete, E., and Schotanus, D.: The use of remote sensing for monitoring environmental indicators: The case of the Incomati estuary, Mozambique, Phys. Chem. Earth, Parts A/B/C, 31, 857–863, doi:10.1016/j.pce.2006.08.023, 2006.

Lennard, C., Coop, L., Morison, D., and Grandin, R.: Extreme events: Past and future changes in the attributes of extreme rainfall and the dynamics of their driving processes, Climate Systems Analysis Group University of Cape Town, 2013.

Love, D., Uhlenbrook, S., Twomlow, S., and van Der Zaag, P.: Changing hydroclimatic and discharge patterns in the northern Limpopo Basin, Zimbabwe, Water SA, 36, 335–350, 2010.

Lynch, S. D.: The Development of a Raster Database of Annual, Monthly and Daily Rainfall for Southern Africa., Water Research Commission, South Africa, Rep. 1156/1/04Rep. 1156/1/04, 2003.

Maingi, J. K. and Marsh, S. E.: Quantifying hydrologic impacts following dam construction along the Tana River, Kenya, J. Arid Environ., 50, 53–79, doi:10.1006/jare.2000.0860, 2002.

Masih, I., Uhlenbrook, S., Maskey, S., and Smakhtin, V.: Streamflow trends and climate linkages in the Zagros Mountains, Iran, Clim. Change, 104, 317–338, doi:10.1007/s10584-009-9793-x, 2011.

Mathews, R. and Richter, B. D.: Application of the Indicators of Hydrologic Alteration Software in Environmental Flow Setting1, JAWRA J. Am. Water Resour. Assoc., 43, 1400–1413, doi:10.1111/j.1752-1688.2007.00099.x, 2007.

Miao, C. Y., Shi, W., Chen, X. H., and Yang, L.: Spatio-temporal variability of streamflow in the Yellow River: possible causes and implications, Hydrol. Sci. J., 57, 1355–1367, doi:10.1080/02626667.2012.718077, 2012.

Milly, P. C. D., Betancourt, J., Falkenmark, M., Hirsch, R. M., Kundzewicz, Z. W., Lettenmaier, D. P., and Stouffer, R. J.: Stationarity Is Dead: Whither Water Management?, Science, 319, 573–574, doi:10.1126/science.1151915, 2008.

Montanari, A., Young, G., Savenije, H. H. G., Hughes, D., Wagener, T., Ren, L. L., Koutsoyiannis, D., Cudennec, C., Toth, E., Grimaldi, S., Blöschl, G., Sivapalan, M., Beven, K., Gupta, H., Hipsey, M., Schaefli, B., Arheimer, B., Boegh, E., Schymanski, S. J., Di Baldassarre, G., Yu, B., Hubert, P., Huang, Y., Schumann, A., Post, D. A., Srinivasan, V., Harman, C., Thompson, S., Rogger, M., Viglione, A., McMillan, H., Characklis, G., Pang, Z., and Belyaev, V.: Panta Rhei – Everything Flows: Change in hydrology and society – The IAHS Scientific Decade 2013–2022, Hydrolog. Sci. J., 58, 1256–1275, doi:10.1080/02626667.2013.809088, 2013.

Moore, D., Dore, J., and Gyawali, D.: The World Commission on Dams + 10: Revisiting the large dam controversy, Water Altern., 3, 3–13, 2010.

Mukororira, F.: Analysis of water allocation in the Komati catchment downstream of Maguga and Driekoppies Dams, MSc Thesis WM 12.18, Water Management, UNESCO-IHE, Delft, 2012.

Mussá, F., Zhou, Y., Maskey, S., Masih, I., and Uhlenbrook, S.: Trend analysis on dry extremes of precipitation and discharge in the Crocodile River catchment, Incomati basin, 14th WaterNet/WARFSA/GWP-SA symposium, 30 October–1 November 2013, Dar es Salaam, Tanzania, 2013,

Nkomo, S. and van der Zaag, P.: Equitable water allocation in a heavily committed international catchment area: the case of the Komati Catchment, Phys. Chem. Earth, Parts A/B/C, 29, 1309–1317, 2004.

Pettitt, A.: A non-parametric approach to the change-point problem., Appl. Stat., 28, 126–135, 1979.

Pollard, S. and du Toit, D.: Integrated water resource management in complex systems: How the catchment management strategies seek to achieve sustainability and equity in water resources in South Africa, Water SA, 34, 671–680, 2009.

Pollard, S. and du Toit, D.: Towards the sustainability of freshwater systems in South Africa: An exploration of factors that enable and constrain meeting the ecological Reserve within the context of Integrated Water Resources Management in the catchments of the lowveld, Water Research Comission, Pretoria, South Africa. Report No. TT 477/10, 2011a.

Pollard, S. and du Toit, D.: Towards Adaptive Integrated Water Resources Management in Southern Africa: The Role of Self-organisation and Multi-scale Feedbacks for Learning and Responsiveness in the Letaba and Crocodile Catchments, Water Resour. Manage., 25, 4019–4035, doi:10.1007/s11269-011-9904-0, 2011b.

Pollard, S., Du Toit, D., and Biggs, H.: River management under transformation: The emergence of strategic adaptive management of river systems in the Kruger National Park, Koedoe, 53, 1011, doi:10.4102/koedoe.v53i2.1011, 2011.

Pollard, S., Mallory, S., Riddell, E., and Sawunyama, T.: Towards improving the assessment and implementation of the reserve: real-time assessment and implementation of the ecological reserve: report to the Water Research Commission, Water Research Commission, Pretoria, South Africa, 2012.

Richter, B. D. and Thomas, G. A.: Restoring environmental flows by modifying dam operations, Ecol. Soc., 12, available at: http://www.ecologyandsociety.org/vol12/iss1/art12/, 2007.

Richter, B. D., Baumgartner, J. V., Powell, J., and Braun, D. P.: A Method for Assessing Hydrologic Alteration within Ecosystems, Conserv. Biol., 10, 1163–1174, 1996.

Richter, B. D., Baumgartner, J. V., Braun, D. P., and Powell, J.: A spatial assessment of hydrologic alteration within a river network, Regulated Rivers: Res. Manage., 14, 329–340, doi:10.1002/(sici)1099-1646(199807/08)14:4< 329::aid-rrr505> 3.0.co;2-e, 1998.

Richter, B. D., Mathews, R., Harrison, D. L., and Wigington, R.: Ecologically sustainable water management: managing river flows for ecological integrity, Ecol. Appl., 13, 206–224, doi:10.1890/1051-0761(2003)013[0206:eswmmr]2.0.co; 2, 2003.

Riddell, E., Pollard, S., Mallory, S., and Sawunyama, T.: A methodology for historical assessment of compliance with environmental water allocations: lessons from the Crocodile (East) River, South Africa, Hydrol. Sci. J., 59, 831–843, doi:10.1080/02626667.2013.853123, 2013.

Rockström, J., Falkenmark, M., Karlberg, L., Hoff, H., Rost, S., and Gerten, D.: Future water availability for global food production: The potential of green water for increasing resilience to global change, Water Resour. Res., 45, W00A12, doi:10.1029/2007wr006767, 2009.

Rouault, M., Fauchereau, N., Pohl, B., Penven, P., Richard, Y., Reason, C., Pegram, G., Phillippon, N., Siedler, G., and Murgia, A.: Multidisciplinary analysis of hydroclimatic variability at the catchment scale, University of Cape Town, CRC, Universite de Dijon, UBO, Universite de Bretagne Occidentale, University of Kwazulu Natal, IMF, University of Kiel, LTHE, Universite de Grenoble, 2010.

Schulze, R. E.: Approaches towards practical adaptive management options for selected water-related sectors in South Africa in a context of climate change, Water SA, 37, 621–645, 2011.

Schulze, R.: A 2011 perspective on climate change and the South African water sector, WRC Report TT 518/12, Gezina, South Africa, 2012.

Shongwe, M. E., van Oldenborgh, G. J., van den Hurk, B. J. J. M., de Boer, B., Coelho, C. A. S., and van Aalst, M. K.: Projected Changes in Mean and Extreme Precipitation in Africa under Global Warming. Part I: Southern Africa, J. Climate, 22, 3819–3837, doi:10.1175/2009jcli2317.1, 2009.

Taylor, V., Schulze, R., and Jewitt, G. P. W.: Application of the Indicators of Hydrological Alteration method to the Mkomazi River, KwaZulu-Natal, South Africa, Afr. J. Aquatic Sci., 28, 1–11, doi:10.2989/16085914.2003.9626593, 2003.

TPTC: Tripartite Permanent Technical Committee (TPTC) between, Moçambique, South Africa, Swaziland, PRIMA: IAAP 3: Consultancy Services for Integrated Water Resources Management. Baseline evaluation and scoping report: Part C. Report No.: IAAP 3: 03C–2010, Prepared for TPTC by Aurecon, Pretoria, 2010.

Tullos, D., Tilt, B., and Liermann, C. R.: Introduction to the special issue: Understanding and linking the biophysical, socioeconomic and geopolitical effects of dams, J. Environ. Manage., 90, Supplement 3, S203–S207, doi:10.1016/j.jenvman.2008.08.018, 2009.

Van den Berg, E., Plarre, C., Van den Berg, H., and Thompson, M.: The South African national land cover 2000, Agricultural Research Council (ARC) and Council for Scientific and Industrial Research (CSIR), Pretoria, Report No. GW/A/2008/86, 2008.

Van der Zaag, P. and Vaz, A. C.: Sharing the Incomati waters: cooperation and competition in the balance, Water Pol., 5, 349–368, 2003.

van Eekelen, M. W., Bastiaanssen, W. G. M., Jarmain, C., Jackson, B., Ferreira, F., van der Zaag, P., Saraiva Okello, A., Bosch, J., Dye, P., Bastidas-Obando, E., Dost, R. J. J., and Luxemburg, W. M. J.: A novel approach to estimate direct and indirect water withdrawals from satellite measurements: A case study from the Incomati basin, Agr. Ecosyst. Environ., 200, 126–142, doi:10.1016/j.agee.2014.10.023, 2015.

Vörösmarty, C. J., McIntyre, P., Gessner, M. O., Dudgeon, D., Prusevich, A., Green, P., Glidden, S., Bunn, S. E., Sullivan, C. A., and Liermann, C. R.: Global threats to human water security and river biodiversity, Nature, 467, 555–561, 2010.

Warburton, M. L., Schulze, R. E., and Jewitt, G. P. W.: Confirmation of *ACRU* model results for applications in land use and climate change studies, Hydrol. Earth Syst. Sci., 14, 2399–2414, doi:10.5194/hess-14-2399-2010, 2010.

Warburton, M. L., Schulze, R. E., and Jewitt, G. P. W.: Hydrological impacts of land use change in three diverse South African catchments, J. Hydrol., 414–415, 118-135, doi:10.1016/j.jhydrol.2011.10.028, 2012.

Zhang, X., Zhang, L., Zhao, J., Rustomji, P., and Hairsine, P.: Responses of streamflow to changes in climate and land use/cover in the Loess Plateau, China, Water Resour. Res., 44, W00A07, doi:10.1029/2007wr006711, 2008.

10

Spatial evapotranspiration, rainfall and land use data in water accounting – Part 1: Review of the accuracy of the remote sensing data

P. Karimi[1] **and W. G. M. Bastiaanssen**[1,2,3]

[1]UNESCO-IHE Institute for Water Education, Delft, the Netherlands
[2]International Water Management Institute, Battaramulla, Sri Lanka
[3]Faculty of Civil Engineering and Geosciences, Water Management Department, Delft University of Technology, Delft, the Netherlands

Correspondence to: P. Karimi (p.karimi@Unesco-ihe.org)

Abstract. The scarcity of water encourages scientists to develop new analytical tools to enhance water resource management. Water accounting and distributed hydrological models are examples of such tools. Water accounting needs accurate input data for adequate descriptions of water distribution and water depletion in river basins. Ground-based observatories are decreasing, and not generally accessible. Remote sensing data is a suitable alternative to measure the required input variables. This paper reviews the reliability of remote sensing algorithms to accurately determine the spatial distribution of actual evapotranspiration, rainfall and land use. For our validation we used only those papers that covered study periods of seasonal to annual cycles because the accumulated water balance is the primary concern. Review papers covering shorter periods only (days, weeks) were not included in our review. Our review shows that by using remote sensing, the absolute values of evapotranspiration can be estimated with an overall accuracy of 95 % (SD 5 %) and rainfall with an overall absolute accuracy of 82 % (SD 15 %). Land use can be identified with an overall accuracy of 85 % (SD 7 %). Hence, more scientific work is needed to improve the spatial mapping of rainfall and land use using multiple space-borne sensors. While not always perfect at all spatial and temporal scales, seasonally accumulated actual evapotranspiration maps can be used with confidence in water accounting and hydrological modeling.

1 Introduction

The demand for fresh water is increasing worldwide due to economic and population growth (Molden et al., 2007; Vörösmarty et al., 2010). Proper planning of such scarce water resources in terms of storage, allocation, return flow and environmental services is vital for optimizing the resource (Chartres and Varma, 2010). There is, however, a lack of fundamental data on vertical and lateral water flows, water stocks, water demand, and water depletion. At the same time, there is a decline in the network density of operational hydrometeorological field stations. The absence of adequate field data sets is an important obstacle for sound, evidence-based water resource management decisions. The consequence of data scarcity is more severe in transboundary river basins where, apart from collection, the accessibility of data is hindered by political issues (Awulachew et al., 2013).

Remotely sensed hydrological data are an attractive alternative to conventional ground data collection methods (Bastiaanssen et al., 2000; Engman and Gurney, 1991; Wagner et al., 2009; Neale and Cosh, 2012). Satellites measure the spatial distribution of hydrological variables indirectly with a high temporal frequency across vast river basins. There are many public data archives where every user can download preprocessed satellite data. Quality flags are often provided, as well as manuals with explanations on how the satellite data have been preprocessed and can be reproduced. These recurrent data sets are highly transparent, politically neutral and consistent across entire river basins, even for large basins such as the Nile and the Ganges. While certain satellite data sets have been processed to a first level of reflectance, emittance, and backscatter coefficients, others will even provide second level products that can be directly explored for water resource planning purposes (e.g., land cover, soil moisture, and rainfall). Evapotranspiration (ET) is one of the parameters that often requires additional processing of the spectral data; only a very few public domain data archives provide preprocessed ET data and, in fact, spatial ET modeling is still underdeveloped. Examples of several remotely sensed ET algorithms that could be applied to interpret raw satellite data into spatial layers of ET are well summarized in a recent book edited by Irmak (2012).

Time series of various hydrological variables such as precipitation, evapotranspiration, snow cover, soil moisture, water levels, and aquifer storage can be downloaded from public domain satellite-based data archives. With the right analytical tools and skills, these abundant data sets of hydrological processes can be used to produce information on water resource conditions in river basins. Tools such as Water Accounting Plus (WA+) (Karimi et al., 2013a, b) are expressly designed to exploit remote sensing estimates of hydrological variables. Water accounting is the process of communicating water related information about a geographical domain, such as a river basin or a country, to users such as policy makers, water authorities, basin managers, and public users. Water accounting information can be key to river basin management policy, especially when administrations are reluctant to share their – sometimes imperfect – in situ data with neighboring states and countries. WA+ can facilitate conflict management in internationally shared river basins. In addition to that, hydrological variables derived from remote sensing can also be used for spatially distributed hydrological modeling. Studies by Houser et al. (1998), Schuurmans et al. (2003), and Immerzeel and Droogers (2008) have, for instance, demonstrated that such inputs have improved hydrological model performance for river basins in Australia, the Netherlands and India, respectively.

A major point of criticism that is commonly laid down on remote sensing data has been the lack of accuracy. With the improvement of technology the accuracy has however improved significantly over the last 30 years; yet it is necessary to remain critical. It is important to note that the conventional methods of measuring hydrological processes (e.g., rainfall and discharge) are not flawless either and, thus, the accuracy of both types of measurements needs to be verified. There are also limitations with what conventional measurement methods can offer especially when spatially distributed data is concerned. For instance, the actual ET of river basins can hardly be measured operationally through ground measurements; therefore, the depletion of water remains difficult to estimate and quantify. Thus, ET is often ignored in water accounting frameworks such as the SEEA-Water system proposed by the United Nations Statistics Division (UN, 2007) and the Australian water accounting system (ABS, 2004). Remote sensing techniques, however, can provide spatially distributed daily estimates of actual ET and this opens new pathways in the accounting of water depletion (Karimi et al., 2013a).

This paper investigates the errors and reliability of remotely sensed ET, rainfall, and land use based on a comprehensive literature review. The choice of the variables that have been investigated in this paper (ET, rainfall, land use/land cover) is based on the common use in hydrological and water resource management studies. Only recent publications on accumulated ET and rainfall for a minimum time period of one growing cycle have been consulted, which implies that some of the well-known reference papers are excluded because they relate to shorter flux observation periods. Elder remote sensing algorithms were also excluded. The companion paper (Karimi et al., 2015) investigates impacts of the errors associated with the satellite measurement for ET, rainfall and land use on the accuracy of WA+ outputs, using a case study from the Awash Basin in Ethiopia. See Appendix D for a glossary of the abbreviations used throughout the paper.

2 Remote sensing data for water accounting (WA+)

2.1 Evapotranspiration

Over the past decades several methods and algorithms to estimate actual ET through satellite measurements have been developed. Most of these estimates are based on the surface energy balance equation. The surface energy balance describes the partitioning of natural radiation absorbed at Earth's surface into physical land surface processes. Evapotranspiration is one of these key processes of the energy balance, because latent heat (energy) is required for evaporation to take place. The energy balance at Earth's surface reads

$$\mathrm{LE} = R_\mathrm{n} - G - H \left(\mathrm{W\,m^{-2}}\right), \qquad (1)$$

where R_n is the net radiation, G is the soil heat flux, H is the sensible heat flux, and LE is the latent heat flux. The sensible heat flux H is a function of the temperature difference between the canopy surface and the lower part of the atmosphere, and the soil heat flux G is a similar function related to the temperature difference between the land surface and the top soil. A rise of surface temperature will thus usually increase H and G fluxes. Evaporative cooling will reduce H and G, and result in a lower surface temperature. The LE is the equivalent energy amount ($\mathrm{W\,m^{-2}}$) of the ET flux ($\mathrm{kg\,m^{-2}\,s^{-1}}$ or $\mathrm{mm\,d^{-1}}$). The net radiation absorbed at the land surface is computed from shortwave and long-wave radiation exchanges. Solar radiation is shortwave and is the most important supplier of energy. More information on the energy balance is provided in general background material such as Brutsaert (1982), Campbell and Norman (1998) or Allen et al. (1998).

Surface temperature is measured routinely by spaceborne radiometers such as the Advanced Very High Resolution Radiometer (AVHRR), Moderate Resolution Imaging Spectrometer (MODIS), Visible Infrared Imager Radiometer Suite (VIIRS), Landsat, Advanced Space borne Thermal Emission and Reflection Radiometer (ASTER), China–Brazil Earth Resources Satellite (CBERS), and the Chinese HJ and Feng Yung satellites. Remotely sensed surface temperature is the major input variable in ET algorithms. Examples of thermal infrared ET algorithms are provided by EARS (Rosema, 1990), SEBAL (Bastiaanssen et al., 1998), TSEB (Norman et al., 1995), SEBS (Su, 2002; Jia et al., 2003), METRIC (Allen et al., 2007), ALEXI (Anderson et al., 1997), and ETWatch (Wu et al., 2012). The differences among these algorithms are often related to the parameterization of H, general model assumptions, and the amount of input data required to operate these models.

Other groups of ET algorithms are based on the vegetation index and its derivatives such as published by Nemani and Running (1989), Guerschman et al. (2009), K. Zhang et al. (2010), Mu et al. (2011), and Miralles et al. (2011). ETLook (Bastiaanssen et al., 2012) is a new ET model that directly computes the surface energy balance using surface soil moisture estimations for the top soil (to feed soil evaporation) and subsoil moisture for the root zone (to feed vegetation transpiration). Soil moisture data can be inferred from thermal measurements (e.g., Scott et al., 2003) or from microwave measurements (e.g., Dunne et al., 2007). Microwave measurements provide a solution for all weather conditions and can be applied at any spatial scale for which moisture data is available.

A different school of remote-sensing-based ET algorithms is built around the derivation of a relative value of ET using trapezoid/triangle methods. Trapezoid/triangle diagrams are constructed from a population of pixel values of surface temperature and vegetation index and used to infer the relative value of ET (e.g., Choudhury, 1995; Moran et al., 1994; Roerink et al., 2000; Wang et al., 2007). In these diagrams, the range of surface temperature values at a given class of vegetation index is the basis for determining relative ET, assuming that the lowest temperature in a certain range of vegetation index represents potential ET. The highest temperature coincides with zero evaporation. The main assumption in triangle/trapezoidal methods is that the variation in vegetation index relation to surface temperature is driven primarily by the variation in soil water content rather than differences in atmospheric conditions.

Merging different global ET products such as MOD16 (Mu et al., 2011) and ERA-Interim (Dee et al., 2011) at global and regional scales into one ET product is another approach that has been used by a group of scientists. This approach mainly uses statistical methods to combine ET products that are based on different methods, algorithms, and origins (e.g., global: Mueller et al., 2013; Africa: Trambauer et al., 2014; US: Velpuri et al., 2013). New ensemble ET products on the basis of several open access and global-scale operational ET products from Earth observations are under development, but are not published yet.

Review papers on advanced algorithms for estimating spatial layers of ET have been published by Moran and Jackson (1991), Kustas and Norman (1996), Bastiaanssen (1998), Courault et al. (2005), Glenn et al. (2007), Gowda et al. (2007), Kalma et al. (2008), Verstraeten et al. (2008), and Allen et al. (2011). While these review papers provide a good understanding of the evolution of ET algorithm development, they rarely report the accuracies attainable, especially at a seasonal or longer time frame.

2.2 Rainfall

There are different algorithms to infer rainfall from satellite data. The four essentially different technologies are (i) indexing the number and duration of clouds (Barrett, 1988), (ii) accumulated cold cloud temperatures (Dugdale and Milford, 1986), (iii) microwave emissivity (Kummerow et al., 1996), and (iv) radar reflectivity (Austin, 1987). Techniques using microwave wavelength information are promising alternatives for measuring rainfall because of the potential for sensing the raindrops themselves and not a surrogate of rain, such as the cloud type. Microwave radiation with wavelengths in the order of 1 mm–5 cm has a strong interaction with raindrops, since the drop size of rain is comparable to this wavelength. This feature makes them suitable to detect rainfall intensity. Active microwave (radar) measurements of rainfall are based on the Rayleigh scattering caused by the interaction of rain and the radar signal (Cracknell and Hayes, 1991). Spaceborne radar measurements of rain intensity are possible with the precipitation radar (PR) aboard the NASA Tropical Rainfall Measuring Mission (TRMM) and Global Precipitation Mission (GPM) satellites, which assesses the attenuation of the radar signal caused by the rain. The PR has a pixel size of 5 km and can oversee a swath of 220 km. Unfortunately, it is usually necessary to evaluate the rainfall radar reflectivity factor empirically on a region-by-region basis over lengthy periods of time. In other words, rain radar systems – both ground-based and satellite-based – need calibration for proper rainfall estimates. We will conclude later that most papers investigated in our review process do apply a certain level of field calibration. Several operational rainfall products based on satellite measurements have been created or improved more recently. Among the new ensemble rainfall products is the Climate Hazards Group InfraRed Precipitation Station (CHIRPS) that provides promising results (Funk et al., 2013).

Review papers on the determination of rainfall from satellite measurements have been prepared, by, for instance, Barrett (1988), Barrett and Beaumont (1994), Petty (1995), Petty and Krajewski (1996), Kummerow et al. (1996), Smith et al. (1998), Kidd (2001), Stephens and Kummerow (2007), and Huffman et al. (2007). A selection of available rainfall

Table 1. Overview of the main existing regional and global-scale satellite-based data sources of rainfall. The column "gauge" indicates whether a calibration against ground data is included.

Product	Main principle data	Resolution	Spatial coverage	Gauge	Minimum time steps interval	Producer
MPE	Meteosat 7, 8, 9 10	3 km	Indian ocean	*N*	15 min	EUMETSAT
CMORPH	Microwave estimates (DMSP F-13, 14 & 15 (SSM/I), NOAA-15, 16, 17 & 18 (AMSU-B), AMSR-E, and TRMM (TMI)), IR motion vectors	8 km	50° N–50° S	*N*	30 min	NOAA/CPC
PERSIANN	Microwave estimates (DMSP F-13, 14, & 15, NOAA-15, 16, 17, and TRMM (TMI))	0.25°	60° N–60° S	*N*	1 h	UC Irvine
GSMap	Microwave estimates (DMSP F-13, 14 & 15 (SSM/I), AMSR, AMSR-E, and TRMM (TMI))	0.1°	60° N–60° S	*N*	1 h	JAXA
NRL-blended	Microwave estimates (DMSP F-13, 14, & 15 (SSM/I), F-16 (SSMIS))	0.25°	60° N–60° S	*N*	3 h	NRL
TCI (3G68)	Microwave estimates (TRMM (TMI)), and PR	0.5°	37° N–37° S	*N*	1 h	NASA
TOVS	HIRS, MSU sounding retrievals	1°	Global	*N*	Daily	NASA
Hydro estimator	GOES IR	4 km	Global	*N*	15 min	NOAA
TRMM 3B42	Microwave estimates (TRMM, SSM/I, AMSR and AMSU), IR estimates from geostationary satellites	0.25°	50° N–50° S	*Y*	3 h	NASA
CPC-RFE2.0	Microwave estimates (SSM/I, AMSU-B), IR estimates from METEOSAT	0.1°	20° W–55° E, 40° S–40° N	*Y*	Daily	FEWS
GPCP 1DD	IR estimates from geostationary satellites, TOVS	1°	50° N–50° S	*Y*	Daily	NASA/GSFC
CMAP	Microwave estimates (SSM/I), GOES IR	2.5°	Global	*Y*	5 days	NOAA
TAMSAT	Meteosat thermal-IR	3 km	Africa	*Y*	10 days	Reading University
TRMM 3B43	Microwave estimates (TRMM, SSM/I, AMSR and AMSU), IR estimates from geostationary satellites	0.25°	40° N–40° S	*Y*	Monthly	NASA
GPCP_V2	Microwave estimates (SSM/I), IR, TOVS	2.5°	Global	*Y*	Monthly	NASA/GSFC

products based on remote sensing techniques – sometimes used in combination with other methodologies – is presented in Table 1.

2.3 Land use

Whereas land cover describes the physical properties of vegetation (e.g., grass, savannah, forest), land use denotes the usage of that land cover (e.g., pasture, crop farming, soccer field). Maps of land use are fundamental to WA+ because it determines the services and processes in a spatial context. Different types of land use provide benefits and services such as food production (agricultural land), economic production (industrial areas), power generation (reservoirs), environmental ecosystems (wetlands), livelihoods etc., and they have an associated water consumptive use. Land use classification based on the use of water, differs from classical land use land cover maps that focus mainly on the description of woody vegetation such as forests and shrubs for ecological and woodland management purposes. WA+ needs land use maps focused on crop types (e.g., rainfed potatoes, irrigated maize) and the source of water consumed (e.g., surface water and groundwater). Some of the first maps dedicated for agricultural water management were prepared by Thenkabail et al. (2005), Cheema and Bastiaanssen (2010), Yalew et al. (2012) and Kiptala et al. (2013). Furthermore, land use classifications for WA+ at river basin scale require a pixel size of 30–100 m that can be delivered by Landsat-8 and Proba-V satellite data, respectively. It is expected that the arrival of Sentinel-2 data during the course of 2014 with pixel sizes ranging between 10–30 m and a short revisit time of 5 days will greatly enhance development of new land use classifications that are tailored for water use and water accounting.

Land use changes affect the water balance of river basins and thus also the amount of water flowing to downstream areas. Bosch and Hewlett (1982) and Van der Walt et al. (2004) discuss for instance how replacing natural vegetation by exotic forest plantations reduced the stream flow in South Africa. Maes et al. (2009) evaluated the effect of land use changes on ecosystem services and water quantity on basins in Belgium and Australia. The role of land use is thus a crucial component of sound water accounting and water resource management (Molden, 2007).

Land use is usually identified on the basis of spectral reflectance and its change with vegetation phonology. The reflectance in the near and middle infrared part of the electromagnetic spectrum, especially, is often related to certain land use classes. The relationship between reflectance and land use is however not unique, and field inspections are usually needed for better interpretation. Soil type, soil moisture and surface roughness all have an influence on reflectance. The health of the vegetation and factors such as the angle and size of leaves also affect the photosynthetic activity of the plants. There is another land use mapping technology that is entirely based on the difference in time profiles of spectral vegetation indices. Fourier analysis of vegetation index can be used to quantify land use classes and crop types (e.g., Roerink et al., 2003), especially when time profiles are linked to existing cropping calendars.

All the land use classification papers we reviewed report on a confusion matrix that describes the overall classification accuracy by showing how often certain land use classes are confused in the remote sensing analysis with other land use classes. Congalton (1991) and Foody (2002) give a full explanation on errors in land use data.

Review papers on the use of remote sensing for land use land cover classification are provided in Bastiaanssen (1998), Smits et al. (1999), Mucher et al. (2000), Cihlar (2000), Franklin and Wulder (2002), Thenkabail et al. (2009b), and García-Mora et al. (2012).

3 Results

3.1 Accuracy of spatial evapotranspiration data

The lack of validation of spatial layers of ET is one of the drawbacks in defining the reliability of remotely sensed ET products. There are no reliable and low-cost ground-based ET flux measurement techniques, although new inventions are always underway (Euser et al., 2014). It is simply too costly to install instruments that have the capacity to measure ET operationally at various locations dispersed across a river basin. The main methods to measure ET at the field scale include lysimeters, Bowen ratio, eddy covariance systems, surface renewal systems, scintillometers, and classical soil water balancing. Lysimeters can be very accurate for in situ measurements of ET at small scale if they are properly maintained. Bowen ratio and eddy covariance flux towers and surface renewal systems are fairly accurate methods for estimating ET at scales of up to 1 km (Rana and Katerji, 2000), although not free of errors (e.g., Teixeira and Bastiaanssen, 2010; Twine et al., 2000). Scintillometers have the capability to measure fluxes across path lengths of 5–10 km (Hartogensis et al., 2010; Meijninger and de Bruin, 2000).

To deal with the problem of measuring ET fluxes in a composite terrain, large-scale field experiments in the African continent (e.g., Sahel: Goutorbe et al., 1997; southern Africa: Otter et al., 2002), the European continent (e.g., France: Andre et al., 1986; Spain: Bolle et al., 2006), the American continent (e.g., Kansas: Smith et al., 1992; Arizona and Oklahoma: Jackson et al., 1993) and the Asian continent (e.g., China: Wang et al., 1992; Korea: Moon et al., 2003) were set up to measure fluxes simultaneously within a certain geographic region at a number of sites with different land use classes. Several remotely sensed ET algorithms were developed and validated using these data sets. The limitation is however that the duration of these field campaigns was for budgetary reasons restricted to several weeks only.

Validation studies with different ET algorithms using the same spatial ground truth data sets are very interesting. The International Water Management Institute (IWMI) undertook for instance a validation study to determine the accuracy of various ET methods for irrigated cotton and grapes in Turkey (Kite and Droogers, 2000). Although here the period was not sufficiently long to encompass one growing season. The Commonwealth Science and Industrial Research Organisation (CSIRO) in Australia studied the predictions of eight different ET products, at a minimum monthly frequency and at a spatial resolution of at least 5 km, using flux tower observations and watershed data across the entire continent as part of the Water Information Research and Development Alliance (WIRADA) project (Glenn et al., 2011). The studied ET products were based on different methods including large-scale water balance modeling, thermal imagery (Mcvicar and Jupp, 1999, 2002), spectral imagery (Guerschman et al., 2009), inferred LAI (leaf area index; Y. Zhang et al., 2010), passive microwave (Bastiaanssen et al., 2012), and the global MODIS reflectance-based algorithm (Mu et al., 2007). The results showed that at annual-scale remote-sensing-based ET estimates, barring the global MODIS product that was at the time an unrefined method that needed improvements (Mu et al., 2011), had an acceptable mean absolute percentage error (MAPE) ranging from 0.6 to 18 % with an average MAPE of 6 % (King et al., 2011). Along similar lines, the Council for Scientific and Industrial Research (CSIR) in South Africa conducted a remote sensing study on a smaller scale to investigate the performance of three ET algorithms (Jarmain et al., 2009).

To assess the overall error in accumulated ET products, a comprehensive literature review was conducted and reported errors by various authors were synthesized. All the papers included in the review were published within the past 13 years (hence from the year 2000 onwards) and they cover a range of in situ measurements and remote sensing ET algorithms. The reviewed papers cover a range of remote sensing methods for ET measurements including SEBAL, METRIC, SEBS, TSEB, ALEXI, ET Watch, and SatDAET. In essence, the spatial ET layers reported in these papers were not a priori calibrated and the authors reported on the validation aspect. Since the primary purpose of this study was to quantify errors in accumulated ET, only papers that report errors on ET estimates over a minimum period of one growing cycle which on average is about 5–6 months, hereafter called seasonal ET, were consulted. Papers dealing with ET over shorter periods were thus excluded in our review (e.g., Anderson et al., 2011; Chávez et al., 2008; Gonzalez-Dugo et al., 2009; Mu et al., 2011). This, also, implies that GEWEX (Global Energy and Water Exchanges Project)-related field experiments could not be used because intensive campaigns with multiple flux covered periods of weeks only. The manifold flux campaigns organized by the US Department of Agriculture (Kustas et al., 2006; JORNEX: Rango et al., 1998; SALSA: Chehbouni et al., 1999) also did not meet our criterion. To be able to compare error levels from different studies only papers that report errors in terms of mean error were included in the review. Thus, some of the valuable papers on this topic that use RMSE (root mean square error) to describe errors without including mean error could not be included in the review (e.g., Batra et al., 2006; Cleugh et al., 2007; Guerschman et al., 2009; Venturini et al., 2008). The data sources consulted are summarized in Appendix A. It reflects the accumulated ET conditions encountered in 11 countries. Thirty-one publications met the criteria specified and were analyzed. One publication often contains more data points due to multiple models, multiple years, and multiple areas. Hence, the total number of points was n=46. Considering this number, the probability density function is unlikely to change if other papers – or more papers – were to be considered in the review.

The probability distribution of mean absolute percentage error in remote sensing ET estimates is presented in Fig. 1.

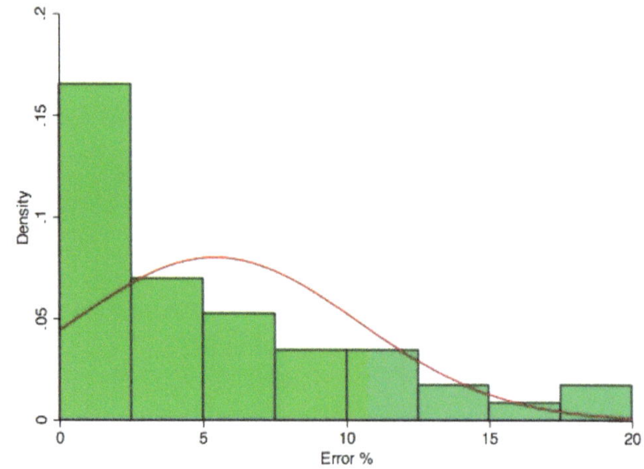

Figure 1. Probability density function of the reported mean absolute percentage error in remotely sensed ET estimates. A season or longer period was considered.

The results demonstrate the absolute error of annual or seasonal ET to vary between 1 and 20 %. The average MAPE is 5.4 %, with a standard deviation of 5.0 %. It is evident from Fig. 1 that the distribution is positively skewed. These results are closely in line with findings by King et al. (2011) in Australia, both in terms of average and the range of error in ET estimates.

Many of the publications reported an error of less than 5 %, a remarkable good and unexpected result. In many cases, the authors of the papers were both the developers and the testers of the algorithms, and parameter tuning was possible. The left-hand bar in Fig. 1 is, we believe, a biased view of the reality. For this reason, the data points were fitted by means of a skewed normal distribution so that less weight is given to the class with exceptionally low errors.

There are seven papers that report a mean absolute percentage error of 1 % for the ET of cropland. Without exception, all these papers are based on the Surface Energy Balance Algorithm for Land (SEBAL) and its related algorithm Mapping ET at High Resolution with Internalized Calibration (METRIC). Apparently, these algorithms work well for crops, which was recognized earlier by Bastiaanssen et al. (2009) and Allen et al. (2011). Another interesting observation is that at river basin scale – i.e., the scale where water accounting is done – all papers report a MAPE of less than 5 %. These case studies include the 3 % difference between the measured ET and remotely sensed ET of selected river basins in Sri Lanka (Bastiaanssen and Chandrapala, 2003), 1.7 % difference observed by Singh et al. (2011) for the Midwest in the USA using the METRIC algorithm, 1.8 and 3 % differences observed by Wu et al. (2012) using ET Watch in the Hai Basin of the North China Plain, 5 % difference observed by Bastiaanssen et al. (2002) for the Indus Basin, 1 % difference observed by Evans et al. (2009) for the Murray–Darling Basin, and 0.6, 2.1, 3.9, and 18 % differences for different algorithms observed by King et al. (2011) for the Australian continent.

At the other end of the spectrum, the largest ET deviations were found by Jiang et al. (2009) for alkali scrubs in south Florida. They used the SatDAET algorithm which is an ET estimation method that uses the contextual relationship between remotely sensed surface temperature and vegetation index to calculate evaporative fraction (EF). They compared the estimated ET using SatDAET for both clear and cloudy days with ET from lysimeters and observed a 19 % difference for 1999.

There is no single preferred ET model. The selection of the algorithm depends on the application, the required spatial resolution, the period for which the ET fluxes should be estimated for, the size of the study area, the land use classes present, etc. A useful distinction is to discern global-scale models (few) and local-scale models (many). Also, the level of validation and application of these models widely differ. Whereas certain models are tested with a single experimental flux site, other models have been applied in more than 30 countries.

Considering this positive evaluation, spatial layers of ET should be encouraged for applications in water accounting and hydrological modeling. Except for Jhorar et al. (2011), Winsemius et al. (2008) and Rientjes et al. (2013), this is rarely done because water managers and hydrologists do not accept ET layers as being sufficiently accurate. This new analysis proves that the science of remote sensing in the last 13 years has advanced and that mapping of ET has become more reliable.

3.2 Accuracy of spatial rainfall data

A comprehensive literature review – similar to ET – was conducted for remote sensing rainfall products. Twenty-four peer reviewed papers that describe the accuracy of annual and seasonal rainfall from satellites, published over the last 5 years were reviewed (see Appendix B). Sixty-eight data points were reconstructed from these publications. The selected papers used various remote sensing rainfall products including TRMM, PERSIANN, RFE, ERA40, CMORPH, and CMAP. A common problem is the scale mismatch between rain gauges and the area integrated rainfall of one single microwave-based pixel of the satellite image.

Several of these papers compared different rainfall algorithms. Some also used the same field data to verify several rainfall algorithms. For example, Asadullah et al. (2008) compared five satellite-based rainfall estimates (SRFEs) with historical average rainfall data from gauges over the period 1960–1990 in Uganda. The difference between gauged data and SRFEs was found to vary between 2 and 19 %. Products such as CMORPH, TRMM 3B42, TAMSAT, and RFE underestimated rainfall by 2, 8, 12, and 19 %, respectively, while PERSIANN overestimated by 8 %. Stisen and Sanholt (2010) compared three global SRFE products,

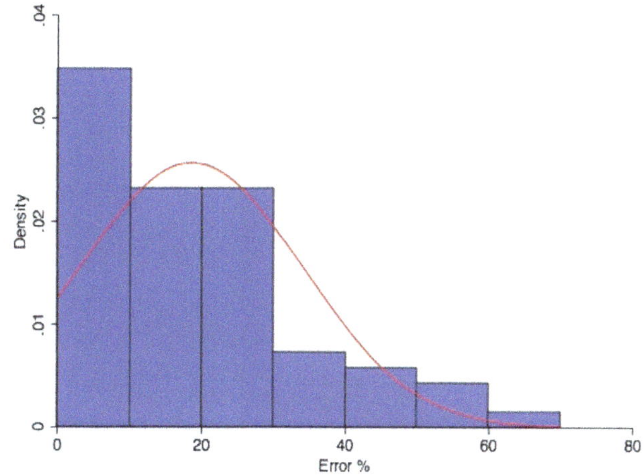

Figure 2. Probability density function of the reported mean absolute percentage error in rainfall estimates from remote sensing. A season or longer period is considered.

i.e., CMORPH, TRMM 3B42 and PERSIANN, and two SRFEs made for Africa, i.e., CPC-FEWS v2 and a locally calibrated product based on TAMSAT data, with the average gauge rainfall in Senegal River basin. They concluded that rainfall estimation methods that are designed for Africa significantly outperform global products. This superior performance is attributed both to the inclusion of local rain gauge data and to the fact that they are made specifically for the atmospheric conditions encountered on the African continent. Of the global products, SRFEs from TRMM were found more accurate, presumably because monthly calibration of the 3B43 product is a default process of the algorithm. The global SRFEs showed an improved performance after bias correction and recalibration. The positive effects of the inclusion of rain gauge data in SRFEs is also reported in the study by Dinku et al. (2011), which compared five SRFEs with rain gauge data in the Blue Nile Basin. Several studies show that local calibration significantly improves the accuracy of satellite-based rainfall estimates: Almazroui et al. (2012) in Saudi Arabia, Cheema and Bastiaanssen (2012) in the Indus Basin, Duan and Bastiaanssen (2013) in the Lake Tana and Caspian Sea regions, and Hunink et al. (2014) in the high-elevation Tungurahua Province in the Andes mountain range of Ecuador.

The error probability distribution function curve reconstructed from the a priori calibrated rainfall data set is shown in Fig. 2. The mean absolute percentage error varies between 0 and 65 %, and the average MAPE for calibrated satellite rainfall estimates is 18.5 %. The standard deviation is 15.4 %, with a positive skewness of 0.9. As with the density function for ET, the curve fitting of the distribution was forced with a skewed normal distribution to ensure that less weight is assigned to the class of 0–10 % deviation. This indicates that for the majority of case studies, the error in calibrated rainfall maps is less than 18.5 %. Large error bands were found

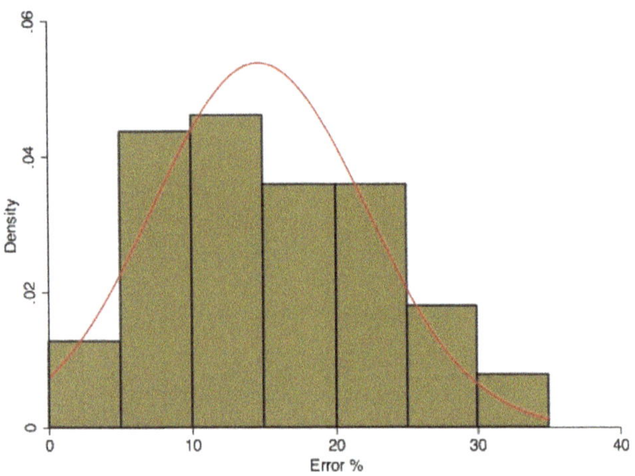

Figure 3. Probability density function of the reported mean absolute percentage error in land use classification using remote sensing.

for all rainfall algorithms, and no particular algorithm performs better in terms of variance. The unresolved problem of the pixel–gauge-scale mismatch is one major source of this problem. The average MAPE is 14, 17, 21, 23, 28, and 29 % for TRMM, ERA40, GPCP 1DD, CMORPH, RFE, and PERSIANN, respectively. These average values represent the average MAPE of each SRFE regardless of the product version.

The interim conclusions are therefore that (i) the processes to derive rainfall from satellite data are more complex than the derivation of ET and (ii) that the performance of existing rainfall products is not satisfactory and requires caution when applied for water accounting and hydrological modeling, despite the fact that most SRFEs have an a priori calibration procedure. More research and development of operational rainfall algorithms using various types of sensors is deemed necessary.

3.3 Accuracy of land use land cover maps

The publications listed in Appendix C were reviewed for land use estimations. Sixty-five papers were reviewed. Seventy-eight data points were reconstructed from these papers. Rather diverging land use classes and data from 35 different countries were included in this comparative data set. The results are presented in Fig. 3. The shape of the probability density function of error differs from the ones obtained for ET and rainfall: it is tending towards a standardized normal distribution, which implies that the number of very good results and very poor results are similar. Table 2 provides a summary of the statistical results. The mean absolute percentage error, defined as 1 minus overall accuracy, for land use classification is 14.6 %, with a standard deviation of 7.4 % and a skewness of 0.35.

The overall performance is rather good, and this can be partially explained by the fact that high-resolution satellites

Table 2. Mean deviation of the input variables and the distribution of the error.

Remote sensing parameter	Calibration	Mean absolute percentage error (%)	Standard deviation error (%)	Skewness (–)	No. of data points
ET	No	5.4	4.9	1.18	41
Rainfall	Yes	18.5	15.4	0.90	69
Land use	Yes	14.6	7.4	0.37	78

were often used for the land use and land cover classification. The spectral measurements of Landsat and Aster satellites were especially often applied because they have bands suitable for the detection of a range of land use classes in the near- and middle-infrared part of the spectrum. To investigate the impact of the spatial resolution of the used imagery on the accuracy of the land use product, we divided the data points into two groups based on the reported resolution. The MAPE for land use classifications that are based on high-resolution images, 30 m and less, is 12.9 %, whereas for those that use moderate- and low-resolution images, more than 200 m, the MAPE is 19.8 %. The number of land use classes shows no significant impact on the overall accuracy of the map. The results reveal that the global-scale land cover maps have lower overall accuracy due to their large pixel size. The overall accuracies of global maps varies between 69 and 87 % with an average of 76.4 %, which is equivalent to a MAPE of 13–31 % and average of 23.4 %. This observation shows that global land cover maps should be used with caution in water accounting applications.

The overall accuracy in the reviewed papers varies between 68 and 98 %. This is in good agreement with the suggested range of 70–90 % by Bach et al. (2006) in their review paper. The review also revealed that Landsat products, with 42 case studies out of the total 78, are the most commonly used imagery for land use land cover classification purposes. The free access Landsat-8 data may thus set the directions for near future development of land use classifications, especially when being complemented with Sentinel data. The Finer Resolution Observation and Monitoring – Global Land Cover (Gong et al., 2013) is an example of that.

Many land use studies are based on ground truth data sets that are used for controlling or supervising the classification process. The data in Appendix C thus have an element of a priori calibration which increases the overall accuracy. Without ground truthing, the overall calibration can be expected to be lower. Also, it must be noted that only the overall accuracy of the confusion matrix is used. While the overall accuracy might be acceptable, it is likely that the error in certain individual land use classes is significantly different.

4 Conclusions and way forward

The increasing number of satellite-based measurements of land and water use data are provided by generally accessible data archives, although evapotranspiration data sets are under development. Satellites provide spatial information with a high temporal frequency over wide areas, which make remotely sensed maps of land use and hydrological variables an attractive alternative to conventionally collected data sets. However, the uncertainty about the possible errors in remote sensing estimates has been an ongoing concern among the users of these products. The goal of this study was to investigate the errors and reliability of some of these remotely sensed hydrological variables created by advanced algorithms through an international literature review. Only recent data sets, not older than 13 years, were reviewed.

The main interest of this review was to understand the measure of error in remote sensing data for water accounting. The review focused on ET, precipitation, and land use classifications. A comprehensive literature review was conducted and for each variable several numbers of post-2000 peer-reviewed publications were consulted for reported differences between satellite-based estimates from conventional ground measurements. It is important to note that conventional ground measurements come with their own errors and uncertainty that should ideally be taken in consideration when used for verifying the accuracy of satellite-based estimates. This holds true for ET where the number of operational flux towers is limited, but also for rainfall that has distinct microscale variability and cannot be measured by a single gauge. However, in most documented studies these ground measurements are treated as "the best available estimates" in the absence of reliable information on their accuracy. As such, they are widely used to validate satellite-based data. The probability distribution functions of the mean absolute percentage errors for all three variables were created, and these functions have more value than a single research paper, with a single algorithm applied to a particular location.

The results show that the average MAPE for satellite-based estimates of annual or seasonal ET, rainfall, and land use classification are 5.4, 18.5, and 14.6 %, respectively. The largest error is thus associated with rainfall. Bias correction and local calibration of global and regional rainfall products seem to improve the quality of the data layers. However, more research is needed to improve remotely sensed rainfall estimation algorithms (e.g., CHIRPS), with a focus on downscaling procedures as the standard pixel size is often too large. Radar-based regional precipitation estimates that offer higher spatiotemporal resolutions are promising and need to be utilized further. Also, the attenuation of microwave signals between cellular communication networks can be used for assessing areal averaged rainfall. In addition, given the differences among reported precipitation measurements by different global and regional products for the same pixels, there is a need for a database that offers an ensemble based on a rigorous and statistically sound method.

In contrast to rainfall, the error in satellite-based ET is relatively small, especially at the aggregation level of a river basin. ET is a vital component of hydrological cycle and reliable estimates of local ET are the essential for modeling river basin hydrology accurately. Remotely sensed ET can be used both as input to distributed hydrological models, and as a means to calibrate the simulations, although locally large errors can occur. Nonetheless, despite its existing potential and accuracy, satellite-based ET is underutilized in hydrological studies. Contributing factors are presumably the difficulty to access and acquire reliable ET data through the public domain, and the difficulty to compare it with reliable field data. Thus, future focus should be on development of open access ET databases. Such efforts are now underway by various organizations such as the US Geological Survey, US Department of Agriculture, the Commonwealth Science and Industrial Research Organization of Australia and the Chinese Academy of Sciences. However, these products have not yet been made fully available to the public, albeit first estimates of an ensemble ET product are under development. There is also a need for higher-resolution ET data in terms of both spatial and temporal resolutions. This is a key factor if satellite-based ET data are to be used extensively in water management and hydrological studies.

Land use mapping was one of the earliest ways in which satellite imagery was used to produce environmental information and it is the most widely studied subject employing remote sensing. The quality of the classifications has improved over time by the availability of high-resolution images and local research projects. The low-resolution and operational land classification mapping product is, however, still the standard method. Global high-resolution land use and land cover databases are conceived as the next generation of information systems for WA+ and other applications, and the product created by Tsinghua University is a first example. The land use classifications come with an overall MAPE of 14.6 %, and accuracy of 85 %. This level of accuracy, although acceptable, calls for improvements given the wide use of these maps. Another important issue is the need for a new type of land use mapping dedicated to agricultural and river basin water management issues. This is of essential value when land use maps are used in hydrological and water management-related studies such as water accounting.

As revealed by the results of this review study, there is a great deal of heterogeneity regarding the accuracy and reliability of remote sensing data and methods. Oftentimes the reliability of remote-sensing-based products is rather case and location specific. Future research could, therefore, aim at cross-comparing remote sensing data and methods on ET, rainfall and land use for different regions. Ensemble mean ET products are currently under development.

Appendix A: Literature review on evapotranspiration

Table A1. Selected ET validation papers that describe experimental data sets covering a season or longer.

Method	Field instrument	Location and year	Land use	No. of images	Source	MAPE (%)
METRIC	Lysimeter	Idhao, US, 1985	Native sedge forage	4	Allen et al. (2005)	4
METRIC	Lysimeter	Idaho, US, 1989	Sugar beet	12	Allen et al. (2007)	1
ALEXI	Eddy covariance	New Mexico, US, 2008	Agricultural areas	6	Anderson et al. (2012)	6.7
SEBAL	Water balance	Sri Lanka	River basin	–	Bastiaanssen and Chandrapala (2003)	1
SEBAL	Water balance	Indus, Pakistan	River basin	20	Bastiaanssen et al. (2002)	5
SEBAL	Lysimeter	California, US, 2002	Alfalfa	7	F. Cassel and M. Robertson (personal communication, 2006)	2
SEBAL	Lysimeter	California, US, 2002	Peaches	7	F. Cassel and M. Robertson (personal communication, 2006)	7
SEBAL	Water balance	Murray–Darling Basin, Australia	River Basin	–	Evans et al. (2009)	1
NDVI-based model	Eddy covariance	New Mexico, US	Cottonwood, saltcedar	10	Groeneveld et al. (2007)	2.2
NDVI-based model	Bowen ratio	Colorado, US, 2006	Greasewood, salt rabbitbrush	5	Groeneveld et al. (2007)	12.2
NDVI-based model	Eddy covariance	California, US, 2000–2002	Salt grass, alkali sacaton	9	Groeneveld et al. (2007)	12.5
SEBAL	Water balance	Central Luzon, 2001	Rice	3	Hafeez et al. (2002)	10.5
SEBAL	Scintillometer	Horana, 1999	Palm trees and rice	5	Hemakumara and Chandrapala (2003)	0.9
METRIC	Bowen ratio	Nebraska, US, 2005	Corn	4	Irmak et al. (2011)	4.3
METRIC	Bowen ratio	Nebraska, US, 2006	Corn	4	Irmak et al. (2011)	4.2
SEBAL	Water balance	Western Cape, South Africa 2004–2006	Grapes	12	Jarmain et al. (2007)	12
ETWatch	Water balance	Hai Basin, China – 2002–2009	Basin	135	Jia et al. (2012)	8.3
SatDAET	Lysimeter	Florida, US, 1998	Alkali scrub	8	Jiang et al. (2009)	14
SatDAET	Lysimeter	Florida, US, 1999	Alkali scrub	3	Jiang et al. (2009)	19
CMRS1	Water balance	Australia	River basin	NA	King et al. (2011)	2.1
CMRS2	Water balance	Australia	River basin	NA	King et al. (2011)	0.6
NDTI	Water balance	Australia	River basin	NA	King et al. (2011)	18
ETLooK	Water balance	Australia	River basin	NA	King et al. (2011)	3.9
SEBAL	Scintillometer	Gediz Basin, Turkey, 1998	Grapes, cotton	4	Kite and Droogers (2000)	16
SEBAL	Surface renewal	Sacramento Valley, US, 2001	Rice	8	Lal et al. (2012)	1
TSEB	Measurements	Yellow River, China, 2004	Wetlands	–	Li et al. (2012)	7.9
SEBS	Measurements	Australia, 2009	Irrigated agriculture	16	Ma et al. (2012)	7.5

Table A1. Continued.

Method	Field instrument	Location and year	Land use	No. of images	Source	MAPE (%)
METRIC/SEBAL	Water balance	India, 2003	Irrigated agriculture	40	Mallick et al. (2007)	11.6
SEBAL	Water balance	Sudd, Sudan, 2000	Wetland	–	Mohamed et al. (2004)	1.8
SEBAL	Water balance	Sobat, Sudan, 2000	Wetland	–	Mohamed et al. (2004)	5.7
SEBAL	Water balance	California, US, 2002	Almonds	7	B. L. Sanden (personal communication, 2005)	1
SEBAL	Bowen ratio	Nebraska, US	Corn	7	Singh et al. (2008)	5
METRIC	Eddy covariance	Nebraska, US	River basin	8	Singh et al. (2011)	1.7
SEBAL	Water balance	California, US	Irrigated agriculture	5	Soppe et al. (2006)	1
SEBAL	Lysimeter	Idaho, US, 1989–1991	Irrigated agriculture	11	Tasumi et al. (2003)	4.3
SEBAL	Eddy covariance	Petrolina, 2001–2007	Mango, grapes	9	Teixeira et al. (2008)	1
SEBAL	Eddy covariance	Brazil	Natural vegetation and irrigated crops	18	Teixeira et al. (2009)	1
SEBAL	Water balance	Imperial Valley, 1997–1998	Several	12	Thoreson et al. (2009)	1
SEBAL	Eddy covariance	Middle Rio Grande, US, 2002–2003	Pecan, alfalfa	7	Wang and Sun (2005)	3
ETWatch	Lysimeter	Hai Basin, China, 2002–2005	Wheat-maize rotation	–	Wu et al. (2012)	9
ETWatch	Eddy covariance	Hai Basin, China, 2002–2005	River Basin	20	Wu et al. (2012)	3
ETWatch	Water balance	Hai Basin, China, 2002–2005	River basin	–	Wu et al. (2012)	1.8
SEBAL	Water balance	North district, China	Regional scale	26	Yang et al. (2012)	5.6
WUE*-based model	Eddy covariance	Jilin Province, China 2003	Mixed forest	45	Zhang et al. (2009)	4
WUE-based model	Eddy covariance	Jilin Province, China 2004	Mixed forest	45	Zhang et al. (2009)	2
WUE-based model	Eddy covariance	Jilin Province, China 2005	Mixed forest	45	Zhang et al. (2009)	0.4

* water use efficiency

Appendix B: Literature review on rainfall

Table B1. Selected validation papers that describe experimental data sets covering a season or longer.

Source	Area	Year	RS data source	Deviation
Almazroui et al. (2011)	Saudi Arabia	1998–2008	TRMM	0
Almazroui et al. (2012)	Saudi Arabia	1998–2008	TRMM	12.05
Asadullah et al. (2010)	Uganda	2003–2007	CMORPH	2
Asadullah et al. (2010)	Uganda	2003–2007	PERSIANN	8
Asadullah et al. (2010)	Uganda	2003–2007	RFE 2.0	19
Asadullah et al. (2010)	Uganda	2003–2007	TRMM 3B42	8
Asadullah et al. (2010)	Uganda	2003–2007	TAMSAT	12
Bitew and Gebremichael (2011)	Gilgel, Ethiopia	2006–2007	CMORPH	29
Bitew and Gebremichael (2011)	Gilgel, Ethiopia	2006–2007	TRMM 3B42RT	29
Bitew and Gebremichael (2011)	Gilgel, Ethiopia	2006–2007	PERSIANN	58
Bitew and Gebremichael (2011)	Gilgel, Ethiopia	2006–2007	TRMM 3B42	64
Cheema and Bastiaanssen (2012)	Indus	2007	TRMM 3B43 V6	6.1
Cheema and Bastiaanssen (2012)	Indus	2007	TRMM 3B43 V6	10.9
Chen et al. (2011)	Dongjing Basin, China	2002–2010	TRMM 3B42RT	22.1
Collischonn et al. (2008)	Tapajo's Basin, Brazil	1997–2006	TRMM 3B42	12
Dinku et al. (2007)	Ethiopian Highlands	1998–2004	TRMM 3B43	8
Dinku et al. (2011)	Blue Nile, Ethiopia	1981–2004	CMAP	3
Dinku et al. (2011)	Blue Nile, Ethiopia	1981–2004	GPCP	5
Dinku et al. (2011)	Blue Nile, Ethiopia	2003–2004	CMORPH	1
Dinku et al. (2011)	Blue Nile, Ethiopia	2003–2004	TRMM 3B42	5
Dinku et al. (2011)	Blue Nile, Ethiopia	2003–2004	RFE	48
Duan and Bastiaanssen (2013)	Lake Tana	1999, 2000, 2004	TRMM 3B43 V7	1
Duan and Bastiaanssen (2013)	Caspian Sea, Iran	2000–2003	TRMM 3B43 V7	20
Feidas (2009)	Greece	1998–2006	TRMM 3B42	4.2
Feidas (2009)	Greece	1998–2007	TRMM 3B43	7.6
Feidas (2009)	Greece	1998–2008	GPCP-1DD	28.7
Fernandes et al. (2008)	Amazon Basin, South America	1980–2002	ERA-40	10
Fernandes et al. (2008)	Amazon Basin, South America	1980–2002	GPCP	7
Fu et al. (2011)	Poyang Basin, China	2003–2006	GSMaP	54
Getirana et al. (2011)	Negro Basin, South America	1998–2002	TMPA	18
Getirana et al. (2011)	Negro Basin, South America	1998–2002	NCEP-2	13
Getirana et al. (2011)	Negro Basin, South America	1998–2002	ERA-40	18
Jiang et al. (2012)	Mishui Basin, China	2003–2008	CMORPH	41
Jiang et al. (2012)	Mishui Basin, China	2003–2008	3B42RT	43
Jiang et al. (2012)	Mishui Basin, China	2003–2008	3B42V6	4.54
Kizza et al. (2012)	Lake Victoria	2001–2004	TRMM 3B43	5
Kizza et al. (2012)	Lake Victoria	2001–2004	PERSIANN	1
Milewski et al. (2009)	Egypt		TRMM	15
Moffitt et al. (2011)	Bangladesh	2000–2005	TRMM 3B42V6	11.6
Pierre et al. (2011)	Sahelian belt	2004–2007	RFE 2.0	23
Pierre et al. (2011)	Sahelian belt	2004–2007	TRMM 3B42	6
Pierre et al. (2011)	Sahelian belt	2004–2007	CMORPH	34
Semire et al. (2012)	Malaysia	2001–2010	TRMM 3B43 V6	15
Stisen and Sandholt (2010)	Senegal River basin	2003–2005	CMORPH	34
Stisen and Sanholt (2010)	Senegal River basin	2003–2005	PERSIANN	47
Stisen and Sanholt (2010)	Senegal River basin	2003–2005	TRMM	23
Stisen and Sanholt (2010)	Senegal River basin	2003–2005	CCD	6
Stisen and Sanholt (2010)	Senegal River basin	2003–2005	CPC-FEWs	21
Su et al. (2008)	La Plata Basin	1998–2006	TRMM	6
Villarini et al. (2009)	Oklahoma, USA	1998–2003	TRMM	10
Voisin et al. (2008)	Amazon	1997–1999	ERA-40	26.5

Table B1. Continued.

Source	Area	Year	RS data source	Deviation
Voisin et al. (2008)	Amazon	1997–1999	GPCP 1DD	24.7
Voisin et al. (2008)	Mississippi, USA	1997–1999	ERA-40	32.3
Voisin et al. (2008)	Mississippi, USA	1997–1999	GPCP 1DD	25.3
Voisin et al. (2008)	Mackenzie, Canada	1997–1999	ERA-40	1.1
Voisin et al. (2008)	Mackenzie, Canada	1997–1999	GPCP 1DD	28.8
Voisin et al. (2008)	Congo, Africa	1997–1999	ERA-40	13.4
Voisin et al. (2008)	Congo, Africa	1997–1999	GPCP 1DD	31
Voisin et al. (2008)	Danube, Europe	1997–1999	ERA-40	29.1
Voisin et al. (2008)	Danube, Europe	1997–1999	GPCP 1DD	17.1
Voisin et al. (2008)	Mekong, SEA	1997–1999	ERA-40	0.4
Voisin et al. (2008)	Mekong, SEA	1997–1999	GPCP 1DD	4.1
Voisin et al. (2008)	Senegal	1997–1999	ERA-40	51.6
Voisin et al. (2008)	Senegal	1997–1999	GPCP 1DD	23.3
Voisin et al. (2008)	Yellow River, China	1997–1999	ERA-40	1.3
Voisin et al. (2008)	Yellow River, China	1997–1999	GPCP 1DD	30.4
Voisin et al. (2008)	Yenisei, Russia	1997–1999	ERA-40	0.7
Voisin et al. (2008)	Yenisei, Russia	1997–1999	GPCP 1DD	26.2
Wilk et al. (2006)	Okavango Basin	1991–1996	TRMM	20

Appendix C: Literature review on land use and land cover

Table C1. Selected validation papers that report on confusion matrices.

Source	Area	Image year	Image source	Overall accuracy (%)
Abd El-Kawy et al. (2011)	Nile Delta, Egypt	2005	Landsat ETM+	96
Aguirre-Gutiérrez et al. (2012)	Sierra Madre, Mexico	2006	Landsat ETM+	87
Bach et al. (2006)	Erda, Germany	1989–1992	CORINE (Landsat TM)	75
Bach et al. (2006)	Erda, Germany	1994	Landsat-5 TM	79
Bach et al. (2006)	Stein, Germany	1989–1992	CORINE (Landsat TM)	69
Bach et al. (2006)	Stein, Germany	1994	Landsat-5 TM	74
Bicheron et al. (2008)	Global	2004–2006	MERIS/Envisat	73
Blanco et al. (2013)	Latin America	2008	Modis-Terra	84
Büttner et al. (2006)	Global	1999–2000	Landsat ETM+/SPOT	87
Cassidy et al. (2013)	Lower Mekong	2005	Landsat TM	85
Cheema and Bastiaanssen (2010)	Indus Basin	2007	SPOT/vegetation	77
Cingolani (2004)	Cordoba, Argentina	1997	Landsat 5 TM	86
Clark et al. (2010)	Dry Chaco, South America	2000–2008	MODIS	80
Colditz et al. (2012)	Mexico	2005	MODIS	83
Hubert-Moy et al. (2001)	Baie de Lannion, France	1996–1997	Landsat 5TM	89
Estes et al. (2012)	Serengeti National Park	2002–2003	Landsat ETM+	83
Friedl et al. (2010)	Global	2000–2001	Modis 5	75
Gamanya et al. (2007)	Central Zimbabwe	2001	ASTER	92
Gamanya et al. (2007)	Central Zimbabwe	2001	Landsat TM	89
Kandrika and Roy (2008)	Orissa, India	2004–2005	AWiFS IRS-P6	87
Kavzoglu and Colkesen (2009)	Kocaeil, Turkey	1997	Landsat ETM+	91
Kavzoglu and Colkesen (2009)	Kocaeil, Turkey	1997	Landsat ETM+	90
Kavzoglu and Colkesen (2009)	Kocaeil, Turkey	2002	Aster	88
Kavzoglu and Colkesen (2009)	Kocaeil, Turkey	2002	Aster	93
Kavzoglu and Colkesen (2009)	Kocaeil, Turkey	2002	Aster	91
Kavzoglu and Colkesen (2009)	Kocaeil, Turkey	1997	Landsat ETM+	87
Kaya et al. (2002)	Kenya	2001	RADARSAT-1	85
Keuchel et al. (2003)	Tenerife, Spain	1988	Landsat 5TM	90

Table C1. Continued.

Source	Area	Image year	Image source	Overall accuracy (%)
Keuchel et al. (2003)	Tenerife, Spain	1988	Landsat 5TM	88
Keuchel et al. (2003)	Tenerife, Spain	1988	Landsat 5TM	93
Klein et al. (2012)	Central Asia	2009	MODIS	91
Kolios and Stylios (2013)	Greece	2009	Landsat 7 ETM+	97
Liu and Yang (2013)	Jilin, China	2009	Landsat TM	95
Liu et al. (2002)	Rondonia, Brazil	1995/1997	Landsat TM/Spot	80
Mayaux et al. (2006)	Global	1999–2000	SPOT-Vegetation	68
Munthali and Murayama (2011)	Dzalanyama, Malawi	2008	ALOS	79
Munthali and Murayama (2011)	Dzalanyama, Malawi	2000	Landsat ETM+	78
Oldeland et al. (2010)	Rehoboth, Namibia	2005	HyMap	98
Otukei and Blaschke (2010)	Pallisa, Uganda	2001	Landsat 7 ETM+	94
Pan et al. (2010)	Honghe Reserve, China	2006	Landsat-5 TM	88
Peña-Barragán et al. (2011)	Yolo County, California	2006	ASTER	79
Pérez-Hoyos et al. (2012)	Regional/Europe	–	Merged-global maps	87
Petropoulos et al. (2012)	Greece	2009	Hyperion	89
Qi et al. (2012)	Panyu, China	2009	RADARSAT-2 PolSAR	87
Ren et al. (2009)	NW Yunnan, China	2000	Landsat ETM+	97
Reno et al. (2011)	Amazon, Brazil	2008	Landsat 5	83
Renó et al. (2011)	Amazon, Brazil	1970	Landsat 2	86
Rodriguez-Galiano and Chica-Olmo (2012)	Granada, Spain	2004	Landsat 5TM	86
Rozenstein and Karnieli (2011)	Israel	2009	Landsat 5 TM	81
Setiawan et al. (2006)	Yogyakarta, Indonesia	1994	Landsat TM	80
Shao and Lunetta (2012)	North Carolina and Virginia, USA	2000–2009	MODIS	91
Shimoni et al. (2009)	Glinska Poljana, Croatia	2001	E-SAR	84
Shrestha and Zinck (2001)	Likhu Basin, Nepal	1988	Landsat TM	94
Song et al. (2005)	Connecticut, USA	2001	Landsat ETM	85
Stavrakoudis et al. (2011)	Lake Koronia, Greece	2005	IKONOS	93
Stefanov et al. (2001)	Arizona, USA	1998	Landsat TM	85
Sulla-Menashe et al. (2011)	Regional/Northern Eurasia	2001–2005	MODIS	73

Table C1. Continued.

Source	Area	Image year	Image source	Overall accuracy (%)
Szuster et al. (2011)	Thai island, Thailand	2004	ASTER	95
Szuster et al. (2011)	Thai island, Thailand	2004	ASTER	94
Szuster et al. (2011)	Thai island, Thailand	2004	ASTER	94
Taşdemir et al. (2012)	Bulgaria	2009	Rapideye	94
Thenkabail et al. (2009a)	Global	1997–1999	AVHHR	79
Tovar et al. (2013)	Cajamarca, Peru	2007	Landsat 5 TM	80
Tseng et al. (2008)	Connecticut, USA	1987	Landsat TM	98
Wang et al. (2010)	Hengshan, China	2003	Hyprion	80
Waske and Braun (2009)	Jena, Germany	2005	ENVISAT/ERS-2	83
Weiers et al. (2002)	Schleswig-Holstein, Germany	1992–1997	Landsat TM	85
Weiers et al. (2002)	Denmark	1992–1997	Landsat TM	70
Whiteside et al. (2011)	Florence Creek, Australia	2000	ASTER	79
Wickham et al. (2013) Wickham et al. (2013)	USA USA	2001 2006	Landsat TM Landsat TM	79 78
Wickham et al. (2013)	USA	2001	Landsat TM	85
Wickham et al. (2013)	USA	2006	Landsat TM	84
Wu et al. (2010)	Dan-Shuei, China	1995	Landsat 5 TM	88
Zhang et al. (2008)	North China plain, China	2003	MODIS_EVI	75
Zhu et al. (2012)	Massachusetts, USA	2007	ALOS	72
Zhu et al. (2012)	Massachusetts, USA	2000–2007	Landsat/ALOS	94
Zhu et al. (2012)	Massachusetts, USA	2000–2002	Landsat	93

Appendix D: Glossary

Table D1. Glossary.

Term	Description
1DD	One degree daily
3B42RT	3B42 real time
ALEXI	Atmosphere–Land Exchange Inverse
ALOS	Advanced Land Observing Satellite
AMSR-E	Advanced Microwave Sounding Radiometer-Earth
AMSU	Advanced Microwave Sounding Unit
ASTER	Advanced Spaceborne Thermal Emission and Reflection Radiometer
AVHRR	Advanced Very High Resolution Radiometer
CBERS	China Brazil Earth Resources Satellite
CMAP	CPC Merged Analysis of Precipitation
CMORPH	CPC Morphing technique
CMRSET	CSIRO MODIS Reflectance-based Scaling ET
CORINE	Coordination of Information on the Environment
CPC	Climate Prediction Center
CSIRO	Commonwealth Science and Industrial Research Organisation
DMSP	Defense Meteorological Satellite Program
EARS	Environmental Analysis and Remote Sensing
EUMETSAT	European Organisation for the Exploitation of Meteorological Satellites
FEWS	Famine Early Warning Systems (FEWS)
GOES	Geostationary Operational Environmental Satellites
GPCC	Global Precipitation Climatology Centre
GPCP	Global Precipitation Climatology Project
GPI	GOES precipitation index
GSFC	Goddard Space Flight Center (GSFC)
GSMaP	Global Satellite Mapping of Precipitation
HIRS	High-resolution Infrared Radiation Sounder
IR	Infrared
IWMI	International Water Management Institute
METRIC	Mapping EvapoTranspiration at high Resolution with Internalized Calibration
MODIS	Moderate Resolution Imaging Spectrometer
MPE	Multi-Sensor Precipitation Estimate
NASA	National Aeronautics and Space Administration
NDTI	Normalized difference temperature index
NOAA	National Oceanic and Atmospheric Administration
PERSIANN	Precipitation Estimation From Remotely Sensed Information using Artificial Neural Networks
PR	Precipitation radar
RFE	Rainfall estimation algorithm
SatDAET	Satellite daily ET
SEBAL	Surface Energy Balance Algorithm for Land
SEBS	Surface Energy Balance System
SEEAW	System of Environmental–Economic Accounts for Water
SPOT	Satellite Pour l'Observation de la Terre
SSM/I	Special Sensor Microwave/Imager
TAMSAT	Tropical Applications of Meteorology using Satellite data
TCI	TRMM Combined Instrument
TMI	TRMM Microwave Imager
TOVS	TIROS Operational Vertical Sounder
TRMM	Tropical Rainfall Measuring Mission
TSEB	Two source energy balance
VIIRS	Visible Infrared Imager Radiometer Suite
WIRADA	Water Information Research and Development Alliance

Acknowledgements. Funds for this research were provided by the CGIAR Research Programme on Water, Land and Ecosystems.

Edited by: B. van den Hurk

References

ABS – Australian Bureau of Statistics: Water Account, Australia – 2000–01, Canberra, 2004.

Abd El-Kawy, O. R., Rød, J. K., Ismail, H. A., and Suliman, A. S.: Land use and land cover change detection in the western Nile delta of Egypt using remote sensing data, Appl. Geogr., 31, 483–494, doi:10.1016/j.apgeog.2010.10.012, 2011.

Aguirre-Gutiérrez, J., Seijmonsbergen, A. C., and Duivenvoorden, J. F.: Optimizing land cover classification accuracy for change detection, a combined pixel-based and object-based approach in a mountainous area in Mexico, Appl. Geogr., 34, 29–37, doi:10.1016/j.apgeog.2011.10.010, 2012.

Allen, R. G., Pereira, L., Raes, D., and Smith, M.: Crop evapotranspiration: guidelines for computing crop water requirements, Issues 56–57, Food and Agriculture Organization of the United Nations, Rome, Italy, 1998.

Allen, R. G., Tasumi, M., Morse, A., and Trezza, R.: A Landsat-based energy balance and evapotranspiration model in Western US water rights regulation and planning, Irrig. Drain. Syst., 19, 251–268, doi:10.1007/s10795-005-5187-z, 2005.

Allen, R. G., Tasumi, M., Morse, A., Trezza, R., Wright, J. L., Bastiaanssen, W., Kramber, W., Lorite, I., and Robison, C. W.: Satellite-Based Energy Balance for Mapping Evapotranspiration with Internalized Calibration (METRIC) – Applications, J. Irrig. Drain. Eng., 133, 395–406, 2007.

Allen, R. G., Irmak, A., Trezza, R., Hendrickx, J. M. H., Bastiaanssen, W. and Kjaersgaard, J.: Satellite-based ET estimation in agriculture using SEBAL and METRIC, Hydrol. Process., 25, 4011–4027, doi:10.1002/hyp.8408, 2011.

Almazroui, M.: Calibration of TRMM rainfall climatology over Saudi Arabia during 1998–2009, Atmos. Res., 99, 400–414, 2011.

Almazroui, M., Islam, M. N., Jones, P. D., Athar, H., and Rahman, M. A.: Recent climate change in the Arabian Peninsula: Seasonal rainfall and temperature climatology of Saudi Arabia for 1979–2009, Atmos. Res., 111, 29–45, doi:10.1016/j.atmosres.2012.02.013, 2012.

Anderson, M. C., Norman, J. M., Diak, G. R., Kustas, W. P., and Mecikalski, J. R.: A Two-Source Time-Integrated Model for Estimating Surface Fluxes Using Thermal Infrared Remote Sensing, Remote Sens. Environ., 60, 195–216, doi:10.1016/S0034-4257(96)00215-5, 1997.

Anderson, M. C., Kustas, W. P., Norman, J. M., Hain, C. R., Mecikalski, J. R., Schultz, L., González-Dugo, M. P., Cammalleri, C., d'Urso, G., Pimstein, A., and Gao, F.: Mapping daily evapotranspiration at field to continental scales using geostationary and polar orbiting satellite imagery, Hydrol. Earth Syst. Sci., 15, 223–239, doi:10.5194/hess-15-223-2011, 2011.

Anderson, M. C., Kustas, W. P., Alfieri, J. G., Gao, F., Hain, C., Prueger, J. H., Evett, S., Colaizzi, P., Howell, T., and Chávez, J. L.: Mapping daily evapotranspiration at Landsat spatial scales during the BEAREX'08 field campaign, Adv. Water Resour., 50, 162–177, doi:10.1016/j.advwatres.2012.06.005, 2012.

Andre, J. C., Goutorbe, J. P., and Penier, A.: HAPEX-MOBILHY, a hydrologic atmospheric pilot experiment for the study of water budget and evaporation flux at the climatic scales, B. Am. Meteorol. Soc., 67, 138–144, 1986.

Asadullah, A., McIntyre, N. E. I. L., and Kigobe, M.: Evaluation of five satellite products for estimation of rainfall over Uganda, Hydrolog. Sci. J., 53, 1137–1150, 2008.

Asadullah, A., Mcintyre, N., and Kigobe, M. A. X.: Evaluation of five satellite products for estimation of rainfall over Uganda/Evaluation de cinq produits satellitaires pour l'estimation des précipitations en Ouganda Evaluation of five satellite products for estimation of rainfall over Uganda, Hydrolog. Sci. J., 53, 1137–1150, 2010.

Austin, P. M.: Relation between measured radar reflectivity and surface rainfall, Mon. Weather Rev., 115, 1053–1070, 1987.

Awulachew, S. B., Molden, D., Smakhtin, V., and Peden, D.: The Nile River Basin: Water, Agriculture, Governance and Livelihoods, Routledge, Oxon, UK, 2013.

Bach, M., Breuer, L., Frede, H. G., Huisman, J. A., Otte, A., and Waldhardt, R.: Accuracy and congruency of three different digital land-use maps, Landsc. Urban Plan., 78, 289–299, doi:10.1016/j.landurbplan.2005.09.004, 2006.

Barrett, E. C.: Precipitation monitoring by satellites. Remote sensing for studies of global environmental changes, ISPRA Courses, RS/88/10, Commission of European communities, Joint Research Center, Ispra Establishment, Ispra, Italy, 1988.

Barrett, E. C. and Beaumont, M. J.: Satellite rainfall monitoring: An overview, Remote Sens. Rev., 11, 23–48, doi:10.1080/02757259409532257, 1994.

Bastiaanssen, W. G. M.: Remote sensing in water resources management: the state of the art, International Water Management Institute (IWMI), Colombo, Sri Lanka, 1998.

Bastiaanssen, W. G. M. and Chandrapala, L.: Water balance variability across Sri Lanka for assessing agricultural and environmental water use, Agric. Water Manage., 58, 171–192, 2003.

Bastiaanssen, W. G. M., Menenti, M., Feddes, R. A., Holtslag, A. A. M., Pelgrum, H., Wang, J., Ma, Y., Moreno, J. F., Roerink, G. J., and van der Wal, T.: A remote sensing surface energy balance algorithm for land (SEBAL), 1. Formulation, J. Hydrol., 212–213, 198–212, 1998.

Bastiaanssen, W. G. M., Molden, D. J., and Makin, I. W.: Remote sensing for irrigated agriculture: examples from research and possible applications, Agric. Water Manage., 46, 137–155, doi:10.1016/S0378-3774(00)00080-9, 2000.

Bastiaanssen, W. G. M., Ahmad, M.-D., and Chemin, Y.: Satellite surveillance of evaporative depletion across the Indus Basin, Water Resour. Res., 38, 9-1–9-9, doi:10.1029/2001WR000386, 2002.

Bastiaanssen, W. G. M., Miltenburg, I., Evans, R., Molloy, R., Bastiaanssen, F. J. M., and van der Pol, E.: An operational satellite-based irrigation monitoring and scheduling tool for saving water in irrigation, in Irrigation and Drainage Conference, Swan Hill, Vic, Australia, 2009.

Bastiaanssen, W. G. M., Cheema, M. J. M., Immerzeel, W. W., Miltenburg, I., and Pelgrum, H.: The surface energy balance and actual evapotranspiration of the transboundary Indus Basin estimated from satellite measurements and the ETLook model,

Water Resour. Res., 48, W11512, doi:10.1029/2011WR010482, 2012.
Batra, N., Islam, S., Venturini, V., Bisht, G., and Jiang, L.: Estimation and comparison of evapotranspiration from MODIS and AVHRR sensors for clear sky days over the Southern Great Plains, Remote Sens. Environ., 103, 1–15, doi:10.1016/j.rse.2006.02.019, 2006.
Bicheron, P., Defourny, P., Brockmann, C., Schouten, L., Vancutsem, C., Huc, M., Bontemps, S., Leroy, M., Achard, F., Herold, M., Ranera, F., and Arino, O.: GLOBCOVER Products Report Description and Products Description and Validation Report, MEDIAS, France, 2008.
Bitew, M. M. and Gebremichael, M.: Evaluation of satellite rainfall products through hydrologic simulation in a fully distributed hydrologic model, Water Resour. Res., 47, W06526, doi:10.1029/2010WR009917, 2011.
Blanco, P. D., Colditz, R. R., López Saldaña, G., Hardtke, L. A., Llamas, R. M., Mari, N. A., Fischer, A., Caride, C., Aceñolaza, P. G., del Valle, H. F., Lillo-Saavedra, M., Coronato, F., Opazo, S. A., Morelli, F., Anaya, J. A., Sione, W. F., Zamboni, P., and Arroyo, V. B.: A land cover map of Latin America and the Caribbean in the framework of the SERENA project, Remote Sens. Environ., 132, 13–31, doi:10.1016/j.rse.2012.12.025, 2013.
Bolle, H. J., Eckardt, M., Koslowsky, D., Maselli, F., Melia-Miralles, J., Menenti, M., Olesen, F.-S., Petkov, L., Rasool, I., and van deGriend, A.: Mediterranean land-surface processes assessed from space, Series: Regional Climate Studies, XXVIII, Springer, Berlin, 760 pp., 2006.
Bosch, J. M. and Hewlett, J. D.: A review of catchment experiments to determine the effect of vegetation changes on water yield and evapotranspiration, J. Hydrol., 55, 3–23, doi:10.1016/0022-1694(82)90117-2, 1982.
Brutsaert, W.: Evaporation into the Atmosphere, Theory, History and Applications, D. Reidel Publishing Company, Dordrecht, the Netherlands, 302 pp., 1982.
Büttner, G., Feranec, J., and Jaffrain, G.: The thematic accuracy of Corine land cover 2000, European Environment Agency, Copenhagen, 2006.
Campbell, G. S. and Norman, J. M.: An introduction to environmental biophysics, Springer, New York, 1998.
Cassidy, L., Southworth, J., Gibbes, C., and Binford, M.: Beyond classifications: Combining continuous and discrete approaches to better understand land-cover change within the lower Mekong River region, Appl. Geogr., 39, 26–45, doi:10.1016/j.apgeog.2012.11.021, 2013.
Chartres, C. and Varma, S.: Out of Water: From Abundance to Scarcity and How to Solve the World's Water Problems, FT Press, New Jersey, 2010.
Chávez, J. L., Neale, C. M. U., Prueger, J. H., and Kustas, W. P.: Daily evapotranspiration estimates from extrapolating instantaneous airborne remote sensing ET values, Irrig. Sci., 27, 67–81, doi:10.1007/s00271-008-0122-3, 2008.
Cheema, M. J. M. and Bastiaanssen, W. G. M.: Land use and land cover classification in the irrigated Indus Basin using growth phenology information from satellite data to support water management analysis, Agric. Water Manage., 97, 1541–1552, 2010.
Cheema, M. J. M. and Bastiaanssen, W. G. M.: Local calibration of remotely sensed rainfall from the TRMM satellite for different periods and spatial scales in the Indus Basin, Int. J. Remote Sens., 33, 2603–2627, 2012.
Chehbouni, A., Kerr, Y. H., Watts, C., Hartogensis, O., Goodrich, D., Scott, R., Schieldge, J., Lee, K., Shuttleworth, W. J., Dedieu, G., and De Bruin, H. A. R.: Estimation of area-average sensible heat flux using a large-aperture scintillometer during the Semi-Arid Land-Surface-Atmosphere (SALSA) Experiment, Water Resour. Res., 35, 2505–2511, doi:10.1029/1999WR900111, 1999.
Chen, C., Yu, Z., Li, L., and Yang, C.: Adaptability Evaluation of TRMM Satellite Rainfall and Its Application in the Dongjiang River Basin, Proced. Environ. Sci., 10, 396–402, doi:10.1016/j.proenv.2011.09.065, 2011.
Choudhury, B. J.: Synergism of optical and microwave observations for land surface studies, in: Passive microwave remote sensing of land–atmosphere interactions, microwave remote sensing of land-atmosphere interactions, edited by: Choudhury, B. J., Kerr, Y. H., Njoku, E. G., and Pampaloni, P., VSP BV, Utrecht, the Netherlands, 155–191, 1995.
Cihlar, J.: Land cover mapping of large areas from satellites?: Status and research priorities, Int. J. Remote Sens., 21, 1093–1114, 2000.
Cingolani, A.: Mapping vegetation in a heterogeneous mountain rangeland using landsat data: an alternative method to define and classify land-cover units, Remote Sens. Environ., 92, 84–97, doi:10.1016/j.rse.2004.05.008, 2004.
Clark, M. L., Aide, T. M., Grau, H. R. and Riner, G.: A scalable approach to mapping annual land cover at 250 m using MODIS time series data: A case study in the Dry Chaco ecoregion of South America, Remote Sens. Environ., 114, 2816–2832, doi:10.1016/j.rse.2010.07.001, 2010.
Cleugh, H. A., Leuning, R., Mu, Q., and Running, S. W.: Regional evaporation estimates from flux tower and MODIS satellite data, Remote Sens. Environ., 106, 285–304, doi:10.1016/j.rse.2006.07.007, 2007.
Colditz, R. R., López Saldaña, G., Maeda, P., Espinoza, J. A., Tovar, C. M., Hernández, A. V., Benítez, C. Z., Cruz López, I., and Ressl, R.: Generation and analysis of the 2005 land cover map for Mexico using 250m MODIS data, Remote Sens. Environ., 123, 541–552, doi:10.1016/j.rse.2012.04.021, 2012.
Collischonn, B., Collischonn, W., and Tucci, C. E. M.: Daily hydrological modeling in the Amazon basin using TRMM rainfall estimates, J. Hydrol., 360, 207–216, doi:10.1016/j.jhydrol.2008.07.032, 2008.
Congalton, R. G.: A Review of Assessing the Accuracy of Classifications of Remotely Sensed Data, Remote Sens. Environ., 46, 35–46, 1991.
Courault, D., Seguin, B., and Olioso, A.: Review on estimation of evapotranspiration from remote sensing data: From empirical to numerical modeling approaches, Irrig. Drain. Syst., 19, 223–249, doi:10.1007/s10795-005-5186-0, 2005.
Cracknell, A. P. and Hayes, L.: Introduction to remote sensing, Taylor & Francis, London, 1991.
Dee, D. P., Uppala, S. M., Simmons, A. J., Berrisford, P., Poli, P., Kobayashi, S., Andrae, U., Balmaseda, M. A., Balsamo, G., Bauer, P., Bechtold, P., Beljaars, A. C. M., van de Berg, L., Bidlot, J., Bormann, N., Delsol, C., Dragani, R., Fuentes, M., Geer, A. J., Haimberger, L., Healy, S. B., Hersbach, H., Hólm, E. V., Isaksen, L., Kållberg, P., Köhler, M., Matricardi, M., McNally,

A. P., Monge-Sanz, B. M., Morcrette, J.-J., Park, B.-K., Peubey, C., de Rosnay, P., Tavolato, C., Thépaut, J.-N., and Vitart, F.: The ERA-Interim reanalysis: configuration and performance of the data assimilation system, Q. J. Roy. Meteorol. Soc., 137, 553–597, doi:10.1002/qj.828, 2011.

Dinku, T., Ceccato, P., Grover-Kopec, E., Lemma, M., Connor, S. J., and Ropelewski, C. F.: Validation of satellite rainfall products over East Africa's complex topography, Int. J. Remote Sens., 28, 1503–1526, doi:10.1080/01431160600954688, 2007.

Dinku, T., Connor, S., and Ceccato, P.: Nile River Basin, in: Nile River Basin, edited by: Melesse, A. M., Springer Netherlands, Dordrecht, 109–127, 2011.

Duan, Z. and Bastiaanssen, W. G. M.: First results from Version 7 TRMM 3B43 precipitation product in combination with a new downscaling–calibration procedure, Remote Sens. Environ., 131, 1–13, doi:10.1016/j.rse.2012.12.002, 2013.

Dugdale, G. and Milford, J. R.: Rainfall estimation over the Sahel using Meteosat thermal infrared data, in: ISLSCP Parameterization of Land-Surface Characteristics: use of Satellite Data in Climate Studies, edited by: Rolfe, E. and Battrick, B., ESA, Paris, 315–319, 1986.

Dunne, S. C., Entekhabi, D., and Njoku, E.: Impact of multiresolution active and passive microwave measurements on soil moisture estimation using the Ensemble Kalman Smoother, IEEE T. Geosci. Remote, 45, 1016–1028, 2007.

Engman, E. T. and Gurney, R. J.: Remote sensing in hydrology, Chapman and Hall Ltd., London, 1991.

Estes, A. B., Kuemmerle, T., Kushnir, H., Radeloff, V. C., and Shugart, H. H.: Land-cover change and human population trends in the greater Serengeti ecosystem from 1984–2003, Biol. Conserv., 147, 255–263, doi:10.1016/j.biocon.2012.01.010, 2012.

Euser, T., Luxemburg, W. M. J., Everson, C. S., Mengistu, M. G., Clulow, A. D., and Bastiaanssen, W. G. M.: A new method to measure Bowen ratios using high-resolution vertical dry and wet bulb temperature profiles, Hydrol. Earth Syst. Sci., 18, 2021–2032, doi:10.5194/hess-18-2021-2014, 2014.

Evans, R., Bastiaanssen, W., Molloy, R., Hulbert, S., and Miltenburg, I.: Improving the picture for irrigation using SEBAL in Australia to measure evapotranspiration (ET), in: Irrigation and Drainage Conference 2009, Swan Hill, Vic, Australia, 2009.

Feidas, H.: Validation of satellite rainfall products over Greece, Theor. Appl. Climatol., 99, 193–216, doi:10.1007/s00704-009-0135-8, 2009.

Fernandes, K., Fu, R., and Betts, A. K.: How well does the ERA40 surface water budget compare to observations in the Amazon River basin?, J. Geophys. Res., 113, D11117, doi:10.1029/2007JD009220, 2008.

Foody, G. M.: Status of land cover classification accuracy assessment, Remote Sens. Environ., 80, 185–201, doi:10.1016/S0034-4257(01)00295-4, 2002.

Franklin, S. E. and Wulder, M. A.: Remote sensing methods in medium spatial resolution satellite data land cover classification of large areas, Prog. Phys. Geogr., 26, 173–205, doi:10.1191/0309133302pp332ra, 2002.

Friedl, M. A., Sulla-Menashe, D., Tan, B., Schneider, A., Ramankutty, N., Sibley, A., and Huang, X.: MODIS Collection 5 global land cover: Algorithm refinements and characterization of new datasets, Remote Sens. Environ., 114(, 168–182, doi:10.1016/j.rse.2009.08.016, 2010.

Fu, Q., Ruan, R., and Liu, Y.: Accuracy Assessment of Global Satellite Mapping of Precipitation (GSMaP) Product over Poyang Lake Basin, China, Proced. Environ. Sci., 10, 2265–2271, doi:10.1016/j.proenv.2011.09.354, 2011.

Funk, C. C., Peterson, P. J., Landsfeld, M. F., Pedreros, D. H., Verdin, J. P., Rowland, J. D., Romero, B. E., Husak, G. J., Michaelsen, J. C., and Verdin, A. P.: A Quasi-Global Precipitation Time Series for Drought Monitoring, US Geological Survey Data Series 832, US Geological Survey, p. 4, doi:10.3133/ds832, 2013.

Gamanya, R., De Maeyer, P., and De Dapper, M.: An automated satellite image classification design using object-oriented segmentation algorithms: A move towards standardization, Expert Syst. Appl., 32, 616–624, doi:10.1016/j.eswa.2006.01.055, 2007.

García-Mora, T. J., Mas, J.-F., and Hinkley, E. A.: Land cover mapping applications with MODIS: a literature review, Int. J. Digit. Earth, 5, 63–87, doi:10.1080/17538947.2011.565080, 2012.

Getirana, A. C. V., Espinoza, J. C. V., Ronchail, J., and Rotunno Filho, O. C.: Assessment of different precipitation datasets and their impacts on the water balance of the Negro River basin, J. Hydrol., 404, 304–322, doi:10.1016/j.jhydrol.2011.04.037, 2011.

Glenn, E. P., Huete, A. R., Nagler, P. L., Hirschboeck, K. K., and Brown, P.: Integrating Remote Sensing and Ground Methods to Estimate Evapotranspiration, CRC, Crit. Rev. Plant Sci., 26, 139–168, doi:10.1080/07352680701402503, 2007.

Glenn, E. P., Doody, T. M., Guerschman, J. P., Huete, A. R., King, E. A., McVicar, T. R., Van Dijk, A. I. J. M., Van Niel, T. G., Yebra, M., and Zhang, Y.: Actual evapotranspiration estimation by ground and remote sensing methods: the Australian experience, Hydrol. Process., 25, 4103–4116, doi:10.1002/hyp.8391, 2011.

Gong, P., Wang, J., Yu, L., Zhao, Y. C., Zhao, Y. Y., Liang, L., Niu, Z. G., Huang, X. M., Fu, H. H., Liu, S., Li, C. C., Li, X. Y., Fu, W., Liu, C. X., Xu, Y., Wang, X. Y., Cheng, Q., Hu, L. Y., Yao, W. B., Zhang, H., Zhu, P., Zhao, Z. Y., Zhang, H. Y., Zheng, Y. M., Ji, L. Y., Zhang, Y. W., Chen, H., Yan, A., Guo, J. H., Yu, L., Wang, L., Liu, X. J., Shi, T. T., Zhu, M. H., Chen, Y. L., Yang, G. W., Tang, P., Xu, B., Ciri, C., Clinton, N., Zhu, Z. L., Chen, J., and Chen, J.: Finer resolution observation and monitoring of global land cover: first mapping results with Landsat TM and ETM+ data, Int. J. Remote Sens., 34, 2607–2654, 2013.

Gonzalez-Dugo, M. P., Neale, C. M. U., Mateos, L., Kustas, W. P., Prueger, J. H., Anderson, M. C., and Li, F.: A comparison of operational remote sensing-based models for estimating crop evapotranspiration, Agr. Forest Meteorol., 149, 1843–1853, doi:10.1016/j.agrformet.2009.06.012, 2009.

Goutorbe, J. P., Lebel, T., Dolman, A. J., Gash, J. H. C., Kabat, P., Kerr, Y. H., Monteny, B., Prince, S. D., Stricker, J. N. M., Tinga, A., and Wallace, J. S.: An overview of HAPEX-Sahel: a study in climate and desertification, J. Hydrol., 188–189, 4–17, doi:10.1016/S0022-1694(96)03308-2, 1997.

Gowda, P. H., Colaizzi, P. D., Evett, S. R., Howell, T. A., and Tolk, J. A.: Remote sensing based energy balance algorithms for mapping ET: current status and future challenges, T. ASABE, 50, 1639–1644, 2007.

Groeneveld, D. P., Baugh, W. M., Sanderson, J. S., and Cooper, D. J.: Annual groundwater evapotranspiration mapped from single satellite scenes, J. Hydrol., 344, 146–156, doi:10.1016/j.jhydrol.2007.07.002, 2007.

Guerschman, J. P., Van Dijk, A. I. J. M., Mattersdorf, G., Beringer, J., Hutley, L. B., Leuning, R., Pipunic, R. C., and Sherman, B. S.: Scaling of potential evapotranspiration with MODIS data reproduces flux observations and catchment water balance observations across Australia, J. Hydrol., 369, 107–119, doi:10.1016/j.jhydrol.2009.02.013, 2009.

Hafeez, M. M., Chemin, Y., Van De Giesen, N., and Bouman, B. A. M.: Estimation of crop water deficit through remote sensing in Central Luzon, Philippines, in: vol. 5, IEEE International Geoscience and Remote Sensing Symposium, 24–28 June, Torento, Canada, 2778–2780, 2002.

Hartogensis, O. K., Weisensee, U., Evans, J., van Kesteren, A. J. H., and Beyrich, F.: First Results of two Optical Millimeter-wave Scintillometer Systems during LITFASS2009, 10th EMS Annu. Meet. 10th Eur. Conf. Appl. Meteorol. Abstr., held: 13–17 September 2010, Zürich, Switzerland, http://meetings.copernicus.org/ems2010/ (last access: 5 October 2013), 2010.

Hemakumara, H. M. and Chandrapala, L.: Evapotranspiration fluxes over mixed vegetation areas measured from large aperture scintillometer, Agr. Water Manage., 58, 109–122, 2003.

Houser, P. R., Shuttleworth, W. J., Famiglietti, J. S., and Goodrich, D. C.: Integration of soil moisture remote sensing and hydrologic modeling using data assimilation, Water Resour. Res., 34, 3405–3420, 1998.

Hubert-Moy, L., Cotonnec, A., Le Du, L., Chardin, A., and Perez, P.: A comparison of parametric classification procedures of remotely sensed data applied on different landscape units, Remote Sens. Environ., 75, 174–187, 2001.

Huffman, G. J., Bolvin, D. T., Nelkin, E. J., Wolff, D. B., Adler, R. F., Gu, G., Hong, Y., Bowman, K. P., and Stocker, E. F.: The TRMM Multisatellite Precipitation Analysis (TMPA): Quasi-Global, Multiyear, Combined-Sensor Precipitation Estimates at Fine Scales, J. Hydrometeorol., 8, 38–55, doi:10.1175/JHM560.1, 2007.

Hunink, J. E., Immerzeel, W. W., and Droogers, P.: A High-resolution Precipitation 2-step mapping Procedure (HiP2P): Development and application to a tropical mountainous area, Remote Sens. Environ., 140, 179–188, doi:10.1016/j.rse.2013.08.036, 2014.

Immerzeel, W. W. and Droogers, P.: Calibration of a distributed hydrological model based on satellite evapotranspiration, J. Hydrol., 349, 411–424, doi:10.1016/j.jhydrol.2007.11.017, 2008.

Irmak, A. (Ed.): Evapotranspiration Remote Sensing and Modeling, InTech, Rijeka, Croatia, 2012.

Irmak, A., Ratcliffe, I., and Hubbard, K.: Estimation of Land Surface Evapotranspiration with a Satellite Remote Sensing Procedure, Great Plains Research 21, University of Nebraska, Lincoln, 73–88, 2011.

Jackson, T. J., Le Vine, D. M., Griffis, A. J., Goodrich, D. C., Schmugge, T. J., Swift, C. T., and O'Neill, P. E.: Soil moisture and rainfall estimation over a semiarid environment with the ESTAR microwave radiometer, IEEE T. Geosci. Remote, 31, 836–841, doi:10.1109/36.239906, 1993.

Jarmain, C., Klaasse, A., Bastiaanssen, W. G. M., and Roux, A. S.: Remote sensing tools for water use efficiency of grapes in the Winelands region, Western Cape, in: 13th Sanciahs symposium, Cape Town, 2007.

Jarmain, C., Everson, C., Savage, M., Mengisto, M., Clulow, A., Walker, S., and Gush, M.: Refining tools for evaporation monitoring in support of water resources management, Water Research Commission, Pretoria, South Africa, 2009.

Jhorar, R. K., Smit, A. A. M. F. R., Bastiaanssen, W. G. M., and Roest, C. W. J.: Calibration of a distributed irrigation water management model using remotely sensed evapotranspiration rates and groundwater heads, Irrig. Drain., 60, 57–69, doi:10.1002/ird.541, 2011.

Jia, L., Su, B., van den Hurk, B., Menenti, M., Moene, A., and de Bruin, H. A. R.: Estimation of sensible heat flux using the Surface Energy Balance System (SEBS) and ATSR measurements, Phys. Chem. Earth, 28, 75–88, 2003.

Jia, Z., Liu, S., Xu, Z., Chen, Y., and Zhu, M.: Validation of remotely sensed evapotranspiration over the Hai River Basin, China, J. Geophys. Res., 117, D13113, doi:10.1029/2011JD017037, 2012.

Jiang, L., Islam, S., Guo, W., Singh Jutla, A., Senarath, S. U. S., Ramsay, B. H., and Eltahir, E.: A satellite-based Daily Actual Evapotranspiration estimation algorithm over South Florida, Global Planet. Change, 67, 62–77, doi:10.1016/j.gloplacha.2008.12.008, 2009.

Jiang, S., Ren, L., Hong, Y., Yong, B., Yang, X., Yuan, F., and Ma, M.: Comprehensive evaluation of multi-satellite precipitation products with a dense rain gauge network and optimally merging their simulated hydrological flows using the Bayesian model averaging method, J. Hydrol., 452–453, 213–225, doi:10.1016/j.jhydrol.2012.05.055, 2012.

Kalma, J. D., McVicar, T. R., and McCabe, M. F.: Estimating land surface evaporation: A review of methods using remotely sensed surface temperature data, Surv. Geophys., 29, 421–469, 2008.

Kandrika, S. and Roy, P. S.: Land use land cover classification of Orissa using multi-temporal IRS-P6 awifs data: A decision tree approach, Int. J. Appl. Earth Obs. Geoinf., 10, 186–193, doi:10.1016/j.jag.2007.10.003, 2008.

Karimi, P., Bastiaanssen, W. G. M., and Molden, D.: Water Accounting Plus (WA+) – a water accounting procedure for complex river basins based on satellite measurements, Hydrol. Earth Syst. Sci., 17, 2459–2472, doi:10.5194/hess-17-2459-2013, 2013a.

Karimi, P., Bastiaanssen, W. G. M., Molden, D., and Cheema, M. J. M.: Basin-wide water accounting based on remote sensing data: an application for the Indus Basin, Hydrol. Earth Syst. Sci., 17, 2473–2486, doi:10.5194/hess-17-2473-2013, 2013b.

Karimi, P., Bastiaanssen, W. G. M., Sood, A., Hoogeveen, J., Peiser, L., Bastidas-Obando, E., and Dost, R.: Spatial evapotranspiration, rainfall and land use data in water accounting – Part 2: Reliability of water accounting results for policy decisions in the Awash basin, Hydrol. Earth Syst. Sci., 19, 533–550, doi:10.5194/hess-19-533-2015, 2015.

Kavzoglu, T. and Colkesen, I.: A kernel functions analysis for support vector machines for land cover classification, Int. J. Appl. Earth Obs. Geoinf., 11, 352–359, doi:10.1016/j.jag.2009.06.002, 2009.

Kaya, S., Pultz, T. J., Mbogo, C. M., Beier, J. C., and Mushinzimana, E.: The Use of Radar Remote Sensing for Identifying Environmental Factors Associated with Malaria Risk in Coastal Kenya, International Geoscience and Remote Sensing Symposium (IGARSS '02), 24–28 June, Toronto, 2002.

Keuchel, J., Naumann, S., Heiler, M., and Siegmund, A.: Automatic land cover analysis for Tenerife by supervised classification using remotely sensed data, Remote Sens. Environ., 86, 530–541, doi:10.1016/S0034-4257(03)00130-5, 2003.

Kidd, C.: Satellite rainfall climatology: a review, Int. J. Climatol., 21, 1041–1066, doi:10.1002/joc.635, 2001.

King, E. A., Van Niel, T. G., Van Dijk, A. I. J. M., Wang, Z., Paget, M. J., Raupach, T., Guerschman, J., Haverd, V., Mcvicar, T. R., Miltenburg, I., Raupach, M. R., and Zhang, Y.: Actual Evapotranspiration Estimates for Australia Inter-comparison and Evaluation, CSIRO: Water for a Healthy Country National Research Flagship Copyright, Canberra, 2011.

Kiptala, J., Mohamed, Y., Mul, M., Cheema, M. J. M., and van der Zaag, P.: Land use and land cover classification using phenological variability from MODIS vegetation in the Upper Pangani River Basin, Eastern Africa, Phys. Chem. Earth A/B/C, 66, 112–122, 2013.

Kite, G. and Droogers, P.: Comparing evapotranspiration estimates from satellites, hydrological models and field data, J. Hydrol., 229, 1–2, doi:10.1016/S0022-1694(99)00193-6, 2000.

Kizza, M., Westerberg, I., Rodhe, A., and Ntale, H. K.: Estimating areal rainfall over Lake Victoria and its basin using ground-based and satellite data, J. Hydrol., 464–465, 401–411, doi:10.1016/j.jhydrol.2012.07.024, 2012.

Klein, I., Gessner, U., and Kuenzer, C.: Regional land cover mapping and change detection in Central Asia using MODIS time-series, Appl. Geogr., 35, 219–234, doi:10.1016/j.apgeog.2012.06.016, 2012.

Kolios, S. and Stylios, C. D.: Identification of land cover/land use changes in the greater area of the Preveza peninsula in Greece using Landsat satellite data, Appl. Geogr., 40, 150–160, doi:10.1016/j.apgeog.2013.02.005, 2013.

Kummerow, C., Olson, W. S., and Giglio, L.: A simplified scheme for obtaining precipitation and vertical hydrometeor profiles from passive microwave sensors, IEEE T. Geosci. Remote, 34, 1213–1232, doi:10.1109/36.536538, 1996.

Kustas, W. P. and Norman, J. M.: Use of remote sensing for evapotranspiration monitoring over land surfaces, Hydrolog. Sci. J., 41, 495–516, doi:10.1080/02626669609491522, 1996.

Kustas, W. P., Anderson, M. C., French, A. N., and Vickers, D.: Using a remote sensing field experiment to investigate flux-footprint relations and flux sampling distributions for tower and aircraft-based observations, Adv. Water Resour., 29, 355–368, doi:10.1016/j.advwatres.2005.05.003, 2006.

Lal, D., Clark, B., Bettner, T., Thoreson, B., and Snyder, R.: Rice evapotranspiration estimates and crop coefficients in Glenn-Colusa Irrigation District, Sacramento Valley, California, in: US-CID Water Management Conference, Austin, 2012.

Li, H., Wang, H., Kong, Y., and Li, L.: Estimation of evapotranspiration in Yellow River Delta wetland based on two-source energy balance (TSEB) model, Remote Sensing Technol. Appl., 27, 58–67, 2012.

Liu, T. and Yang, X.: Mapping vegetation in an urban area with stratified classification and multiple endmember spectral mixture analysis, Remote Sens. Environ., 133, 251–264, doi:10.1016/j.rse.2013.02.020, 2013.

Liu, X.-H., Skidmore, A. K., and Van Oosten, H.: Integration of classification methods for improvement of land-cover map accuracy, ISPRS J. Photogramm. Remote Sens., 56, 257–268, doi:10.1016/S0924-2716(02)00061-8, 2002.

Ma, W., Hafeez, M., Rabbani, U., Ishikawa, H., and Ma, Y.: Retrieved actual ET using SEBS model from Landsat-5 TM data for irrigation area of Australia, Atmos. Environ., 59, 408–414, doi:10.1016/j.atmosenv.2012.05.040, 2012.

Maes, W. H., Heuvelmans, G., and Muys, B.: Assessment of Land Use Impact on Water-Related Ecosystem Services Capturing the Integrated Terrestrial-Aquatic System, Environ. Sci. Technol., 43, 7324–7330, doi:10.1021/es900613w, 2009.

Mallick, K., Bhattacharya, B. K., Chaurasia, S., Dutta, S., Nigam, R., Mukherjee, J., Banerjee, S., Kar, G., Rao, V. U. M., Gadgil, A. S., and Parihar, J. S.: Evapotranspiration using MODIS data and limited ground observations over selected agroecosystems in India, Int. J. Remote Sens., 28, 2091–2110, 2007.

Mayaux, P., Eva, H., Gallego, J., Strahler, A. H. A. H., Herold, M., Agrawal, S., Naumov, S., De Miranda, E. E., Di Bella, C. M., Ordoyne, C., Kopin, Y., Roy, P. S. S., Member, S., De Miranda, E. E., and Di Bella, C. M.: Validation of the global land cover 2000 map, IEEE T. Geosci. Remote, 44, 1728–1739, doi:10.1109/TGRS.2006.864370, 2006.

Mcvicar, T. R. and Jupp, D. L. B.: Estimating one-time-of-day meteorological data from standard daily data as inputs to thermal remote sensing based energy balance models, Agr. Forest Meteorol., 96, 219–238, 1999.

Mcvicar, T. R. and Jupp, D. L. B.: Using covariates to spatially interpolate moisture availability in the Murray – Darling Basin A novel use of remotely sensed data, Remote Sens. Environ., 79, 199–212, 2002.

Meijninger, W. M. L. and de Bruin, H. A. R.: The sensible heat fluxes over irrigated areas in western Turkey determined with a large aperture scintillometer, J. Hydrol., 229, 42–49, doi:10.1016/S0022-1694(99)00197-3, 2000.

Milewski, A., Sultan, M., Jayaprakash, S. M., Balekai, R., and Becker, R.: RESDEM, a tool for integrating temporal remote sensing data for use in hydrogeologic investigations, Comput. Geosci., 35, 2001–2010, doi:10.1016/j.cageo.2009.02.010, 2009.

Miralles, D. G., Holmes, T. R. H., De Jeu, R. A. M., Gash, J. H., Meesters, A. G. C. A., and Dolman, A. J.: Global land–surface evaporation estimated from satellite-based observations, Hydrol. Earth Syst. Sci., 15, 453–469, doi:10.5194/hess-15-453-2011, 2011.

Moffitt, C. B., Hossain, F., Adler, R. F., Yilmaz, K. K., and Pierce, H. F.: Validation of a TRMM-based global Flood Detection System in Bangladesh, Int. J. Appl. Earth Obs. Geoinf., 13, 165–177, doi:10.1016/j.jag.2010.11.003, 2011.

Mohamed, Y. A., Bastiaanssen, W. G. M., and Savenije, H. H. G.: Spatial variability of evaporation and moisture storage in the swamps of the upper Nile studied by remote sensing techniques, J. Hydrol., 289, 145–164, doi:10.1016/j.jhydrol.2003.11.038, 2004.

Molden, D.: Water for food, water for life: A comprehensive assessment of water management in agriculture, Earthscan, London, 2007.

Molden, D., Dong, B., Loeve, R., Barker, R., and Tuong, T. P.: Agricultural water productivity and savings: policy lessons from two diverse sites in China, Water Policy, 9, 29–44, 2007.

Moon, B., Hong, J., Lee, B., Yun, J. I., Park, E. W., and Kim, J.: CO_2 and energy exchange in a rice paddy for the growing season of 2002 in Hari, Korea Byung-Kwan, Korean J. Agric. For. Meteorol., 5, 51–60, 2003.

Moran, M. S. and Jackson, R. D.: Assessing the spatial distribution of evapotranspiration using remotely sensed inputs, J. Environ. Qual., 20, 725–737, 1991.

Moran, M. S., Clarke, T. R., Inoue, Y., and Vidal, A.: Estimating crop water deficit using the relation between surface-air temperature and spectral vegetation index, Remote Sens. Environ., 49, 246–263, doi:10.1016/0034-4257(94)90020-5, 1994.

Mu, Q., Heinsch, F. A., Zhao, M., and Running, S. W.: Development of a global evapotranspiration algorithm based on MODIS and global meteorology data, Remote Sens. Environ., 111, 519–536, 2007.

Mu, Q., Zhao, M., and Running, S. W.: Improvements to a MODIS global terrestrial evapotranspiration algorithm, Remote Sens. Environ., 115, 1781–1800, doi:10.1016/j.rse.2011.02.019, 2011.

Mucher, C. A., Steinnocher, K. T., Kressler, F. P., and Heunks, C.: Land cover characterization and change detection for environmental monitoring of pan-Europe, Int. J. Remote Sens., 21, 1159–1181, doi:10.1080/014311600210128, 2000.

Mueller, B., Hirschi, M., Jimenez, C., Ciais, P., Dirmeyer, P. A., Dolman, A. J., Fisher, J. B., Jung, M., Ludwig, F., Maignan, F., Miralles, D. G., McCabe, M. F., Reichstein, M., Sheffield, J., Wang, K., Wood, E. F., Zhang, Y., and Seneviratne, S. I.: Benchmark products for land evapotranspiration: LandFlux-EVAL multi-data set synthesis, Hydrol. Earth Syst. Sci., 17, 3707–3720, doi:10.5194/hess-17-3707-2013, 2013.

Munthali, K. G. and Murayama, Y.: Land use/cover change detection and analysis for Dzalanyama forest reserve, Lilongwe, Malawi, Procedia – Social and Behavioral Sciences, 21, 203–211, doi:10.1016/j.sbspro.2011.07.035, 2011.

Neale, C. M. U. and Cosh, M. H.: Remote sensing and hydrology, IAHS Red Book Series, Publ. 352, IAHS Wallingford, UK, 482 pp., 2012.

Nemani, R. R. and Running, S. W.: Estimation of Regional Surface Resistance to Evapotranspiration from NDVI and Thermal-IR AVHRR Data, J. Appl. Meteorol., 28, 276–284, doi:10.1175/1520-0450(1989)028<0276:EORSRT>2.0.CO;2, 1989.

Norman, J. M., Kustas, W. P., and Humes, K. S.: Source approach for estimating soil and vegetation energy fluxes in observations of directional radiometric surface temperature, Agr. Forest Meteorol., 77, 263–293, doi:10.1016/0168-1923(95)02265-Y, 1995.

Oldeland, J., Dorigo, W., Lieckfeld, L., Lucieer, A., and Jürgens, N.: Combining vegetation indices, constrained ordination and fuzzy classification for mapping semi-natural vegetation units from hyperspectral imagery, Remote Sens. Environ., 114, 1155–1166, doi:10.1016/j.rse.2010.01.003, 2010.

Otter, L. B., Guenther, A., and Greenberg, J.: Seasonal and spatial variations in biogenic hydrocarbon emissions from southern African savannas and woodlands, Atmos. Environ., 36, 4265–4275, 2002.

Otukei, J. R. and Blaschke, T.: Land cover change assessment using decision trees, support vector machines and maximum likelihood classification algorithms, Int. J. Appl. Earth Obs. Geoinf., 12, S27–S31, doi:10.1016/j.jag.2009.11.002, 2010.

Pan, X., Zhang, S., Zhang, H., Na, X., and Li, X.: A variable precision rough set approach to the remote sensing land use/cover classification, Comput. Geosci., 36, 1466–1473, doi:10.1016/j.cageo.2009.11.010, 2010.

Peña-Barragán, J. M., Ngugi, M. K., Plant, R. E., and Six, J.: Object-based crop identification using multiple vegetation indices, textural features and crop phenology, Remote Sens. Environ., 115, 1301–1316, doi:10.1016/j.rse.2011.01.009, 2011.

Pérez-Hoyos, A., García-Haro, F. J., and San-Miguel-Ayanz, J.: A methodology to generate a synergetic land-cover map by fusion of different land-cover products, Int. J. Appl. Earth Obs. Geoinf., 19, 72–87, doi:10.1016/j.jag.2012.04.011, 2012.

Petropoulos, G. P., Kalaitzidis, C., and Prasad Vadrevu, K.: Support vector machines and object-based classification for obtaining land-use/cover cartography from Hyperion hyperspectral imagery, Comput. Geosci., 41, 99–107, doi:10.1016/j.cageo.2011.08.019, 2012.

Petty, G. W.: The status of satellite-based rainfall estimation over land, Remote Sens. Environ., 51, 125–137, doi:10.1016/0034-4257(94)00070-4, 1995.

Petty, G. W. and Krajewski, W. F. W. F.: Satellite estimation of precipitation over land, Hydrolog. Sci. J., 41, 433–452, doi:10.1080/02626669609491519, 1996.

Pierre, C., Bergametti, G., Marticorena, B., Mougin, E., Lebel, T., and Ali, A.: Pluriannual comparisons of satellite-based rainfall products over the Sahelian belt for seasonal vegetation modeling, J. Geophys. Res., 116, D18201, doi:10.1029/2011JD016115, 2011.

Qi, Z., Yeh, A. G.-O., Li, X., and Lin, Z.: A novel algorithm for land use and land cover classification using RADARSAT-2 polarimetric SAR data, Remote Sens. Environ., 118, 21–39, doi:10.1016/j.rse.2011.11.001, 2012.

Rana, G. and Katerji, N.: Measurement and estimation of actual evapotranspiration in the field under Mediterranean climate: a review, Eur. J. Agron., 13, 125–153, doi:10.1016/S1161-0301(00)00070-8, 2000.

Rango, A., Ritchie, J. C., Kustas, W. P., Schmugge, T. J., Humes, K. S., Hipps, L. E., Prueger, J. H., and HavstadRango, K. M.: JORNEX: A multidisciplinary remote sensing campaign to quantify plant community/atmospheric interactions in the northern Chihuahuan desert of New Mexico, in: Hydrology in a Changing Environment, edited by: Wheater, H. and Kirby, C., John Wiley, London, UK, 585–590, 1998.

Ren, G., Zhu, A.-X., Wang, W., Xiao, W., Huang, Y., Li, G., Li, D., and Zhu, J.: A hierarchical approach coupled with coarse DEM information for improving the efficiency and accuracy of forest mapping over very rugged terrains, Forest Ecol. Manage., 258, 26–34, doi:10.1016/j.foreco.2009.03.043, 2009.

Renó, V. F., Novo, E. M. L. M., Suemitsu, C., Rennó, C. D., and Silva, T. S. F.: Assessment of deforestation in the Lower Amazon floodplain using historical Landsat MSS/TM imagery, Remote Sens. Environ., 115, 3446–3456, doi:10.1016/j.rse.2011.08.008, 2011.

Rientjes, T. H. M., Muthuwatta, L. P., Bos, M. G., Booij, M. J., and Bhatti, H. A.: Multi-variable calibration of a semi-distributed hydrological model using streamflow data and satellite-based evapotranspiration, J. Hydrol., 505, 276–290, 2013.

Rodriguez-Galiano, V. and Chica-Olmo, M.: Land cover change analysis of a Mediterranean area in Spain using different sources of data: Multi-seasonal Landsat images, land surface temperature, digital terrain models and texture, Appl. Geogr., 35, 208–218, doi:10.1016/j.apgeog.2012.06.014, 2012.

Roerink, G. J., Su, Z., and Menenti, M.: S-SEBI: A simple remote sensing algorithm to estimate the surface energy balance, Phys. Chem. Earth B, 25, 147–157, doi:10.1016/S1464-1909(99)00128-8, 2000.

Roerink, G. J., Menenti, M., Soepboer, W., and Su, Z.: Assessment of climate impact on vegetation dynamics by using remote sensing, Phys. Chem. Earth, A/B/C, 28, 103–109, doi:10.1016/S1474-7065(03)00011-1, 2003.

Rosema, A.: Comparison of Meteosat-based rainfall and evapotranspiration mapping in the Sahel region, Int. J. Remote Sens., 11, 2299–2309, doi:10.1080/01431169008955176, 1990.

Rozenstein, O. and Karnieli, A.: Comparison of methods for land-use classification incorporating remote sensing and GIS inputs, Appl. Geogr., 31, 533–544, doi:10.1016/j.apgeog.2010.11.006, 2011.

Schuurmans, J. M., Troch, P. A., Veldhuizen, A. A., Bastiaanssen, W. G. M., and Bierkens, M. F. P.: Assimilation of remotely sensed latent heat flux in a distributed hydrological model, Adv. Water Resour., 26, 151–159, 2003.

Scott, C. A., Bastiaanssen, W. G. M., and Ahmad, M.-D.: Mapping Root Zone Soil Moisture Using Remotely Sensed Optical Imagery, J. Irrig. Drain. Eng., 129, 326–335, doi:10.1061/(ASCE)0733-9437(2003)129:5(326), 2003.

Semire, F. A., Mohd-Mokhtar, R., Ismail, W., Mohamad, N., and Mandeep, J. S. S.: Ground validation of space-borne satellite rainfall products in Malaysia, Adv. Space Res., 50, 1241–1249, doi:10.1016/j.asr.2012.06.031, 2012.

Setiawan, H., Mathieu, R., and Thompson-Fawcett, M.: Assessing the applicability of the V-I-S model to map urban land use in the developing world: Case study of Yogyakarta, Indonesia, Comput. Environ. Urban Syst., 30, 503–522, doi:10.1016/j.compenvurbsys.2005.04.003, 2006.

Shao, Y. and Lunetta, R. S.: Comparison of support vector machine, neural network, and CART algorithms for the land-cover classification using limited training data points, ISPRS J. Photogramm. Remote Sens., 70, 78–87, doi:10.1016/j.isprsjprs.2012.04.001, 2012.

Shimoni, M., Borghys, D., Heremans, R., Perneel, C., and Acheroy, M.: Fusion of PolSAR and PolInSAR data for land cover classification, Int. J. Appl. Earth Obs. Geoinf., 11, 169–180, doi:10.1016/j.jag.2009.01.004, 2009.

Shrestha, D. P. and Zinck, J. A.: Land use classification in mountainous areas: integration of image processing, digital elevation data and field knowledge (application to Nepal), Int. J. Appl. Earth Obs. Geoinf., 3, 78–85, doi:10.1016/S0303-2434(01)85024-8, 2001.

Singh, R. K., Irmak, A., Irmak, S., and Martin, D. L.: Application of SEBAL Model for Mapping Evapotranspiration and Estimating Surface Energy Fluxes in South-Central Nebraska, J. Irrig. Drain. Eng., 134, 273–285, doi:10.1061/(ASCE)0733-9437(2008)134:3(273), 2008.

Singh, R. K., Liu, S., Tieszen, L. L., Suyker, A. E., and Verma, S. B.: Estimating seasonal evapotranspiration from temporal satellite images, Irrig. Sci., 30, 303–313, doi:10.1007/s00271-011-0287-z, 2011.

Smith, D. M., Kniveton, D. R., and Barrett, E. C.: A Statistical Modeling Approach to Passive Microwave Rainfall Retrieval, J. Appl. Meteorol., 37, 135–154, 1998.

Smith, E. A., Cooper, H. J., Xiang, X., Mugnai, A., and Tripoli, G. J.: Foundations for Statistical-Physical Precipitation Retrieval from Passive Microwave Satellite Measurements. Part I: Brightness-Temperature Properties of a Time-dependent Cloud-Radiation Model, J. Appl. Meteorol., 31, 506–531, doi:10.1175/1520-0450(1992)031<0506:FFSPPR>2.0.CO;2, 1992.

Smits, P. C., Dellepiane, S. G., and Schowengerdt, R. a.: Quality assessment of image classification algorithms for land-cover mapping: A review and a proposal for a cost-based approach, Int. J. Remote Sens., 20, 1461–1486, doi:10.1080/014311699212560, 1999.

Song, M., Civco, D. L., and Hurd, J. D.: A competitive pixel-object approach for land cover classification, Int. J. Remote Sens., 26, 4981–4997, doi:10.1080/01431160500213912, 2005.

Soppe, R. W., Bastiaanssen, W., Keller, A., Clark, B., Thoreson, B., Eckhardt, J. and Davids, G.: Use of High Resolution Thermal Landsat Data to Estimate Evapotranspiration Within the Imperial Irrigation District in Southern California, Am. Geophys. Union, 2006.

Stavrakoudis, D. G., Theocharis, J. B., and Zalidis, G. C.: A Boosted Genetic Fuzzy Classifier for land cover classification of remote sensing imagery, ISPRS J. Photogramm. Remote Sens., 66, 529–544, doi:10.1016/j.isprsjprs.2011.01.010, 2011.

Stefanov, W. L., Ramsey, M. S., and Christensen, P. R.: Monitoring urban land cover change?: An expert system approach to land cover classification of semiarid to arid urban centers, Remote Sens. Environ., 77, 173–185, 2001.

Stephens, G. L. and Kummerow, C. D.: The Remote Sensing of Clouds and Precipitation from Space: A Review, J. Atmos. Sci., 64, 3742–3765, doi:10.1175/2006JAS2375.1, 2007.

Stisen, S. and Sandholt, I.: Evaluation of remote-sensing-based rainfall products through predictive capability in hydrological runoff modelling, Hydrol. Process., 24, 879–891, doi:10.1002/hyp.7529, 2010.

Su, F., Hong, Y., and Lettenmaier, D. P.: Evaluation of TRMM Multisatellite Precipitation Analysis (TMPA) and Its Utility in Hydrologic Prediction in the La Plata Basin, J. Hydrometeorol., 9, 622–640, doi:10.1175/2007JHM944.1, 2008.

Su, Z.: The Surface Energy Balance System (SEBS) for estimation of turbulent heat fluxes, Hydrol. Earth Syst. Sci., 6, 85–100, doi:10.5194/hess-6-85-2002, 2002.

Sulla-Menashe, D., Friedl, M. A., Krankina, O. N., Baccini, A., Woodcock, C. E., Sibley, A., Sun, G., Kharuk, V., and Elsakov, V.: Hierarchical mapping of Northern Eurasian land cover using MODIS data, Remote Sens. Environ., 115, 392–403, doi:10.1016/j.rse.2010.09.010, 2011.

Szuster, B. W., Chen, Q., and Borger, M.: A comparison of classification techniques to support land cover and land use analysis in tropical coastal zones, Appl. Geogr., 31, 525–532, doi:10.1016/j.apgeog.2010.11.007, 2011.

Taşdemir, K., Milenov, P., and Tapsall, B.: A hybrid method combining SOM-based clustering and object-based analysis for identifying land in good agricultural condition, Comput. Electron. Agric., 83, 92–101, doi:10.1016/j.compag.2012.01.017, 2012.

Tasumi, M., Trezza, R., Allen, R. G., and Wright, J. L.: U.S. Validation Tests on the SEBAL Model for Evapotranspiration via Satellite, in: Vol. 17, 54th IEC meeting of the international commission on irrigation and drainage (ICID) Workshop remote sensing of ET for large regions, 17 September, Montpellier, France, 1–14, 2003.

Teixeira, A. H. D. C. and Bastiaanssen, W. G. M.: Five methods to interpret field measurements of energy fluxes over a micro-sprinkler-irrigated mango orchard, Irrigation Sci., 30, 13–28, doi:10.1007/s00271-010-0256-y, 2010.

Teixeira, A. H. D. C., Bastiaanssen, W. G. M., Moura, M. S. B., Soares, J. M., Ahmad, M. D., and Bos, M. G.: Energy and water balance measurements for water productivity analysis in irrigated mango trees, Northeast Brazil, Agr. Forest Meteorol., 148, 1524–1537, doi:10.1016/j.agrformet.2008.05.004, 2008.

Teixeira, A. H. D. C., Bastiaanssen, W. G. M., Ahmad, M. D., and Bos, M. G.: Reviewing SEBAL input parameters for assessing evapotranspiration and water productivity for the Low-Middle São Francisco River basin, Brazil, Agr. Forest Meteorol., 149, 462–476, 2009.

Thenkabail, P. S., Schull, M., and Turral, H.: Ganges and Indus river basin land use/land cover (LULC) and irrigated area mapping using continuous streams of MODIS data, Remote Sens. Environ., 95, 317–341, doi:10.1016/j.rse.2004.12.018, 2005.

Thenkabail, P. S., Biradar, C. M., Noojipady, P., Dheeravath, V., Li, Y., Velpuri, M., Gumma, M., Gangalakunta, O. R. P., Turral, H., Cai, X., Vithanage, J., Schull, M. A., and Dutta, R.: Global irrigated area map (GIAM), derived from remote sensing, for the end of the last millennium, Int. J. Remote Sens., 30, 3679–3733, doi:10.1080/01431160802698919, 2009a.

Thenkabail, P. S., Turral, H., Biradar, C., and Lyon, J. G.: Remote sensing of global croplands for food security, CRC Press, Boca Raton, FL, 2009b.

Thoreson, B., Clark, B., Soppe, R., Keller, A., Bastiaanssen, W. G. M., and Eckhardt, J.: Comparison of evapotranspiration estimates from remote sensing (SEBAL), water balance, and crop coefficient approaches, in: ASCE World Environmental and Water Resources Congress 2009: Great Rivers, 17–21 May, Kansas City, 1–15, doi:10.1061/41036(342)437, 2009.

Tovar, C., Seijmonsbergen, A. C., and Duivenvoorden, J. F.: Monitoring land use and land cover change in mountain regions: An example in the Jalca grasslands of the Peruvian Andes, Landsc. Urban Plan., 112, 40–49, doi:10.1016/j.landurbplan.2012.12.003, 2013.

Trambauer, P., Dutra, E., Maskey, S., Werner, M., Pappenberger, F., van Beek, L. P. H., and Uhlenbrook, S.: Comparison of different evaporation estimates over the African continent, Hydrol. Earth Syst. Sci., 18, 193–212, doi:10.5194/hess-18-193-2014, 2014.

Tseng, M.-H., Chen, S.-J., Hwang, G.-H., and Shen, M.-Y.: A genetic algorithm rule-based approach for land-cover classification, ISPRS J. Photogramm. Remote Sens., 63, 202–212, doi:10.1016/j.isprsjprs.2007.09.001, 2008.

Twine, T. E., Kustas, W. P., Norman, J. M., Cook, D. R., Houser, P. R., Meyers, T. P., Prueger, J. H., Starks, P. J., and Wesely, M. L.: Correcting eddy-covariance flux underestimates over a grassland, Agr. Forest Meteorol., 103, 279–300, doi:10.1016/S0168-1923(00)00123-4, 2000.

UN: System of Environmental Economic Accounting for Water, Geneva, 2007.

Van der Walt, I. J., Struwig, A., and van Rensburg, J. R. J.: Forestry as a streamflow reduction activity in South Africa: Discussion and evaluation of the proposed procedure for the assessment of afforestation permit applications in terms of water sustainability, GeoJournal, 61, 173–181, doi:10.1007/s10708-004-2872-7, 2004.

Velpuri, N. M., Senay, G. B., Bohms, R. K., and Verdin, J. P.: A comprehensive evaluation of two MODIS evapotranspiration products over the conterminous United States: Using point and gridded FLUXNET and water balance ET, Remote Sens. Environ., 139, 35–49, 2013.

Venturini, V., Islam, S., and Rodriguez, L.: Estimation of evaporative fraction and evapotranspiration from MODIS products using a complementary based model, Remote Sens. Environ., 112, 132–141, doi:10.1016/j.rse.2007.04.014, 2008.

Verstraeten, W. W., Veroustraete, F., and Feyen, J.: Assessment of Evapotranspiration and Soil Moisture Content Across Different Scales of Observation, Sensors, 8, 70–117, doi:10.3390/s8010070, 2008.

Villarini, G., Krajewski, W. F., and Smith, J. A.: New paradigm for statistical validation of satellite precipitation estimates: Application to a large sample of the TMPA 0.25° 3-hourly estimates over Oklahoma, J. Geophys. Res., 114, D12106, doi:10.1029/2008JD011475, 2009.

Voisin, N., Wood, A. W., and Lettenmaier, D. P.: Evaluation of Precipitation Products for Global Hydrological Prediction, J. Hydrometeorol., 9, 388–407, doi:10.1175/2007JHM938.1, 2008.

Vörösmarty, C. J., McIntyre, P. B., Gessner, M. O., Dudgeon, D., Prusevich, A., Green, P., Glidden, S., Bunn, S. E., Sullivan, C. A., Liermann, C. R., and Davies, P. M.: Global threats to human water security and river biodiversity, Nature, 467, 555–561, doi:10.1038/nature09440, 2010.

Wagner, W., Verhoest, N. E. C., Ludwig, R., and Tedesco, M.: Editorial "Remote sensing in hydrological sciences", Hydrol. Earth Syst. Sci., 13, 813–817, doi:10.5194/hess-13-813-2009, 2009.

Wang, J., Chen, Y., He, T., Lv, C., and Liu, A.: Application of geographic image cognition approach in land type classification using Hyperion image: A case study in China, Int. J. Appl. Earth Obs. Geoinf., 12, S212–S222, doi:10.1016/j.jag.2009.06.003, 2010.

Wang, J. R., Gogineni, S. P., and Ampe, J.: Active and passive microwave measurements of soil moisture in FIFE, J. Geophys. Res., 97, 18979, doi:10.1029/92JD00848, 1992.

Wang, K., Wang, P., Li, Z., Cribb, M., and Sparrow, M.: A simple method to estimate actual evapotranspiration from a combination of net radiation, vegetation index, and temperature, J. Geophys. Res., 112, D15107, doi:10.1029/2006JD008351, 2007.

Wang, Y. and Sun, D.: The ET estimation from ASTER image based on SEBAL and TSEB method, Proceedings of SPIE 6045, MIPPR 2005: Geospatial Information, Data Mining, and Applications, 604532, doi:10.1117/12.651845, 2005.

Waske, B. and Braun, M.: Classifier ensembles for land cover mapping using multitemporal SAR imagery, ISPRS J. Photogramm. Remote Sens., 64, 450–457, doi:10.1016/j.isprsjprs.2009.01.003, 2009.

Weiers, S., Groom, G., and Wissen, M.: Comparability and subjectivity of land cover maps pro- duced with digital image classification techniques?: some recent experiences from Denmark and northern Germany, Geogr. Tidsskr. Danish J. Geogr., 102, 59–77, 2002.

Whiteside, T. G., Boggs, G. S., and Maier, S. W.: Comparing object-based and pixel-based classifications for mapping savannas, Int. J. Appl. Earth Obs. Geoinf., 13, 884–893, doi:10.1016/j.jag.2011.06.008, 2011.

Wickham, J. D., Stehman, S. V., Gass, L., Dewitz, J., Fry, J. A., and Wade, T. G.: Accuracy assessment of NLCD 2006 land cover and impervious surface, Remote Sens. Environ., 130, 294–304, doi:10.1016/j.rse.2012.12.001, 2013.

Wilk, J., Kniveton, D., Andersson, L., Layberry, R., Todd, M. C., Hughes, D., Ringrose, S., and Vanderpost, C.: Estimating rainfall and water balance over the Okavango River Basin for hydrological applications, J. Hydrol., 331, 18–29, doi:10.1016/j.jhydrol.2006.04.049, 2006.

Winsemius, H. C., Savenije, H. H. G., and Bastiaanssen, W. G. M.: Constraining model parameters on remotely sensed evaporation: justification for distribution in ungauged basins?, Hydrol. Earth Syst. Sci., 12, 1403–1413, doi:10.5194/hess-12-1403-2008, 2008.

Wu, B., Yan, N., Xiong, J., Bastiaanssen, W. G. M., Zhu, W., and Stein, A.: Validation of ETWatch using field measurements at diverse landscapes: A case study in Hai Basin of China, J. Hydrol., 436–437, 67–80, 2012.

Wu, C., Cheng, C., Lo, H., and Chen, Y.: Study on estimating the evapotranspiration cover coefficient for stream flow simulation through remote sensing techniques, Int. J. Appl. Earth Obs. Geoinf., 12, 225–232, doi:10.1016/j.jag.2010.03.001, 2010.

Yalew, S., Teferi, E., and Van Griensven, A.: Land Use Change and Suitability Assessment in the Upper Blue Nile Basin Under Water Resources and Socio-economic Constraints: A Drive Towards a Decision Support System, in: International Congress on Environmental Modelling and Software Managing Resources of a Limited Planet, Sixth Biennial Meeting, edited by: Seppelt, R., Voinov, A. A., Lange, S., and Bankamp, D., 1–5 July, Leipzig, Germany, 2124–2131, 2012.

Yang, Y., Shang, S., and Jiang, L.: Remote sensing temporal and spatial patterns of evapotranspiration and the responses to water management in a large irrigation district of North China, Agr. Forest Meteorol., 164, 112–122, doi:10.1016/j.agrformet.2012.05.011, 2012.

Zhang, J., Hu, Y., Xiao, X., Chen, P., Han, S., Song, G., and Yu, G.: Satellite-based estimation of evapotranspiration of an old-growth temperate mixed forest, Agr. Forest Meteorol., 149, 976–984, doi:10.1016/j.agrformet.2008.12.002, 2009.

Zhang, K., Kimball, J. S., Nemani, R. R., and Running, S. W.: A continuous satellite-derived global record of land surface evapotranspiration from 1983 to 2006, Water Resour. Res., 46, W09522, doi:10.1029/2009WR008800, 2010.

Zhang, X., Sun, R., Zhang, B., and Tong, Q.: Land cover classification of the North China Plain using MODIS_EVI time series, ISPRS J. Photogramm. Remote Sens., 63, 476–484, doi:10.1016/j.isprsjprs.2008.02.005, 2008.

Zhang, Y., Leuning, R., Hutley, L. B., Beringer, J., McHugh, I., and Walker, J. P.: Using long-term water balances to parameterize surface conductances and calculate evaporation at 0.05° spatial resolution, Water Resour. Res., 46, W05512, doi:10.1029/2009WR008716, 2010.

Zhu, Z., Woodcock, C. E., Rogan, J., and Kellndorfer, J.: Assessment of spectral, polarimetric, temporal, and spatial dimensions for urban and peri-urban land cover classification using Landsat and SAR data, Remote Sens. Environ., 117, 72–82, doi:10.1016/j.rse.2011.07.020, 2012.

11

Spatial evapotranspiration, rainfall and land use data in water accounting – Part 2: Reliability of water acounting results for policy decisions in the Awash Basin

P. Karimi[1], **W. G. M. Bastiaanssen**[1,2,3], **A. Sood**[2], **J. Hoogeveen**[4], **L. Peiser**[4], **E. Bastidas-Obando**[5], **and R. J. Dost**[5]

[1]UNESCO-IHE Institute for Water Education, Delft, the Netherlands
[2]International Water Management Institute, Battaramulla, Sri Lanka
[3]Faculty of Civil Engineering and Geosciences, Water Management Department, Delft University of Technology, Delft, the Netherlands
[4]Land and Water Division, FAO, Rome, Italy
[5]eLEAF Competence Centre, Wageningen, the Netherlands

Correspondence to: P. Karimi (p.karimi@unesco-ihe.org)

Abstract. Water Accounting Plus (WA+) is a framework that summarizes complex hydrological processes and water management issues in river basins. The framework is designed to use satellite-based measurements of land and water variables and processes as input data. A general concern associated with the use of satellite measurements is their accuracy. This study focuses on the impact of the error in remote sensing measurements on water accounting and information provided to policy makers. The Awash Basin in the central Rift Valley in Ethiopia is used as a case study to explore the reliability of WA+ outputs, in the light of input data errors. The Monte Carlo technique was used for stochastic simulation of WA+ outputs over a period of 3 yr. The results show that the stochastic mean of the majority of WA+ parameters and performance indicators are within 5 % deviation from the original WA+ values based on one single calculation. Stochastic computation is proposed as a standard procedure for WA+ water accounting because it provides the uncertainty bandwidth for every WA+ output, which is essential information for sound decision-making processes. The majority of WA+ parameters and performance indicators have a coefficient of variation (CV) of less than 20 %, which implies that they are reliable and provide consistent information on the functioning of the basin. The results of the Awash Basin also indicate that the utilized flow and basin closure fraction (the degree to which available water in a basin is utilized) have a high margin of error and thus a low reliability. As such, the usefulness of them in formulating important policy decisions for the Awash Basin is limited. Other river basins will usually have a more accurate assessment of the discharge in the river mouth.

1 Introduction

Water Accounting Plus (WA+) is a novel analytical framework that summarizes complex hydrological processes and water management issues in vast river basins by means of four simple sheets (Bastiaanssen, 2009; Karimi et al., 2013a), although the accounting system is expanded continuously. WA+ has the ability to accommodate satellite measurements to quantify land use and hydrological variables. WA+ is a successor of the water accounting (WA) system initiated by the International Water Management Institute (IWMI) that was introduced by Molden (1997) and Molden and Sakthivadivel (1999) for describing the depletion of water resources in river basins. Whereas the IWMI WA is based on piezometers, water levels, discharge measurement, rain gauges and reference evapotranspiration to assess water stocks, water usage, and depletion in river basins, WA+ is also designed to allow for the use of remote sensing data. Remote sensing information can replace hydrometeorological data sets measured in situ, especially when administrations are reluctant to share data, and also where the data quality from field observatories is questionable.

WA+ facilitates the understanding of the water resource situation and the use of water by riparian administrations. The number of river basins under water stress is rapidly growing (Vörösmarty et al., 2010; Wolf et al., 2003), and there is a growing need for transparent and independent water-related data (CA, 2007; FAO, 2003; UN-Water, 2013). WA+ meets this need by quantifying the resources and their depletion by all the agroecological land use units in the river basin. WA+ provides policy makers with data for water (re-) allocation, withdrawal permits, flows to sustain ecosystems, and for soil and water conservation, among others.

The art of using remote sensing to derive hydrological variables is well established (e.g., Neale et al., 2012; Stewart et al., 1996). A recent literature review by Karimi and Bastiaanssen (2015) showed that the average errors in land use mapping, and annual or seasonal precipitation and evapotranspiration estimates on the basis of multispectral remote sensing data were 14.5, 18.5, and 5.4 %, respectively. These figures are based on a comprehensive literature review in which for each variable several numbers of post-2000 peer-reviewed publications were consulted for reported differences of satellite-based estimates from conventional ground measurements. Results of the study show that errors in satellite-based estimates are within an acceptable range and comparable to errors reported in conventional ground-based observations. They are thus suitable for application in WA+ for any river basin, including ungauged basins. Bastiaanssen and Chandrapala (2003), Bastiaanssen et al. (2014), Karimi et al. (2012, 2013b), Drooger et al. (2010), Shilpakar et al. (2012) and Dost et al. (2012) used remote sensing data for the water accounts of ungauged river basins in Sri Lanka, and for the Nile, Okavango, East Rapti, Indus, and Awash Basins, respectively. While this is great for basin planning, arbiters may raise concerns on the reliability of the accounts if they have not been verified on the ground, especially if the water accounts are not favorable for the water manager that is responsible for the operational distribution of water resources. While field devices are considered reliable measurement instruments, the radiometer onboard a satellite is often interpreted as futuristic, and not having accurate measurement capabilities. This is not correct, as many in situ devices also measure variables indirectly. In situ soil moisture sensors for instance measure the soil dielectric properties, and not soil moisture; measured river and canal discharge are sometimes based on the sound of water flow, rather than being direct measurements; leaf area index is based on intercepted solar radiation, and not a direct measurement of total leaf area.

By demonstrating the accuracy of satellite measurements they will become more acceptable for use in water accounting. Water-resources-related court cases in the USA and Spain have already used remote sensing data in dealing with conflicts between competing water users (e.g., Allen et al., 2007). This created a precedent for more frequent usage of satellite measurements to alleviate international water conflicts. However, certain critical scientists only trust their own devices and measurements obtained in a particular location, preferably operated by themselves and not by their colleagues. It is already known for quite some time that the quantification of water stocks, fluxes and flows in river basins will not necessarily be better if conventional point measurements are used. Pelgrum and Bastiaanssen (1996) demonstrated for instance that the regional scale actual evapotranspiration (ET) for an area of 10 000 km^2 cannot be predicted accurately even if 15 advanced flux towers are installed. Hence, in situ measurements are not the ultimate solution for determining water flows at river basin scale, although they are needed to verify local model predictions. The core issue is to determine the reliability of WA+ accounts if remote sensing input data is used.

This paper investigates the impact of the errors in remote sensing measurements on water accounting and the information provided to policy makers. The water accounting exercise in this study has been done at an annual scale because the monthly storage chances were not know with sufficient accuracy. Future studies will focus on monthly water accounts though. The degree of inaccuracies in remote sensing data is based on the comprehensive review of Karimi and Bastiaanssen (2015). The objective of the current paper is to study the impact of these errors on the water resources of two WA+ reporting sheets.

2 Background information

2.1 Awash Basin

The Awash River is located in the central Rift Valley in Ethiopia. The river emerges from the central highlands 150 km west of Addis Ababa and flows via the central Rift Valley to Lake Abbe on the border with Djibouti (Edossa et al., 2010). The mean annual rainfall is 530 mm and varies from about 1600 mm yr^{-1} at Ankober, in the highlands northeast of Addis Ababa to 160 mm yr^{-1} at Asayita on the northern border of the basin. The drainage area of the Awash River basin is 116 449 km^2 (Fig. 1). Lake Abbe is located in the downstream end of the basin and has an average size of 340 km^2 open water, surrounded by 110 km^2 of salt flats. The lake surface area and water depth fluctuates with rainfall and runoff. The water level can drop as much as 5 m. The maximum depth of the lake is 36 m.

The Awash Basin is located in the tectonically active East African Rift System and it has a complex geology. The complexity of geology of the basin has a direct impact on its hydrogeological characteristics and geohydrological flows. Groundwater flow is of key importance in the Awash Basin both as a major source of water supply for people and be-

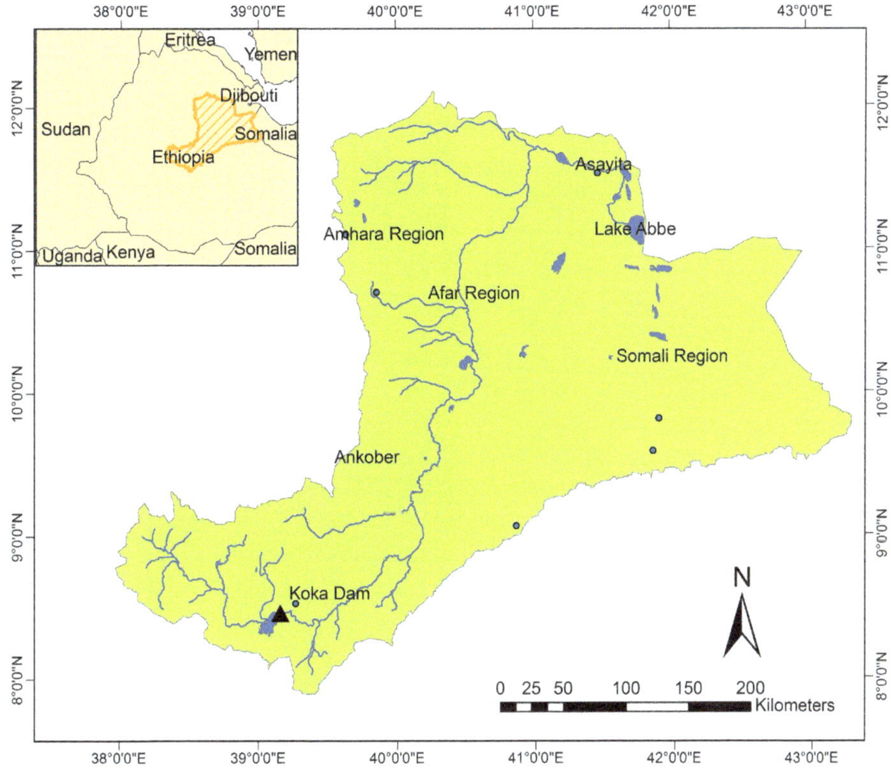

Figure 1. Location of the Awash River basin in the central Rift Valley of Ethiopia.

cause of its impact on hydrographs, especially during the dry period. The highland's fractured volcanic cover is favorable for groundwater recharge processes (Ayenew et al., 2008). Thus, groundwater recharge from the highlands is substantial. Groundwater gradually percolates into the lower aquifers through large marginal faults before it reaches the rift floor (Ayenew, 2001). In the upper and middle parts of the valley, the groundwater levels range between 30 and 70 m. The levels drop to lower than 200 m in some areas in the southern corner of the Awash Valley. In the upper basin, upstream of the Koka dam, the Awash River is hydraulically linked to the aquifers. However, this link weakens downstream of the dam. The major and deeper aquifers in this region are fractured basalt and ignimbrites. The axial faults together with the thickness and the extent of Quaternary deposits control groundwater occurrence below the pediment slopes. In the southern Afar plains the thick alluvio–colluvial deposits and the underlying Mesozoic limestones, dolomites and sandstones form highly productive aquifers. These aquifers are recharged by seasonal floods in wadis and wide river beds that are often highly permeable Quaternary deposits (Ayenew et al., 2008; Meskale, 1982). These aquifers are recharged by the streams that originate from the eastern highlands and seasonal floods that occur in summer.

The Awash Basin has an irrigation potential of 205 400 ha (FAO, 2003). Agriculture, providing livelihood for 85 % of the population, contributes to 45 % of Ethiopia's GDP. According to the FAO's AQUASTAT country fact sheet for Ethiopia, the country has an estimated 2.7×10^6 ha of irrigable land, yet only about 289 000 ha (11 %) are presently irrigated and only provides approximately 3 % of the country's food crop requirements. Most of the irrigation developed to date in Ethiopia is located in the Awash Basin.

The basin has been selected by the FAO as a case study for testing its approach in coping with water scarcity (FAO, 2012). Awash is experiencing water shortage for irrigated agriculture and for the wetlands and natural lakes along the riparian corridor of the river. The salt flows at the downstream end of the system are also suffering from water shortage, and there is a threat of salt storms when these flows dry up. It is therefore necessary to understand the hydrological processes and ecosystem services better, and summarize the management options. WA+ is an ideal framework for such a situation and has been applied to 3 consecutive years with rainfall varying from an average (510 mm yr^{-1}) in 2009, a high (862 mm yr^{-1}) in 2010, and a low (364 mm yr^{-1}) in 2011. Table 1 shows long-term average rainfall and potential ET (PET) in the Awash Basin.

2.2 Remote sensing input data used

Annual actual ET for the Awash Basin was computed by means of the two-layer ETLook surface energy balance model, using input data from MODIS (Moderate-

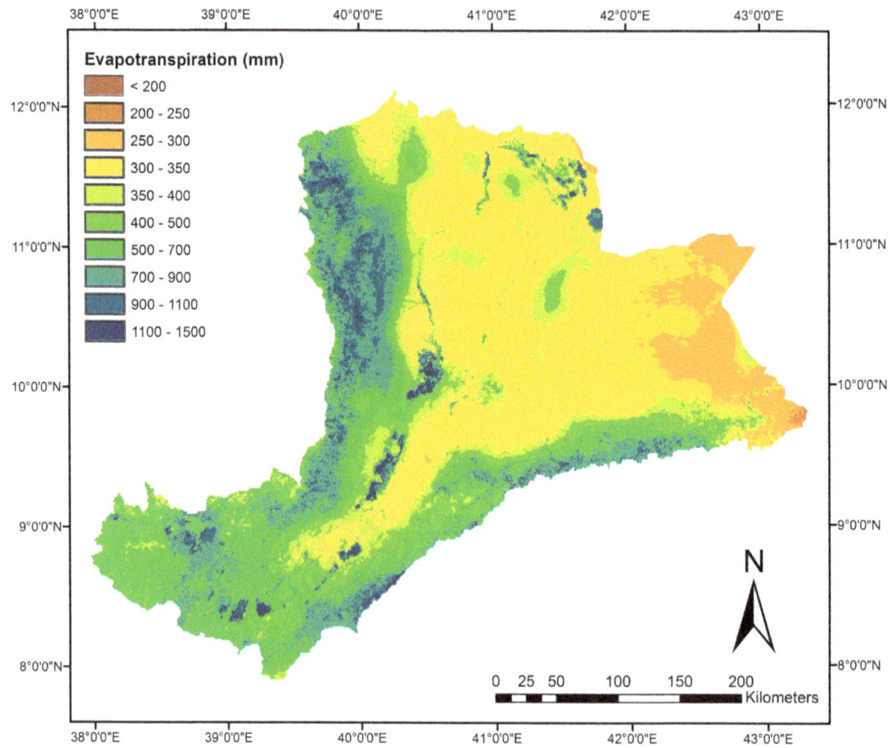

Figure 2. Spatial distribution of the annual ET of the Awash Basin for 2009 computed with the ETLook model (after Dost et al., 2013).

Table 1. Long-term average (1961–1990) rainfall and PET in the Awash Basin.

Month	Rainfall ($mm\,month^{-1}$)	PET ($mm\,day^{-1}$)
Jan	5.2	4.8
Feb	15.1	5.2
Mar	38.4	5.6
Apr	56.3	5.4
May	40.5	5.4
Jun	30.2	5.8
Jul	117.6	5.5
Aug	142.1	5.3
Sep	65.3	5.2
Oct	13.7	4.7
Nov	4	4.4
Dec	1.5	4.4

resolution Imaging Spectroradiometer; albedo, vegetation index), AMSR-E (Advanced Microwave Scanning Radiometer for EOS; top soil moisture) and Meteosat Second Generation (cloud cover). ETLook is based on a two-layer Penman–Monteith equation that describes soil evaporation and plant transpiration as separated physical processes (Bastiaanssen et al., 2012). Evaporation from wet leaves (i.e., interception) and open water is also computed. An interval of 8 days was applied based on recurrent MODIS measurements, and the accumulated ET value for 2009 is presented in Fig. 2.

Daily rainfall maps were acquired from the US Agency for International Development (USAID) Famine Early Warning Systems Network (FEWS NET). FEWS NET is an information system designed to identify problems in the food supply system that can potentially lead to famine or other food-insecure conditions in the Horn of Africa, amongst other regions. FEWS NET provides daily rainfall with a spatial resolution of 8 km × 8 km. The FEWS RFE (rainfall estimate) 2.0 algorithm is implemented by NOAA's Climate Prediction Center and uses an interpolation method to combine Meteosat and global telecommunication system (GTS) data. More background information on the FEWS rainfall algorithm can be found in Herman et al. (1997). Figure 3 shows the spatial distribution of annual rainfall in 2009.

A new land use map customized for application of water accounting in the Awash Basin was generated by Dost et al. (2013). The basis for the new land use map is the existing GlobCover map (Bicheron et al., 2008). The new additions are related to the separation of rainfed and irrigated agriculture, and the temporal changes in the size of the open water body. The institute of Physical Geography of the Goethe University of Frankfurt developed the MIRCA (Monthly Irrigated and Rainfed Crop Areas) data set, containing monthly maps of growing areas and crop calendars of 26 irrigated and rainfed

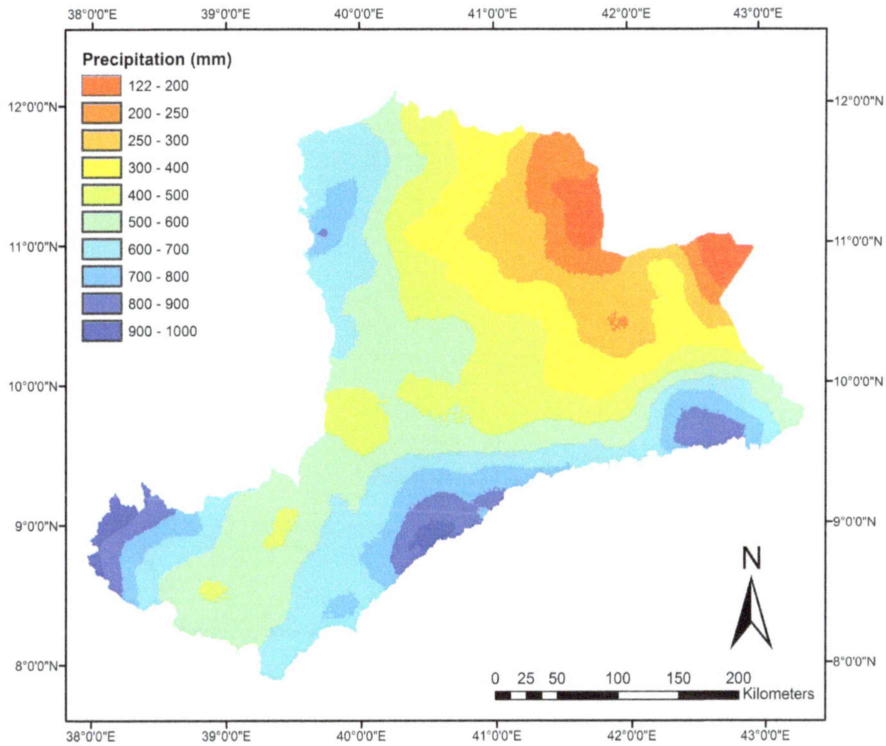

Figure 3. Spatial distribution of annual rainfall of the Awash Basin for the average rainfall year 2009 taken from FEWS Net (after Dost et al., 2013).

crops (documented at http://www.geo.uni-frankfurt.de/jpg/ag/dl/forschun/MIRCA/index.html). MIRCA contains data for 1999–2002 and has a spatial resolution of 5 arcmin (+10 km). The cropped area is based on the period with maximum rainfed crop acreage. Areas equipped for irrigation are extracted from the irrigated area map of the FAO and the university of Kassel (Döll and Siebert, 2000). Since these data sets are to some extent outdated, a time series of the normalized difference vegetation index (NDVI) during 2009 was used to verify the crop phenology. Fallow land was identified and reclassified. Figure 4 shows the resulting locations of irrigated and rainfed cropland in the Awash Basin. The area of irrigated croplands is 216 900 ha and the area of rainfed croplands is 2 258 500 ha. The irrigated acreage is close to the irrigation potential of 205 400 ha, which suggests that most potential land for irrigation is exploited already. While the alluvial soils and flat topography are suitable for irrigation, the unreliability of water resources due to the overall water scarcity is the constraint for further land reclamation.

2.3 Water Accounting (WA+)

The latest version of the WA+ framework provides four sheets including (i) a resource base sheet, (ii) an evapotranspiration sheet, (iii) a productivity sheet, and (iv) a withdrawal sheet (Karimi et al., 2013a). The resource base sheet deals with water volumes and provides information on water availability, water depletion and outflow processes. The evapotranspiration sheet distinguishes beneficial water depletion from non-beneficial depletion by partitioning total evapotranspiration (ET) into evaporation (E), transpiration (T), and interception (I). The productivity sheet links water depletion with benefits gained through biomass production. It extends to carbon sequestration, crop production and water productivity. The withdrawal sheet presents information on water withdrawals, depletions, and returns.

Each sheet has a set of indicators that are used to summarize the overall water resources situation. WA+ explicitly recognizes the influence of land use on the water cycle. To provide the link between land use and water use, land use classes with common management characteristics were defined. These are conserved land use (CLU), utilized land use (ULU), modified land use (MLU), and managed water use (MWU). CLU comprises environmentally sensitive land uses and natural ecosystems which are set aside for environmental protection. ULU represents a low to moderate resource utilization, such as savannah, woodland and mixed pastures which provide ecosystem services. MLU represents areas where the original vegetation was replaced for increased utilization of land resources or treatment of the soil. Rainfed crop land, plantations and biofuel crops are examples of replacement cover. The soil treatment can for instance be plowing, mulching and tilling. MWU represents landscapes that receive withdrawals by means of man-made

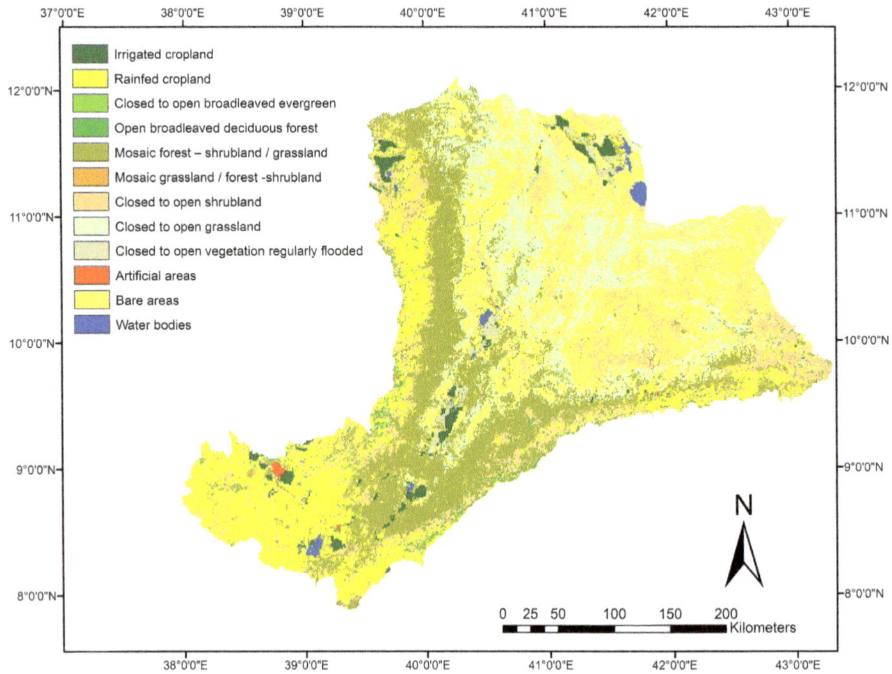

Figure 4. Updated spatial distribution of land use in the Awash Basin (after Dost et al., 2013).

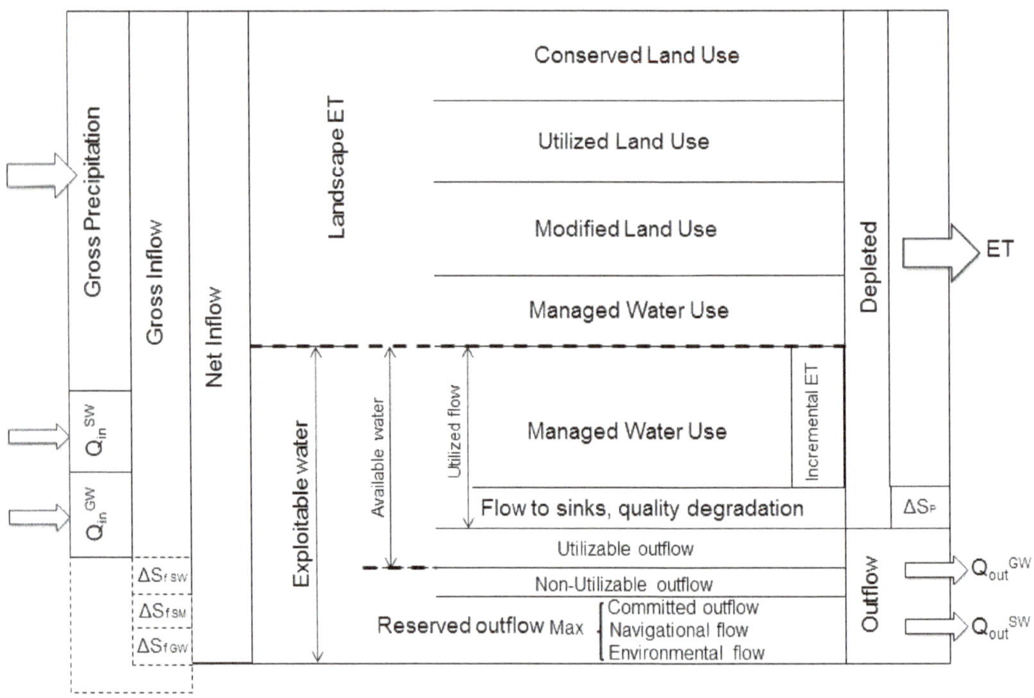

Figure 5. Resources base sheet for WA+ (after Karimi et al., 2013a).

infrastructure (diversion dams, canals, ditches, pumping stations, gates, weirs, pipes, etc.). This is also known as blue water usage (Falkenmark and Rockström, 2006).

The resource base sheet's main components are gross inflow, storage change, net inflow, landscape ET, exploitable water, available water, utilized flow, utilizable outflow, incremental ET, reserved outflow, non-utilizable outflow, and surface and groundwater outflows (see Fig. 5). Gross inflow is the total amount of water that flows into the domain, including precipitation and any inflow of surface or ground water

Table 2. Key performance indicators of the resource base sheet.

Indicator	Definition	What does it indicate?
Exploitable water fraction	$\frac{\text{Exploitable water}}{\text{Net inflow}}$	The part of the net inflow that is not depleted by landscape ET, and thus exploitable
Storage change fraction	$\frac{\Delta S_{\text{fw}}}{\text{Exploitable water}}$	The dependency of exploitable water on fresh water storage change
Available water fraction	$\frac{\text{Available water}}{\text{Exploitable water}}$	The portion of exploitable water that is actually available for withdrawals
Basin closure fraction	$\frac{\text{Utilized water}}{\text{Available water}}$	The extent to which available water is depleted in a basin
Reserved outflow fraction	$\frac{\text{Reserved outflow}}{\text{Outflow}}$	The degree of meeting the flows set aside for interbasin transfer, navigation and environmental purposes

from adjacent basins. Net inflow is the gross inflow after correction for annual storage change (ΔS) and represents water available for landscape ET, and exploitable water. Landscape ET is the water that evaporates directly from the soil surface and water intercepted by the vegetation cover, as well as water taken up by plant roots and transpired into the atmosphere. Exploitable water represents water in reservoirs, rivers, lakes and groundwater that can be partitioned further into utilized, utilizable, non-utilizable and reserved outflows. Available water, the part of water that can be allocated to various water use sectors, would be a good definition for the often used term "renewable water resources". Reserved outflow is the water that has to be reserved to meet the committed outflow, navigational flows, and environmental flow. Available water is the exploitable water minus reserved outflows and non-utilizable outflow. The latter is the part of water that cannot be utilized due to the lack of required infrastructure; e.g., a flash flood in utilized flow is the part of available water that is depleted by uses and hence is no longer available for downstream usage. Utilized flow is the difference between the withdrawals and the return flow from these withdrawals. Utilizable outflow is the available water for resources development and defined as the difference between available and utilized flow (see Fig. 5).

The resource base sheet indicators include the exploitable water fraction, storage change fraction, available water fraction, basin closure fraction, and the reserved outflows fraction. Exploitable water fraction is the part of the net inflow that is not depleted by landscape ET processes. Storage change fraction defines the degree of dependency of exploitable water on fresh storage change (ΔS_{fw}). Available water fraction relates available water to exploitable water. It describes the portion of exploitable water that is actually available for withdrawals within a basin because certain water resources have to be committed to sustain minimal environmental flows, navigation or should be allocated to users outside the basin. Basin closure fraction describes to what extent available water is already depleted in a basin or domain. Reserved outflows fraction relates the reserved outflows to outflow via streams and aquifers. It indicates whether the committed outflows are being met. A summary is provided in Table 2.

The WA+ evapotranspiration sheet (Fig. 6) relates ET to the generated benefits. ET processes are classed as managed, manageable, and non-manageable, which indicate the level of human influence on water consumption. The sheet provides a breakdown of ET into its components: interception evaporation, and transpiration. Knowing the proportion that each of these components contributes to total ET of each land use class makes it possible to determine the proportion of ET that has beneficial use, called *beneficial* ET, and the portion that does not have a beneficial use, called *non-beneficial* ET.

Beneficial ET comprises beneficial T and beneficial evaporation E. T is generally considered as beneficial. However, it can be considered non-beneficial in some cases such as for weed infestations in cropland or in degraded landscapes, or when it originates from non-desirable plants. E is usually considered as non-beneficial. However, the E in some cases such as evaporation from natural water surfaces is considered beneficial as these water bodies serve their purpose for fishing, aquatic birds, buffering floods, water sports, leisure, etc. In short, beneficial ET would be transpiration by usable vegetation cover (crops for example), and also evaporation from natural water surfaces and from cooling towers. Non-beneficial ET would be transpiration and evaporation from unwanted vegetation such as weeds and invasive species and evaporation from wet surfaces such as bare soil, buildings, and roads.

The evapotranspiration sheet indicators, summarized in Table 3, include a transpiration fraction, beneficial ET fraction, managed fraction, agricultural ET fraction, and irrigated ET fraction. The transpiration fraction is the proportion of ET

Table 3. Key performance indicators of the evapotranspiration sheet.

Indicator	Definition	What does it indicate?
Transpiration fraction	$\frac{T}{\mathrm{ET}}$	The part of ET that is transpired by plants and it reflects a biophysical process.
Beneficial ET fraction	$\frac{E\ \text{beneficial}+T\ \text{beneficial}}{\mathrm{ET}}$	Relates beneficial E and T to the total ET in a basin.
Managed ET fraction	$\frac{\mathrm{ET\ managed}}{\mathrm{ET}}$	The ET processes in a basin that is manipulated by land use change, cultivation practices and water withdrawals.
Agricultural ET fraction	$\frac{\mathrm{Agricultural\ ET}}{\mathrm{ET}}$	The part of ET that is from agricultural activities.
Irrigated ET fraction	$\frac{\mathrm{Irrigated\ agricultural\ ET}}{\mathrm{Agricultural\ ET}}$	Irrigated ET fraction describes the portion of agricultural ET that is related to irrigated agriculture

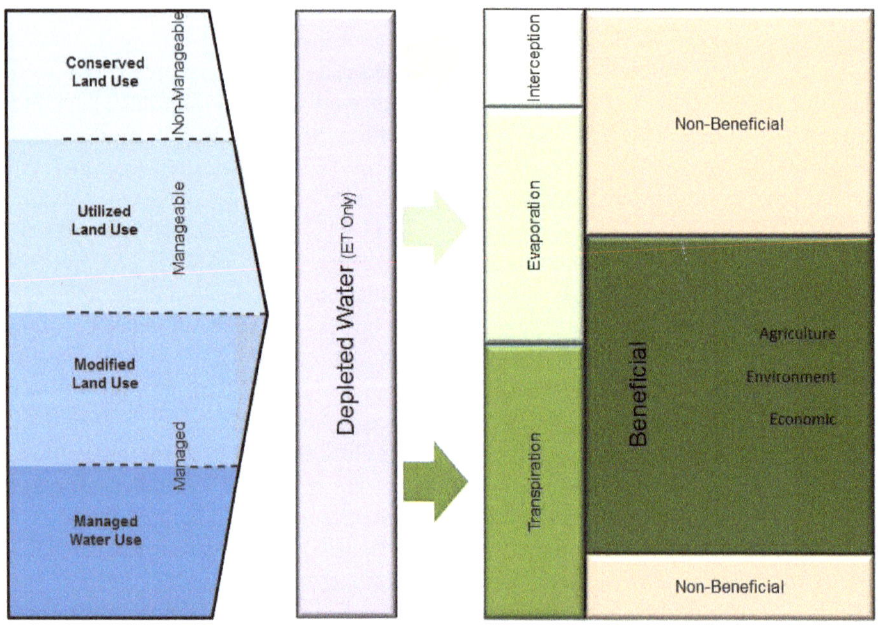

Figure 6. Schematic representation of the evapotranspiration sheet.

that is transpired by plants and relates to net carbon assimilation of vegetation. The beneficial ET fraction relates beneficial E and T to the total ET in a basin. Managed ET fraction indicates the ET processes in a basin that are manipulated by land use change, soil treatment, cultivation practices and water withdrawals. This includes ET from managed water use, e.g., irrigated areas and urban parks, and ET from modified land use, e.g., rainfed areas. Agricultural ET fraction is the part of ET attributable to the agricultural production from rainfed and irrigated crops. Lastly, irrigated ET fraction describes the portion of agricultural ET that is related to irrigated agriculture.

2.4 Methodology to express the reliability of the WA+ framework

The Monte Carlo (MC) technique was used to validate the WA+ outputs. The MC technique involves selecting numbers randomly from a predefined probabilistic distribution and applying them in stochastic simulation. MC computes the variability of the WA+ output parameters by defining the variability of the input parameters. The variability in this case expresses the accuracy and thus confidence that can be attached to the outputs, because the variability of the input parameter space expresses error in the remotely sensed hydrological variables. The space of input parameters in this

Table 4. Statistics of the probability density function of variation for each remote sensing input parameter into WA+.

Remote sensing parameter	Shape α	Shape Δ	Skewness γ (–)	Scale ω	Variance (%)	Standard deviation error (%)	Location ζ	Mean (%)
ET	25	1.000	1.18	2.444	2.17	4.7	3.5	5.4
Rainfall	6.4	0.988	0.90	3.218	3.92	15.4	16.0	18.5
Land use	1.66	0.856	0.35	2.258	2.72	7.4	13.1	14.6

Table 5. Annual total precipitation and ET in the Awash Basin averaged over the 3 yr: 2009, 2010 and 2011. Rainfall and ET data are based on remote sensing. The actual evapotranspiration is partitioned into evaporation, transpiration and interception following ETLook principles.

Year	Rainfall (mm)	ET (mm)	Interception (mm)	Evaporation (mm)	Transpiration (mm)	Biomass ($kg\,ha^{-1}$) production
2009	515	480	18	310	152	5744
2010	865	554	26	308	220	8570
2011	366	486	18	293	175	6455
Average	582	507	21	304	182	6923

MC study is defined by a skewed normal distribution as explained by Karimi and Bastiaanssen (2015). The statistical input data are specified in Table 4. A program was developed to generate random numbers from a positively skewed normal distribution based on mean, variance and skewness. This code handled only skewness smaller or equal to 1.0 and hence this number has been modified accordingly. The skewness γ is defined as the third standardized moment (γ_3):

$$\gamma_3 = \frac{4-\pi}{2}\frac{(\delta\sqrt{2/\pi})^3}{\left(1-\frac{2\delta^2}{\pi}\right)^{3/2}}, \quad (1)$$

where δ and α are shape parameters:

$$\delta = \frac{\alpha}{\sqrt{1+\alpha^2}}. \quad (2)$$

The variance is described by means of the scale ω and δ as

$$\text{variance} = \omega^2\left(1-\frac{2\delta^2}{\pi}\right). \quad (3)$$

The means value of the population can be computed from the location ζ:

$$\text{mean} = \varepsilon + \omega\delta\sqrt{\frac{2}{\pi}}. \quad (4)$$

The results of this exercise is a set of 1000 WA+ resource base sheets and evapotranspiration sheets, each of them based on a unique combination of ET, rainfall and land use. Care has been taken that the total basin area is conserved and that the mass balance of water flows applies. While in simulations the distribution of different land use classes was flexible and a function of randomly chosen error, a constant correction factor was applied to all land use classes to match the total basin area of 116 449 km^2 and hence keep the total physical area constant. The 1000 WA indicators were then analyzed to determine their accuracy and thus reliability.

3 Awash Basin results

3.1 Baseline hydrology and water accounting

Rainfall and ET are the two most important hydrological variables for WA+. The average rainfall from FEWS NET for the 3 yr investigated is 582 $m\,yr^{-1}$ (see Table 5). The average ET computed with ETLook is 507 $mm\,yr^{-1}$, which compares well with the average rainfall. Note that ETLook is based on an energy balance and is computed independent from rainfall. The magnitude of the annual ET for the different years is apparently dampened, which could be ascribed to compensating effects of atmospheric demand and soil moisture availability: dry years have a high potential ET but the ET reduction due to soil moisture stress is high as well, yielding an ET value similar to lower potential ET but lower reductions due to soil moisture stress. This behavior is also observed for ET from other surface energy balance models in the Nile Basin (e.g., Yilmaz et al., 2014). Dry years also partially compensate the lack of infiltrated water by consuming moisture from the unsaturated zone that is carried over from a previous wetter year.

Another interesting observation is that soil and water evaporation (304 mm) exceed transpiration (182 mm). The relatively low values of transpiration and interception are due to the reduced fractional vegetation cover in the Awash Basin, especially during the dry season. A large portion of the basin has barren land and the vegetation is senescent during elongated dry periods. The ETLook results show that transpiration from the vegetation (rainfed crops and hillslope forests) in the western and southern parts of the basin and the irrigated croplands are the major contributing factors to evapotranspiration in the river basin during the dry winters. During the rainy season, transpiration is higher due to the increased photosynthesis and biomass production of natural vegetation. In the eastern plains, evaporation values rise as the soil is saturated with water during the wet summer period, while the transpiration remains low due to the low vegetation cover.

Many national and international sources report a mean annual surface runoff of 4.6–4.9 $km^3\,yr^{-1}$ for the Awash Basin (e.g., Behailu, 2004; Edossa et al., 2010). This annual runoff data is based on measured discharge rates. This surface water flow is withdrawn by irrigation systems, wetlands, inundation areas and lakes. The long-term average annual flow at the Awash station in the middle of the basin is 1.7 $km^3\,yr^{-1}$, revealing that a substantial part is withdrawn in the upstream part of the basin (approximately 3 $km^3\,yr^{-1}$). The non-utilized water from Awash River flows into the saline depressions of Afar at the downstream end of the basin, where it is exposed to evaporation. In 2009, the average rainfall year of this study, total evaporation from all natural lakes amounted to $622 \times 10^6\,m^3\,yr^{-1}$, while the rainfall over these lakes was only 278 $Mm^3\,yr^{-1}$. This difference of $344 \times 10^6\,m^3\,yr^{-1}$ must be the inflow to the lakes from the Awash River, which matches the flow measured near the Awash station. This finding shows that all Awash Basin surface water resources are consumed and that no surface water outflow takes place. Awash is an example of a basin in which the available water is depleted (Molden, 1997).

Hence, all river flow that is not recharging the aquifer evaporates inside the basin either as a result of withdrawals or due to evaporation from the sink at the downstream end of the system. The evaporation from terminal lakes is included in the total ET value of 507 $mm\,yr^{-1}$ (see Table 5). Hence, the rainfall surplus of 75 mm (582–507 mm) or 8.7 $km^3\,yr^{-1}$ is not related to surface runoff and has to go somewhere else. The only possible outlet is underground basin discharge. Taddese et al. (2003) refer to a study of UNDP (1973) that estimates the total groundwater recharge in Awash to be 3.8 $km^3\,yr^{-1}$, while EVDSA (1989) estimated 4.1 $km^3\,yr^{-1}$. Ayenew et al. (2008) reported a basin-wide average recharge of 30 mm, which is equivalent to 3.5 $km^3\,yr^{-1}$. These estimates are mutually close, and the average number is 3.8 $km^3\,yr^{-1}$. Groundwater flows towards the downstream end of the basin at Lake Abbe, where the elevation is only 240 m. Ayenew et al. (2008) describe a regional groundwater flow in the direction of the Afar Depression. While detailed groundwater studies were not available, a regional flow of 3.8 $km^3\,yr^{-1}$ is likely. This assumes that all groundwater recharge will flow across the basin boundaries into deep depressions in the Horn of Africa. Since deep regional groundwater flow is usually rather stationary, variability of rainfall will have a limited impact on this interbasin transfer process. As such, in the absence of data and information on groundwater flows in the basin, it is assumed – without any scientific underpinning computations – that the basin outflow should not fluctuate more than 50 %. This assumption is used to define lower and upper boundaries for changes in annual groundwater outflow, allowing for a range of between 2.5 and 5.7 $km^3\,yr^{-1}$. This is a basis for computing the storage changes for the 3 different rainfall years analyzed (see Table 6), which must have significantly more dynamics than a deep quasi-stationary groundwater flow. Since the storage change is calculated as a residual term for P minus ET minus the underground recharge, it is important to note that it collects the errors of all three parameters including the error that assumed underground recharge might have. Thus, to have better and more accurate estimates of these flows, further research is needed to understand groundwater flows and outflows from the Awash Basin. A large uncertainty will be associated with groundwater outflow in the stochastic analysis described in the next section.

Table 6. Annual water balance of the Awash Basin for the selected hydrological years. The basin area is 116 449 km^2.

Year	Rainfall (km^3)	ET (km^3)	Basin outflow (km^3)	Storage change (km^3)
2009	59.8	56.4	3.8	−0.4
2010	100.5	65.1	5.7	+29.7
2011	42.4	57.2	2.5	−17.3
Average	67.6	59.6	3.9	4.1

There is an unexplained difference between the 8.7 $km^3\,yr^{-1}$ basin rainfall surplus and a groundwater recharge of 3.8 $km^3\,yr^{-1}$ that requires more detailed discussion. It is possible that some groundwater seeps to deeper levels via faults and tectonic plates. Another possible explanation is the change in storage. The storage changes among years must be significant and this is confirmed by reported changes in water levels of lakes and reservoirs. These storage changes occur in lakes and reservoirs, as well as in the deep aquifers of the Awash Basin. The average area of water bodies is 754 km^2 and wetlands cover 1078 km^2. If we assume that one-third of the 2010 storage (29.7 $km^3\,yr^{-1}$) takes place in the aquifer and unsaturated zone (i.e., 9.9 $km^3\,yr^{-1}$), then two-thirds of the storage change had to be stored as surface water (19.8 $km^3\,yr^{-1}$), which over 1832 km^2 (754 + 1078 km^2) signifies a rise in surface water level of 10.8 m. Reported changes in water levels are of 5 m, meaning that the area of open water bodies and wetlands expands with a factor of 2 during a wet year

Table 7. Rainfall and ET by land use class for 2009, 2010 and 2011. CLU is conserved land use, ULU is utilized land use, MLU is modified land use, and MWU is managed water use.

Land use class		Area	Precipitation (mm)			ET (mm)		
		km^2	2009	2010	2011	2009	2010	2011
Bare areas	CLU	1270	352	757	222	340	433	340
Closed to open grassland	CLU	1639	362	779	217	336	425	335
Closed/open vegetation regularly flooded	CLU	17	356	745	238	392	447	372
Rainfed croplands	CLU	39	520	727	340	364	407	376
Closed to open shrubland	CLU	173	343	698	225	326	342	308
Mosaic forest–shrubland/grassland	CLU	778	631	818	418	370	418	382
Irrigated cropland	CLU	24	674	718	425	795	864	810
Bare areas	ULU	30579	387	728	255	340	382	343
Closed to open grassland	ULU	16132	413	740	266	347	388	345
Closed to open shrubland	ULU	12936	557	935	399	489	551	484
Mosaic forest–shrubland/grassland	ULU	24414	608	930	420	507	618	525
Open broadleaved deciduous forest	ULU	1376	678	1017	505	705	771	686
Mosaic grassland/forest–shrubland	ULU	327	690	1026	528	841	883	797
Closed/open vegetation regularly flooded	ULU	1078	426	784	309	931	1142	963
Closed to open broadleaved evergreen or semideciduous forest	ULU	102	637	960	524	969	945	889
Water bodies	ULU	746	373	681	340	833	878	953
Rainfed croplands	MLU	22546	638	1034	504	687	796	697
Water bodies	MWU	8	667	691	415	499	533	414
Irrigated cropland	MWU	2145	550	854	428	826	924	867
Artificial areas	MWU	120	703	1130	587	533	520	493

for hosting a storage of 19.8 km^3 yr^{-1}. Expansion and level changes do happen and, together with some unknown deep seepage of groundwater, explain the total water balance.

A basic component of water accounting is the distribution of the rainfall and ET across all land use classes. Table 7 shows that the classes open broadleaved deciduous forests, mosaic forest shrubland/grassland and rainfed cropland receive more rainfall than the other land use classes, which indeed is the source of the existence of these types of vegetation. The highest ET is found in the land use class of 'regularly flooded closed and open vegetation. These are the wetlands in the riparian corridor of Awash River system. ET is highest in the wet year when a larger contiguous layer of water is ponded in these wetlands. The average ET for this particular land use class in 2010 is 1142 mm, and the rainfall is 784 mm. Hence, this inundation water must come from upstream drainage areas. The evaporation from water bodies is lower than for wetlands because the saline sinks of the Afar depression are also included in this data set, and these brines evaporate significantly less than wetlands.

The WA+ framework was applied for the average rainfall year, 2009, using Table 7 as input and the basin outflows as specified in Table 6. The flow to sink has been assigned a zero value because all surface flow is assumed to be depleted by evaporation and it thus included already in the ULU class. Reserved flow, which is the required flow to maintain a specific constant river flow, was fixed in accordance with the general guidelines for environmental flow requirements (Smakhtin and Eriyagama, 2008). Environmental flows were estimated to be 622 × 10^6 m^3 yr^{-1}, being the river flow required to meet the evaporation from natural lakes. The calculation is based on the assumption that this volume water is necessary to maintain the lakes and consequent conservation of aquatic ecosystems. The basin has no surface outflow and, since evaporation from the lakes is already accounted for, the outflow from the basin is through underground flows. Theses flows recharge the aquifers and leave the basin through underground interbasin transfers as outlined in the previous section. This outflow could be utilized by installing deep pumping stations that withdraw from this water before it flows away. We thus assume this portion of water to be utilizable, although in reality the abstraction should be an economic discussion. The resulting resource base sheet is presented in Fig. 7.

The results show that the ULU class is with 37.7 km^3 yr^{-1} depleting the majority of the net inflow of 60.2 km^3 yr^{-1}. This contributes to ecosystem services and grazing. The benefits and value of these depletion processes are moderate to low, especially considering that the majority is bare soil evaporation. The largest value is related to the biodiversity of flora and fauna. The MLU class depletes 15.5 km^3 yr^{-1}, and this contributes to a better food security in the basin. MLU consists of rainfed crops such as wheat and teff that occupy an area of 2 254 600 ha. Depletion from surface water with-

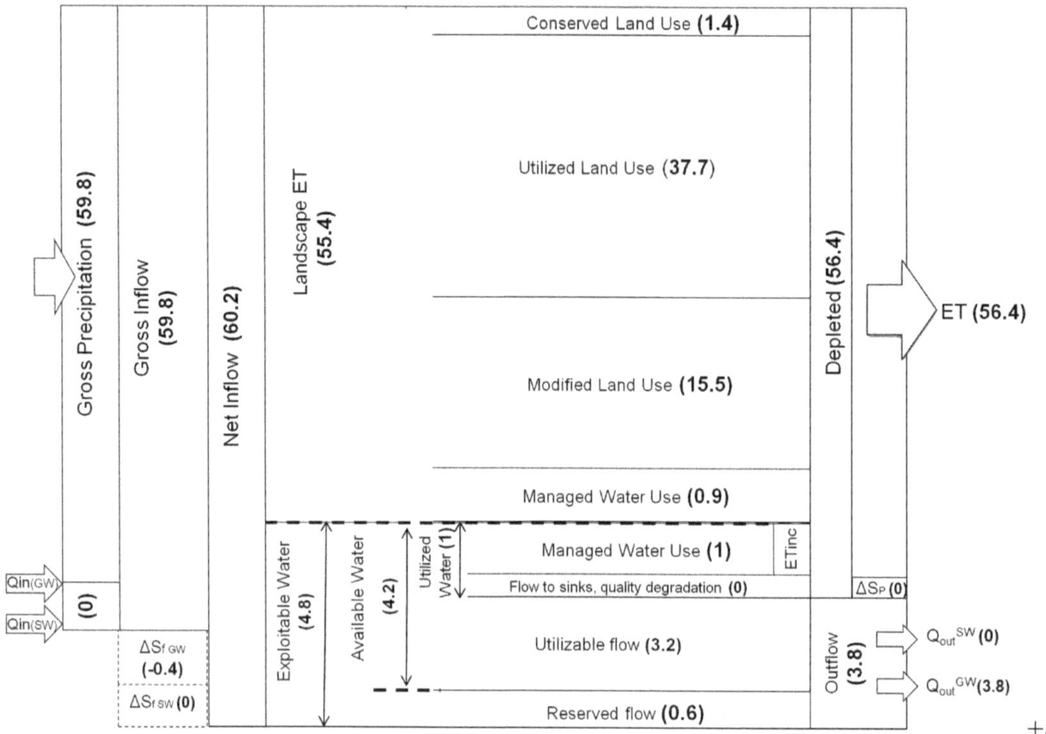

Figure 7. Resource base sheet of WA+for the Awash Basin during 2009. *All units are* km^3 *(adjusted after Dost et al., 2013).*

drawals to irrigated land, industry and domestic water use is with 1 $km^3\,yr^{-1}$ minimal. While the depletion of this water provides many benefits in terms of energy, the economy and domestic services, the amount of water being depleted is very low compared to the significant amount of water depleted by utilized land use. Land use planning is thus crucial for improving the benefits from water depletion in the Awash Basin. The introduction of agroforestry systems and short-duration low-water consuming crops could generate more benefits (e.g., Baudron et al., 2014). A volume of 3.2 km^3 is utilizable flow. This is groundwater that is not utilized. Options for groundwater abstraction and expansion of irrigated areas could be appraised.

3.2 Probability distribution of WA+ for 2009

The goal of this paper is to investigate the difference between the reference data of Table 7 and the results if the remote sensing input data is made variable according to the errors identified by Karimi and Bastiaanssen (2015). For this purpose the average rainfall year 2009 has been analyzed. The frequency distribution of the input parameters, randomly generated through the Monte Carlo technique based on their levels of uncertainty, is demonstrated in Fig. 8.

While precipitation and ET follow a similar unimodal normal distribution, the area of each land use class follows a bimodal distribution. This different result for land use is related to two factors: firstly, the error is an absolute error and, secondly, the skewness of error probability distribution is low. The error probability distribution function (PDF) of both precipitation and ET are highly skewed to the right (see Table 4). The implication is that the majority of cases have an error that is less than the mean value. As such, randomly generated inputs tend to be more concentrated. For example, the mean absolute error for ET is 5.4 % with a high positive skewness of 1.18. This implies that the majority of randomly generated error levels are less than 5.4 %, with a higher proportion between 0 and 4 % which is the median. Therefore, the generated input data are concentrated around one peak maximum between −4 and +4 %, which creates a unimodal distribution. For land use area, the low skewness of the error probability distribution function would imply that the randomly generated inputs are concentrated around the absolute mean of 14.5 %, which generates two peaks of −14.5 and +14.5 %. Because the error is absolute, the observed distribution in the randomly generated input follows a mirrored shaped of the error PDF for each parameter (see Karimi and Bastiaanssen, 2015).

In addition to variability of remote sensing input data, outflow and reserved flow have also been made variable. Outflow was allowed to vary between 2.5 and 5.7 $km^3\,yr^{-1}$ (see Table 6). The reserved flow variability was taken as equal to the observed variability of lake evaporation. The water balance of the Awash Basin was closed by mass conservation on the storage change. An example of the variability of two output parameters is demonstrated in Fig. 9.

The results are 1000 versions of the WA+ sheets. Table 8 shows the mean value of all 1000 different versions, referred to as the stochastic mean. The differences between the original results – using the reference values – and the stochastic mean are often within a few percent, except for a few interesting cases where the differences are 10.6 % (storage change), 6.3 % (utilizable flow), 7.6 % (beneficial ET fraction), 9.4 % (basin closure fraction) and 10.6% (reserved flow fraction). These differences are mainly a result of the larger variability that each of these parameters have. A few numerical outliers in the population of the output data distribution of a given parameter can yield a different mean value of the 1000 water accounts. The large uncertainty of (groundwater) outflow and its translation into utilizable flow (outflow is utilizable plus reserved flow) is the root cause of these differences and is in agreement with the general difficulty of estimating the groundwater flow of hydrological systems. Since the storage change is the residual of the water balance, it will automatically also get a large variability. The resource base performance indicators follow the same trend as the absolute values. Hence, the absence of reliable outflow data has, in this specific case study, impacts on the uncertainty of utilizable flow and storage change and thus also on the basin closure fraction.

The stochastically generated data sets of error ranges can be described by an interval around the mean value. This gives an indication of the error probability and accuracy of each of the parameters in a standardized way; it will allow comparing the variability of different parameters. The band widths surrounding the mean value at 95 % confidence intervals for the main input and output parameters of WA+ resource base and evapotranspiration sheets are presented in Table 8.

The interim conclusion is that MC simulations provide a slightly different result to standard modeling which, for certain parameters, can exceed 5 %. Since the consideration of larger variability of certain terms is realistic, it is recommended to run the WA+ sheets always in the MC mode.

3.3 Temporal variability and error probabilities for multiple years

To understand the temporal variability of the error band width in WA+, the MC analysis was extended to multiple years. The covered period was 2009–2011. As explained earlier, the period includes an average rainfall year, 2009, a wet year, 2010, and a dry year, 2011. For every year the MC model was run 1000 times and a stochastic mean for each WA+ parameter was calculated. To normalize the variability of the error component, their CVs (coefficients of variation) were calculated for all the parameters. Table 9 summarizes the results of this exercise.

The CV is an indication of the variability of the population of output values of one particular parameter. In this case, a larger variability can be attributed to a larger uncertainty of the MC results. A CV of 10 % or less is generally considered indicative of a very good accuracy because the variability is within an acceptable value range of a particular variable. CV values in the range of 10–20 % are deemed acceptable and are close to the accuracy that is generally achieved through field measurements. Estimates with CVs of more than 20 % require caution and those with CVs of 40 % are unreliable. However, in all the cases the mean value must also be considered because a low mean value with a high CV may represent a smaller variation range compared to a high mean with a low CV.

Figure 10 illustrates the temporal variability of CVs for all the WA+ parameters and performance indicators in the study period. A low average CV value implies that the frequency distribution of a single WA parameter for a given year has a relatively minor variability, meaning that the results are stable and accurate. The CV for multiple years of the CV of the stochastic distribution of a single parameter is indicated by the height of the bar. The latter reflects how vulnerable a decision is on the time frame considered for the water accounting. A certain WA+ parameter can be more accurate in a low rainfall year than during a high rainfall year. The results indicate that CVs for available water, exploitable water, utilized flow, and the outflow, vary from year to year, which is a mere consequence of the combination of the temporal variable rainfall and temporal constant ET values. Consequently, the performance indicators that are related to these parameters have varying CVs. These include exploitable water fraction, basin closure fraction, and the reserved flow fraction. In general, more temporal variability implies lower accuracy.

An important observation is that the majority of the performance indicators show low sensitivity to the input data. Four out of eight indicators, i.e., available water fraction, T fraction, managed fraction, and beneficial fraction, have CV's of less than, or close, to 10 % in all 3 yr. Another two indicators, exploitable water fraction and irrigated ET fraction, have average CVs of close to 20 % which fall in the acceptable range. All the parameters and indicators that have CVs of less than 20 % are deemed to be reliable and can be used in policy formulation processes. However, some of the indicators such as basin closure fraction have a high average CV and a high temporal variability. The same applies to utilized flow, utilizable flows, and the reserved flow fraction. This shows that these parameter and indicators should be treated with caution and should not be used to formulate policy decisions. Most inaccurate indicators are related to the outflows and their low reliability is directly linked to the lack of accurate information on groundwater outflows. The uncertainty of groundwater flows is a general problem in ungauged basins (Hrachowitz et al., 2013) and should not interpreted as being typical for the Awash Basin. Hence, extra allocation of water and exploitation of utilizable flows are highly unreliable, and should not be done without some cross-examination and execution of advanced groundwater studies. First priority should

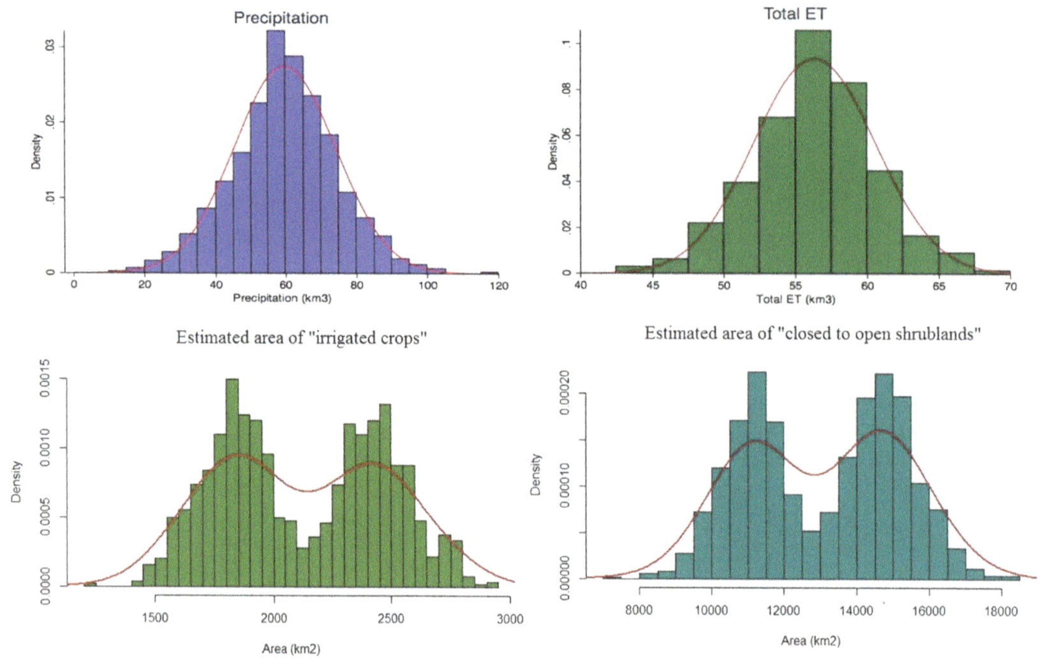

Figure 8. Example of the frequency distribution of the randomly simulated input parameters ET, rainfall and land use into the Monte Carlo simulations for 2009 (for land use estimated area of two classes, i.e., irrigated crops and closed to open shrublands, are demonstrated).

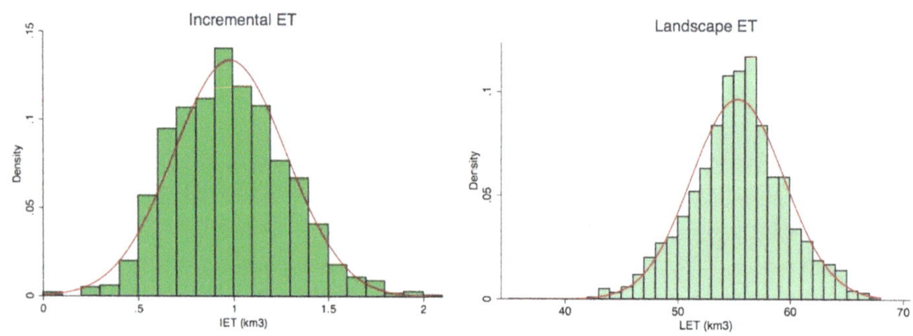

Figure 9. Variability of two selected output parameters for 2009, i.e., incremental ET and landscape ET, following from the Monte Carlo simulation of 1000 runs.

be given to understanding spatial and temporal variability of the less-known flows in any basin. In other basins, however, outflow could be measured or modeled with a much more comfortable accuracy, which will make it feasible to make decisions on re-allocation of available water resources. Hence, the uncertain outflow is specific for the Awash Basin.

4 Summary and conclusions

WA+ is a novel analytical framework that summarizes complex hydrological processes and water management issues in river basins. The framework uses state of the art satellite-based measurements of land and water consumption to quantify hydrological variables and water accounts. This makes WA+ to a large extent independent from conventional hydrological measurements. Such independence is necessary to apply WA+ on any river basin, including poorly gauged and ungauged basins. However, the use of satellite-based measurements for water accounting may raise concerns about the reliability of the accounts if they have not been verified on the ground. To address this concern, this paper examined the impact of the errors in satellite-based input data to WA+ on the confidence that policy makers can have in the outputs and information provided.

The focus of the study was on the WA+ resource base sheet and evapotranspiration sheet. ET, precipitation, and land use are the three main satellite-based spatial data sets used for these two sheets. The Awash Basin in the central Rift Valley in Ethiopia was used to demonstrate the influence that errors in the input data could have on the confidence in the outputs. The analysis covered a period of 3 yr which included

Table 8. Difference between standard and stochastic modeling of the WA+ outputs for 2009 (RB – resource base).

Parameter	Reference computation (km^3)	Stochastic mean (km^3)	Confidence interval (0.95) (km^3)	Difference between standard and stochastic mean
Resource base sheet				
Precipitation	59.80	59.96	±28.55	0.3 %
Storage change	−0.40	−0.45	±26.98	10.6 %
Net inflow	60.20	60.41	±7.76	0.3 %
Landscape ET	55.37	55.36	±7.74	0.0 %
Exploitable water	4.83	5.05	±2.77	4.4 %
Available water	4.21	4.42	±2.80	4.9 %
Utilized flow	0.98	0.98	±0.59	0.0 %
Reserved flows	0.62	0.62	±0.24	0.5 %
Outflow	3.84	4.07	±3.10	5.4 %
Utilizable flow	3.22	3.44	±3.14	6.3 %
Evapotranspiration sheet				
Total ET	56.36	56.35	±7.98	0.0 %
ET managed	17.34	17.33	±5.14	0.0 %
T total	18.00	18.00	±3.42	0.0 %
E total	36.31	36.30	±5.03	0.0 %
I total	2.05	2.05	±0.37	0.0 %
Beneficial depletion	20.08	18.66	±3.50	7.6 %
Non-benef. depletion	36.28	37.69	±5.20	3.7 %
RB sheet indicators*				
Exploitable water fraction	0.08	0.08	±0.05	4.1 %
Available water fraction	0.87	0.87	±0.09	0.6 %
Basin closure fraction	0.23	0.26	±0.26	9.4 %
Reserved flow fraction	0.16	0.18	±0.16	10.6 %
Evapotranspiration sheet indicators*				
T fraction	0.32	0.32	±0.03	0.1 %
Managed fraction	0.31	0.31	±0.07	0.2 %
Beneficial fraction	0.36	0.33	±0.03	7.7 %
Irrigated ET fraction	0.10	0.11	±0.05	5.4 %

* Indicators are dimensionless.

an average rainfall year (510 mm yr^{-1}) 2009, a wet year (862 mm yr^{-1}) 2010, and a dry year (364 mm yr^{-1}) 2011. Spatial ET data for the Awash Basin was computed by means of the ETLook model. Daily rainfall maps were acquired from the FEWS NET, and a land use map, customized by Dost et al. (2013) was also used for the application of water accounting in the Awash Basin. The errors in these satellite-based land and water use measurements are based on a comprehensive review by Karimi and Bastiaanssen (2015). The Monte Carlo technique that is based on selecting numbers randomly from a predefined probabilistic distribution was used for stochastic simulation of WA+ outputs. The simulation was repeated 1000 times for all 3 yr.

The results of this exercise show that the stochastic mean of the majority of WA+ parameters and performance indicators (13 out of 25) are within a 1 % deviation from the original value. Out of 25, 19 are within a 5 % deviation. The maximum deviation of 10 % was observed for the storage change and reserved flow fraction. This shows that stochastic simulation can be used as part of a standard procedure to produce water accounts with WA+. There are two main advantages related to the MC technique. Firstly, it allows for the incorporation and acknowledgement of input data errors in producing water accounts. Secondly, it provides the possibility to estimate and report on the error bandwidth that surrounds every WA+ output. The latter is of essential value to informed decision making, as it enables users to better understand the error margin that is associated with the generated information. The goal is to separate reliable information from that with low reliability. In such a way, outputs with a

Table 9. Temporal variability of WA+ across a longer period with low and high rainfall years.

Parameter	2009		2010		2011	
	Mean	CV	Mean	CV	Mean	CV
			Resource base sheet			
Precipitation	59.96	24 %	101.51	23 %	42.84	24 %
Storage change	−0.45	–	30.81	77 %	−17.25	61 %
Net inflow	60.41	6.8 %	70.70	6.7 %	60.09	6.8 %
Landscape ET	55.36	7.4 %	64.60	7.2 %	56.20	7.2 %
Exploitable water	5.05	27 %	6.10	12 %	3.89	15 %
Utilized flow	0.98	30 %	0.71	52 %	1.26	23 %
Available water	4.42	5.3 %	5.44	14 %	3.17	18 %
Reserved flows	0.62	18 %	0.66	19 %	0.71	20 %
Outflow	4.07	38 %	5.40	17 %	2.62	22 %
Utilizable flow	3.44	45 %	4.74	20 %	1.91	30 %
			Evapotranspiration sheet			
Total ET	56.35	7.5 %	65.30	7.3 %	57.46	7 %
ET managed	17.33	15 %	20.01	15 %	17.65	15 %
T total	18.00	10 %	25.95	9 %	20.76	9.3 %
E total	36.30	7.3 %	36.25	7 %	34.61	6.8 %
I total	2.05	9.6 %	3.11	8.8 %	2.10	9 %
Beneficial depletion	18.66	9.8 %	26.64	8.9 %	21.49	9.2 %
Non-benef. depletion	37.69	7.3 %	38.66	6.9 %	35.97	6.8 %
			Indicators			
Exploitable water fraction	0.08	27 %	0.09	13 %	0.06	15 %
Available water fraction	0.87	5 %	0.89	3 %	0.81	5 %
Basin closure fraction	0.26	51 %	0.14	73 %	0.40	20 %
Reserved flow fraction	0.18	45 %	0.13	39 %	0.28	22 %
T fraction	0.32	5 %	0.40	3 %	0.36	4 %
Managed fraction	0.31	12 %	0.31	12 %	0.31	12 %
Beneficial fraction	0.33	5 %	0.41	3 %	0.37	4 %
Irriated ET fraction	0.11	21 %	0.10	21 %	0.11	20 %

high error margin and low reliability will be identified and it is recommended that they should not be used to formulate policy decisions. This reliability can be normalized and quantified by calculating coefficients of variation for all the WA+ parameters.

Results of the multiyear analysis for the Awash Basin, after incorporating input data error, showed that the majority of WA+ parameters and performance indicators have CVs of less than 20 % which implies that they are reliable. The results also indicate that parameters and indicators such as utilized flow, utilizable flow, and basin closure fraction have a high margin of error and thus have low reliability. This implies, for instance, that despite the fact that accounting results show that the utilizable flow is on average about $3.4\,km^3\,yr^{-1}$, this estimate has low reliability. The same applies for the figures related to basin closure fraction and utilized flow. In other words, although the accounting outputs, i.e., utilized flow, utilizable flow, and basin closure fraction, suggest that more water can be utilized in the basin, the high margin of error associated with these outputs means they are not reliable enough to be used for formulating policy decisions. As such, more research with more accurate input data is required to verify and endorse such possibilities which, in this case, are related to uncertain groundwater flows in deeper geological layers. This finding applies to the Awash Basin and cannot be generalized because for many basins, the discharge at the river mouth is not properly measured or modeled.

Every measurement, regardless of the method used, has some level of uncertainty. In many instances, hydrologists and engineers know the uncertainty associated with in situ measurements such as runoff, canal water levels, etc., yet these estimates are used in studies and in policy decision making. It is a fact that compared to ground measurements, our knowledge on the accuracy of remote-sensing-based estimates of hydrological parameters is less complete. The review conducted by Karimi and Bastiaanssen (2015) provides essential information in this regard for annual and seasonal

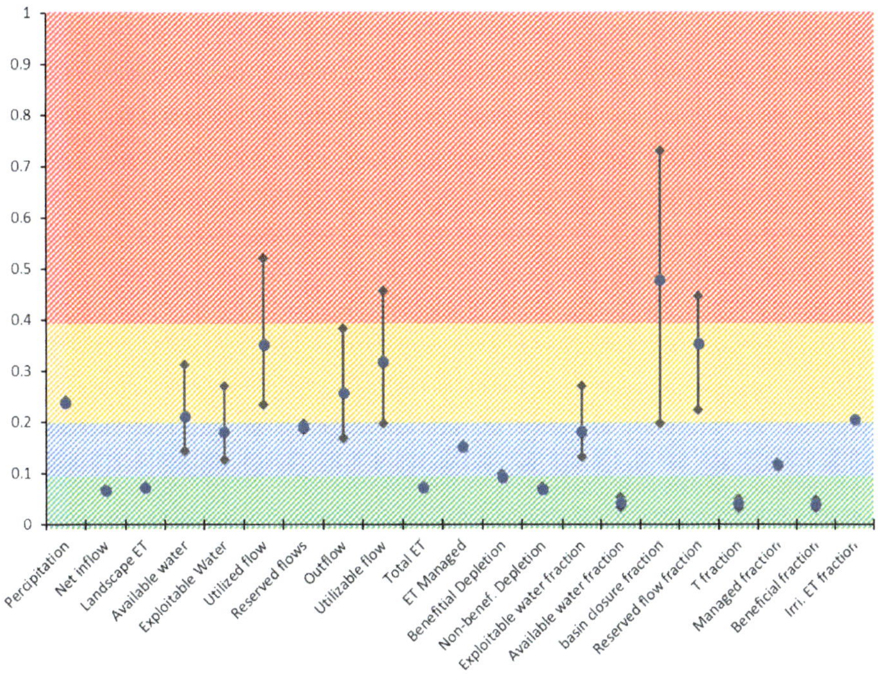

Figure 10. The level of inaccuracy expressed as a coefficient of variation for a dry, a wet and an average rainfall year. The height of the bars expresses temporal variability. The background colors indicate where a certain parameter should be considered in the water management decision process.

estimates. At this timescale, remote-sensing-based model performance outweighs monthly, weekly, and daily scale estimates that are known to have larger uncertainties. The important point is to acknowledge these uncertainties while processing information and inform the users accordingly. Nevertheless, remote-sensing-based information can be very valuable in data-scarce areas of the world and can contribute to bridging spatial scales in hydrology (Stewart et al., 1996).

Acknowledgements. Funds for this research were provided by the CGIAR Research Programme on Water Land and Ecosystem.

Edited by: B. van den Hurk

References

Allen, R. G., Tasumi, M., Morse, A., Trezza, R., Wright, J. L., Bastiaanssen, W., Kramber, W., Lorite, I., and Robison, C. W.: Satellite-Based Energy Balance for Mapping Evapotranspiration with Internalized Calibration (METRIC) – Applications, J. Irrig. Drain. Eng., 133, 395–406, 2007.

Ayenew, T.: Numerical groundwater flow modelling of the Central Main Ethiopian Rift lakes basin, SINET Ethiop. J. Sci., 24, 167–184, 2001.

Ayenew, T., Kebede, S., and Alemyahu, T.: Environmental isotopes and hydrochemical study applied to surface water and groundwater interaction in the Awash, Hydrol. Process., 22, 1548–1563, doi:10.1002/hyp.6716, 2008.

Bastiaanssen, W. G. M.: Water accountants: de nieuwe generatie waterbeheercontroleurs, intreerede, in: Inaugural Address, Delft University of Technology, Delft, the Netherlands, p. 16, 2009.

Bastiaanssen, W. G. M. and Chandrapala, L.: Water balance variability across Sri Lanka for assessing agricultural and environmental water use, Agr. Water Manage., 58, 171–192, 2003.

Bastiaanssen, W. G. M., Cheema, M. J. M., Immerzeel, W. W., Miltenburg, I., and Pelgrum, H.: The surface energy balance and actual evapotranspiration of the transboundary Indus Basin estimated from satellite measurements and the ETLook model, Water Resour. Res., 48, W11512, doi:10.1029/2011WR010482, 2012.

Bastiaanssen, W. G. M., Karimi, P., Rebelo, L.-M., Duan, Z., Senay, G., Muttuwatte, L., and Smakhtin, V.: Earth observation-based assessment of the water production and water consumption of Nile Basin Agro-Ecosystems, Remote Sensing, 6, 10306–10334, 2014.

Baudron, F., Jaleta, M., Okitoi, O., and Tegegn, A.: Conservation agriculture in African mixed crop-livestock systems: Expanding the niche, Agr. Ecosyst. Environ., 187, 171–182, 2014.

Behailu, S.: Stream flow simulation for the Upper Awash Basin, MSc Thesis, Addia Ababa University, Addis Ababa, 2004.

Bicheron, P., Defourny, P., Brockmann, C., Schouten, L., Vancutsem, C., Huc, M., Bontemps, S., Leroy, M., Achard, F., Herold, M. F., Ranera, F., and Arino, O.: GLOBCOVER Products Report Description and Products Description and Validation Report, MEDIAS, Toulouse, France, 2008.

CA: Water for Food, Water for Life: A Comprehensive Assessment of Water Management in Agriculture, Earthscan, London and International Water Management Institute, Colombo, 2007.

Döll, P. and Siebert, S.: A digital global map of irrigated areas, ICID J., 49, 55–66, 2000.

Dost, R., Obando, E. B., Bastiaanssen, W., and Hoogeveen, J.: Background report: Water accounting+ in the Awash River Basin, eLEAF Report, Wageningen, the Netherlands, 56 pp., 2013.

Drooger, P., Simons, G. W. H. G., Bastiaanssen, W. G. M., and Hoogeveen, J.: Water accounting Okavango; Coping with water scarcity – developing national water audits Africa, WaterWatch, Wageninegen, the Netherlands, 58 pp., 2010.

Edossa, D. C., Babel, M. S., and Das Gupta, A.: Drought Analysis in the Awash River Basin, Ethiopia, Water Resour. Manage., 24, 1441–1460, doi:10.1007/s11269-009-9508-0, 2010.

EVDSA – Ethiopian Valleys Development Studies Authority: Master plan for the development of Surface water resources in the Awash Basin, Addis Ababa, Ethiopia, 1989.

Falkenmark, M. and Rockström, J.: The new Blue and Green water paradigm: Breaking new ground for water resources planning and management, J. Water Resour. Pl. Manage., 132, 129–132, 2006.

FAO: Review of world water resources by country, Water Report No. 23, Rome, Italy, 2003.

FAO: Coping with water scarcity; an action framework for agriculture and food security, Water Reports No. 38, FAO, Rome, Italy, 78 pp., 2012.

Herman, A., Kumar, V. B., Arkin, P. A., and Kousky, J. V.: Objectively determined 10-day African rainfall estimates created for famine early warning systems, Int. J. Remote Sens., 18, 2147–2159, doi:10.1080/014311697217800, 1997.

Hrachowitz, M., Savenije, H. H., Bloschl, G., McDonnell, J. J., Sivapalan, M., Pomeroy, J., Arheimer, B., Blume, T., Clark, M. P., Ehret, U., Fenicia, F., Freer, J. E., Gelfan, A., Gupta, H. V., Hughes, D. A., Hut, R., Montanari, A., Pande, S., Tetzlaff, D., Troch, P. A., Uhlenbrook, S., Wagener, T., Winsemius, H., Woods, R. A., Zehe, E., and Cudennec, C.: A decade of Predictions in Ungauged Basins (PUB) – a review, Hydrolog. Sci. J., 58, 1198–1255, 2013.

Karimi, P., Molden, D., Bastiaanssen, W. G. M., and Cai, X.: Water accounting to assess use and productivity of water: evolution of a concept and new frontiers, in Water accounting: International approaches to policy and decision-making, edited by: Godfrey, J. M. and Chalmers, K., Edward Elgar Publishing, Cheltenham, UK, 76–88, 2012.

Karimi, P., Bastiaanssen, W. G. M., and Molden, D.: Water Accounting Plus (WA+) – a water accounting procedure for complex river basins based on satellite measurements, Hydrol. Earth Syst. Sci., 17, 2459–2472, doi:10.5194/hess-17-2459-2013, 2013a.

Karimi, P., Bastiaanssen, W. G. M., Molden, D., and Cheema, M. J. M.: Basin-wide water accounting based on remote sensing data: an application for the Indus Basin, Hydrol. Earth Syst. Sci., 17, 2473–2486, doi:10.5194/hess-17-2473-2013, 2013b.

Karimi, P. and Bastiaanssen, W. G. M.: Spatial evapotranspiration, rainfall and land use data in water accounting – Part 1: Review of the accuracy of the remote sensing data, Hydrol. Earth Syst. Sci., 19, 507–532, doi:10.5194/hess-19-507-2015, 2015.

Meskale, M.: Hydrogeology of South Afar and adjacent areas, MSc Thesis, International Inst. for Aerial Survey and Earth Sciences, Enschede, the Netherlands, 1982.

Molden, D.: Accounting for water use and productivity, SWIM paper 1, International Water Management Institute, Colombo, Sri Lanka, 1997.

Molden, D. and Sakthivadivel, R.: Water Accounting to Assess Use and Productivity of Water, Int. J. Water Resour. Dev., 15, 55–71, doi:10.1080/07900629948934, 1999.

Neale, C. M. U., Geli, H. M. E., Kustas, W. P., Alfieri, J. G., Gowda, P. H., Evett, S. R., Prueger, J. H., Hipps, L. E., Dulaney, W. P., Chávez, J. L., French, A. N., and Howell, T. a.: Soil water content estimation using a remote sensing based hybrid evapotranspiration modeling approach, Adv. Water Resour., 50, 152–161, doi:10.1016/j.advwatres.2012.10.008, 2012.

Pelgrum, H. and Bastiaanssen, W. G. M.: An Intercomparison of Techniques to Determine the Area-Averaged Latent Heat Flux from Individual in Situ Observations: A remote Sensing Approach Using the European Field Experiment in a Desertification-Threatened Area Data, Water Resour. Res., 32, 2775–2786, doi:10.1029/96WR01396, 1996.

Shilpakar, R. L., Bastiaanssen, W. G. M., and Molden, D. J.: A remote sensing-based approach for water accounting in the East Rapti River Basin, Nepal, Himal. J. Sci., 7, 15–30, 2012.

Smakhtin, V. U. and Eriyagama, N.: Developing a software package for global desktop assessment of environmental flows, Environ. Model. Softw., 23, 1396–1406, 2008.

Stewart, J. B., Engman, E. T., Feddes, R. A., and Kerr, Y.: Scaling up in hydrology using remote sensing, Wiley, New York, 270 pp., 1996.

Taddese, G., Sonder, K., and Peden, D.: The water of the Awash River basin a future challenge to Ethiopia, Addis Ababa, Ethiopia, 2003.

UNDP: Geology, geochemistry and hydrology of hot springs of the East African Rift system within Ethiopia, New York, 1973.

UN-Water: UN-Water country briefs project, Geneva, 2013.

Vörösmarty, C. J., McIntyre, P. B., Gessner, M. O., Dudgeon, D., Prusevich, A., Green, P., Glidden, S., Bunn, S. E., Sullivan, C. A., Liermann, C. R., and Davies, P. M.: Global threats to human water security and river biodiversity, Nature, 467, 555–561, doi:10.1038/nature09440, 2010.

Wolf, A. T., Yoffe, S. B., and Giordano, M.: International waters?: identifying basins at risk, Water Policy, 5, 29–60, 2003.

Yilmaz, M. T., Anderson, M. C., Zaitchik, B., Hain, C. R., Crow, W. T., Ozdogan, M., Chun, J. A., and Evans, J.:Comparison of prognostic and diagnostic surface flux modeling approaches over the Nile River basin, Water Resour. Res., 50, 386–408, doi:10.1002/2013WR014194, 2014.

12

Uncertainty analysis of a spatially explicit annual water-balance model: case study of the Cape Fear basin, North Carolina

P. Hamel[1] **and A. J. Guswa**[2]

[1]Natural Capital Project, Woods Institute for the Environment, Stanford University, Stanford, CA, USA
[2]Picker Engineering Program, Smith College, Northampton, MA, USA

Correspondence to: P. Hamel (perrine.hamel@stanford.edu)

Abstract. There is an increasing demand for assessment of water provisioning ecosystem services. While simple models with low data and expertise requirements are attractive, their use as decision-aid tools should be supported by uncertainty characterization. We assessed the performance of the InVEST annual water yield model, a popular tool for ecosystem service assessment based on the Budyko hydrological framework. Our study involved the comparison of 10 subcatchments ranging in size and land-use configuration, in the Cape Fear basin, North Carolina. We analyzed the model sensitivity to climate variables and input parameters, and the structural error associated with the use of the Budyko framework, a lumped (catchment-scale) model theory, in a spatially explicit way. Comparison of model predictions with observations and with the lumped model predictions confirmed that the InVEST model is able to represent differences in land uses and therefore in the spatial distribution of water provisioning services. Our results emphasize the effect of climate input errors, especially annual precipitation, and errors in the ecohydrological parameter Z, which are both comparable to the model structure uncertainties. Our case study supports the use of the model for predicting land-use change effect on water provisioning, although its use for identifying areas of high water yield will be influenced by precipitation errors. While some results are context-specific, our study provides general insights and methods to help identify the regions and decision contexts where the model predictions may be used with confidence.

1 Introduction

The interactions between hydrology and land use and land-management decisions have received increased attention in recent years. The International Association of Hydrological Sciences (IAHS) recently declared this decade Panta Rhei – everything flows – to focus on the changing dynamics of the water cycle in connection with changing human systems (Montanari et al., 2013). Socio-hydrology has recently been proposed as a "use-inspired" discipline to focus on understanding the human-modified water cycle (Sivapalan et al., 2014). The quantification of water services, or the value that humans derive from natural processes, is also increasingly seen a means of elucidating the interactions between people and water. Examples of this approach abound globally: through its Grain-to-Green program, China, incentivizes land owners to convert annual crops to perennial species or natural forests (Liu et al., 2008). In South America, there now exist dozens of water funds, which invest in upstream conservation measures to ensure the downstream provision of clean water (Martin-Ortega et al., 2013). In the United States, federal investments in water resources projects now require an assessment of impacts to ecosystem services (Council on Environmental Quality, 2013).

To quantify the impact of land-use and land-management decisions on ecosystem services, a number of tools have been developed by researchers and practitioners (Bagstad et al., 2013). Typical applications of these tools include the development of spatial planning policies, such as the delineation of priority areas for conservation or for agricultural development (Guswa et al., 2014). These applications often (i) occur in data-scarce environments; (ii) require spatially explicit information, at the scale of individual land holdings and parcels; and (iii) integrate a range of ecosystem services rather than focus on the precise quantification of a particular service. Accordingly, state-of-the-art models requiring extensive data and expertise are generally not appropriate for such applications. Instead, models for ecosystem-service valuation often focus on ease of use, using globally available data, accepting spatially explicit input and producing spatially explicit output, and limiting the model structure to key biophysical processes involved in land-use change (Guswa et al., 2014).

The InVEST annual water yield model was developed in line with this philosophy (Sharp et al., 2014). It includes a biophysical component, computing the provision of freshwater, or water yield, by different parts of the landscape, and a valuation component, representing the benefits of water provisioning to people. The biophysical module, the focus of this paper, is based on the Budyko theory, which has a long history and continues to receive interest in the hydrological literature (Budyko, 1961; Zhou et al., 2012; Zhang et al., 2001, 2004; Donohue et al., 2012; Xu et al., 2013; Wang and Tang, 2014). The InVEST model applies a one-parameter formulation of the theory (Zhang et al., 2004) in a spatially explicit way. This raises two issues. First, application of the model to ungauged basins or to future land-use scenarios requires a methodology for determining the value of the model parameter from known characteristics of the climate and basin, since it cannot be determined via calibration. Second, the Budyko approaches have been developed for long-term water balances at the catchment scale, rather than at the scale of individual land parcels, which is required for ecosystem-service decisions. The effect of this change in spatial scale is unclear, and calls for a rigorous analysis of the model and its uncertainties.

Uncertainty analyses remain rare or incomplete in ecosystem service assessments, where the focus is on analyzing trade-offs and valuation of multiple services, often at the expense of characterizing uncertainty of individual modeling components. For example, in reviewing the literature using the InVEST annual water yield model, we found the following common limitations: absence of or inadequate comparison with observed data, calibration of the model without prior identification of sensitive parameters, and lack of validation of the predictive capabilities in the context of land-use and land-cover (LULC) change (Bai et al., 2012; Su and Fu, 2013; Terrado et al., 2014; Nelson et al., 2010). To varying degrees, these limitations jeopardize the production of credible assessments of ecosystem services.

Recent work paved the way for understanding the uncertainties in the InVEST model predictions. Sánchez-Canales et al. (2012) analyzed the sensitivity of the model in their case study of the Llobregat catchment, in Spain. They found that the model was sensitive to climate variables, but less so to the Z parameter (see model description). Similarly, Boithias et al. (2014) and Terrado et al. (2014) reflect on the sensitivity of the model to climate inputs and calibrate the model based on the climate parameters and return flows. The conclusions of these studies are often context-specific and lack a quantitative estimate of the model's structural uncertainties. In particular, they assess the effect of the climate variables' uncertainty but do not examine the ability of the model to represent land-use change.

This paper aims to extend this work by characterizing the uncertainty in the InVEST annual water-yield model, and assess its utility to inform ecosystem-service decisions. As indicated above, the focus on water services implies a focus on decisions related to land use and land management, thus requiring spatially explicit descriptions of the landscape and associated hydrologic parameters (Guswa et al., 2014). Ecosystem-service decisions may be based on spatially aggregated output (e.g., which landscape scenario provides the greatest water yield at the base of the catchment), or may require spatially explicit output (e.g., which parcels in this catchment are of highest priority for conservation). While the proposed model is capable of providing output to inform the latter, this paper focuses on the former, since typical measurements of water yield (streamflow) are inherently aggregated. Using a case study in the Cape Fear region of North Carolina (NC), our study (i) quantifies the effect of parameter uncertainty on model outputs through sensitivity analyses, (ii) compares the distributed application of the water balance to the catchment-scale application, and (iii) quantifies the accuracy of calibrated and uncalibrated versions of the model by comparing model predictions to observations. From a practical standpoint, this work helps InVEST model users identify modeling uncertainties and proposes simple and replicable methods that can be used to quantify the reliability of water-service decisions.

2 InVEST annual water yield model

2.1 Background theory

The Budyko curve is a unique empirical function that relates the ratio of actual evapotranspiration to precipitation (averaged over a catchment and over many years) to the ratio of potential evapotranspiration to precipitation (Budyko, 1961). The function is bounded by two limits – an energy limit in which actual evapotranspiration is equal to potential, and a water limit for which actual evapotranspiration is equal to precipitation. Due to spatial and temporal variability in climate forcing, the asynchronicity of water supply (P) and demand (PET), the imperfect capacity of the root zone to buffer that asynchronicity, and lateral redistribution of water within the catchment, the Budyko curve lies below those two limits (Fig. 1).

To describe the degree to which long-term catchment water balances deviate from the parameterless Budyko curve, a number of scholars have proposed one-parameter functions

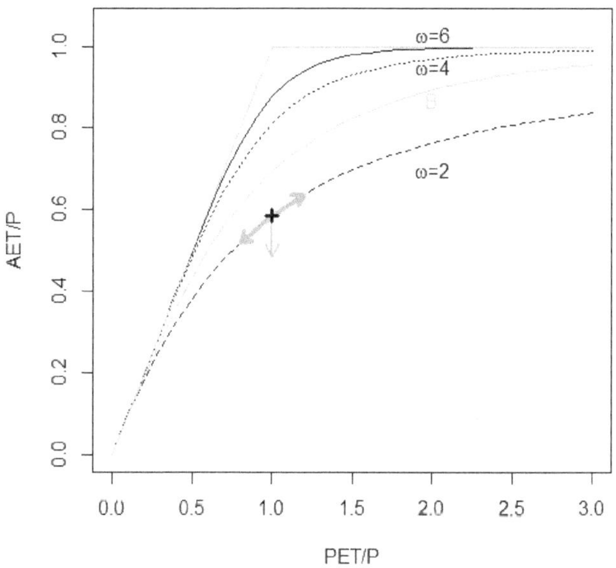

Figure 1. Original Budyko curve (B) and its variations used in the lumped model (Eq. 1), shown for ω values of 2, 4, and 6. Grey lines represent the energy and water limits. Arrows illustrate the effect of a change in the climate forcing (thick arrows) and a change in the ω parameter, a function of Z, precipitation, and soil and vegetation properties (thin arrow, see Eq. (3) for details).

that are similar (e.g., Fu, 1981; Choudhury, 1999; Zhang et al., 2004; Wang and Tang, 2014). The InVEST water yield model employs the formulation by Zhang et al. (2004), which incorporates a catchment parameter, ω:

$$\frac{\mathrm{AET}}{P} = 1 + \frac{\mathrm{PET}}{P} - \left[1 + \left(\frac{\mathrm{PET}}{P}\right)^{\omega}\right]^{1/\omega}, \tag{1}$$

where AET is the actual evapotranspiration (mm), P is precipitation (mm), and PET is the potential evapotranspiration (mm). ω affects the partitioning of precipitation between evapotranspiration and runoff, and is a function of climate and physical factors. Larger values of ω indicate those basins that are more "efficient" in converting precipitation to transpiration, e.g., those with precipitation synchronous with PET and/or with deeper root zones. Gentine et al. (2012) and Troch et al. (2013) have shown that the natural coevolution of vegetation, climate, and topography may lead to basins for which the effects implicitly captured by ω counterbalance each other, offering an explanation for the observed convergence of data close to the original Budyko curve. The intent of the InVEST model, however, is to predict the effects of human-induced changes, i.e., to examine catchments for which natural coevolution is disrupted.

2.2 Spatially explicit application to land-use change

Model overview

To represent parcel-level changes to the landscape, InVEST represents explicitly the spatial variability in precipitation and PET, soil depth, and vegetation. The model is GIS-based, using rasters of climate and soil properties as inputs (see Sharp et al., 2014, for full details).

For vegetated land uses, InVEST applies the Zhang formulation in a spatially explicit way at the pixel scale (10–100 m on a side):

$$\frac{\mathrm{AET}_i}{P_i} = 1 + \frac{\mathrm{PET}_i}{P_i} - \left[1 + \left(\frac{\mathrm{PET}_i}{P_i}\right)^{\omega_i}\right]^{1/\omega_i}. \tag{2}$$

In contrast to Eq. (1), P, PET, ω, and AET are all functions of the local position, indicated by the subscript i.

The parameter ω is further deconstructed to separate the effects of soil depth, rainfall frequency, and other factors, following an approach proposed by Donohue et al. (2012):

$$\omega_i = Z\frac{\mathrm{AWC}_i}{P_i} + 1.25, \tag{3}$$

where AWC_i is the plant-available water content (depth), and Z is an empirical parameter. The constant, 1.25, in Eq. (2) reflects the minimum value of ω corresponding to bare soil, following Donohue et al. (2012). In this representation, differences in land use and land cover affect both PET, through changes to the crop factor, K_c, and Z, through changes to the root depth and plant-available water content.

For open water, wetlands, and urban land uses, InVEST computes AET_i directly as a user-defined proportion of PET_i, with classical approaches such as the FAO 56 guidelines (Allen et al., 1998) or local knowledge used to determine the appropriate proportion (Sharp et al., 2014). The simple representation of these LULCs, compared to the vegetated land uses modeled with Eqs. (2) and (3), reflects the focus of the model on vegetation-dominated landscapes.

Total evapotranspiration from a catchment is computed as the sum of AET_i attributed to each cell, and water yield is obtained by subtracting this value from the total precipitation.

Selection of the Z parameter

The empirical constant Z captures catchment-wide characteristics of climate seasonality, rainfall intensity, and topography that are not described by the plant-available water content (AWC) and annual P. Given the empirical nature of the model, the value of the Z parameter remains uncertain. In this work, we examine the three methods for the determination of Z that are proposed in the InVEST user's guide (Sharp et al., 2014). The first draws upon recent work that suggests that Z is positively correlated with the average annual number of rain events per year, N, and that Z may be approximated by $N/5$ (Donohue et al., 2012). This implies that Z captures rainfall patterns, distinguishing between catchments with similar annual precipitation but different intensity. The second method relies on globally available estimates of ω (Xu et al., 2013; Liang and Liu, 2014). Z is inferred from these published values of ω by inverting Eq. (2) with values of AWC and P averaged over the catchment. In the third method, Z is determined via calibration to streamflow data (see Sect. 2.5).

3 Methods

The goal of the InVEST model is not to reproduce observations with a high degree of accuracy and precision, but to provide reliable information to inform decisions. Therefore, utility or acceptability of the model should be couched in terms of relative uncertainty. That is, the uncertainty associated with the model (due to its simple structure or challenge of parameter identification) should be on par with or less than the irreducible predictive uncertainty that arises due to uncertainty in the forcing variables – in this case, precipitation and potential evapotranspiration. To assess the relative importance of the three sources of error (structural error, parameter selection, climate variables), we applied the InVEST annual model to 10 subcatchments in the Cape Fear basin, NC. Their colocation implies a similarity in climate and seasonality and facilitates a focus on variations in land use, size and topography (Hrachowitz et al., 2013). In the following sections, we describe the study area, the methods for the sensitivity analyses and uncertainty assessment of input parameters and forcing variables, and our approach to assess the structural error of the model: comparison with observations, and with the (classical) lumped model predictions.

3.1 Cape Fear study area

The Cape Fear catchment is a 23 600 km^2 area in North Carolina. Its major land uses are forest (40 %), wetland (15 %), grassland (14 %), and agriculture (12 %), mainly in the lower parts of the catchment and including intensive swine and poultry farms. Urban and agricultural development has generated significant groundwater extraction throughout the catchment.

The climate is humid subtropical, with a precipitation average of $\sim$ 1200 mm over the 2002–2012 study period (Table A1 in the Appendix). This period was used for the analyses based on the longest period available for climate data, observed streamflow, and matching LULC map. The available precipitation data comprise the PRISM data set (Parameter-elevation Regressions on Independent Slopes Model; Gilliland, 2003) and a network of eight rain gauges maintained by the USGS (USGS, 2014). For our analyses, we use the PRISM data and two additional rasters interpolated from the USGS point data (rain gauges) via spline and inverse-distance weighting (IDW). The three input rasters (hereafter referred to as PRISM, IDW, and spline) were used separately to compute the average precipitation over each of the 10 subcatchments and assess the error introduced by the input data selection. The variability in average annual precipitation among the PRISM, IDW, and spline rasters (averaging 1118, 975, and 966 mm, respectively; Table 1) represents the uncertainty that may arise when precipitation data are limited, a situation that is common in many places around the world (McGlynn et al., 2012).

Potential evapotranspiration is represented by reference evapotranspiration (ET_0) times a crop factor K_c (Sharp et al., 2014). ET_0 was obtained from three sources: FAO data, representing a long-term average from 1961 to 1990 (FAO, 2001), MODIS (Moderate Resolution Imaging Spectroradiometer) data (Mu et al., 2012), and interpolation (IDW) from a network of 13 weather stations maintained by the Climate Office of North Carolina (hereafter referred to as Climate Office; NCSU, 2014). These three sources indicate average annual PET for the Cape Fear region to be 1240 mm (FAO), 1160 mm (MODIS), and 1310 mm (NCSU). These climate data indicate an aridity index (P / PET) of approximately 0.9 for the Cape Fear catchment. A summary of InVEST inputs is given in the Appendix (Tables A1, A2).

Streamflow observations were obtained from the USGS monitoring network (USGS, 2014). A total of 10 stations with a minimum of 10 years of data were used for the analyses (Fig. 2, Table 2). Subcatchments draining to each of these points were delineated based on the 30 m DEM (digital elevation model).

Water withdrawal data were obtained from governmental agencies (NC Department of Environment and Natural Resources, 2014). Due to the lack of spatially explicit information for water withdrawals (reported by county, which do not follow the subcatchment boundaries), and on the magnitude of return flow, we represented their effect as homogeneous over the entire catchment. We think this decision has a limited effect on model testing since the value of water withdrawals is small compared to water yields (see Sect. 4). In addition, we explicitly accounted for this uncertainty by examining the effect of a 50 % error on the water withdrawal – a magnitude consistent with the variance among the county withdrawals. The average withdrawal rate, 39 mm yr^{-1}, was subtracted from the predicted water yields for comparison with observations.

3.2 Sensitivity analyses

Sensitivity to Z and K_c

Step one in our assessment of the InVEST model was a local sensitivity analysis of water yield to the Z parameter and K_c for forest – the dominant LU class. The sensitivity of the

Table 1. Precipitation and evapotranspiration in Cape Fear according to different data sources. Mean and standard deviation values are obtained from the 10 subcatchments. The relative difference between baseline data (i.e., PRISM and FAO sources, respectively, for P and ET_0) and the alternative data sources is given as the mean and the range for the 10 subcatchments.

	Annual P (mm)			Annual ET_0 (mm)		
	PRISM	Spline	IDW	FAO	Climate Office	MODIS
Mean (± SD*)	1118 ± 11	966 ± 81	975 ± 38	1200 ± 18	1189 ± 56	1459 ± 19
Relative difference from baseline data (mean difference and range)		−14 % (−23; 2)%	−13 % (−17; −4)%		−1 % (−8; 5)%	+22 % (14; 24)%

* SD: standard deviation

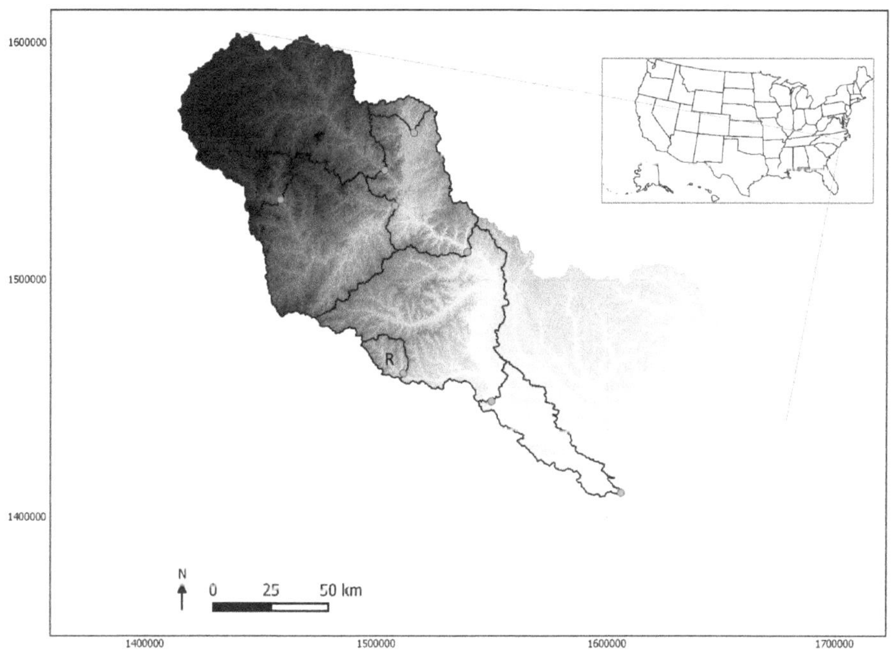

Figure 2. Cape Fear catchment showing locations of the stream gauges and subcatchments used in the study. The Rockfish catchment, discussed in the text, is indicated as "R".

model to Z can also be interpreted as the sensitivity to AWC, when the raster values are varied homogeneously over the catchment, since these parameters play a similar role in the model structure (Eq. 3).

As previously noted, large uncertainties surround the selection of the Z parameter (Sharp et al., 2014). For what we term the "baseline" case, we set Z equal to one-fifth of the number of rain days per year ($Z = N/5$). Based on historic precipitation data (SERCC, 2014), the average number of rain days per year is approximately 110, yielding a Z value of 22. We used this value as a baseline for all subcatchments, and allowed the parameter to vary between 1 and 30 for the sensitivity analyses. This range was estimated from Eq. (3) used with extreme values of P and AWC found in our catchments, and extreme values of ω (2.1 and 3.75) found in the study by Zhang et al. (2004).

Forest was the dominant LULC in all basins, with its cover ranging from 43 to 72 % of subcatchments. We therefore decided to use the crop factor K_c-forest for the sensitivity analyses, and a baseline value of 1 for K_c-forest was obtained from the FAO 56 guidelines (Allen et al., 1998). Uncertainties on this value are large since it remains difficult to provide accurate estimates of the actual evapotranspiration from forests (McMahon et al., 2013). We set the upper bound to 1.1, because values greater than this are unlikely (McMahon et al., 2013), and the lower bound to 0.7.

For the two parameters, we performed sensitivity analyses with the ranges defined above. The results are presented as a change in predicted water yield compared to the baseline run, thus assessing absolute sensitivity. Precipitation and reference evapotranspiration used for these runs were from the PRISM (1118 mm) and the FAO (1240 mm) data sets, respectively (see Sects. 2.5 and 5 for insights into the error introduced by climate data).

Table 2. Summary of mean flow, precipitation, reference evapotranspiration, and land-use characteristics of the 10 study subcatchments. LULC classes shrubland, swine farm, open water and barren represented $\leq 2\,\%$ and are not reported here. Predicted mean flow values are results from the InVEST model with Z set to 22 (the difference with the calibrated run, with $Z = 14$, is shown in parentheses). P and ET_0 are precipitation and reference evapotranspiration, respectively.

ID	Name	Area (km^2)	Observed flow (mm)	Predicted flow (mm)[1]	P (mm)	ET_0 (mm)	% Forest	% Grassland	% Agriculture	% Pasture	% Wetland	% Urban
2105769	Cape Fear at Kelly	13 567	278	208 (−31)	1112	1212	49	13	9	6	6	13
2105500	Cape Fear at Tarheel	12 535	265	218 (−32)	1109	1207	51	13	9	6	3	14
2102500	Cape Fear at Lillington	8973	236	225 (−29)	1110	1196	55	10	9	8	1	14
2104220	Rockfish CR at Raeford	237	368	174 (−53)	1118	1240	62	18	1	0	7	8
2102000	Deep River at Moncure	3727	250	210 (−39)	1113	1203	58	9	7	11	0	11
2097314	New Hope CR at Blands	197	357	322 (−14)	1143	1199	49	5	2	2	3	39
2100500	Deep River at Ramseur	913	289	287 (−27)	1112	1177	43	9	9	10	0	27
2096960	Haw River at Bynum	3294	278	264 (−23)	1110	1181	48	10	14	9	0	17
2097464	Morgan CR at White Cross	22	177	176 (−26)	1133	1198	72	7	10	5	0	5
2096846	Cane CR at Orange CR	20	202	163 (−20)	1123	1192	71	6	11	6	0	4

[1] In parentheses, we report the difference in corrected water yield (mm) between the baseline and calibrated runs ($Z = 22$, and $Z = 14$, respectively)

Sensitivity to climate inputs

To provide a context for the uncertainty in the predictions of water yield from the InVEST model, we compared the prediction error to the uncertainty in water yield that arises from uncertainty in climate (i.e., variability in the rasters of P and ET_0). Uncertainties in climatic data and their impact on rainfall-runoff models are commonly cited in the literature (McMahon et al., 2013; McGlynn et al., 2012). To be an effective decision-support tool, errors attributed to model structure and parameter selection should be on par with or less than the irreducible error associated with uncertainty in the climate.

As illustrated in Table 1, the average precipitation differed significantly across subcatchments depending on the data source: the mean differences between the PRISM and USGS data sets with the spline or IDW interpolation methods, respectively, were −14 and −13 %. Catchment-by-catchment differences were more spatially heterogeneous with the spline method, with some subcatchments receiving less precipitation relative to the baseline (PRISM data set) and others receiving more. The reference evapotranspiration data also showed significant differences across sources, although the FAO and Climate Office sources showed good agreement. The MODIS values were on average 22 % higher than those from the other two sources (Table 1). Differences between the Climate Office and FAO data were also spatially variable, ranging from −8 to 5 % across catchments.

To assess the uncertainty in water yield due to variability in climate inputs (precipitation and reference evapotranspiration), we examined the sensitivity of the baseline model results to spatially homogeneous increases and decreases in climate forcing. We considered climate inputs that are 10 % greater and 10 % less than the baseline, applied uniformly across the landscape.

3.3 Comparison of spatially explicit and lumped models

Although the InVEST annual water yield model is based on the well-studied Budyko framework, it departs from its classical application by applying the partitioning model at the pixel scale. To our knowledge, the effect of the pixel-by-pixel calculation performed by InVEST has not been previously studied. In such an application, three issues arise related to lateral flows of water, the spatial variability in climate variables, and the covariance of climate and soil in the prediction of the parameter ω.

In the catchment-scale application of Budyko-type models, lateral inflows and outflows across the catchment boundary are presumed negligible, resulting in a simple water budget based on catchment precipitation, evapotranspiration, and water yield. This assumption will not hold for a parcel-based application of Eq. (2). Thus, error in the catchment-scale water balance will arise by ignoring the excess water generated at one spot that is later evaporated at a downgradient location. Such explicit routing is not included in the InVEST model.

Additionally, even if lateral flows are negligible, applying the nonlinear Budyko curve locally and aggregating the yield will lead to different results than applying Eq. (2) to average values of P and PET. The concave nature of the function indicates that application over a range of climates will produce an average water yield that is higher than what would be predicted if applied at the catchment scale (Fig. 1).

Finally, since local values of both available water content and precipitation combine to affect the local values of omega (Eq. 3), average values of omega from the spatially explicit model will be different from what one would obtain if average values of AWC and P were used to compute an average value of omega.

To investigate these effects, we compared the model predictions to those obtained by applying the lumped model (Zhang et al., 2004) at the catchment scale. Application of the lumped model requires a value of ω, which we derived from Eq. (3) with average values of P and AWC and with Z set to the baseline value of 22, as would be done in a typical ungauged application. We thus obtained, for each subcatchment, an estimate of areal AET and water yields for the vegetated areas. AET for urban areas and wetlands was calculated separately, following the same method as InVEST, and total water yield was calculated as the area-weighted average of water yield from the vegetated and urban areas.

3.4 Testing the spatially explicit model with observed data

To quantify the accuracy and precision associated with the InVEST water yield model, we assessed model performance by comparison with observed data for each of the 10 subcatchments in the Cape Fear area. Our method aims to measure the aggregated value of water yields at the subcatchment scale, not to test whether the water yield predicted by each pixel is accurate. We measured performance with the model bias, i.e., the relative difference between predicted and observed water yields, and with the subcatchment ranking by water yields. The ability of the model to predict ranking is important for applications where prioritization of areas of low and high water yields is needed (Guswa et al., 2014).

Uncalibrated model

We first examined the performance of the model when Z was determined without calibration. We calculated Z both from the number of rain days and from a global value of ω, to evaluate the appropriateness of these recommended methods. In addition to assessing overall model performance, we also assessed the correlation between model performance and the proportion of forest in the catchment. These analyses aimed to identify a potential bias that may be corrected by modifying the LULC-specific K_c.

Calibrated model

To separate the effects of error associated with model structure from error attributed to parameter estimation, we also determined the value of Z via calibration. We calibrated to individual subcatchments, identifying for each the Z value that resulted in zero error in the water yield. We examined the similarity of Z values across the 10 basins, allowing us to assess the robustness of the model structure since we expect Z to depend on larger-scale climate and geology and not on local-scale land use. We also considered the performance of the model with a single value of Z applied to all subcatchments, determined by minimizing the average bias for all basins. This allowed us to assess the uncertainty in prediction of water yield due to model structure, i.e., the inherent uncertainty to applying Eqs. (2) and (3) to different basins when the parameter, Z, is chosen by best fit for the entire region.

4 Results

In the baseline case, we applied Eqs. (2) and (3) in a spatially explicit way with a precipitation field from the PRISM data and potential evapotranspiration data from the FAO. The value of Z in Eq. (3) was set to 22, as mentioned above. In this baseline case, predicted water yields ranged from 163 to 322 mm across the 10 subcatchments. Results are presented in Table 2.

4.1 Sensitivity analyses

Water yield predictions are very sensitive to climate inputs. The sensitivity is higher for precipitation than ET_0. Relative to the baseline case, a 10 % increase in precipitation resulted in a 30 % increase in water yield (Fig. 3), while the same increase in ET_0 resulted in a 15 % decrease in water yield.

In contrast to the climate variables, water yield is less sensitive to values of Z: for example, a change in Z from the baseline value of 22 to a value of 10 results in an increase in water yield of approximately 27 % (Fig. 3). However, given the large uncertainties in the Z parameter, potential errors in water yield can be large: for example, the water yield is 155 % higher when Z is set to 1, relative to the baseline case with $Z = 22$. The sensitivity to Z is catchment-specific, as expected, since its effect on yield is modulated by AWC and P, both of which are spatially variable. In addition, the relative sensitivity of water yield to Z decreased with increasing values of Z and increased with increasing values of the aridity index (PET / P, results not shown).

The model was found to be more sensitive to K_c (Fig. 3) with a 30 % change in K_c resulting in a 41 % change in the water yield. However, given the small range of K_c values, the effect of parameter uncertainty on the water yield prediction is lower than for Z.

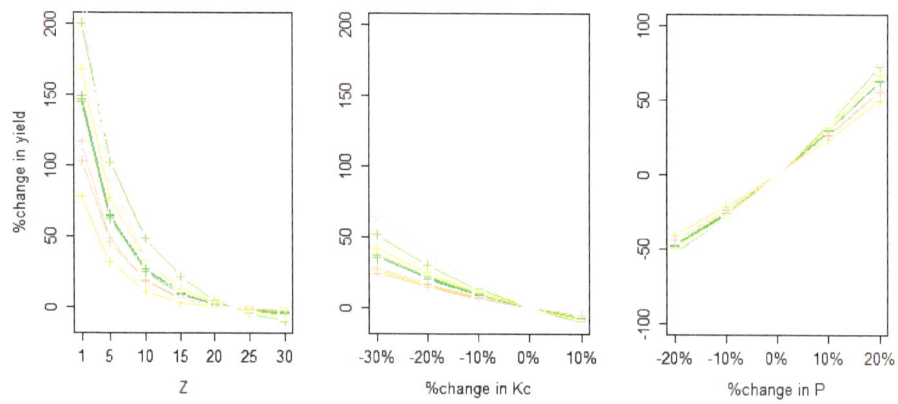

Figure 3. Sensitivity of the water yield output to the Z coefficient and crop coefficient for forest LULC (K_c). Changes are relative to the baseline run (where $Z = 22$ and $K_c = 1$). On the left-hand-side plot, absolute Z values are plotted on the x axis to facilitate the discussion on the Z coefficient. Each curve represents a subcatchment.

4.2 Comparison of spatially explicit and lumped models

Across the 10 subcatchments, the water yields predicted by the spatially explicit InVEST model were on average 10 % lower than the outputs from the lumped model. For 8 of the 10 catchments, the spatially explicit model predicted lower water yields than the lumped model, and differences ranged from −24 to 14 % (Table 3). The two catchments for which the lumped model predicted a lower water yield than the InVEST model were the Morgan Creek and Cane Creek catchments, which have the highest proportions of forest and the lowest proportions of urbanized area across the 10 catchments (Table 2).

The ω values computed for the lumped model ranged from 4.29 to 6.25 across the 10 catchments. These values are in the higher range of the values obtained by Zhang et al. (2004), as discussed in Sect. 5.2.

4.3 Testing the spatially explicit model with observed data

Uncalibrated model

Figure 4 shows the spatially explicit output from the InVEST model. The figure is for illustrative purposes only; as indicated above, we aggregate the pixel values of water yield to the subcatchment scale to compare with observations. Such comparison is presented in Fig. 5a, where the Z parameter for the InVEST model is determined from the number of rain days ($Z = 22$). Open triangles represent results from the InVEST model. To contextualize the error, grey bars represent the uncertainty in predicted water yield due to a 10 % uncertainty in precipitation and black bars represent the uncertainty in water yield due to a 50 % uncertainty in water withdrawals.

The performance of the model for this baseline run is fair. Across all basins, predicted water yields range from 163 to 322 mm yr^{-1} versus an observed range of 177–368 mm yr^{-1}. The bias between predicted and observed values averages −16 % across the 10 subcatchments, ranging from −53 to −1 % (Table 3). This indicates that the model structure combined with this choice of Z leads to a systematic underestimation of water yield. With the exception of one catchment, the biases ranged from −25 to −1 %. The outlier with an error of −53 %, Rockfish catchment, is relatively small (237 km^2), and the observed water yield is also an outlier, being the highest in the data set (367 mm). This area is also characterized by sandy soils; the plant available water content averages 0.11, compared to values between 0.17 and 0.20 for the other subcatchments. This suggests that the catchment may exhibit a unique behavior, which we will highlight in the following analyses.

Figure 5b presents the ranking of catchments in terms of their observed and predicted water yields. Discarding the outlier catchment, the figure indicates that the model accurately predicts the high- and low-ranking catchments, while there is some dispersion in ranks for the five midrange water yields, which vary from 236 to 289 mm yr^{-1}.

When Z was determined from published values of ω, the average value across the 10 catchments was 6 (compared to 22 for the baseline case). Model performance was not satisfactory in this case and the model bias averaged 68 %.

Calibrated model

In the first approach to the calibration of Z, we determined the value for which the predicted water yield exactly matched the observations. In this case, values of Z range from 6 to 20 across the 10 catchments. Not including the Rockfish catchment, the range is narrower (10–20) and the average across the nine remaining catchments is 14.5. The narrow range of

Table 3. Bias between the water yields obtained from the InVEST model (baseline value $Z = 22$), the lumped model, and observed data. The average, minimum, and maximum bias values for all the subcatchments are reported. Note that comparison with observations discards the Rockfish subcatchment which was identified as an outlier (see text for details).

	Average	Min.	Max.
InVEST/Lumped model	−0.10	−0.24	0.14
InVEST/Observations	−0.16	−0.53	−0.01
Lumped model/Observations	0.04	−0.36	0.29

variability translates into relatively small changes in water yield – the average difference among the basins is 27 %.

In a second approach, we determined a single value of Z for all 10 catchments by minimizing the average bias. This gives a value of $Z = 14$, and the error in yield for all subcatchments ranges from −38 to 14 % with a median value of −3 %. Predicted water yields range from 183 to 336 mm yr^{-1} versus an observed range from 177 to 368 mm yr^{-1}. The open circles in Fig. 5a present predictions from the calibrated model of water yield versus the observed values.

Model bias is not correlated with forest cover ($R^2 = 0.01$), nor with any other LULC (Table 1). The absence of systematic bias suggests that K_c values are in a realistic range, with no significant error due to LULC parameter selection. No significant bias was detected with regard to catchment size, suggesting that this characteristic did not systematically influence the model predictions either.

5 Discussion

5.1 Sensitivity analyses

Variability in the Z parameter, which is linearly related to ω, results in a shift of the Budyko curve, which affects water yield predictions (Fig. 1). Our results in Cape Fear suggest that the sensitivity of water yield to Z is low compared to the climate inputs, and decreases for larger values of Z (Fig. 3). This is consistent with the lumped model for which the sensitivity to ω decreases with increasing values of ω (Fig. 1). Due to this low sensitivity, small errors in estimating Z are likely to have limited impact on the reliability of water yield predictions. In particular, we note that the range investigated in the study (from 1 to 30) is greater than the typical uncertainty associated with Z: irrespectively of the selection method, values smaller than 5 are unlikely.

The sensitivity to Z also provides a sense of the sensitivity to AWC, which is a function of the local ecohydrological properties: plant available water content, root depth and soil depth (cf. Sharp et al., 2014 for details). Examination of Eq. (3) suggests that a relative change in Z has the same effect as a relative change in these ecohydrological parameters. The confidence interval for these physical parameters may be large but is reducible by measurements.

When analyzing the model sensitivity to K_c, two things are to be considered. First, the K_c value affects only the portion of the landscape covered with forest and this reduces its effect. Because total water yield is the sum of the yields from the different parts of the landscape, parameters affecting only a portion of the landscape will have a smaller effect. Second, it is worth noting that the K_c coefficient directly affects PET for a given LULC, since the latter is the product of K_c by ET_0. Examining the sensitivity of the model to K_c is therefore equivalent to a displacement along the Budyko curve, rather than a shift of this curve (Fig. 1).

In summary, the sensitivity analyses showed that, for expected and reasonable ranges of parameter variability, precipitation and potential evapotranspiration have the greatest influence on water yield. These are followed by Z and then K_c.

5.2 Comparison of spatially explicit and lumped models

Comparison of the model predictions with the classical lumped model application suggests three insights. First, it provides a sense of the effect of the pixel-by-pixel application of the Budyko theory. Because of its nonlinear nature, the average response of Eq. (2) applied across the landscape in a spatially explicit way is not equivalent to the response of the function applied to the entire catchment, characterized by average parameters. Our results suggest that this discretization effect is not large for the Cape Fear subcatchments, with the difference between the lumped and explicit models ranging from −24 to +14 %. This range is consistent with the typical errors expected from the application of simple empirical models. This point can be illustrated by the performance of the lumped model when compared with the observations: bias ranges from −36 to 29 % (Table 3). It is worth noting that the larger, positive biases (> 22 %), i.e., when the lumped model largely overestimated observed water yields, were obtained for the two subcatchments that had > 25 % urban cover, and the three basins with the least urban cover (Cane Creek, Rockfish, and Morgan Creek) had the largest underestimates of water yield. These results suggest that the contribution from urban areas was overestimated by the simple model.

The second point is related to the first one, focusing on the observation that water yields predicted by the spatially explicit model were consistently less than those predicted by the lumped model. As stated in Sect. 3.3 (Methods) , this difference can be expected from the differences in average climate values or average ω values, due to the nonlinearity in Eq. (2). In our case, the average ω values were high for the lumped model (ranging from 4.29 to 6.25). This indicates that the empirical expression for Z, developed for a lumped application (e.g., Donohue et al., 2012), may give values of

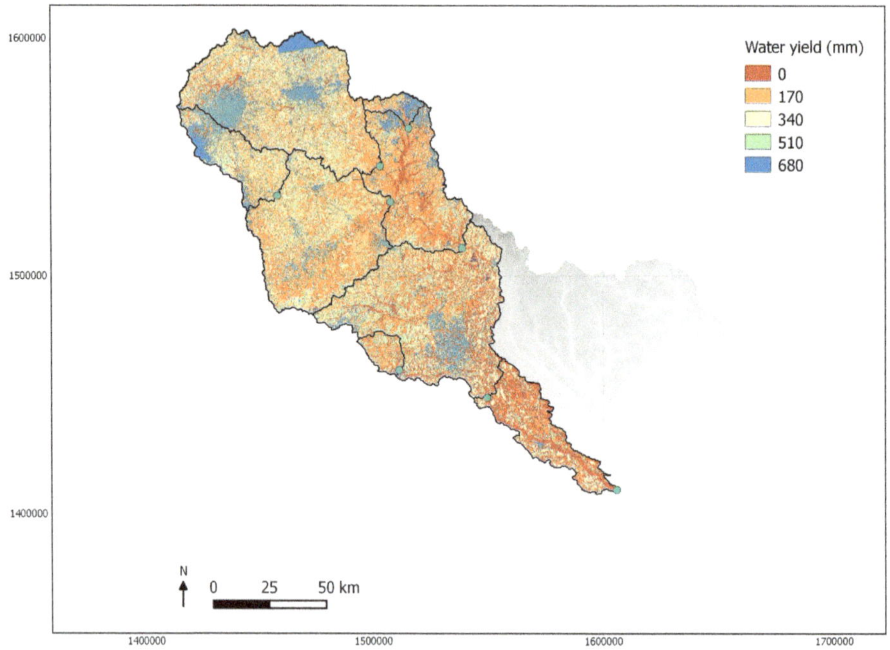

Figure 4. Spatially explicit output of the InVEST model, showing the water yield computed on a pixel scale. Model outputs are aggregated at the subcatchment scale, delineated by black lines, to be compared with observations at the gauging stations (green circles).

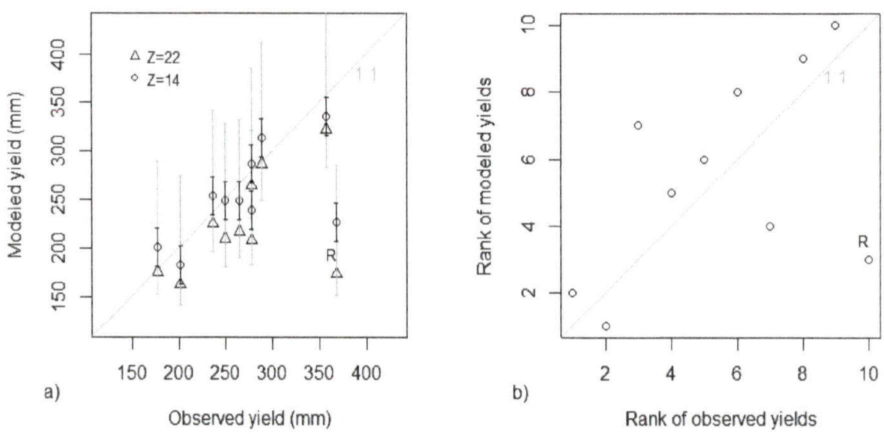

Figure 5. (a) Comparison between modeled yields (corrected for water withdrawal) and observed yields, both for the baseline run ($Z = 22$), and the calibrated run ($Z = 14$). Black error bars represent the uncertainty on the value for water withdrawal, while grey bars represent a 10 % error in the precipitation input. **(b)** Comparison of subcatchment ranks. The outlier (Rockfish) subcatchment is noted as "R" on each figure (see text for details).

Z (and, therefore, ω) that are too large for our case study; this effect is emphasized when used in a spatially explicit model. Calibration of the model based on Z allows correcting this error in the empirical expression, although further studies would be necessary to gain insights into the extrapolation of the Z parameter to spatially explicit models like InVEST.

Finally, the good agreement between the InVEST model and the lumped model allows drawing from the large body of work investigating the performance of the latter model. For example, Zhou et al. (2012) report a bias of less than 20 % in a long-term study of 150 large basins worldwide; similarly, Zhang et al. (2004) report a mean absolute error of < 60 mm in their study of over 470 catchments worldwide, corresponding to a bias of < 10 % for the majority of the catchments. Other local examples may be drawn by users to understand how the Budyko theory may apply locally (e.g., Yang et al., 2007, in China). Overall, there is a large ongoing effort to improve the parameterization and predictive use of the Budyko framework (Donohue et al., 2012; Xu et al., 2013; Liang and Liu, 2014). Future progress may therefore be used to refine the InVEST model interpretation in different geographic contexts. In particular, ongoing research on linking patch-scale and catchment-scale hydrology should provide critical insights into the effect of the simple spatial disaggregation used in the InVEST model (Thompson et al., 2011). We note, however, that the agreement between the lumped model and the catchment model is context specific. As illustrated in Table 2, the differences between the lumped model and the InVEST model vary among the catchments, such that extrapolation of the results from such studies will need to be done cautiously.

5.3 Model performance with and without calibration

Calibrated model

Our results indicate a fair performance of the calibrated model for multiple catchments ranging in size and LULC. The bias ranged from −38 to 14 % for all subcatchments, and from −14 to 14 % when discarding the Rockfish catchment. This narrow range suggests that the calibrated model was able to explain the variability in observed water yields. While it is possible that such variability is explained by climate more than LULC, this is not the case in Cape Fear since the average values of P and PET varied by less than 3 % between subcatchments (Table 2).

Further consideration of the Z values obtained by calibrating it for each subcatchment provides insights into the interpretation of this parameter. With the exception of the Rockfish catchment, a value between 10 and 20 is able to characterize the nine other subcatchments. This suggests that the parameter captures the topography and climate of the area, as intended by the model. The calibrated value of Z for the Rockfish catchment was much lower ($Z = 6$), producing a higher water yield. This difference could be due to the inadequacy of Eq. (3) to relate ω to soil characteristics (since the soils in the Rockfish catchment are particularly sandy). It could also be attributed to errors in the treatment of water withdrawals and return flows, especially since the entire subcatchment lies within Hoke County, which has minimal water withdrawals.

Despite the uncertainties around the outlier, the multi-catchment analyses allowed us to assess the model performance in representing LULC change. Use of the model for evaluation of LULC change is crucial in ecosystem service assessments, where scenario analyses of LULC development are common (Guswa et al., 2014). Validating the use of models in such contexts is extremely challenging since it is rare for modelers to have sufficient pre- and post-LULC-change data (Hrachowitz et al., 2013). In our study, the length of the precipitation and streamflow data did not allow conducting such temporal analyses. Regional analyses where space is substituted for time thus represent a powerful way to assess the ability of the model to capture differences in LULC configuration.

Uncalibrated model

Another important lesson from the analyses is that the calibrated Z value is relatively close to the baseline value, which was derived independently from the average annual number of rain events. Based on Fig. 3, using one value or the other would result in a difference in water yield of approximately 10 %. This error is small compared to other model uncertainties, suggesting that this method for determining Z is robust. The underprediction of water yield for ungauged catchments could be explained by errors in the precipitation raster, the Z parameter, and the treatment of water withdrawals. Based on Eq. (2), the negative bias implies the underestimation of the precipitation data or overestimation of the Z coefficient. As already noted, errors in precipitation data are difficult to characterize. However, precipitation was more likely underestimated in this study since it did not include snowfall.

Conversely, the method relying on a constant ω value was not found satisfactory for this case study, since it resulted in large overestimation of the water yields. Using $\omega = 4$, the Z value found for individual subcatchments ranged from 4 to 8, averaging 6, a value that results in a large model bias (averaging 68 %).

With regard to relative water yield values, the model was able to predict subcatchment ranks fairly accurately (Fig. 4b), which means that priority areas would be correctly identified. The uncertainties in ranking for medium water yield catchments (ranking from 3 to 6) could be partly explained by their similarity (observed water yields range from 236 to 278 mm) and the uncertainty in the water abstraction, as suggested by the overlapping error bars in Fig. 4a. Interestingly, although these results were obtained with the calibrated value of Z, they are only slightly sensitive to the value of Z, since the ranking of subcatchments is largely maintained when the value of Z changes. The ranking of subcatchments based on the baseline run, for example, was identical to the one with $Z = 14$.

5.4 Practical implications

In this final section, we discuss the results with a focus on practical implications for model users.

Our analyses suggest that the uncertainty introduced by "variability in the precipitation inputs" is high, comparable or higher than the uncertainty introduced by the parameter Z and the use of the lumped model theory on a pixel-by-pixel basis. Importantly, the sensitivity observed in Cape Fear (e.g., that a 10 % change in precipitation may result in a 30 % change in water yield) is specific to the climate; for example, in arid climates where evapotranspiration is water limited, an error in precipitation may have a lower effect on water yield since the precipitation surplus or deficit will be mostly converted to evapotranspiration. In Cape Fear, comparison of three climate input data sources suggested that large errors may occur when using point data or data sets obtained with different modeling assumptions. These results confirm a wide body of research that highlights the importance of precipitation inputs for rainfall runoff models (Zhou et al., 2012; McGlynn et al., 2012) and in particular for the InVEST model (Boithias et al., 2014; Sánchez-Canales et al., 2012). Although it was outside the scope of this study to investigate which climate data sets are less prone to errors, our results also draw attention to spatially heterogeneous errors. If model users are interested in the relative ranking of subcatchments, the spatial distribution of errors should be specifically investigated (e.g., probability of a systematic bias in mountainous areas).

The model is not very sensitive to "uncertainty in Z" over a modest range (e.g., 14–22). This is consistent with the conclusions from Sánchez-Canales et al. (2012), who report a low sensitivity to Z in a Mediterranean catchment, for which Z varied between 7 and 9. Since the viable range of Z is quite wide, however, it is possible that large uncertainties in that parameter will translate to significant uncertainty in water yield (Fig. 3). Our analyses provided further insights into the methods for Z selection and highlighted that the sensitivity of the model to Z decreased with increasing values of Z. Based on the examination of Eq. (2), this property will apply generally. Therefore, in temperate climates where values of Z are high (based on the interpretation of Z as the number of annual rain events), the model outputs are likely to be less sensitive to this parameter.

Our study also presented a method to detect a "bias related to the LULC parameters" when multiple observations are available in a catchment. Because K_c values are LULC-specific, the correlation between model performance and K_c values can be used to identify a possible error in the parameter and rectify the values accordingly. No bias was found in this study, bringing confidence in the ability of the model to capture the differences in LULC. We note that these correlation analyses rely on nested subcatchments that are not independent from each other, which decreases the significance of the relationship: five subcatchments are independent, while the other five partially overlap. However, proportions of forest cover varied widely between all subcatchments (from 43 to 72 %), which justifies our interpretation of the analyses.

We conclude this section with a perspective on the model performance assessment, highlighting key limitations in the calibration/testing exercise. First, we note that some water transfers are missing in the model, including irrigation and water abstraction. The model represents agriculture in the same way that it does natural vegetation, and irrigation is not included explicitly. Second, in the Cape Fear catchment, the magnitudes of the water withdrawals are small but this aspect of the modeling may be improved in future applications. In particular, distinction between uses of groundwater (crop irrigation or drinking water) are necessary to account for the fate of water extraction: evapotranspiration in the case of irrigation water, or return flow to the river in the case of drinking water (e.g., Terrado et al., 2014). Additionally, performance was evaluated at the catchment scale. A potential benefit of a spatially explicit model, however, is the ability to predict patterns of water yield within a basin. To properly evaluate that capability, further work should focus on comparing the InVEST model to more sophisticated fully distributed models.

6 Conclusion

Our study aimed to assess the performance of the InVEST annual water yield, a tool that is gaining interest in the ecosystem services community. While such simple models with low requirements for data and level of expertise are needed for practical applications, greater attention should be paid to characterizing the modeling uncertainties. Our assessment of the potential input errors, sensitivity analyses and comparison with observations in the Cape Fear catchment add to this body of work. Key results of the analyses are as follows.

- In the Cape Fear catchment, the InVEST model was most sensitive to uncertainty in the precipitation forcing.
- Errors in climate input data may be significant and non-spatially homogeneous, resulting in uncertainties in the assessment of zones of high and low water yields.
- The study supports the recommendations for setting the Z parameter based on the number of rain events, or via calibration with available observed data.
- Based on the average bias and the explained variance in water yield among the subcatchments, the model performance was fair to high, suggesting that the effects of land use and land cover are adequately captured by the model.
- The errors potentially introduced by a pixel-level application of the Budyko theory will depend on catchment configuration; in Cape Fear, they remained small, comparable to the climate and parameter errors of the empirical model.
- Water abstractions and irrigation processes that are not represented in simple models may confuse the calibration exercise, especially in data-scarce environments where the ecosystem services approach is gaining momentum.

Rigorous uncertainty analyses have not been the norm in the ecosystem service community; however, such work is essential to help users interpret models correctly to inform land-management decisions appropriately.

Appendix A

Table A1. Data sources and statistics for model inputs. Raster statistics are for the entire Cape Fear catchment delineated in Fig. 2.

Data	Type	Value (mean and range)	Source	Range for sensitivity analyses
Precipitation	Raster	1180 mm (1030; 1450) mm	PRISM* (Gilliland, 2003) (USGS, 2014)	±20 %
Reference evapotranspiration	Raster	1240 mm (1160; 1310) mm	FAO, 2001* MODIS (Mu et al., 2012) Climate Office (NCSU, 2014)	±10 %
DEM	Raster	90 (0; 250) m	USGS, 2013a	n/a
LULC	Raster	see Table A2	NASS, 2013	n/a
Soil depth	Raster	1710 m (0; 2110) mm	USGS, 2013b	n/a
PAWC	Raster	0.18 (0.07; 0.52)	USGS, 2013b	n/a
Root depth	Per LULC class	see Table A2	Allen et al., 1998	n/a
K_c	Per LULC class	see Table A2	Allen et al., 1998	(−30 %; +10 %)
Z	Constant	22*	Sharp et al., 2014	(1; 30)

* Indicates the data source used for the baseline run (see Sect. 3.2)

Table A2. Biophysical table used for the baseline InVEST model run, giving the root depth and K_c for each LULC class (values from Allen et al., 1998; Ag – agriculture)

LULC	Root depth (mm)	K_c
Ag-corn	1500	0.75
Ag-other	1100	0.7
Grass	1100	0.9
Forest	5000	1
Wetland	n/a	1.1
Urban	n/a	0.4

Acknowledgements. We thank Rebecca Benner and Margaret Fields, from the Nature Conservancy, and Chris Cook for their support and help with data preprocessing.

Edited by: R. Uijlenhoet

References

Allen, R. G., Pereira, L. S., Raes, D., and Smith, M.: Crop evapotranspiration – Guidelines for computing crop water requirements, FAO Irrigation and drainage paper no. 56, Rome, Italy, 1998.

Bagstad, K. J., Semmens, D. J., Waage, S., and Winthrop, R.: A comparative assessment of decision-support tools for ecosystem services quantification and valuation, Ecosystem Services, 5, 27–39, 2013.

Bai, Y., Zheng, H., Ouyang, Z., Zhuang, C., and Jiang, B.: Modeling hydrological ecosystem services and tradeoffs: a case study in Baiyangdian watershed, China, Environ. Earth Sci., 70, 709–718, 2012.

Boithias, L., Acuña, V., Vergoñós, L., Ziv, G., Marcé, R., and Sabater, S.: Assessment of the water supply: demand ratios in a Mediterranean basin under different global change scenarios and mitigation alternatives, Sci. Total Environ,, 470–471, 567–77, 2014.

Budyko, M.: The Heat Balance of the Earth's Surface, US Department of Commer., Washington, D.C., USA, 259 pp., 1961.

Choudhury, B. J.: Evaluation of an empirical equation for annual evaporation using field observations and results from a biophysical model, J. Hydrol., 216, 99–110, 1999.

Council on Environmental Quality: Principles and Requirements for Federal Investments in Water Resources, available at: http://www.whitehouse.gov/sites/default/files/final_principles_and_requirements_march_2013.pdf (last access: July 2014), 2013.

Donohue, R. J., Roderick, M. L., and McVicar, T. R.: Roots, storms and soil pores?, Incorporating key ecohydrological processes into Budyko's hydrological model, J. Hydrology, 436–437, 35–50, 2012.

FAO: Global agro-ecological zones, FAO Land and Water Digital Media Series No. 11, Rome, Italy, available at: http://www.fao.org/geonetwork/srv/en/metadata.show?id=5017&currTab=summary (last access: July 2014), 2001.

Fu, B. P.: On the calculation of the evaporation from land surface, Sci. Atmos. Sin., 5, 23–31, 1981 (in Chinese).

Gentine, P., Odorico, P. D., Lintner, B. R., Sivandran, G., and Salvucci, G.: Interdependence of climate, soil, and vegetation as constrained by the Budyko curve, Geophys. Res. Lett., 39, L19404, doi:10.1029/2012GL053492, 2012.

Gilliland, B. K.: Technical Report 04-013, Idaho Water Resources Research Institute, University of Idaho, 14 pp., 2003.

Guswa, A., Brauman, K. A., Brown, C., Hamel, P., Keeler, B. L., and Sayre, S. S.: Ecosystem Services: Challenges and Opportunities for Hydrologic Modeling to Support Decision Making, Water Resour. Res., 50, 4535–4544, 2014.

Hrachowitz, M., Savenije, H. H. G., Blöschl, G., McDonnell, J. J., Sivapalan, M., Pomeroy, J. W., Arheimer, B., Blume, T., Clark, M. P., Ehret, U., Fenicia, F., Freer, J. E., Gelfan, A., Gupta, H. V., Hughes, D. A., Hut, R. W., Montanari, A., Pande, S., Tetzlaff, D., Troch, P. A., Uhlenbrook, S., Wagener, T., Winsemius, H. C., Woods, R. A., Zehe, E., and Cudennec, C.: A decade of Predictions in Ungauged Basins (PUB) – a review, Hydrol. Sci. J., 58, 1198–1255, 2013.

Liang, L. and Liu, Q.: Streamflow sensitivity analysis to climate change for a large water-limited basin, Hydrol. Process., 28, 1767–1774, 2014.

Liu, J., Li, S., Ouyang, Z., Tam, C., and Chen, X.: Ecological and socioeconomic effects of China's policies for ecosystem services, P. Natl. Acad. Sci. USA, 105, 9477–9482, 2008.

Martin-Ortega, J., Ojea, E., and Roux, C.: Payments for Water Ecosystem Services in Latin America: A literature review and conceptual model, Ecosystem Services, 6, 122–132, 2013.

McGlynn, B. L., Blöschl, G., Borga, M., Bormann, H., Hurkmans, R., Komma, J., Nandagiri, L., Uijlenhoet, R., and Wagener, T.: A data acquisition framework for prediction of runoff in ungauged basins, in: Predictions in Ungauged Basins, IAHS, 29–52, doi:10.1017/CBO9781139235761.006, 2012.

McMahon, T. A., Peel, M. C., Lowe, L., Srikanthan, R., and McVicar, T. R.: Estimating actual, potential, reference crop and pan evaporation using standard meteorological data: a pragmatic synthesis, Hydrol. Earth Syst. Sci., 17, 1331–1363, doi:10.5194/hess-17-1331-2013, 2013.

Montanari, A., Young, G., Savenjie, H. H. G., Hughes, D., Wagener, T., Ren, L. L., Koutsoyiannis, D., Cudennec, C., Toth, E., Grimaldi, S., Blöschl, G., Sivapalan, M., Beven, K., Gupta, H., Hipsey, M., Schaefli, B., Arheimer, B., Boegh, E., Schymanski, S. J., Di Baldassarre, G., Yu, B., Hubert, P., Huang, Y., Schumann, A., Post, D., Srinivasan, V., Harman, C., Thompson, S., Rogger, M., Viglione, A., McMillan, H., Characklis, G., Pang, Z., and Beylaev, V.: "Panta Rhei – Everything Flows": Change in hydrology and society – The IAHS Scientific Decade 2013–2022, Hydrol. Sci. J., 58, 1256–1275, 2013.

Mu, Q., Zhao, M., and Running, S. W.: Brief Introduction to MODIS Evapotranspiration Data Set (MOD16), available at: ftp://ftp.ntsg.umt.edu/pub/MODIS/NTSG_Products/MOD16/MOD16A3.105_MERRAGMAO/Geotiff (last access: July 2014), 2012.

NASS: National Agricultural Statistics Service – Cropland data layers 2012, available at: http://www.nass.usda.gov/research/Cropland/SARS1a.htm (last access: July 2014), 2013.

NC Department of Environment and Natural Resources: Reported daily water withdrawals, available at: http://www.ncwater.org/Water_Withdrawals/ResultsTabJS.php?wsrc=gwandfip=1andtab=dataandbyRegionType=basinandbasinid=2, last access: July 2014.

NCSU: NC Climate Office, available at: www.nc-climate.ncsu.edu/, last access: July 2014.

Nelson, E., Sander, H., Hawthorne, P., Conte, M., Ennaanay, D., Wolny, S., Manson, S., and Polasky, S.: Projecting Global Land-Use Change and Its Effect on Ecosystem Service Provision and Biodiversity with Simple Models, PLoS ONE, 5, e14327, doi:10.1371/journal.pone.0014327, 2010.

Sánchez-Canales, M., López Benito, A., Passuello, A., Terrado, M., Ziv, G., Acuña, V., Schuhmacher, M., and Elorza, F. J.: Sensitivity analysis of ecosystem service valuation in a Mediterranean watershed, Sci. Total Environ., 440, 140–153, 2012.

SERCC: Number of days of precipitation Equal to or above 0.01 inches in selected cities in the Southeast, The Southeast Regional Climate Center, available at: https://www.sercc.com/climateinfo/historical/meanprecip.html, last access: July 2014.

Sharp, R., Tallis, H. T., Ricketts, T., Guerry, A. D., Wood, S. A., Nelson, E., Ennaanay, D., Wolny, S., Olwero, N., Vigerstol, K., Pennington, D., Mendoza, G., Aukema, J., Foster, J., Forrest, J., Cameron, D., Arkema, K., Lonsdorf, E., Kennedy, C., Verutes, G., Kim, C. K., Guannel, G., Papenfus, M., Toft, J., Marsik, M., Bernhardt, J., Griffin, R., Glowinski, K., Chaumont, N., Perelman, A., Lacayo, M., Mandle, L., Hamel, P., and Chaplin-Kramer, R.: InVEST 3.0 User's Guide, available at: http://ncp-dev.stanford.edu/~dataportal/invest-releases/documentation/current_release/, last access: June 2014.

Sivapalan, M., Konar, M., Srinivasan, V., Chhatre, A., Wutich, A., Scott, C. A., and Wescoat, J. L.: Socio-hydrology: Use-inspired water sustainability science for the Anthropocene, Earth's Future, 2, 225–230, 2014.

Su, C. and Fu, B.: Evolution of ecosystem services in the Chinese Loess Plateau under climatic and land use changes, Global Planet. Change, 101, 119–128, 2013.

Terrado, M., Acuña, V., Ennaanay, D., Tallis, H., and Sabater, S.: Impact of climate extremes on hydrological ecosystem services in a heavily humanized Mediterranean basin, Ecol. Indicators, 37, 199–209, 2014.

Thompson, S. E., Harman, C. J., Troch, P. A., Brooks, P. D., and Sivapalan, M.: Spatial scale dependence of ecohydrologically mediated water balance partitioning: A synthesis framework for catchment ecohydrology, Water Resour. Res., 47, W00J03, doi:10.1029/2010WR009998 , 2011.

Troch, P. A., Carrillo, G., Sivapalan, M., Wagener, T., and Sawicz, K.: Climate-vegetation-soil interactions and long-term hydrologic partitioning: signatures of catchment co evolution, Hydrol. Earth Syst. Sci., 17, 2209–2217, doi:10.5194/hess-17-2209-2013, 2013.

USGS: Digital Elevation Model, The National Map, available at: http://nationalmap.gov/viewer.html (last access: July 2014), 2013a.

USGS: STATSGO and SSURGO datasets, available at: http://water.usgs.gov/GIS/metadata/usgswrd/XML/ussoils.xml (last access: July 2014), 2013b.

USGS: USGS Real time water data for North Carolina, available at: http://nc.water.usgs.gov/realtime/real_time_cape_fear.html, last access: July 2014.

Wang, D. and Tang, Y.: A one-parameter Budyko model for water balance captures emergent behavior in darwinian hydrologic models, Geophys. Res. Lett., 41, 4569–4577, doi:10.1002/2014GL060509, 2014.

Xu, X., Liu, W., Scanlon, B. R., Zhang, L., and Pan, M.: Local and global factors controlling water-energy balances within the Budyko framework, Geophys. Res. Lett., 40, 6123–6129, 2013.

Yang, D., Sun, F., Liu, Z., Cong, Z., Ni, G., and Lei, Z.: Analyzing spatial and temporal variability of annual water-energy balance in nonhumid regions of China using the Budyko hypothesis, Water Resour. Res., 43, W04426, doi:10.1029/2006WR005224, 2007.

Zhang, L., Dawes, W. R., and Walker, G. R.: Response of mean annual evapotranspiration to vegetation changes at catchment scale, Water Resour. Res., 37, 701–708, 2001.

Zhang, L., Hickel, K., Dawes, W. R., Chiew, F. H. S., Western, A. W., and Briggs, P. R.: A rational function approach for estimating mean annual evapotranspiration, Water Resour. Res., 40, W02502, doi:10.1029/2003WR002710, 2004.

Zhou, X., Zhang, Y., Wang, Y., Zhang, H., Vaze, J., Zhang, L., Yang, Y., and Zhou, Y.: Benchmarking global land surface models against the observed mean annual runoff from 150 large basins, J. Hydrol., 470–471, 269–279, 2012.

13

Model study of the impacts of future climate change on the hydrology of Ganges–Brahmaputra–Meghna basin

M. Masood[1,2], P. J.-F. Yeh[3], N. Hanasaki[4], and K. Takeuchi[1]

[1]International Centre for Water Hazard and Risk Management (ICHARM), PWRI, Tsukuba, Japan
[2]National Graduate Institute for Policy Studies (GRIPS), Tokyo, Japan
[3]National University of Singapore, Singapore
[4]National Institute for Environmental Studies, Tsukuba, Japan

Correspondence to: M. Masood (masood35bd@yahoo.com)

Abstract. The intensity, duration, and geographic extent of floods in Bangladesh mostly depend on the combined influences of three river systems, the Ganges, Brahmaputra and Meghna (GBM). In addition, climate change is likely to have significant effects on the hydrology and water resources of the GBM basin and may ultimately lead to more serious floods in Bangladesh. However, the assessment of climate change impacts on the basin-scale hydrology by using well-calibrated hydrologic modeling has seldom been conducted in the GBM basin due to the lack of observed data for calibration and validation. In this study, a macroscale hydrologic model H08 has been applied over the basin at a relatively fine grid resolution (10 km) by integrating the fine-resolution DEM (digital elevation model) data for accurate river networks delineation. The model has been calibrated via the analysis of model parameter sensitivity and validated based on long-term observed daily streamflow data. The impacts of climate change (considering a high-emissions path) on runoff, evapotranspiration, and soil moisture are assessed by using five CMIP5 (Coupled Model Intercomparison Project Phase 5) GCMs (global circulation models) through three time-slice experiments; the present-day (1979–2003), the near-future (2015–2039), and the far-future (2075–2099) periods. Results show that, by the end of 21st century, (a) the entire GBM basin is projected to be warmed by $\sim 4.3\,^{\circ}$C; (b) the changes of mean precipitation (runoff) are projected to be +16.3 % (+16.2 %), +19.8 % (+33.1 %), and +29.6 % (+39.7 %) in the Brahmaputra, Ganges, and Meghna, respectively; and (c) evapotranspiration is projected to increase for the entire GBM (Brahmaputra: +16.4 %, Ganges: +13.6 %, Meghna: +12.9 %) due to increased net radiation as well as warmer temperature. Future changes of hydrologic variables are larger in the dry season (November–April) than in the wet season (May–October). Amongst the three basins, the Meghna shows the highest increase in runoff, indicating higher possibility of flood occurrence. The uncertainty due to the specification of key model parameters in model predictions is found to be low for estimated runoff, evapotranspiration and net radiation. However, the uncertainty in estimated soil moisture is rather large with the coefficient of variation ranging from 14.4 to 31 % among the three basins.

1 Introduction

Bangladesh is situated in the active delta of three of the world's major rivers, the Ganges, Brahmaputra and Meghna. Due to its unique geographical location, the occurrence of water-induced disasters is a regular phenomenon. In addition, the anticipated change in climate is likely to lead to an intensification of the hydrological cycle and to have a major impact on the overall hydrology of these basins and ultimately lead to the increase in the frequency of water-induced disasters in Bangladesh. However, the intensity, duration and geographic extent of floods in Bangladesh mostly depend on the combined influences of these three river systems. Previous studies indicated that flood damages have become more severe and devastating when more than one flood peak in these three river basins coincide (Mirza, 2003; Chowdhury, 2000).

The Ganges–Brahmaputra–Meghna (hereafter referred to as GBM) basin with a total area of about 1.7 million km^2 (FAO-AQUASTAT, 2014; Islam et al., 2010) is shared by a number of countries (Fig. 1). The Brahmaputra River begins in the glaciers of the Himalayas and travels through China, Bhutan, and India before emptying into the Bay of Bengal in Bangladesh. It is snow-fed braided river and it remains a natural stream with no major hydraulic structures built along its reach. The Ganges River originates at the Gangotri glaciers in the Himalayas and it passes through Nepal, China and India and empties into the Bay of Bengal at Bangladesh. It is snowmelt-fed river and its natural flow is controlled by a number of dams constructed by the upstream countries. The Meghna River is a comparatively smaller, rain-fed, and relatively flashier river that runs through a mountainous region in India before entering Bangladesh. Major characteristics of the GBM Rivers are presented in Table 1. This river system is the world's third largest freshwater outlet to the oceans (Chowdhury and Ward, 2004). During extreme floods, over 138 700 $m^3 s^{-1}$ of water flows into the Bay of Bengal through a single outlet, which is the world largest intensity, even exceeding that of the Amazon discharge by about 1.5 times (FAO-AQUASTAT, 2014). The GBM basin is unique in the world in terms of diversified climate. For example, the Ganges River basin is characterized by low precipitation (760–1020 mm $year^{-1}$) in the northwestern upper region and high precipitation (1520–2540 mm $year^{-1}$) along the coastal areas. High precipitation zones and dry rain shadow areas are located in the Brahmaputra River basin, whereas the world's highest precipitation (11 871 mm $year^{-1}$ at Mawsynram, Meghalaya state, India) area is situated in the Meghna River basin (FAO-AQUASTAT, 2014).

Several studies have focused on the rainfall and discharge relationships in the GBM basin by (1) identifying and linking the correlation between basin discharge and the El Niño–Southern Oscillation (ENSO) and sea surface temperature (SST) (Chowdhury and Ward, 2004; Mirza et al., 1998; Nishat and Faisal, 2000), (2) analyzing available observed or reanalysis data (Chowdhury and Ward, 2004, 2007; Mirza et al., 1998; Kamal-Heikman et al., 2007), and (3) evaluating historical data of flood events (Mirza, 2003; Islam et al., 2010). Various statistical approaches were used in the above studies instead of using hydrologic model simulations. In recent years, a number of global-scale hydrologic model studies (Haddeland et al., 2011, 2012; Pokhrel et al., 2012) have been reported. Although their modeling domains include the GBM basin, these global-scale simulations are not fully reliable due to the lack of model calibration at both the global and basin scales.

Few studies have been conducted to investigate the impact of climate change on the hydrology and water resources of the GBM basin (Immerzeel, 2008; Kamal et al., 2013; Biemans et al., 2013; Gain et al., 2011; Ghosh and Dutta, 2012). In most of these studies, future streamflow is projected on the basis of linear regression between rainfall and streamflow derived from historical data (Immerzeel, 2008; Chowdhury and Ward, 2004; Mirza et al., 2003). Immerzeel (2008) used the multiple regression technique to predict streamflow at the Bahadurabad station (the outlet of Brahmaputra Basin) under future temperature and precipitation conditions based on a statistically downscaled GCM (global circulation model) output. However, since most hydrologic processes are nonlinear, they cannot be predicted accurately by extrapolating empirically derived regression equations to the future projections. The alternative for the assessment of climate change impacts on basin-scale hydrology is via well-calibrated hydrologic modeling, but this has rarely been conducted for the GBM basin due to the lack of observed data for model calibration and validation. Ghosh and Dutta (2012) applied a macroscale distributed hydrologic model to study the change of future flood characteristics at the Brahmaputra Basin, but their study domain is only focused on the regions inside India. Gain et al. (2011) estimated future trends of the low and high flows in the lower Brahmaputra Basin using outputs from a global hydrologic model (grid resolution: 0.5°) forced by multiple GCM outputs. Instead of model calibration, the simulated future streamflow is weighted against observations to assess the climate change impacts.

In this study, a hydrologic model simulation is conducted in which the calibration and validation is based on a rarely obtained long-term (1980–2001) observed daily streamflow data set in the GBM basin provided by the Bangladesh Water Development Board (BWDB). Relative to previous GBM basin studies, it is believed that the availability of this unique long-term streamflow data can lead to more precise estimation of model parameters and hence more accurate hydrological simulations and more reliable future projection of the hydrology over the GBM basin.

The objective of this study is to (1) setup a hydrologic model for the GBM basin and calibrate and validate the model with the long-term observed daily streamflow data, and (2) to study the impact of future climate changes on the basin-scale hydrology. A global hydrologic model H08 (Hanasaki et al., 2008, 2014) is applied regionally over the GBM basin at a relatively fine grid resolution (10 km) by integrating the fine-resolution ($\sim$ 0.5 km) DEM (digital elevation model) data for the accurate river networks delineation. The hourly atmospheric forcing data from the Water and Global Change (WATCH) model-intercomparison project (Weedon et al., 2011) (hereafter referred to as WFD, i.e., WATCH forcing data set) are used for the historical simulations. WFD is considered as one of the best available global climate forcing data sets to provide accurate representation of meteorological events, synoptic activity, seasonal cycles and climate trends (Weedon et al., 2011). The studies by Lucas-Picher et al. (2011) and Siderius et al. (2013) found that for southern Asia and the Ganges, respectively, the WFD rainfall is consistent with the APHRODITE (Yatagai et al., 2012), a gridded (0.25°) rainfall product for the southern Asia region developed based on a large number of rain gauge

Table 1. Major characteristics of the Ganges, Brahmaputra and Meghna river basins.

Item		Brahmaputra	Ganges	Meghna
Basin area (km^2)		583 000[b] 530 000[f,g] 543 400[h]	907 000[b] 1 087 300[h] 1 000 000[c]	65 000[b] 82 000[h]
River length (km)		1800[b] 2900[f] 2896[a]	2000[b] 2510[c] 2500[a]	946[b]
Elevation (m a.s.l.)[e]	Range	8 ~ 7057	3 ~ 8454	−1 ~ 2579
	Average	3141	864	307
	Area below 500 m:	20 %	72 %	75 %
	Area above 3000 m:	60 %	11 %	0 %
Discharge ($m^3 s^{-1}$)	Station	Bahadurabad	Hardinge Bridge	Bhairab Bazar
	Lowest	3430[d]	530[d]	2[d]
	Highest	102 535[d]	70 868[d]	19 900[d]
	Average	20 000[g]	11 300[d]	4600[d]
Land use (% area)[i]	Agriculture	19 %	68 %	27 %
	Forest	31 %	11 %	54 %
Basin-averaged normalized difference vegetation index (NDVI)[j]		0.38	0.41	0.65
Total number of dams (both for hydropower and irrigation purpose)[k]		6	75	–

[a] Moffitt et al. (2011). [b] Nishat and Faisal (2000). [c] Abrams (2003). [d] BWDB (2012). [e] Estimated from SRTM DEM data by Lehner et al. (2006). [f] Gain et al. (2011). [g] Immerzeel (2008). [h] FAO-AQUASTAT (2014). [i] Estimated from Tateishi et al. (2014). [j] Estimated from NEO (2014). [k] Lehner et al. (2008).

data. For future simulations, the H08 model is forced by climate model output under the high-emissions scenario (RCP 8.5 – representative concentration pathway) from five different coupled atmosphere–ocean GCMs, all of which participate in the Coupled Model Intercomparison Project Phase 5 (CMIP5) (Taylor et al., 2012). In order to be consistent with the historical data, for each basin the monthly correction factor (i.e., the ratio between the monthly precipitation of the WFD data and that of the GCM data for each month) is applied to the GCM's future precipitation outputs. Three time-slice experiments are performed for the present-day (1979—2003), the near-future (2015–2039), and the far-future (2075—2099) periods.

Our present modeling study makes advances over previous similar studies in three aspects. First, the H08 model has been demonstrated as a suitable tool for large-scale hydrologic modeling (Hanasaki et al., 2008) and, in this study, it is first calibrated via analyzing model parameter sensitivity in the GBM basin before being validated against the observed long-term daily streamflow data set. Second, the uncertainty due to the determination of model parameters in hydrologic simulations, which is seldom considered in previous studies, is analyzed intensively in this study. Third, three large GBM basins and their spatial variability are studied, respectively, in this study via an integrated model framework which benefits the analysis of the combined influences of the three rivers on the large-scale floods and droughts in Bangladesh, as extensively reported in literature (Chowdhury, 2000; Mirza, 2003). Finally, the impacts of climate change not only on streamflow, but also on other hydro-meteorological variables, including evapotranspiration, soil moisture and net radiation, are also assessed in this study, unlike in most previous studies where the climate change impact on streamflow is often the only focus.

The paper is organized into five sections as follows. A brief description of the data and hydrologic model used is presented in Sect. 2. Section 3 presents the model setup as well as the results from the model parameter sensitivity analysis. Results and discussion are presented in Sect. 4, and important conclusions of this study are summarized in Sect. 5.

2 Data and tools

Basic information and characteristics (type, source, resolution and periods of data) of input data used in this study are summarized in Table 2.

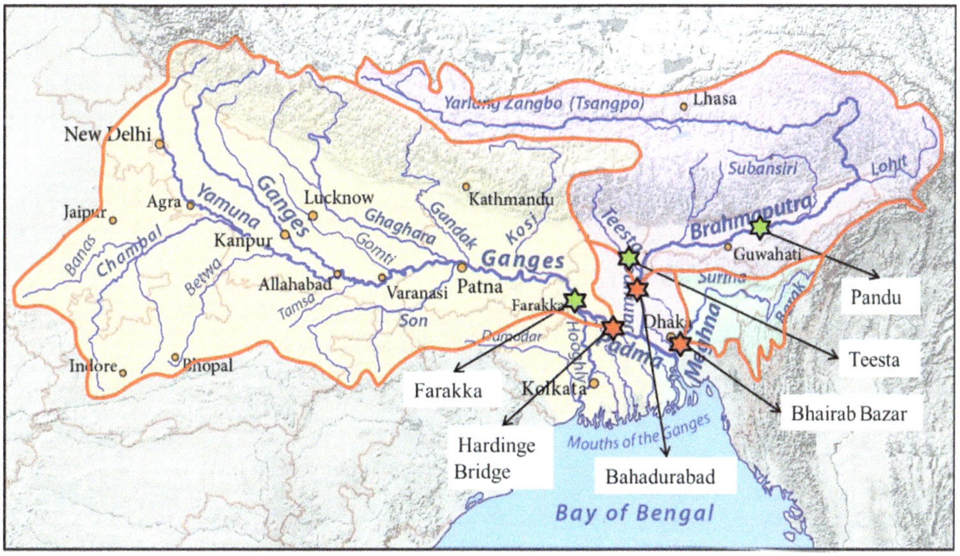

Figure 1. The boundary of the Ganges–Brahmaputra–Meghna (GBM) basin (thick red line), the three outlets (red star): Hardinge Bridge, Bahadurabad and Bhairab Bazar for the Ganges, Brahmaputra and Meghna river basins, respectively. Green stars indicate the locations of three additional upstream stations: Farakka, Pandu and Teesta (modified from Pfly, 2011).

2.1 Meteorological forcing data sets

The WFD (Weedon et al., 2011) is used to drive the H08 model for the historical simulation. The WFD variables, including rainfall, snowfall, surface pressure, air temperature, specific humidity, wind speed, long-wave downward radiation, and shortwave downward radiation were taken from the ERA-40 reanalysis product of the European Centre for Medium Range Weather Forecasting (ECMWF). The ERA reanalysis data with the 1° resolution were interpolated into the 0.5° resolution on the Climate Research Unit of the University of East Anglia (CRU) land mask, adjusted for elevation changes where needed and bias-corrected using monthly observations. For detailed information on the WFD, see Weedon et al. (2011, 2010). The albedo values are based on the monthly albedo data form the Second Global Soil Wetness Project (GSWP2).

2.2 Hydrologic data

Observed river water level (daily) and discharge (weekly) data from 1980 to 2012 for the hydrological stations located inside the Bangladesh (the outlets of three basins shown in Fig. 1, i.e., the Ganges Basin at Hardinge Bridge, the Brahmaputra Basin at Bahadurabad, and the Meghna Basin at Bhairab Bazar) were provided by the Hydrology Division, Bangladesh Water Development Board (BWDB). River water levels were regularly measured 5 times a day (at 06:00, 09:00, 12:00, 15:00 and 18:00 local time) and discharges were measured weekly by the velocity–area method. Since the Brahmaputra River is highly braided, the discharge measurements at Bahadurabad were carried out on multiple channels. In contrast, the Meghna River at Bhairab Bazar is seasonally tidal – after withdrawal of the monsoon the river near this station becomes tidal, and from December to May the river shows both a horizontal and a vertical tide (Chowdhury and Ward, 2004). Under this condition, during the dry season tidal discharge measurements were made at this station once per month. Daily discharges of Ganges and Brahmaputra rivers were calculated from the daily water level data by using the rating equations developed by the Institute of Water Modelling (IWM) (IWM, 2006). A rating equation for the Meghna River is not reported in literature. In this study an attempt was made to develop the rating equation for the Meghna Basin. Discharge (monthly) data of another three stations (Farakka, Pandu, Teesta) located upstream of these basins (Fig. 1) were collected from the Global Runoff Data Centre (GRDC) and were also useful for model validation purposes.

2.3 Topographic data

DEM data were collected from HydroSHEDS (Hydrological data and maps based on SHuttle Elevation Derivatives at multiple Scales) (HydroSHEDS, 2014). It offers a suite of georeferenced data sets (vector and raster), including stream networks, watershed boundaries, drainage directions, and ancillary data layers such as flow accumulations, distances and river topology information (Lehner et al., 2006). The HydroSHEDS data were derived from the elevation data of the Shuttle Radar Topography Mission (SRTM) at a ~0.5 km resolution. Preliminary quality assessments indicate that the accuracy of HydroSHEDS significantly exceeds that of existing global watershed and river maps (Lehner et al., 2006).

2.4 GCM data

Climate data from five CMIP5 climate models – MIROC5, MIROC-ESM, MRI-CGCM3, HadGEM2-ES (under the RCP 8.5) and MRI-AGCM3.2S (under the SRES A1B – Special Report on Emissions Scenarios) – are used in this study as the forcing data for future hydrological simulations (see Appendix B, Table B1). The climate data have been interpolated from their original climate model resolutions (ranging from 0.25 × 0.25° to 2.8 × 2.8°) to 5′ × 5′ (∼ 10 km-mesh) using linear interpolation (nearest four-point). In order to be consistent with the historical simulation forced by WFD, the precipitation forcing data in each GBM basin from each GCM are corrected by multiplying a monthly correction factor, which is equal to the ratio between the basin-averaged long-term mean precipitation from WFD and that from each GCM for all the months. Among these GCMs, MRI-AGCM3.2S (where the "S" refers to the "super-high resolution") provides higher-resolution (20 km) atmospheric forcing data which shows improvements in simulating heavy precipitation, global distribution of tropical cyclones, and the seasonal march of the east Asian summer monsoon (Mizuta et al., 2012). The MRI-AGCM3.2S forcing data set has been used in several recent climate change impact studies focused on southern Asia (Rahman et al., 2012; Endo et al., 2012; Kwak et al., 2012).

2.5 Hydrologic model: H08

H08 is a macroscale hydrological model developed by Hanasaki et al. (2008) which consists of six main modules: land surface hydrology, river routing, crop growth, reservoir operation, environmental flow requirement estimation, and anthropogenic water withdrawal. For this study, only two modules, the land surface hydrology and the river routing are used. The land surface hydrology module calculates the energy and water budgets above and beneath the land surface as forced by the high-temporal-resolution meteorological data.

The runoff scheme in H08 is based on the bucket model concept (Manabe, 1969) but differs from the original formulation in certain important aspects. Although runoff is generated only when the bucket is overfilled as in the original bucket model, H08 uses a "leaky bucket" formulation in which subsurface runoff occurs continually as a function of soil moisture. Soil moisture is expressed as a single-layer reservoir with the holding capacity of 15 cm for all the soil and vegetation types. When the reservoir is empty (full), soil moisture is at the wilting point (the field capacity). Evapotranspiration is expressed as a function of potential evapotranspiration and soil moisture (Eq. 2). Potential evapotranspiration and snowmelt are calculated from the surface energy balance (Hanasaki et al., 2008).

Potential evaporation E_{P} is expressed in this model as

$$E_{\mathrm{P}}(T_{\mathrm{S}}) = \rho C_D U (q_{\mathrm{SAT}}(T_{\mathrm{S}}) - q_a), \quad (1)$$

where ρ is the density of air, C_D is the bulk transfer coefficient, U is the wind speed, $q_{\mathrm{SAT}}(T_{\mathrm{S}})$ is the saturated specific humidity at surface temperature, and q_a is the specific humidity. Evaporation from a surface (E) is expressed as

$$E = \beta E_{\mathrm{P}}(T_{\mathrm{S}}), \quad (2)$$

where

$$\beta = \begin{cases} 1 & 0.75W_{\mathrm{f}} \le W \\ W/W_{\mathrm{f}} & W < 0.75W_{\mathrm{f}} \end{cases}, \quad (3)$$

where W is the soil water content and W_{f} is the soil water content at field capacity (fixed at 150 kg m^{-2}).

Surface runoff (Q_{s}) is generated whenever the soil water content exceeds the field capacity:

$$Q_{\mathrm{s}} = \begin{cases} W - W_{\mathrm{f}} & W_{\mathrm{f}} < W \\ 0 & W \le W_{\mathrm{f}} \end{cases}. \quad (4)$$

Subsurface runoff (Q_{sb}) is incorporated to the model as

$$Q_{\mathrm{sb}} = \frac{W_{\mathrm{f}}}{\tau}\left(\frac{W}{W_{\mathrm{f}}}\right)^{\gamma}, \quad (5)$$

where τ is a time constant and γ is a parameter characterizing the degree of nonlinearity of Q_{sb}. These two parameters are calibrated in this study as described later in Sect. 3.1.

The river module is identical to the Total Runoff Integrating Pathways (TRIP) model (Oki and Sud, 1998). The module has a digital river map covering the whole globe at a spatial resolution of 1° (∼ 111 km). The land–sea mask is identical to the GSWP2 meteorological forcing input. Effective flow velocity and meandering ratio are set as the default values at 0.5 m s^{-1} and 1.5, respectively. The module accumulates runoff generated by the land surface model and routes it downstream as streamflow. However, for this study a new digital river map of the GBM basin with the spatial resolution of ∼ 10 km is prepared. Effective flow velocity and meandering ratio have been calibrated, respectively, for the three basins.

3 Methodology: model setup and simulation

Figure 2 presents the methodology used in this study from model setup to the historical and future simulations. The H08 simulation with the 10 km (5′) resolution is calibrated to find the optimal parameter sets by using the parameter-sampling simulation technique, and validated with observed daily streamflow data. The default river module of H08 uses the digital river map from TRIP (Oki and Sud, 1998) with the global resolution of 1° (∼ 111 km) which is too coarse for the regional simulation in this study, which has the 10 km resolution. Therefore, a new digital river map of the 10 km resolution is prepared for this purpose by integrating the finer-resolution (∼ 0.5 km) DEM data.

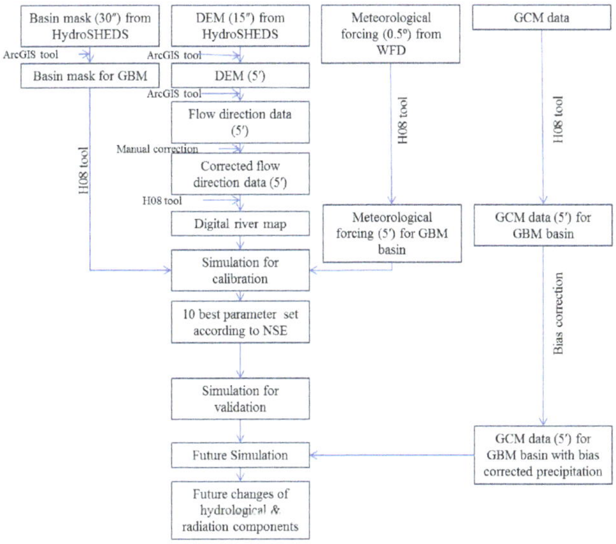

Figure 2. Flow chart of the methodology used in this study.

3.1 Parameter sensitivity

The parameter-sampling simulation is conducted to investigate the sensitivity of the H08 model parameters to simulation results. The most sensitive parameters in H08 include the root-zone depth d (m), the bulk transfer coefficient C_D (–) controlling the potential evaporation (Eq. 1), and the parameters sensitive to subsurface flow, that is, τ (day) and γ (–) (Eq. 5) (Hanasaki et al., 2014); hence, they are treated as calibration parameters in this study. The parameter τ is a time constant determining the daily maximum subsurface runoff. The parameter γ is a shape parameter controlling the relationship between subsurface flow and soil moisture (Hanasaki et al., 2008). Their default parameter values in H08 are 1 m for d, 0.003 for C_D, 100 days for τ, and 2 for γ. For each of these four parameters, five different values are selected from their feasible physical ranges. The parameter-sampling simulations of the H08 model were run by using all the combinations of four parameters, which consist of a total of 5^4 (625) simulations conducted by using the same 11-year (1980– 1990) atmospheric forcing data of WFD.

Figure 3 plots the 11-year long-term average seasonal cycles of simulated total runoff, surface runoff and subsurface runoff of the Brahmaputra Basin. Each of the five lines in each panel represents the average of 5^3 (125) runs with one of the four calibration parameters fixed at a given value. As shown, the overall sensitivity of the selected model parameters to the flow partitioning is high. When d is low, surface runoff is high (due to higher saturated fractional area) (Fig. 3b). As d increases, subsurface runoff increases and surface runoff decreases (Fig. 3c and b). Due to these compensating effects, the effect of d on the total runoff becomes more complex: from March to August higher d causes lower total runoff, but the trend is reversed from August on for the Brahmaputra Basin. Similar behaviors can be observed for the other two basins (figure not shown).

The parameter C_D is the bulk transfer coefficient in the calculation of potential evaporation (Eq. 1); thus, its effect on runoff is relatively small (Fig. 3d–f). However, higher C_D causes more evaporation and hence lower (both surface and subsurface) runoff (Eqs. 1, 2). The sensitivity of parameter γ to runoff is also smaller than d and τ. As γ increases, surface runoff increases and subsurface runoff decreases (Fig. 3h, i). The overall sensitivity of γ to the total runoff becomes negligible due to the compensating effects (Fig. 3g).

As shown in Eq. (5) and Fig. 3k–l, the parameter τ has a critical impact on the surface and subsurface flow partitioning. A larger τ corresponds to larger surface runoff and hence smaller subsurface runoff (Fig. 3k–l), but it has relatively a small impact on total runoff (Fig. 3j).

These four calibration parameters have a combined influence on total runoff partitioning as well as on the simulations of other hydrologic variables. To summarize, (1) the sensitivity of d on the total runoff is complex, i.e., the trend is reversed between the two halves of a year; (2) parameters d and τ have a significant impact on flow partitioning whereas C_D and γ have less sensitivity to runoff simulation; and(3) the influence of d and τ is reversed between surface and subsurface runoff: surface runoff increases as d decreases and τ increases.

Figure 4e plots the uncertainty bands of the simulated discharges by using 10 optimal parameter combinations according to the Nash–Sutcliffe coefficient of efficiency (NSE) (Nash and Sutcliffe, 1970). It is observed that the spread of the uncertainty band is located mainly around the low flow period (dry season from November to March) over the Brahmaputra Basin (Fig. 4e). No surface runoff is generated in the dry season when the soil moisture is lower than the field capacity (Eq. 4 and Fig. 3b). It is noted from the 10 optimal parameter combinations that the optimal τ is 150, C_D is 0.001, d and γ range from 3 to 5 and 1.0 to 2.5, respectively. The spread of the uncertainty bands is mainly due to the variations of d and γ. As d increases, the subsurface runoff increases (Figs. 3c, 4e). On the other hand, in the case of the Ganges and Meghna basins the spread of uncertainty bands are observed through the entire period of a year (in low flow as well as in peak flow regimes). Among the 10 optimal parameter combinations for the Ganges (Meghna) it is found that parameter C_D is 0.008 (0.008), τ is 150 (50), d and γ range from 4 to 5 (4 to 5) and 2.5 to 4 (1.5 to 2), respectively. In the dry period, when surface runoff is nearly zero, subsurface runoff increases as d increases. A higher C_D causes higher evaporation which influences runoff as well (Eq. 1). As discussed earlier, the influence of d on the total runoff is complex, which results in the variation of simulated runoff throughout the year. The spread of the uncertainty bands is large in the peak flow period as the sensitivity of both surface and subsurface runoff is also large with respect to the value of d (not shown).

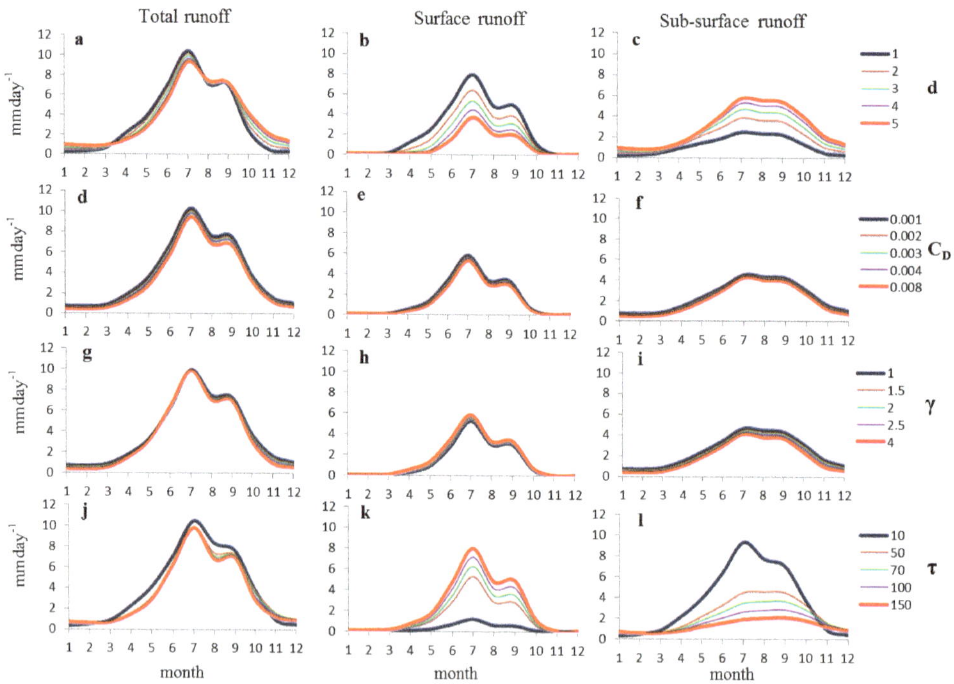

Figure 3. The 11-year (1980–1990) mean seasonal cycles of the simulated total runoff, surface runoff and subsurface runoff (mm day^{-1}) in the Brahmaputra Basin. Each of the five lines in each panel represents the average of 5^3 (125) runs with one of the four calibration parameters fixed at a given reasonable value.

3.2 Calibration and validation

The historical simulation from 1980 to 2001 is divided into two periods with the first half (1980–1990) as the calibration period and the second half (1991–2001) as validation. Basic information and characteristics (location, drainage area, and periods of available observed data) of the six validation stations in the GBM basin are summarized in Table 3. Model performance is evaluated by comparing observed and simulated daily streamflow by the NSE (Nash and Sutcliffe, 1970), the optimal objective function for assessing the overall fit of a hydrograph (Sevat and Dezetter, 1991). A series of sensitivity analysis of H08 parameters was conducted from which 10 sets of optimal parameters are determined by using the parameter-sampling simulation as discussed earlier; these parameter sets are used to quantify the uncertainty in both historical and future simulations in the following. Figure 4 plots the daily hydrograph comparisons at the outlets of the three river basins with the corresponding daily observations for both calibration and validation periods. The obtained NSE for the calibration (validation) period is 0.84 (0.78), 0.80 (0.77), and 0.84 (0.86), while the percent bias (PBIAS) is 0.28 % (6.59 %), 1.21 % (2.23 %) and −0.96 % (3.15 %) for the Brahmaputra, Ganges, and Meghna basins, respectively. For all basins, the relative root-mean-square error (RRMSE), the correlation coefficient (cc), and the coefficient of determination (R^2) for the calibration (validation) period range from 0.32 to 0.60 (0.32 to 0.59), 0.91 to 0.93 (0.89 to 0.94) and 0.82 to 0.86 (0.79 to 0.88), respectively. These statistical indices (Table 4) suggest that the model performance is overall satisfactory. To further evaluate model performance at upstream stations, the monthly discharge data at three upstream stations (Farakka, Pandu, Teesta) collected from the GRDC are used to compare with model simulations, and the result shows that the mean seasonal cycle of simulated streamflow matches well with the corresponding GRDC observations in these three upstream stations (see Appendix A).

4 Results and discussion

The calibrated H08 model is applied to the simulations for the following three time-slice periods, the present (1979–2003), the near-future (2015–2039), and the far-future (2075–2099) period. For the present simulation, both WFD and GCM climate forcing data are used. For the future simulation, only GCM forcing data are used. Simulation results for the two future periods are then compared with the present period (1979–2003) simulation forced by a GCM to assess the effect of climate change on the hydrology and water resources of the GBM basin in terms of precipitation, air temperature, evapotranspiration, soil moisture and net radiation. The results are presented in the following.

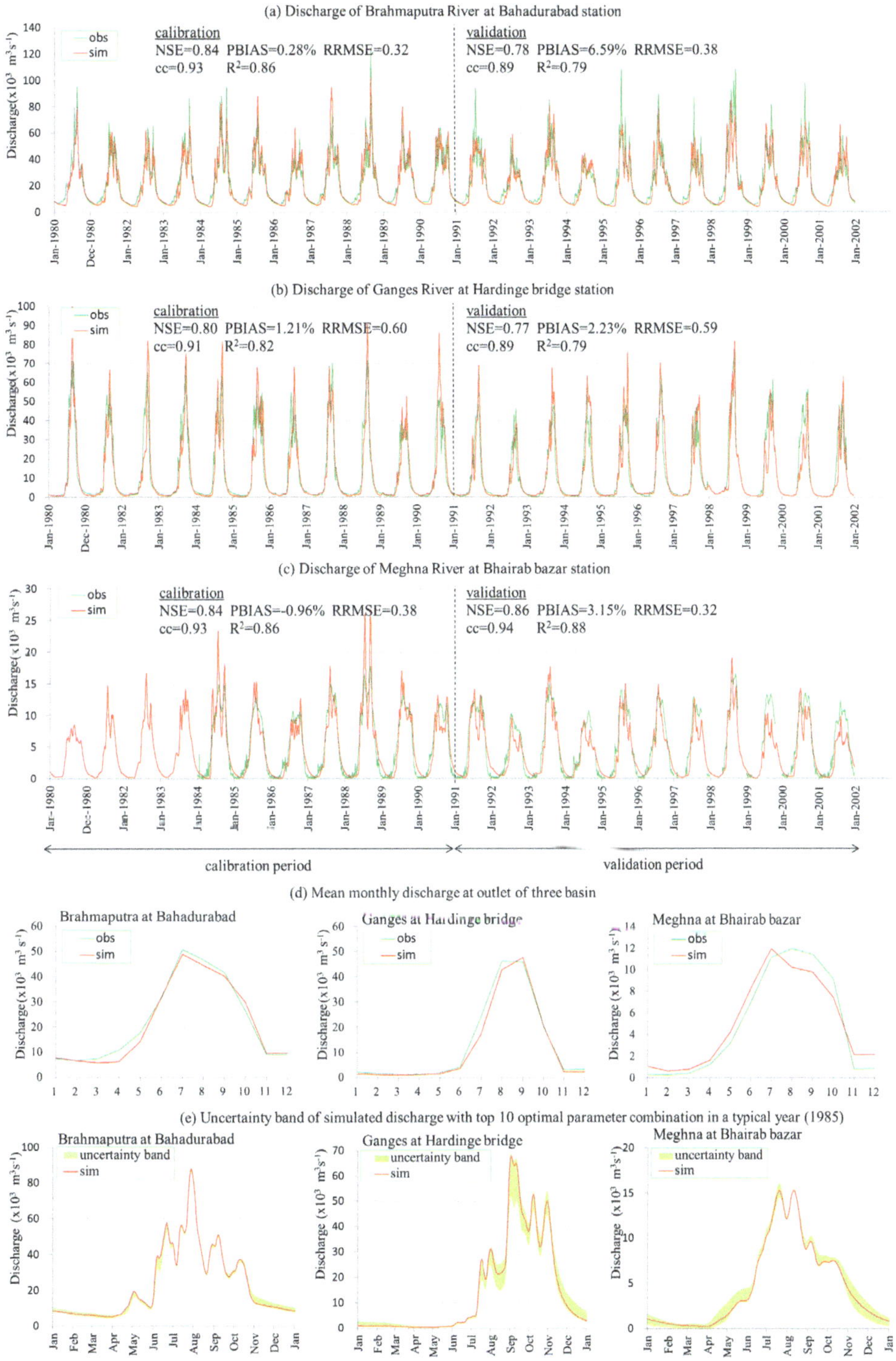

Figure 4. The simulated discharges (red line) using the WFD forcing data (both calibration and validation period) compared with observations (green line) at outlets of the **(a)** Brahmaputra, **(b)** Ganges, and **(c)** Meghna rivers, **(d)** mean monthly (1980–2001) simulated discharges compared with those of observations at outlets; **(e)** simulated discharges by using the 10 optimal parameter sets (red line) and the associated uncertainty bands (green shading) in a typical year (1985). NSE, PBIAS, RRMSE, cc and R^2 for both calibration and validation periods are noted at subplots **(a)**, **(b)** and **(c)**.

Table 2. Basic input data used in this study.

Type	Description	Source/reference(s)	Original spatial resolution	Period	Remarks
Physical data	Digital elevation map (DEM)	HydroSHEDS[a] (HydroSHEDS, 2014)	15″ (∼ 0.5 km)	–	Global data
	Basin mask	HydroSHEDS[a] (HydroSHEDS, 2014)	30″ (∼ 1 km)	–	
Meteorological data	Rainfall, snowfall, surface pressure, air temperature, specific humidity, wind speed, long-wave downward radiation, shortwave downward radiation	WFD[b] (Weedon et al., 2010, 2011)	0.5°	1980–2001	5′ (∼ 10 km mesh) data has been prepared by linear interpolating for this study.
	albedo	GSWP2[c]	1°	1980–1990	Mean monthly 5′ (∼ 10 km mesh) data has been prepared for this study.
Hydrologic data	Water level discharge	Bangladesh Water Development Board (BWDB)	Gauged	1980–2012	Water level (daily), discharge (weekly) data at outlets of three basins, i.e., the Ganges Basin at Hardinge Bridge, the Brahmaputra Basin at Bahadurabad, and the Meghna Basin at Bhairab Bazar obtained from BWDB.
	Discharge	Global Runoff Data Centre (GRDC)	Gauged	1949–1973 (Farakka), 1975–1979 (Pandu), 1969–1992 (Teesta) with missing data	Discharge (monthly) data at three upstream stations, i.e., at Farakka (Ganges), Pandu (Brahmaputra) and Teesta (Brahmaputra).
GCM data	Rainfall, snowfall, surface pressure, air temperature, specific humidity, wind speed, long-wave downward radiation, shortwave downward radiation	MRI-AGCM3.2S[d]	0.25° (∼ 20 km-mesh)	1979–2003, 2015–2039, 2075–2099	Bias of precipitation data set has been corrected by multiplying by the monthly correction coefficient (ratio between basin-averaged long-term monthly mean precipitation from WFD and that from each GCM) for each GBM basins.
		MIROC5	1.41 × 1.39°		
		MIROC-ESM	2.81 × 2.77°		
		MRI-CGCM3	1.125 × 1.11°		
		HadGEM2-ES	1.875 × 1.25°		

[a] HydroSHEDS is Hydrological data and maps based on SHuttle Elevation Derivatives at multiple Scales, [b] WFD is WATCH forcing data, [c] GSWP2 is Second Global Soil Wetness Project, [d] MRI-AGCM is Meteorological Research Institute-Atmospheric General Circulation Model.

Table 3. Basic information of the streamflow validation stations in the GBM basin.

Basin name	Brahmaputra			Ganges		Meghna
Station name	Bahadurabad	Pandu	Teesta	Hardinge Bridge	Farakka	Bhairab Bazar
Latitude	25.18° N	26.13° N	25.75° N	24.08° N	25° N	25.75° N
Longitude	89.67° E	91.7° E	89.5° E	89.03° E	87.92° E	89.5° E
Drainage area (km^2)	583 000	405 000	12 358	907 000	835 000	65 000
Available observed data period (with missing)	1980–2001	1975–1979	1969–1992	1980–2001	1949–1973	1980–2001

4.1 Seasonal cycle

Figure 5 plots the 22-year (1980–2001) mean seasonal cycles of the climatic (from WFD forcing) and hydrologic (from model simulations) quantities averaged over the three basins (the corresponding mean annual amounts of these variables are presented in Table 5). Also given in Fig. 5 is the box-and-whisker plot showing the range of variability for each month. The interannual variation of precipitation in the Brahmaputra and Meghna basins is high from May to September (Fig. 5a, c), whereas in the Ganges it is from June to October. However, the magnitude of precipitation differs substantially among the three basins. The Meghna Basin has a significantly higher precipitation than the other two basins

Table 4. Statistical indices that measure the model performance in the three basins (GBM) during both calibration and validation periods.

Statistical indices	Brahmaputra		Ganges		Meghna	
	Calibration	Validation	Calibration	Validation	Calibration	Validation
Nash–Sutcliffe efficiency (NSE)	0.84	0.78	0.80	0.77	0.84	0.86
Percent bias (PBIAS)	0.28 %	6.59 %	1.21 %	2.23 %	0.96 %	3.15 %
Root-Mean Square Error (RRMSE)	0.32	0.38	0.60	0.59	0.38	0.32
Correlation coefficient (cc)	0.93	0.89	0.91	0.89	0.93	0.94
Coefficient of determination (R^2)	0.86	0.79	0.82	0.79	0.86	0.88

Table 5. The 22-year (1980–2001) averages of the meteorological (from the WFD forcing data) and hydrologic variables in the GBM river basins.

	Unit	Brahmaputra	Ganges	Meghna
(a) Meteorological variables				
Precipitation (Prcp)	mm year^{-1}	1609	1157	3212
Temperature (T_{air})	°C	9.1	21.7	23.0
Net radiation (Net rad)	W m^{-2}	31	74	84
Specific humidity	g kg^{-1}	9.3	11.8	14.4
(b) Hydrological variables				
Runoff	mm year^{-1}	1360	406	2193
Evapotranspiration (ET)	mm year^{-1}	251	748	1000
Potential Evapotranspiration (PET)	mm year^{-1}	415	2359	1689

(Table 5); also, the maximum (monthly) precipitation during 1980–2001 occurs in May with a magnitude of 32 mm day^{-1}, while that in the Brahmaputra and Ganges occur in July with the magnitudes of 15 and 13 mm day^{-1}, respectively. Moreover, the seasonality of runoff in all three basins corresponds well with that of precipitation. Runoff (Fig. 5j–l) in the Ganges Basin is much lower (the monthly maximum of 4.3 mm day^{-1} in August) than in the other two basins (a monthly maximum of 9.3 mm day^{-1} in the Brahmaputra and 15.9 mm day^{-1} in the Meghna, both in July). In addition, ET (evapotranspiration) in the Brahmaputra Basin is significantly lower (251 mm year^{-1}) than in the other two basins (748 mm year^{-1} in Ganges and 1000 mm year^{-1} in Meghna). The contrasting ET magnitudes among the three basins are due to multiple reasons: differences in elevation, amounts of surface water to evaporate, air temperature, and possibly wind and solar irradiance situations. Lower ET in the Brahmaputra Basin is likely due to its cooler air temperature, higher elevation and less vegetated area. The basin-average normalized difference vegetation index (NDVI) in the Brahmaputra is 0.38, whereas in the Ganges and Meghna NDVI is 0.41 and 0.65, respectively (NEO, 2014). However, the patterns of seasonal ET variability in the Brahmaputra and Meghna basins are quite similar, except there is a drop in July in the Brahmaputra (Fig. 5m–o). ET is relatively stable from May to October in the Brahmaputra and Meghna basins in contrast to that in the Ganges where ET does not reach the peak until September. Finally, both the pattern and magnitude of seasonal soil moisture variations are rather different among the three basins (Fig. 5p–r). However, the peak of soil moisture occurs consistently in August in all three basins.

Figure 5d–f present the 22-year mean seasonal cycle of basin-average air temperature (T_{air}). The Brahmaputra Basin is much cooler (mean temperature 9.1 °C) than the Ganges (21.7 °C) and the Meghna (23.0 °C). Figure 5g–i plot the mean seasonal cycle of net radiation averaged over the three basins. The seasonal pattern of net radiation is similar, but the magnitudes differ significantly among the three basins: the average net radiation is ∼ 31, 74 and 84 W m^{-2} in the Brahmaputra, Ganges and Meghna basins, respectively, while the maximum (monthly-average) net radiation is ∼ 47, 100 and 117 W m^{-2}, respectively, in these three basins (Table 5).

4.2 Correlation between meteorological and hydrological variables

Figure 6 presents the scatter plots and correlation coefficients between monthly meteorological and hydrological variables in three river basins. Three different colors represent three different seasons: dry/winter (November–March), pre-monsoon (April–June), and monsoon (July–October). From this plot, the following summary can be drawn. Total runoff and surface runoff of the Brahmaputra Basin have

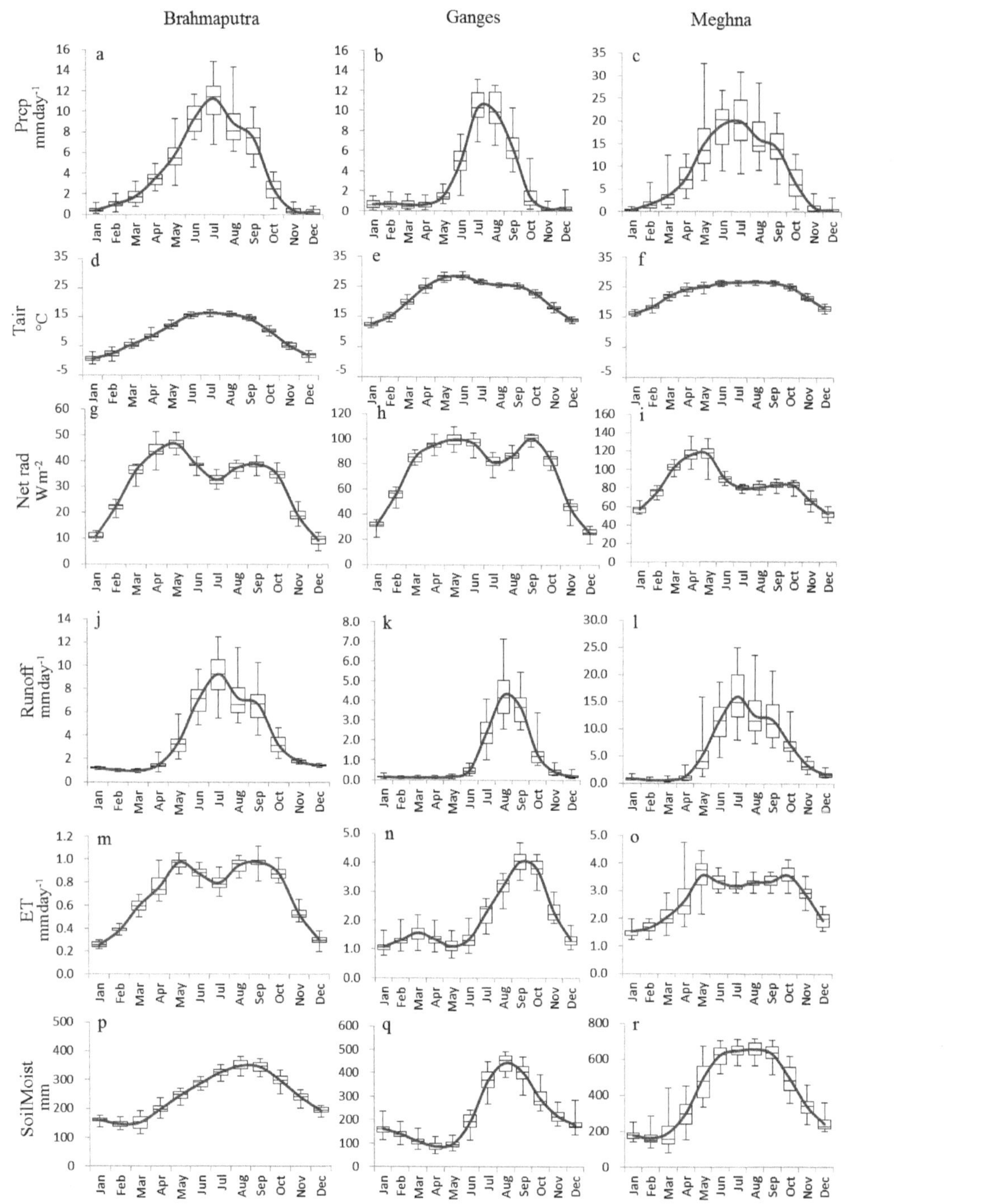

Figure 5. (a)–(r) Seasonal cycle of climatic and hydrologic quantities during 1980–2001. Box-and-whisker plots indicate minimum and maximum (whiskers), 25th and 75th percentiles (box ends), and median (black solid middle bar). Solid curve line represents the interannual average value. All abbreviated terms here refer to Table 5.

stronger correlation (cc = 0.95 and 0.97, both are statistically significant at $p < 0.05$) with precipitation than the other two basins. However, subsurface runoff in the Brahmaputra Basin has weaker correlation (cc = 0.62, $p < 0.05$) with precipitation than that in the Ganges (cc = 0.75, $p < 0.05$) and Meghna (cc = 0.77, $p < 0.05$). These relationships imply that the deeper soil depths enhance the correlation between subsurface runoff and precipitation. The deeper root-zone soil depth (calibrated $d = 5$ m) in the Meghna Basin generates more subsurface runoff (69 % of total runoff) than other two basins. Soil moisture in the Meghna Basin also shows stronger correlation (cc = 0.87, $p < 0.05$) with pre-

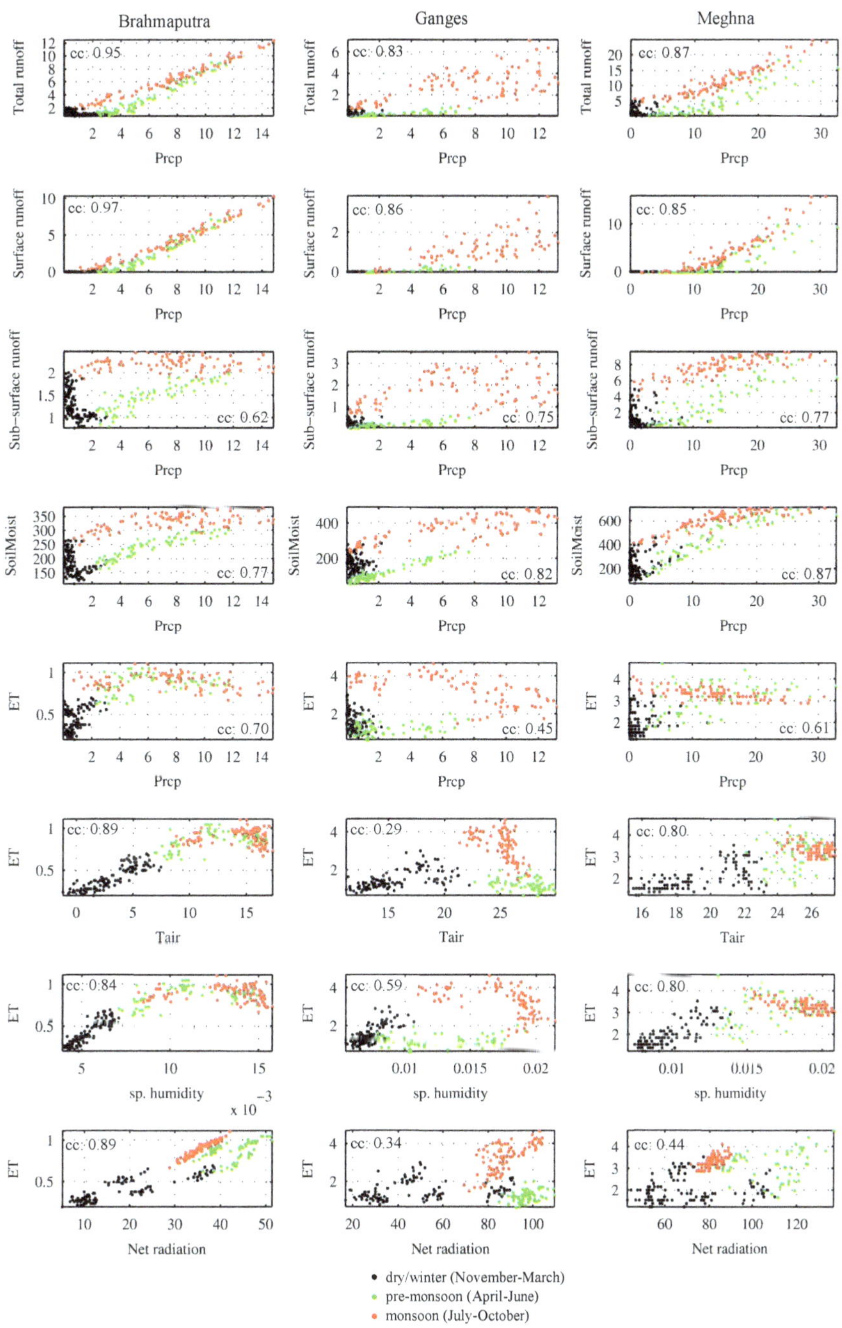

Figure 6. The correlation between the monthly means of meteorological variables (WFD) and that of hydrological variables for the Brahmaputra, Ganges and Meghna basins. Three different colors represent the data in three different seasons. Black: dry/winter (November–March); green: pre-monsoon (April–June); red: monsoon (July–October). The cc for each pair (all three seasons together) is noted at each subplot. The units are millimeters per day for Prcp, ET, and runoff; millimeters for SoilMoist; degrees Celcius for T_{air}; and watts per square meter for net radiation. All abbreviated terms here are referred to in Table 5.

cipitation than that in the Brahmaputra (cc = 0.77, $p < 0.05$) and Ganges (cc = 0.82, $p < 0.05$).

The relationships of evapotranspiration with various atmospheric variables (radiation, air temperature) and soil water availability are rather complex (Shaaban et al., 2011). Different methods for estimating PET in different hydrological models may also be a source of uncertainty (Thompson et al., 2014). However, the ET scheme in the H08 model uses the bulk formula where the bulk transfer coefficient is used to calculate turbulent heat fluxes (Haddeland et al., 2011). In estimating PET (and hence ET), H08 uses humidity, air temperature, wind speed and net radiation. Figure 6 presents the correlation of ET with different meteorological variables in three basins. The ET in the Brahmaputra has a significant correlation with precipitation, air temperature, specific humidity and net radiation with the cc ranging from 0.70 to 0.89 (all of which are statistically significant at $p < 0.05$). The correlations of ET in the Meghna Basin with the meteorological variables are also relatively strong (cc ranges from 0.61 to 0.80, $p < 0.05$) except for the net radiation (cc = 0.44, $p < 0.05$). However, ET in the Ganges Basin has a weak correlation with the meteorological variables (cc from 0.29 to 0.59, $p < 0.05$). A weaker correlation of ET with the meteorological variables is likely attributed to the overestimation of actual ET in the Ganges Basin, because the upstream water use (which is larger in the Ganges) may be incorrectly estimated as ET by the H08 model to ensure water balance.

4.3 Interannual variability

Figure 7 presents the interannual variability of meteorological and hydrologic variables from simulations driven by using five different GCMs and that of the multimodel mean (shown by the thick blue line) for three basins. It can be seen from the figure that the magnitude of interannual variations of variables corresponding to individual GCMs are noticeably larger than that of the multimodel mean. However, the long-term trends in the meteorological and hydrologic variables of the multimodel mean are generally similar to that of each of the GCMs. Figure 7a1–a3 shows that the long-term trend in precipitation is not pronounced in the Brahmaputra and Meghna basins, but its interannual variability is rather large for each GCM. Among the five GCMs used, the precipitation of MRI-AGCM3 has the largest interannual variability (particularly in the Ganges and Meghna basins). A clear increasing trend in air temperature can be observed for all three basins. As there is strong correlation between precipitation and runoff (Fig. 6), the interannual variabilities of them are similar. There is no clear trend for ET in each basin from the present to the near-future periods. However, in the far-future period a notable increasing trend is observed for all basins (Fig. 7e1–e3). Figure 7f1–f3 plots the interannual variability of soil moisture. Since there are no clear trends (from the present to the near-future period) identified for precipitation and evapotranspiration, the effect of climate change on soil moisture is not pronounced.

4.4 Projected mean changes

The long-term average seasonal cycles of hydro-meteorological variables in the two projected periods (2015–2039 and 2075–2099) were compared with that in the reference period (1979–2003). All the results presented here are from the multimodel mean of all simulations driven by the climate forcing data from five GCMs for both reference and future periods. The solid lines in Fig. 8 represent the monthly averages and the dashed lines represent the upper and lower bounds of the uncertainty bands as determined from the 10 simulations using the 10 optimal parameter sets (identified by ranking the NSE). Figure 9 plots the corresponding percentage changes and Table 6 summarizes these relative changes in the hydro-meteorological variables over the three basins on the annual and 6-month (dry season and wet season) basis.

4.4.1 Precipitation

Considering a high-emissions scenario, by the end of 21st century the long-term mean precipitation is projected to increase by 16.3, 19.8 and 29.6 % in the Brahmaputra, Ganges and Meghna basins, respectively (Table 6), in agreement with previous studies which compared GCM simulation results over these regions. For example, Immerzeel (2008) estimated the increase of precipitation in the Brahmaputra Basin as 22 and 14 % under the SRES A2 and B2 scenarios, respectively. Endo et al. (2012) considered the SRES A1B scenario and estimated the country-wise increase in precipitation as 19.7 and 13 % for Bangladesh and India, respectively. Based on the present study, for the Brahmaputra and Meghna basins, the changes in precipitation in the dry season (November–April) are of 23 and 33.6 %, respectively; both are larger than the change in wet season (May–October) (Brahmaputra: 15.1 %; Meghna: 29 %) (Fig. 9b, c). However, the change of precipitation in the dry season in the Ganges Basin (3.6 %) is lower than that in wet season (21.5 %).

4.4.2 Air temperature

The GBM basin will be warmer by about 1 °C in the near-future period (Brahmaputra: 1.2 °C; Ganges: 1.0 °C; Meghna: 0.7 °C) and by about 4.3 °C in the far-future period (Brahmaputra: 4.8 °C; Ganges: 4.1 °C; Meghna: 3.8 °C) (Table 6). According to the projected changes, the cooler Brahmaputra Basin will be significantly warmer, with the maximum increase of up to 5.9 °C in February (Fig. 9d). In Immerzeel (2008), the increase of air temperature in the Brahmaputra Basin is projected (under the SRES A2 and B2 scenarios) as 2.3 ~ 3.5 °C by the end of 21st century. However, the rate of increase over the year is not uniform for all these basins. Temperature will increase more in winter than in summer (Fig. 9d–f). Therefore, a shorter winter and an extended spring can be expected in the future of the GBM basin, which may significantly affect the crop growing season as well.

4.4.3 Runoff

Long-term mean runoff is projected to increase by 16.2, 33.1 and 39.7 % in the Brahmaputra, Ganges and Meghna basins, respectively, by the end of the century (Table 6). Percentage increase of runoff in the Brahmaputra will be quite large in May (about 36.5 %), which may be due to the increase of precipitation and also smaller evapotranspi-

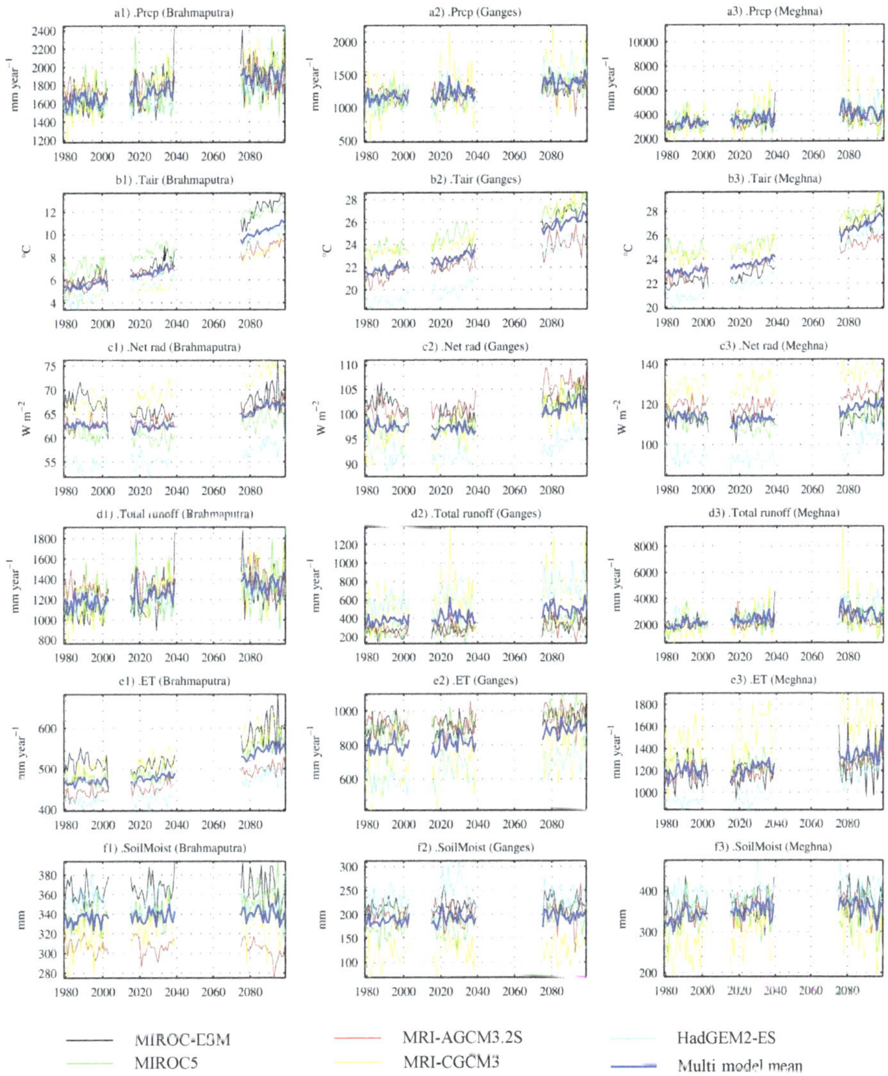

Figure 7. (a1)–(f3) Interannual variation of mean of meteorological and hydrological variables of five GCMs for the present-day (1979–2003), near-future (2015–2039) and far-future (2075–2099) periods. Thick blue lines represent the means of five GCMs.

ration caused by lower net radiation (Fig. 9g, m). In response to seasonally varying degrees of changes in air temperature, net radiation and evaporation, the changes of runoff in the wet season (May–October) (Brahmaputra: 20.3 %; Ganges: 36.3 %; Meghna: 41.8 %) are larger than in the dry season (November–April) (Brahmaputra: 2.9 %, Ganges: −2.3 %, Meghna: 24.2 %) (Fig. 9j, k). Runoff in the Meghna Basin shows a larger response to precipitation increase, which could lead to higher possibility of floods in this basin and prolonged flooding conditions in Bangladesh. These findings are in general consistent with previous findings. Mirza (2002) reported that the probability of occurrence of 20-year floods is expected to be higher in the Brahmaputra and Meghna rivers than in Ganges river. However, Mirza et al. (2003) found that future change in the peak discharge of the Ganges River (as well as the Meghna River) is expected to be larger than that of the Brahmaputra River.

4.4.4 Evapotranspiration

It can be seen from Fig. 9m–o that the change of ET in the near-future period is relatively low but increases to be quite large by the end of the century (Brahmaputra: 16.4 %; Ganges: 13.6 %; Meghna: 12.9 %). This is due to the increase of net radiation (Brahmaputra: 5.6 %; Ganges: 4.1 %; Meghna: 4.4 %) as well as the higher air temperature. Following the seasonal patterns of radiation (Fig. 9g–i) and air temperature (Fig. 9d–f), the change of ET is expected to be considerably larger in the dry season (November–April) (Brahmaputra: 25.6 %; Ganges: 19.3 %; Meghna: 18.2 %) than that in wet season (May–October) (Brahmaputra: 12.9 %; Ganges: 10.9 %; Meghna: 10.5 %).

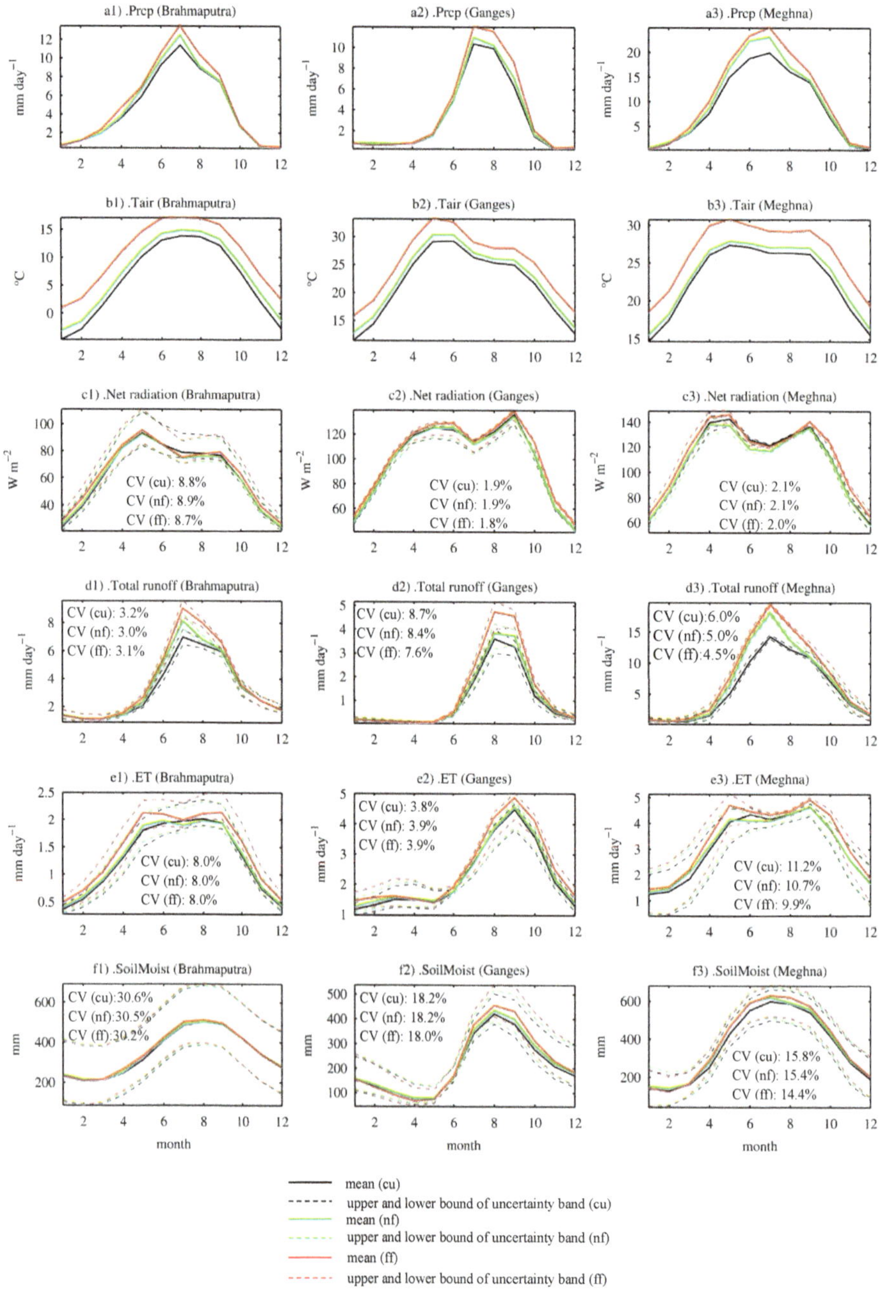

Figure 8. (a1)–(f3) The mean (solid line), upper and lower bounds (dashed line) of the uncertainty band of the hydrological quantities and net radiation components for the present-day (black), near-future (green) and far-future (red) simulations as determined found from 10 simulation results considering 10 optimal parameter sets according to NSE (cu: present-day, nf: near-future, ff: far-future). Coefficients of variations (CVs) for all periods (Table 7) are noted in each subplot.

4.4.5 Soil moisture

Soil moisture is expressed in terms of the water depth per unit area within the spatially varying soil depths (3–5 m). The change of soil moisture (ranges from 1.5–6.9 % in the far-future) is lower compared to other hydrological quantities, except for the Meghna Basin in April where the soil moisture is projected to increase by 22 %. However, the associated uncertainties for all seasons are relatively high compared to other variables (Fig. 8f1–f3).

4.4.6 Net radiation

Net radiation is projected to increase by > 4 % for all seasons except summer in the entire GBM basin by the end of the century (Fig. 9g–i). Due to the increase in the future air

Table 6. The 10-simulation average of annual mean and percentage changes of hydrological and meteorological variables.

Variable	Period	Brahmaputra % change (T_{air}: °C)				Ganges % change (T_{air}: °C)				Meghna % change (T_{air}: °C)			
		annual mean	dry season (November–April)	wet season (May–October)	annual	annual mean	dry season (November–April)	wet season (May–October)	annual	annual mean	dry season (November–April)	wet season (May–October)	annual
(a) Meteorological variables													
Precipitation (mm year^{-1})	present-day (1979–2003)	1632	–	–	–	1154	–	–	–	3192	–	–	–
	near-future (2015–2039)	1720	4.2	5.6	5.4	1218	−0.1	6.2	5.6	3598	11.4	12.9	12.7
	far-future (2075–2099)	1897	23.0	15.1	16.3	1383	3.6	21.5	19.8	4139	33.6	29.0	29.6
T_{air} (°C)	present-day (1979–2003)	5.5	–	–	–	21.7	–	–	–	23.0	–	–	–
	near-future (2015–2039)	6.7	1.4	1.0	1.2	22.8	1.1	0.9	1.0	23.7	0.8	0.6	0.7
	far-future (2075–2099)	10.3	5.5	4.1	4.8	25.9	4.6	3.7	4.1	26.8	4.3	3.4	3.8
Net radiation (W m^{-2})	present-day (1979–2003)	63	–	–	–	97	–	–	–	114	–	–	–
	near-future (2015–2039)	62	2.0	−1.6	−0.4	97	−0.2	−0.9	−0.7	112	−0.4	−2.2	−1.5
	far-future (2075–2099)	66	10.3	3.1	5.6	101	5.3	3.4	4.1	119	6.5	3.0	4.4
(b) Hydrological variables													
Total runoff (mm year^{-1})	present-day (1979–2003)	1166	–	–	–	372	–	–	–	1999	–	–	–
	near-future (2015–2039)	1244	0.5	8.6	6.7	414	2.5	12.1	11.3	2380	10.5	20.2	19.1
	far-future (2075–2099)	1355	2.9	20.3	16.2	495	−2.3	36.3	33.1	2793	24.2	41.8	39.7
ET (mm year^{-1})	present-day (1979–2003)	467	–	–	–	785	–	–	–	1193	–	–	–
	near-future (2015–2039)	477	5.5	0.9	2.1	808	4.9	2.1	3.0	1216	5.2	0.4	1.9
	far-future (2075–2099)	513	25.6	12.9	16.4	892	19.3	10.9	13.6	1347	18.2	10.5	12.9
Soil moisture (mm)	present-day (1979–2003)	335	–	–	–	186	–	–	–	336		–	–
	near-future (2015–2039)	338	0.4	1.2	0.9	192	2.7	3.4	3.1	354	6.6	5.1	5.5
	far-future (2075–2099)	340	0.2	2.3	1.5	197	0.4	8.3	5.8	359	6.7	6.9	6.9

Table 7. Statistical indices (the CV and SD) of the uncertainty in model simulations due to the uncertainty in model parameters.

Variable	Period	Brahmaputra		Ganges		Meghna	
		Coefficient of variation (CV) of mean (Fig. 8) (%)	Standard deviation (SD) of mean (Fig. 8)	Coefficient of variation (CV) of mean (Fig. 8) (%)	Standard deviation (SD) of mean (Fig. 8)	Coefficient of variation (CV) of mean (Fig. 8) (%)	Standard deviation (SD) of mean (Fig. 8)
Net radiation	present-day	8.6	5.4	2.0	2.0	2.1	2.4
	near-future	8.6	5.4	1.9	1.9	2.1	2.3
	far-future	8.4	5.6	1.8	1.8	2.0	2.4
Total runoff	present-day	3.2	0.1	7.6	0.1	6.7	0.4
	near-future	3.0	0.1	7.2	0.1	5.4	0.4
	far-future	3.1	0.1	6.6	0.1	4.6	0.4
ET	present-day	7.9	0.1	3.6	0.1	11.3	0.4
	near-future	7.9	0.1	3.7	0.1	10.6	0.4
	far-future	7.8	0.1	3.7	0.1	9.7	0.4
Soil moisture	present-day	31.0	103.7	18.5	34.5	15.9	53.5
	near-future	30.8	104.1	18.5	35.5	15.4	54.5
	far-future	30.5	103.7	18.3	36.1	14.4	51.6

temperature, the downward long-wave radiation would increase accordingly and lead to the increase in net radiation. However, the change of net radiation in the far-future period is larger in the dry season (Brahmaputra: 10.3 %; Ganges: 5.3 %; Meghna: 6.5 %) than in the wet season (Brahmaputra: 3.1 %; Ganges: 3.4 %; Meghna: 3 %). For the near-future period, net radiation is projected to decrease by < 1 % through almost all seasons due to the smaller increase in air temperature (∼ 1 °C) as well as decreased incoming solar radiation (not shown) in this basin.

4.5 Uncertainty in projection due to model parameters

In recent decades, along with the increasing computational power there has been a trend towards increasing complexity

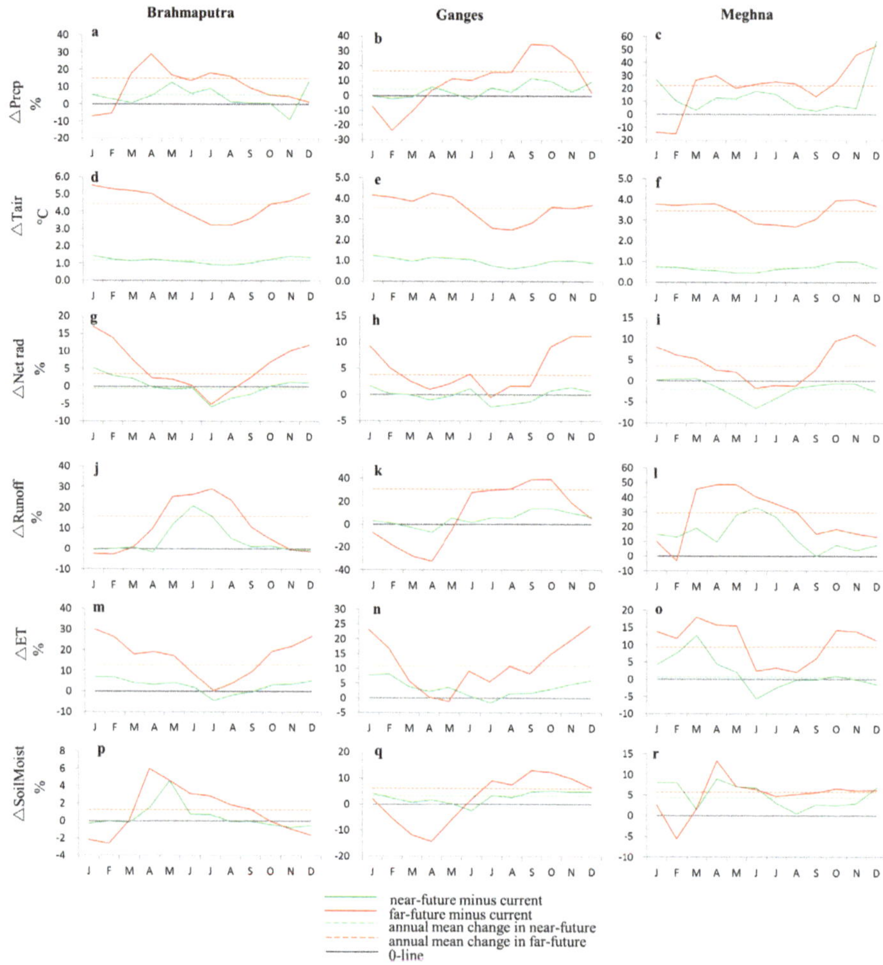

Figure 9. (a)–(r) Percentage changes in the monthly means of the climatic and hydrologic quantities from the present-day period to the near-future and far-future periods. The dashed lines represent the annual mean changes.

of hydrological models to capture natural phenomena more precisely. However, the increased complexity of hydrological models does not necessarily improve their performance for unobserved conditions due to the uncertainty in the model parameter values (Carpenter and Georgakakos, 2006; Tripp and Niemann, 2008). An increase in complexity may improve the calibration performance due to the increased flexibility in the model behavior, but the ability to identify correct parameter values is typically reduced (Wagener et al., 2003). Model simulations with multiple combinations of parameter sets can perform equally well in reproducing the observations. Another source of uncertainty comes from the assumption of stationary model parameters, which is one of the major limitations in modeling the effects of climate change. Model parameters are commonly estimated under the current climate conditions as a basis for predicting future conditions, but the optimal parameters may not be stationary over time (Mirza and Ahmad, 2005). Therefore, the uncertainty in future projections due to model parameter specification can be critical (Vaze et al., 2010; Merz et al., 2011; Coron et al., 2012), although it is usually ignored in most climate change impact studies (Lespinas et al., 2014). Results obtained by Vaze et al. (2010) indicated that the model parameters can generally be used for climate impact studies when a model is calibrated using more than 20 years of data and where the future precipitation is not more than 15 % lower or 20 % higher than that in the calibration period. However, Coron et al. (2012) found a significant number of errors in simulations due to this uncertainty and suggested further research to improve the methods of diagnosing parameter transferability under the changing climate. For the purpose of minimizing this parameter uncertainty, the average results from the 10 simulations using 10 optimal parameter sets are considered as the simulation result for the two future periods in this study. Also, the propagating uncertainty in simulation results due to the uncertainty in mode parameters will be quantified and compared among various hydrologic variables in this study.

The upper and lower bounds of the uncertainty of hydrometeorological variables are plotted in Fig. 8 for all the simulation periods. It can be seen from the figure that the uncertainty band of runoff is relatively narrow, which indicates that future runoff is well predictable through model simulations. The uncertainty due to model parameters in runoff projection is lower (the CV ranges between 3 and 7.6 % among three basins) than that of other hydrologic variables (Table 7, Fig. 8d1–d3). In addition, in Fig. 4e it is observed that there is no significant uncertainty in simulated peak discharge for the Brahmaputra and Meghna rivers. Lower uncertainty in simulating runoff is highly desirable for climate change impact studies; for instance, the flood risk assessment where the runoff estimate (especially the peak flow) is the main focus. However, a relatively wide uncertainty band of runoff can be found in the Ganges Basin during the wet season (Fig. 8d2), which might be due to the fact that the upstream water use (diversion) in the Ganges was not well represented in the model. Notice that the lower uncertainty in runoff projection relative to other variables could be expected as the model was calibrated and validated against observed streamflow at the basin outlet. The uncertainty in ET projection is also lower (CV: 3.6–11.3 %; SD: 0.1–0.4), which can be related to the narrower uncertainty band of net radiation (CV: 1.8–8.6 %; SD: 1.8–5.6). On the other hand, the projection of soil moisture is rather uncertain for all three basins (CV: 14.4–31 %; SD: 35–104). Large uncertainty in predicting soil moisture can be a serious issue which is significant in land use management and agriculture. This emphasizes the critical significance of (1) suitable parameterization of soil water physics in the model, (2) a reliable regional soil map for the specification of model parameters, and (3) soil moisture observations for model calibration and validation.

5 Conclusions

This study presents model analyses of the climate change impact on the Ganges–Brahmaputra–Meghna (GBM) basin focusing on (1) the setup of a hydrologic model by integrating the fine-resolution (~ 0.5 km) DEM data for the accurate river network delineation to simulate at relatively fine grid resolution (10 km) (2) the calibration and validation of the hydrologic model with long-term observed daily discharge data and (3) the impacts of future climate changes in the basin-scale hydrology. The uncertainties in the future projection stemming from model parameters were also assessed. The time-slice numerical experiments were performed using the model forced by the climatic variables from five GCMs (all participating in the CMIP5) for the present-day (1979–2003), near-future (2015–2039) and the far-future (2075–2099) periods.

The following findings and conclusions were drawn from the model analysis:

- (a) The entire GBM basin is projected to be warmer, in the range of 1–4.3 °C in the near-future and far-future periods. The cooler Brahmaputra Basin will be warmer than the Ganges and Meghna. (b) Considering a high-emissions scenario, by the end of 21st century the long-term mean precipitation is projected to increase by +16.3, +19.8 and +29.6 %, and the long-term mean runoff is projected to increase by +16.2, +33.1 and +39.7 % in the Brahmaputra, Ganges and Meghna basins, respectively. (c) The change of ET in the near-future is relatively low, but increases to be quite large by the end of the century due to the increase of net radiation as well as the higher air temperature. However, the change will be considerably larger in the dry season than in the wet season. (d) The change of soil moisture is lower compared to other hydrological quantities.
- Overall, it is observed that the climate change impact on the hydrological processes of the Meghna Basin is larger than that in the other two basins. For example, in the near-future period runoff in the Meghna Basin is projected to increase by 19.1 % whereas it is projected to increase by 6.7 and 11.3 % for the Brahmaputra and Ganges, respectively. In the far-future period, a larger increase of precipitation (29.6 %), lower increase of ET (12.9 %) and consequently larger increase of runoff (39.7 %) lead to a higher possibility of floods in this basin.
- The uncertainty due to model parameters in runoff projection is lower than that of other hydrologic variables. The uncertainty in ET projection is also lower, which can be related to the narrower uncertainty band of net radiation. On the other hand, the projection of soil moisture is rather uncertain in all three basins, which can be significant in land use management and agriculture in particular. This emphasizes the significance of (1) suitable parameterization of soil water physics in the model, (2) a reliable regional soil map for the specification of model parameters, and (3) soil moisture observations for model calibration and validation.

However, this study still has some limitations which can be addressed in future research. (a) All results presented here are basin-averaged. The basin-averaged large-scale changes and trends are difficult to translate to regional- and local-scale impacts. Moreover, the changes in averages do not reflect the changes in variability and extremes, (b) anthropogenic and industrial water use upstream are important factors in altering the hydrologic cycle; however, they were not considered in the present study due to data constraints, (c) urbanizing watersheds are characterized by rapid land use changes and the associated landscape disturbances can shift the rainfall–runoff relationships away from natural processes. Hydrological changes in future can also be amplified by changing land uses. However, in our study future changes of demography and land uses were not considered.

Appendix A: Model validation at three upstream station

The model performance was further evaluated by comparing the simulated monthly streamflow with the observed data from the Global Runoff Data Centre (GRDC) at three upstream gauging stations (Farakka, Pandu and Teesta) in the GBM basin. The locations and drainage areas of these three stations are summarized in Table 3. Although the available data period does not cover the study period 1980–2001 (except for the Teesta which has the data from 1985 to 1991), the mean seasonal cycle and the mean, maximum, minimum, and the standard deviation of the streamflow are compared in Fig. A1 and Table A1. It can be seen that the mean seasonal cycle of simulated streamflow matches well with the corresponding GRDC data (Fig. A1d–f). Also, the agreement of the simulated and observed 1985–1991 monthly streamflow at the Teesta station of the Brahmaputra Basin is excellent (Fig. A1c).

Table A1. Comparison between observed (data source: GRDC) and simulated discharge ($m^3\ s^{-1}$) at the Farakka gauging station in the Ganges Basin, and Pandu and Teesta stations in the Brahmaputra Basin.

Basin Station	Ganges Farakka		Brahmaputra Pandu		Brahmaputra Teesta	
Data type	observed	simulated	observed	simulated	observed	simulated
Data period (with missing)	1949–1973	1980–2001	1975–1979	1980–2001	1969–1992	1980–2001
Mean	12 037	11 399	18 818	15 868	915	920
Maximum	65 072	69 715	49 210	46 381	3622	4219
Minimum	1181	414	4367	3693	10	122
Standard deviation	14 762	15 518	12 073	11 709	902	948

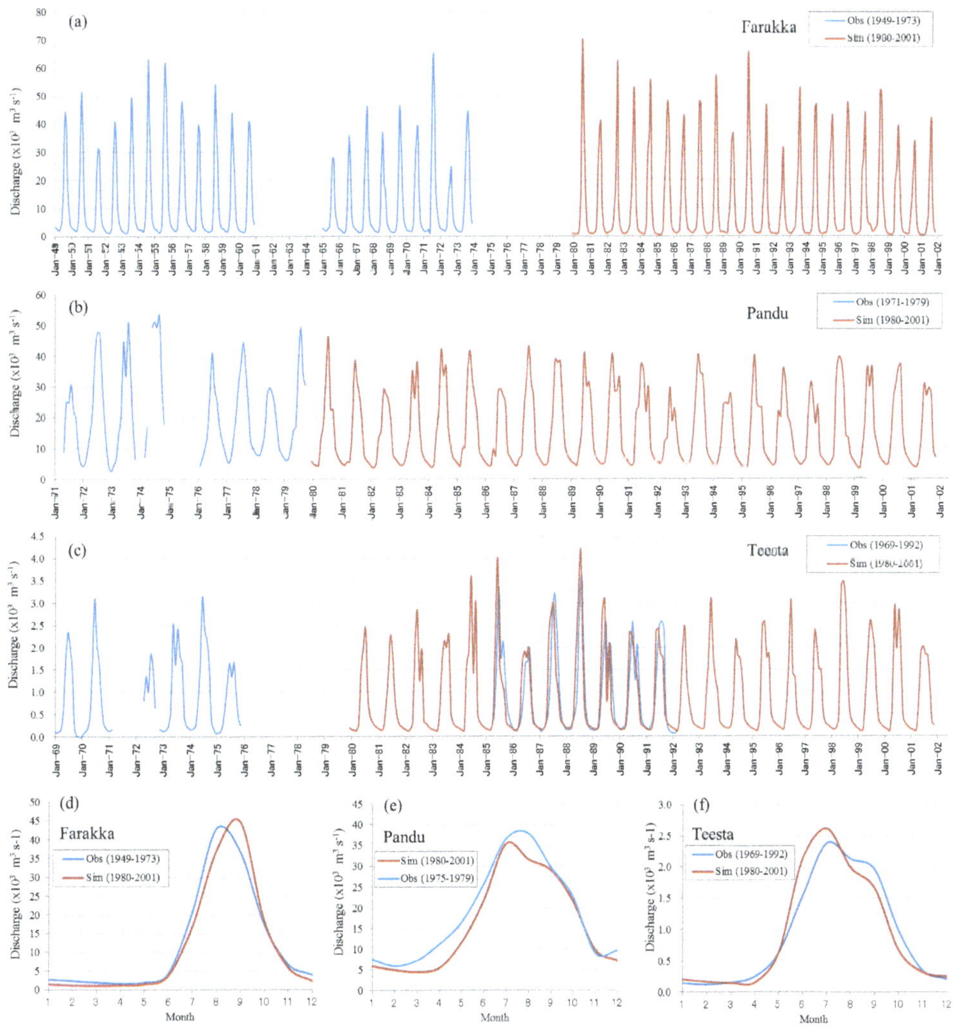

Figure A1. Comparisons between simulated (magenta line) and observed GRDC (blue line) data for **(a)–(c)** the monthly time series of discharges and **(d)–(f)** long-term mean seasonal cycles at the Farakka gauging station in the Ganges Basin and the Pundu and Teesta stations in the Brahmaputra Basin.

Appendix B

Table B1. Salient features of CMIP5 climate models used in the analysis.

Model name	MIROC-ESM	MIROC5	MRI-AGCM3.2S	MRI-CGCM3	HadGEM2-ES
Modeling center	Japan Agency for Marine-Earth Science and Technology, Atmosphere and Ocean Research Institute (The University of Tokyo), and National Institute for Environmental Studies	Atmosphere and Ocean Research Institute (The University of Tokyo), National Institute for Environmental Studies, and Japan Agency for Marine-Earth Science and Technology	Meteorological Research Institute (MRI), Japan and Japan Meteorological Agency (JMA), Japan	Meteorological Research Institute (MRI), Japan	Met Office Hadley Centre, UK
Scenario	RCP 8.5	RCP 8.5	SRES A1B	RCP 8.5	RCP 8.5
Nominal horizontal resolution	$2.81 \times 2.77°$	$1.41 \times 1.39°$	$0.25 \times 0.25°$	$1.125 \times 1.11°$	$1.875 \times 1.25°$
Model type	ESM[a]	ESM[a]	AMIP[b]	ESM[a]	ESM[a]
Aerosol component name or type	SPRINTARS	SPRINTARS	Prescribed		Interactive
Atmospheric Chemistry	Not implemented	Not implemented	Not implemented	Not implemented	Included
Land surface component	MATSIRO	MATSIRO	SiB0109	HAL	Included
Ocean Biogeochemistry	NPZD-type	Not implemented	Not implemented	Not implemented	Included
Sea ice	Included	Included	Not implemented	Included	Included

[a] ESM is Earth system model. Atmosphere–ocean general circulation models (AOGCMs) with representation of biogeochemical cycles. [b] AMIP refers to models with atmosphere and land surface only, using observed sea surface temperature and sea ice extent.

Acknowledgements. This study is supported by the Public Works Research Institute (PWRI), Japan. The first author is indebted to the authority of Nippon Koei Co., Ltd., Japan, for the grant from The Kubota Fund. Also, thanks are given to A. Hasegawa, and T. Sayama for help in data preparation and for suggestions, and to the Bangladesh Water Development Board (BWDB) for providing observed hydrological data. Finally, the authors wish to thank the editor and five anonymous reviewers for their constructive comments and suggestions, which greatly improved the quality of the paper.

Edited by: F. Tian

References

Abrams, P.: River Ganges, available at: http://www.africanwater.org/ganges.htm (last access date: 13 July 2014), 2003.

Biemans, H., Speelman, L. H., Ludwig, F., Moors, E. J., Wiltshire, A. J., Kumar, P., Gerten, D., and Kabat, P.: Future water resources for food production in five South Asian river basins and potential for adaptation – A modeling study, Sci. Total Environ., 468–469, Supplement, S117–S131, doi:10.1016/j.scitotenv.2013.05.092, 2013.

BWDB: Rivers of Bangladesh, Bangladesh Water Development Board, Dhaka, 2012.

Carpenter, T. M. and Georgakakos, K. P.: Intercomparison of lumped versus distributed hydrologic model ensemble simulations on operational forecast scales, J. Hydrol., 329, 174–185, doi:10.1016/j.jhydrol.2006.02.013, 2006.

Chowdhury, M. R.: An Assessment of Flood Forecasting in Bangladesh: The Experience of the 1998 Flood, Nat. Hazards, 139–163, 2000.

Chowdhury, M. R. and Ward, M. N.: Hydro-meteorological variability in the greater Ganges-Brahmaputra-Meghna basins, Int. J. Climatol., 24, 1495–1508, doi:10.1002/joc.1076, 2004.

Chowdhury, M. R. and Ward, M. N.: Seasonal flooding in Bangladesh – variability and predictability, Hydrol. Process., 21, 335–347, doi:10.1002/hyp.6236, 2007.

Coron, L., Andréassian, V., Perrin, C., Lerat, J., Vaze, J., Bourqui, M., and Hendrickx, F.: Crash testing hydrological models in contrasted climate conditions: An experiment on 216 Australian catchments, Water Resour. Res., 48, W05552, doi:10.1029/2011wr011721, 2012.

Endo, H., Kitoh, A., Ose, T., Mizuta, R., and Kusunoki, S.: Future changes and uncertainties in Asian precipitation simulated by multiphysics and multi–sea surface temperature ensemble experiments with high-resolution Meteorological Research Institute atmospheric general circulation models (MRI-AGCMs), J. Geophys. Res., 117, D16118, doi:10.1029/2012jd017874, 2012.

FAO-AQUASTAT: Ganges–Brahmaputra–Meghna River Basin, available at: http://www.fao.org/nr/water/aquastat/basins/gbm/index.stm, last access: 19 April 2014.

Gain, A. K., Immerzeel, W. W., Sperna Weiland, F. C., and Bierkens, M. F. P.: Impact of climate change on the stream flow of the lower Brahmaputra: trends in high and low flows based on discharge-weighted ensemble modelling, Hydrol. Earth Syst. Sci., 15, 1537–1545, doi:10.5194/hess-15-1537-2011, 2011.

Ghosh, S. and Dutta, S.: Impact of climate change on flood characteristics in Brahmaputra basin using a macro-scale distributed hydrological model, J. Earth Syst. Sci., 121, 637–657, doi:10.1007/s12040-012-0181-y, 2012.

Haddeland, I., Clark, D. B., Franssen, W., Ludwig, F., Voß, F., Arnell, N. W., Bertrand, N., Best, M., Folwell, S., Gerten, D., Gomes, S., Gosling, S. N., Hagemann, S., Hanasaki, N., Harding, R., Heinke, J., Kabat, P., Koirala, S., Oki, T., Polcher, J., Stacke, T., Viterbo, P., Weedon, G. P., and Yeh, P.: Multimodel Estimate of the Global Terrestrial Water Balance: Setup and First Results, J. Hydrometeorol., 12, 869–884, doi:10.1175/2011jhm1324.1, 2011.

Haddeland, I., Heinke, J., Voß, F., Eisner, S., Chen, C., Hagemann, S., and Ludwig, F.: Effects of climate model radiation, humidity and wind estimates on hydrological simulations, Hydrol. Earth Syst. Sci., 16, 305–318, doi:10.5194/hess-16-305-2012, 2012.

Hanasaki, N., Kanae, S., Oki, T., Masuda, K., Motoya, K., Shirakawa, N., Shen, Y., and Tanaka, K.: An integrated model for the assessment of global water resources – Part 1: Model description and input meteorological forcing, Hydrol. Earth Syst. Sci., 12, 1007–1025, doi:10.5194/hess-12-1007-2008, 2008.

Hanasaki, N., Saito, Y., Chaiyasaen, C., Champathong, A., Ekkawatpanit, C., Saphaokham, S., Sukhapunnaphan, T., Sumdin, S., and Thongduang, J.: A quasi-real-time hydrological simulation of the Chao Phraya River using meteorological data from the Thai Meteorological Department Automatic Weather Stations, Hydrol. Res. Lett., 8, 9–14, doi:10.3178/hrl.8.9, 2014.

Hydrological data and maps based on SHuttle Elevation Derivatives at multiple Scales, available at: http://hydrosheds.cr.usgs.gov/hydro.php, last access: 19 April 2014.

Immerzeel, W.: Historical trends and future predictions of climate variability in the Brahmaputra basin, I. J. Climatol., 28, 243–254, doi:10.1002/joc.1528, 2008.

Inomata, H., Takeuchi, K., and Fukami, K.: Development of a Statistical Bias Correction Method For Daily Precipitation Data of GCM20, J. Japan Soc. Civil Eng., 67, I_247–I_252, doi:10.2208/jscejhe.67.I_247, 2011.

Islam, A. S., Haque, A., and Bala, S. K.: Hydrologic characteristics of floods in Ganges-Brahmaputra-Meghna (GBM) delta, Nat. Hazards, 54, 797–811, 2010.

IWM: Updating and Validation of North West Region Model (NWRM), Institute of Water Modelling, Bangladesh, 2006.

Kamal-Heikman, S., Derry, L. A., Stedinger, J. R., and Duncan, C. C.: A Simple Predictive Tool for Lower Brahmaputra River Basin Monsoon Flooding, Earth Inter., 11, 1–11, doi:10.1175/ei226.1, 2007.

Kamal, R., Matin, M. A., and Nasreen, S.: Response of River Flow Regime to Various Climate Change Scenarios in Ganges-Brahmaputra- Meghna Basin, J. Water Resour. Ocean Sci., 2, 15–24, doi:10.11648/j.wros.20130202.12, 2013.

Kwak, Y., Takeuchi, K., Fukami, K., and Magome, J.: A new approach to flood risk assessment in Asia-Pacific region based on MRI-AGCM outputs, Hydrol. Res. Lett., 6, 70–75, doi:10.3178/HRL.6.70, 2012.

Lehner, B., Liermann, C. R., Revenga, C., Vörösmarty, C., Fekete, B., Crouzet, P., Döll, P., Endejan, M., Frenken, K., Magome, J., Nilsson, C., Robertson, J. C., Rödel, R., Sindorf, N., and Wisser, D.: High-resolution mapping of the world's reservoirs and dams for sustainable river-flow management, Frontiers in Ecology and the Environment, 9, 494–502, doi:10.1890/100125, 2011.

Lehner, B., Verdin, K., and Jarvis, A.: HydroSHEDS technical documentation v1.0, World Wildlife Fund US, Washington, DC, 1–27, 2006.

Lespinas, F., Ludwig, W., and Heussner, S.: Hydrological and climatic uncertainties associated with modeling the impact of climate change on water resources of small Mediterranean coastal rivers, J. Hydrol., 511, 403–422, doi:10.1016/j.jhydrol.2014.01.033, 2014.

Lucas-Picher, P., Christensen, J. H., Saeed, F., Kumar, P., Asharaf, S., Ahrens, B., Wiltshire, A. J., Jacob, D., and Hagemann, S.: Can Regional Climate Models Represent the Indian Monsoon?, J. Hydrometeorol., 12, 849–868, doi:10.1175/2011jhm1327.1, 2011.

Manabe, S.: Climate and the ocean circulation – 1: The atmospheric circulation and the hydrology of the Earth's surface, Mon. Weather Rev., 97, 739–774, 1969.

Merz, R., Parajka, J., and Blöschl, G.: Time stability of catchment model parameters: Implications for climate impact analyses, Water Resour. Res., 47, W02531, doi:10.1029/2010wr009505, 2011.

Mirza, M. M. Q., Warrick, R. A., Ericksen, N. J., and Kenny, G. J.: Trends and persistence in precipitation in the Ganges, Brahmaputra and Meghna river basins, Hydrol. Sci. J., 43, 345–858, 1998.

Mirza, M. M. Q.: Global warming and changes in the probability ofoccurrence of floods in Bangladesh and implications, Global Environ. Change, 12, 127–138, 2002.

Mirza, M. M. Q.: Three Recent Extreme Floods in Bangladesh: A Hydro-Meteorological Analysis, Nat. Hazards, 28, 35–64, doi:10.1023/A:1021169731325, 2003.

Mirza, M. M. Q., Warrick, R. A., and Ericksen, N. J.: The implications of climate change on floods of the Ganges, Brahmaputra and Meghna rivers in Bangladesh, Clim. Change, 57, 287–318, 2003.

Mirza, M. M. Q. and Ahmad, Q. K.: Climate Change And Water Resources, in: South Asia, edited by: Monirul Qader Mirza, M., Q. K. A., A. A. Balkema Publishers, Leiden, the Netherlands, 2005.

Mizuta, R., Yoshimura, H., Murakami, H., Matsueda, M., Endo, H., Ose, T., Kamiguchi, K., Hosaka, M., Sugi, M., Yukimoto, S., Kusunoki, S., and Kitoh, A.: Climate Simulations Using MRI-AGCM3.2 with 20-km Grid, J. Meteorol. Soc. Japan, 90A, 233–258, doi:10.2151/jmsj.2012-A12, 2012.

Nash, J. E. and Sutcliffe, J. V.: River flow forecasting through conceptual models part I – a discussion of principles, J. Hydrol., 10, 282–290, 1970.

NEO: Vegetation Index [NDVI] (1 Month - Terra/Modis), available at: http://neo.sci.gsfc.nasa.gov/view.php?datasetId=MOD13A2_M_NDVI&year=2000, last access: 19 April 2014.

Nishat, A. and Faisal, I. M.: An assessment of the Institutional Mechanism for Water Negotiations in the Ganges–Brahmaputra–Meghna system, Int. Negotiations, 5, 289–310, 2000.

Nishat, B. and Rahman, S. M. M.: Water Resources Modeling of the Ganges-Brahmaputra-Meghna River Basins Using Satellite Remote Sensing Data, JAWRA J. Am. Water Resour. Assoc., 45, 1313–1327, doi:10.1111/j.1752-1688.2009.00374.x, 2009.

Oki, T. and Sud, Y. C.: Design of Total Runoff Integrating Pathways (TRIP)-A Global River Channel Network, Earth Inter., 2, EI013, 1998.

Pfly: Ganges-Brahmaputra-Meghna basins.jpg, available at: http://en.wikipedia.org/wiki/File:Ganges-Brahmaputra-Meghna_basins.jpg# (last access: April 2014), 2011.

Pokhrel, Y., Hanasaki, N., Koirala, S., Cho, J., Yeh, P. J. F., Kim, H., Kanae, S., and Oki, T.: Incorporating Anthropogenic Water Regulation Modules into a Land Surface Model, J. Hydrometeorol., 13, 255–269, doi:10.1175/jhm-d-11-013.1, 2012.

Rahman, M. M., Ferdousi, N., Sato, Y., Kusunoki, S., and Kitoh, A.: Rainfall and temperature scenario for Bangladesh using 20 km mesh AGCM, International J. Clim. Change Strat. Manage., 4, 66–80, doi:10.1108/17568691211200227, 2012.

Sevat, E. and Dezetter, A.: Selection of calibration objective functions in the context of rainfall-runoff modeling in a Sudanese savannah area, Hydrological Sci. J., 36, 307–330, 1991.

Shaaban, A. J., Amin, M. Z. M., Chen, Z. Q., and Ohara, N.: Regional Modeling of Climate Change Impact on Peninsular Malaysia Water Resources, J. Hydrol. Eng., 16, 1040–1049, doi:10.1061/(ASCE)HE.1943-5584.0000305, 2011.

Siderius, C., Biemans, H., Wiltshire, A., Rao, S., Franssen, W. H., Kumar, P., Gosain, A. K., van Vliet, M. T., and Collins, D. N.: Snowmelt contributions to discharge of the Ganges, The Sci. Total Environ., 468–469 Supplement, S93–S101, doi:10.1016/j.scitotenv.2013.05.084, 2013.

Taylor, K. E., Stouffer, R. J., and Meehl, G. A.: An overview of CMIP5 and the experimental design, B. Am. Meteorol. Soc., 93, 485–498, doi:10.1175/BAMS-D-11-00094.1, 2012.

Tateishi, R., Hoan, N. T., Kobayashi, T., Alsaaideh, B., Tana, G., and Phong, D. X.: Production of Global Land Cover Data – GLCNMO2008, J. Geogr. Geol., 6, 99–122, doi:10.5539/jgg.v6n3p99, 2014.

Thompson, J. R., Green, A. J., and Kingston, D. G.: Potential evapotranspiration-related uncertainty in climate change impacts on river flow: An assessment for the Mekong River basin, J. Hydrol., 510, 259–279, doi:10.1016/j.jhydrol.2013.12.010, 2014.

Tripp, D. R. and Niemann, J. D.: Evaluating the parameter identifiability and structural validity of a probability-distributed model for soil moisture, J. Hydrol., 353, 93–108, doi:10.1016/j.jhydrol.2008.01.028, 2008.

Vaze, J., Post, D. A., Chiew, F. H. S., Perraud, J. M., Viney, N. R., and Teng, J.: Climate non-stationarity – Validity of calibrated rainfall–runoff models for use in climate change studies, J. Hydro., 394, 447–457, doi:10.1016/j.jhydrol.2010.09.018, 2010

Wagener, T., McIntyre, N., Lees, M. J., Wheater, H. S., and Gupta, H. V.: Towards reduced uncertainty in conceptual rainfall-runoff modelling: dynamic identifiability analysis, Hydrol. Process., 17, 455–476, doi:10.1002/hyp.1135, 2003.

Weedon, G. P., Gomes, S., Viterbo, P., Österle, H., Adam, J. C., Bellouin, N., Boucher, O., and Best, M.: The watch forcing data 1958–2001: A meteorological forcing dataset for land surface- and hydrological-models, WATCH Technical Report No. 22, available at: url:http://www.eu-watch.org/media/default.aspx/emma/org/10376311/WATCH+Technical+Report+Number+22+The+WATCH+forcing+data+1958-2001+A+meteorological+forcing+dataset+for+land+surface-+and+hydrological-models.pdf, 22, 1–41, 2010.

Weedon, G. P., Gomes, S., Viterbo, P., Shuttleworth, J., Blyth, E., Osterle, H., Adam, J. C., Bellouin, N., Boucher, O., and Best, M.: Creation of the WATCH Forcing Data and its use to assess global and regional reference crop evaporation over land during the twentieth century, J. Hydrometeorol., 12, 823–848, doi:10.1175/2011JHM1369.1, 2011.

Yatagai, A., Kamiguchi, K., Arakawa, O., Hamada, A., Yasutomi, N., and Kitoh, A.: APHRODITE: Constructing a Long-Term Daily Gridded Precipitation Dataset for Asia Based on a Dense Network of Rain Gauges, B. Am. Meteorol. Soc., 93, 1401–1415, doi:10.1175/BAMS-D-11-00122.1, 2012.

Permissions

All chapters in this book were first published in HESS, by Copernicus Publications; hereby published with permission under the Creative Commons Attribution License or equivalent. Every chapter published in this book has been scrutinized by our experts. Their significance has been extensively debated. The topics covered herein carry significant findings which will fuel the growth of the discipline. They may even be implemented as practical applications or may be referred to as a beginning point for another development.

The contributors of this book come from diverse backgrounds, making this book a truly international effort. This book will bring forth new frontiers with its revolutionizing research information and detailed analysis of the nascent developments around the world.

We would like to thank all the contributing authors for lending their expertise to make the book truly unique. They have played a crucial role in the development of this book. Without their invaluable contributions this book wouldn't have been possible. They have made vital efforts to compile up to date information on the varied aspects of this subject to make this book a valuable addition to the collection of many professionals and students.

This book was conceptualized with the vision of imparting up-to-date information and advanced data in this field. To ensure the same, a matchless editorial board was set up. Every individual on the board went through rigorous rounds of assessment to prove their worth. After which they invested a large part of their time researching and compiling the most relevant data for our readers.

The editorial board has been involved in producing this book since its inception. They have spent rigorous hours researching and exploring the diverse topics which have resulted in the successful publishing of this book. They have passed on their knowledge of decades through this book. To expedite this challenging task, the publisher supported the team at every step. A small team of assistant editors was also appointed to further simplify the editing procedure and attain best results for the readers.

Apart from the editorial board, the designing team has also invested a significant amount of their time in understanding the subject and creating the most relevant covers. They scrutinized every image to scout for the most suitable representation of the subject and create an appropriate cover for the book.

The publishing team has been an ardent support to the editorial, designing and production team. Their endless efforts to recruit the best for this project, has resulted in the accomplishment of this book. They are a veteran in the field of academics and their pool of knowledge is as vast as their experience in printing. Their expertise and guidance has proved useful at every step. Their uncompromising quality standards have made this book an exceptional effort. Their encouragement from time to time has been an inspiration for everyone.

The publisher and the editorial board hope that this book will prove to be a valuable piece of knowledge for researchers, students, practitioners and scholars across the globe.

List of Contributors

A. Nazemi and H. S. Wheater
Global Institute for Water Security, University of Saskatchewan, 11 Innovation Boulevard, Saskatoon, SK, S7N 3H5, Canada

T. Antofie, G. Naumann, J. Spinoni and J. Vogt
European Commission, Joint Research Centre, Institute for Environment and Sustainability (IES), Climate Risk Management Unit, Ispra, Italy

T. R. Marthews, S. J. Dadson and S. Abele
School of Geography and the Environment, University of Oxford, Oxford, OX1 3QY, UK

B. Lehner
Department of Geography, McGill University, Montreal, H3A 0B9, Quebec, Canada

N. Gedney
Met. Office, Hadley Centre for Climate Prediction and Research, (JCHMR), Wallingford, OX10 8BB, UK

L. E. Condon
Bureau of Reclamation, Technical Service Center, Denver, Colorado, USA
Hydrologic Science and Engineering Program and Department of Geology and Geological Engineering, Colorado School of Mines, Golden, Colorado, USA

S. Gangopadhyay and T. Pruitt
Bureau of Reclamation, Technical Service Center, Denver, Colorado, USA

Y. B. Sulistioadi
Hydrological Sciences Laboratory, NASA Goddard Space Flight Center, Greenbelt, MD, USA
Department of Forest Science, University of Mulawarman, Samarinda, Indonesia

K.-H. Tseng
Center of Space&Remote Sensing Research, National Central University, Taoyuan, Taiwan

C. K. Shum
Division of Geodetic Science, School of Earth Sciences, the Ohio State University, Columbus, OH, USA
Institute of Geodesy and Geophysics, Chinese Academy of Sciences, Wuhan, China

H. Hidayat
Hydrology and Quantitative Water Management Group, Wageningen University, Wageningen, the Netherlands
Research Center for Limnology, Indonesian Institute of Sciences, Cibinong, Indonesia

M. Sumaryono
Department of Forest Science, University of Mulawarman, Samarinda, Indonesia

A. Suhardiman
Department of Forest Science, University of Mulawarman, Samarinda, Indonesia
Department of Global Agricultural Sciences, the University of Tokyo, Tokyo, Japan

F. Setiawan
Research Center for Limnology, Indonesian Institute of Sciences, Cibinong, Indonesia

S. Sunarso
PT Vale Indonesia, Tbk, Sorowako, Indonesia

D. Siergieiev and A. Lundberg
Division of Geosciences and Environmental Engineering, Luleå University of Technology, Luleå, Sweden

L. Ehlert
Division of Geosciences and Environmental Engineering, Luleå University of Technology, Luleå, Sweden
Institute for Groundwater Management, Faculty of Environmental Sciences, Technische Universtität Dresden, Dresden, Germany

T. Reimann and R. Liedl
Institute for Groundwater Management, Faculty of Environmental Sciences, Technische Universtität Dresden, Dresden, Germany

M. C. Demirel, M. J. Booij and A. Y. Hoekstra
Water Engineering and Management, Faculty of Engineering Technology, University of Twente, P.O. Box 217, 7500 AE Enschede, the Netherlands
Portland State University, Department of Civil&Environmental Engineering, 1930 S. W. 4th Avenue, Suite 200, Portland, OR 97201, USA

A. M. L. Saraiva Okello
UNESCO-IHE, Institute for Water Education, P.O. Box 3015, 2601 DA Delft, the Netherlands
Centre for Water Resources Research, School of Agriculture, Earth and Environmental Science, University of KwaZulu-Natal, Private Bag X01, Scottsville, 3209, South Africa

I. Masih
UNESCO-IHE, Institute for Water Education, P.O. Box 3015, 2601 DA Delft, the Netherlands

S. Uhlenbrook and P. van der Zaag
UNESCO-IHE, Institute for Water Education, P.O. Box 3015, 2601 DA Delft, the Netherlands
Delft University of Technology, Department of Water Resources, P.O. Box 5048, 2600 GA Delft, the Netherlands

G. P. W. Jewitt and E. Riddell
Centre for Water Resources Research, School of Agriculture, Earth and Environmental Science, University of KwaZulu-Natal, Private Bag X01, Scottsville, 3209, South Africa

P. Karimi
UNESCO-IHE Institute for Water Education, Delft, the Netherlands

W. G. M. Bastiaanssen
UNESCO-IHE Institute for Water Education, Delft, the Netherlands
International Water Management Institute, Battaramulla, Sri Lanka
Faculty of Civil Engineering and Geosciences, Water Management Department, Delft University of Technology, Delft, the Netherlands

P. Karimi
UNESCO-IHE Institute for Water Education, Delft, the Netherlands

W. G. M. Bastiaanssen
UNESCO-IHE Institute for Water Education, Delft, the Netherlands
International Water Management Institute, Battaramulla, Sri Lanka
Faculty of Civil Engineering and Geosciences, Water Management Department, Delft University of Technology, Delft, the Netherlands

A. Sood
International Water Management Institute, Battaramulla, Sri Lanka

J. Hoogeveen and L. Peiser
Land and Water Division, FAO, Rome, Italy

E. Bastidas-Obando and R. J. Dost
eLEAF Competence Centre, Wageningen, the Netherlands

P. Hamel
Natural Capital Project, Woods Institute for the Environment, Stanford University, Stanford, CA, USA

A. J. Guswa
Picker Engineering Program, Smith College, Northampton, MA, USA

M. Masood
International Centre for Water Hazard and Risk Management (ICHARM), PWRI, Tsukuba, Japan
National Graduate Institute for Policy Studies (GRIPS), Tokyo, Japan

P. J.-F. Yeh
National University of Singapore, Singapore

N. Hanasaki
National Institute for Environmental Studies, Tsukuba, Japan

K. Takeuchi
International Centre for Water Hazard and Risk Management (ICHARM), PWRI, Tsukuba, Japan